11th Pacific Rim Conference on Stellar Astrophysics 2015

Physics and Chemistry of the Late Stages of Stellar Evolution

Journal of Physics: Conference Series Volume 728

Hong Kong, China
14 – 17 December 2015

Editors:

Sun Kwok
Ching Leung

ISBN: 978-1-5108-2815-5
ISSN: 1742-6588

Printed from e-media with permission by:

Curran Associates, Inc.
57 Morehouse Lane
Red Hook, NY 12571

Some format issues inherent in the e-media version may also appear in this print version.

Printed by Curran Associates, Inc. (2016)

Institute of Physics
Dirac House, Temple Back
Bristol BS1 6BE UK

Phone: 44 1 17 929 7481
Fax: 44 1 17 920 0979

techtracking@iop.org

Additional copies of this publication are available from:

Curran Associates, Inc.
57 Morehouse Lane
Red Hook, NY 12571 USA
Phone: 845-758-0400
Fax: 845-758-2633
Email: curran@proceedings.com
Web: www.proceedings.com

Table of contents

Volume 728

July 2016

Preface

SESSION I: Mass Loss from Evolved Stars

Invited

Contributed

Invited

Contributed

SESSION II: Evolution of Planetary Nebulae

Invited

Contributed

Invited

Contributed

Invited

Contributed

Contributed

POSTERS

Posters

Journal of Physics: Conference Series **728** (2016) 011002 doi:10.1088/1742-6596/728/1/011002

History of the Pacific Rim Conference on Stellar Astrophysics

The First Conference was sponsored by the US Science Foundation (NSF) and the Chinese Academy of Science (CAS), as a Workshop between the USA and China. The NSF grant supported a delegation of 8 astronomers to China. It was decided to turn the Workshop into an international conference to attract international researchers to China. The Conference was named Critical Observations versus Physical Models for Close Binary Systems, Beijing, China 1985. It was the first international conference in Astronomy held in China with the People's Republic. Professor Wang Shou Guan served as the Co-Chair of the SOC.

After several visits with my former college mate, Professor Il-Seong Nha at Yonsei University, Korea, he agreed to host their first international meeting on Astronomy in Seoul and Daejeon, Korea in 1990. Dr. Nha served as the Co-Chair of SOC. The title of the Conference was New Frontiers in Binary Star Research. Again, the Conference was sponsored by the US NSF with the Korean Academy of Science and Engineering as a Workshop between the two countries and expanded to an international conference invitees.

During frequent visits to China in the 80s and 90s I usually stopped by Chiang Mai, Thailand and Seoul, Korea. Professor Boonrucksar Soonthornthum and his colleagues were working to obtain photometric observations with a 20 inch refractor. He had studied in New Zealand and had befriended many binary stars researchers. For the third time US NSF supported a joint workshop with the Thai Academy of Science and Engineering. Thus, we had the first international meeting in Thailand at Chiang Mai University on Recent Developments on Binary Stars Research in 1995. The Conference was Co-Chaired by Boonrucksar.

The Conferences in China, Korea and Thailand were very successful in attracting international researchers to Asia. Many Asian colleagues were interested in having more conferences in Asia. As a result, we renamed the conferences the Pacific Rim Conference on Stellar Astrophysics and shortened the frequency of the series from every five years to about three years.

The 4th Pacific Rim Conference on Stellar Astrophysics was held at Hong Kong University of Science and Technology in 1997. Professor Kwing-Lam Chan served as a Co-Chair of the SOC. 1997 was chosen for the conference to coincide with the year the British Government handed back Hong Kong to China.

The 5th Pacific Rim Conference on Stellar Astrophysics was again held in Hong Kong but at Hong Kong University in 1990. Professor K. S. Cheng was the Co-Chair of the SOC.

The 6th Pacific Rim Conference on Stellar Astrophysics, PRCSA, was held in Xi'an, China in 2002. Shaanxi Observatory of Chinese Academy of Science helped to organize the conference. The SOC decided to have the 6th Conference be a Tribute the Dr. Helmut A. Abt.

Journal of Physics: Conference Series **728** (2016) 011002 doi:10.1088/1742-6596/728/1/011002

The 7th PRSCA went back to Seoul, Korea at the King Sejong University in 2005. Professor, Yoong Woon Kang was the Co-Chair of the SOC.

The 8th PRSCA was held at the resort city of Phuket, Thailand in 2008. Boonrucksar Soonthornthum of Chiang Mai University was the Co-Chair of the SOC. About a month or so prior to the conference I developed shingles on half my face so had to wear a blindfold to limit double vision. Perfect for a double star person. Unknown to me, the SOC had already decided to dedicate the conference to me. I have been wondering ever since if they decided to honor me because I was about to die!

The 9th PRCSA went back the China the second time. It was held in Lijiang, Yunnan in 2011. The conference was planned to coincide with the 3rd week of the IAU International School for Young Astronomers, ISYA. The main purpose was to give the Young Astronomers an exposure to an international conference and opportunity to meet many professional astronomers. Some students presented poster papers to the Conference. Professor Shengbang Qian was the Co-Chair of the SOC.

The 10th PRCSA went back to Kora for the 3rd time at the Sejong University in 2013. Professor Yoon Woon Kang again served as the Co-Chair of the SOC.

This current Pacific Rim Conference on Stellar Astrophysics returns to Hong Kong. Many of us on the SOC understand Professor Sun Kwok will be leaving his post as Dean of Sciences in 2016 and feel strongly he should be honored for his long career in astrophysics and his service to the astronomical community. We dedicate the 11th PRCSA to Professor Sun Kwok. We congratulate him for his significant achievements in stellar astrophysics,

Kam-Ching Leung

Journal of Physics: Conference Series **728** (2016) 011003 doi:10.1088/1742-6596/728/1/011003

Preface

Stellar mass loss is now widely recognized to have a significant impact on stellar evolution. Mass loss on the asymptotic giant branch (AGB) allows stars with initial masses under 8 solar masses to avoid the fate of going supernovae. Over 95% of stars in our Galaxy will evolve through the planetary nebulae phase to end up as white dwarfs instead of neutron stars or black holes. Massive stars undergo mass loss both in the blue and red phases of evolution and create new classes of stars such as Wolf-Rayet stars and luminous blue variables. The circumstellar matter ejected by these mass loss processes becomes new laboratories to study the physical and chemical processes of interstellar matter. The interaction between different phases of mass loss (with variable mass loss rates, ejection speeds, and directions) leads to spectacular morphological transformation of the circumstellar nebulae. The circumstellar nebulae are also sites of molecular and solid-state synthesis. Close to 100 molecular species and a variety of solids, including minerals and complex organics, have been detected in circumstellar envelopes. Since the dynamical time scale of the ejection puts an upper limit on the chemical time scale, we are witnessing a rapid synthesis of chemical species in an extremely low-density environment, creating new challenges to our understanding of chemical reactions.

Effects of mass loss are not limited to single stars. Mass loss by one component of a binary system allows mass transfer to occur at separations beyond the Roche Lobe limit. Accreted wind materials on the surface of a degenerate star can lead to periodic outbursts through H-shell burning. When both components are losing mass, we have interesting dynamical systems such as symbiotic novae.

The theme of this conference is "Physics and Chemistry of the Late Stages of Stellar Evolution". We try to bring together experts in different fields to exchange ideas in the hope of solving the many unsolved problems in the late stages of stellar evolution. The size of the conference was limited to about 100 participants to allow for sufficient time for discussions and personal exchanges of opinions. Although the conference is part of the Pacific Rim series, astronomers from Asia, Europe, North and South America participated in the conference. We hope that these proceedings will serve as a record of our current understanding of the late stages of stellar evolution and be a useful resource for future generation of scientists working in the field.

Sun Kwok

Kam Ching Leung

Editors

Journal of Physics: Conference Series **728** (2016) 011004 doi:10.1088/1742-6596/728/1/011004

Words from the LOC

The University of Hong Kong (HKU) is honored to have the opportunity to host the 11th Pacific Rim Conference on Stellar Astrophysics. The conference was attended by 100 researchers from 20 countries, including Australia, Belgium, Brazil, Chile, China, Czech Republic, France, Germany, Hong Kong, Italy, Japan, Korea, Mexico, Poland, Spain, Sweden, Taiwan, UK, USA, and Vietnam.

The meeting began on the evening of December 13 with the reception held in the foyer of the Faculty of Science office. The 4-day (December 14-17) scientific sessions were held in the Centennial Campus, a new extension of the HKU campus constructed in 2011. On December 16, we had an organized afternoon tour of the south side of the island of Hong Kong, making stops at Repulse Bay and Stanley and ending at the Victoria Peak. The conference banquet was held in the Tien Yi Restaurant in the Peak Tower, which commands a gorgeous view of the Victoria Harbor. Before the banquet, we were treated to the magnificent traditional Chinese face-changing show, where the artist was able to change her face masks in fractions of a second right in front of the audience. After the conference, there was a self-guided tour to Macau on December 19, cumulating with a Portuguese dinner at the historical building of the Clube Militar de Macau.

I would like to thank the many people who helped in the organization of the conference. First I would like to acknowledge the contributions of Anisia Tang, who ably oversaw the logistics of the entire conference, including the venue, budget, registration, web site, and tours. Not only she had to deal with the various vendors and suppliers, she also managed to satisfy the many complex university regulations. After the conference, Anisia was instrumental in the editing and production of these Proceedings. To Anisia, I want to express my deepest personal thanks. Linda Lee was in charge of the organization of the travel and lodgings of the participants. Venus Chu made the arrangements for the banquet as well as the Macau tour and she chose the menus that we so much enjoyed. Y.L. Lee provided essential support of many aspects of the organization, in particular the reception. Jenny Cheng was the official photographer and Emily Wong did the video taping of the proceedings. I am in debt to Angela Tsang, the Faculty Secretary, whose unwavering support of this conference was crucial to its success.

We are grateful for the financial support provided by the Croucher Foundation, K.C. Wong Education Foundation, and the Faculty of Science of the University of Hong Kong. David Foster, Director of the Croucher Foundation, has been a good friend of HKU astronomy and his continued support is greatly appreciated.

I want to express my deepest gratitude to my old friend Kam Ching Leung. Kam Ching was the person who got the Pacific Rim series started 30 years ago. His perseverance and organization skills are the reason why the conferences have been so successful over the years. I thank him in particular for proposing to hold the 11th meeting in Hong Kong and the hard work that he has put in to help this conference achieve all its goals.

Journal of Physics: Conference Series **728** (2016) 011004 doi:10.1088/1742-6596/728/1/011004

Finally, I want to thank the participants. Many of you came from distant parts of the world for this conference and your talks and posters are the reasons behind the scientific success of this conference. I hope that these proceedings will serve as a useful landmark for the development of evolved stars research and that this field will continue to prosper for a long time to come.

Sun Kwok

Chair, Local Organizing Committee

Journal of Physics: Conference Series **728** (2016) 011005 doi:10.1088/1742-6596/728/1/011005

Tribute to Sun Kwok

Sun Kwok was born in Hong Kong in 1949. He did all his early schooling in Hong Kong and went to the same high school, Pui Ching Middle School, as I did but he was more than a decade later. There are two Education Systems in Hong Kong; the Chinese Language Schools and English Language School. Pui Ching was started by Christian missionaries in China and has a long history of providing quality education. Pui Ching is a Chinese Language School, and during colonial times, school entrance was difficult for students as we were not eligible to apply for admission to the University of Hong Kong, nor were we able to join the civil service. In spite of these handicaps, the school still managed to produce many excellent academics, including one Nobel Prize winner in physics and one Field's medalist in mathematics. Most of its graduates who sought further education went to the U.S. Or Canada as Sun Kwok did. Sun graduated from McMaster University and then went to the University of Minnesota for graduate studies.

In the early 1970s, the University of Minnesota had just built one of the world's first infrared bolometers and the astronomers there (Nick Woolf and Ed Ney) were able to make some of the first infrared observations in the mid-infrared region. Through these observations, circumstellar dust was discovered, leading to the realization the evolved stars are losing mass. Sun wrote his PhD thesis on the mass loss mechanism of red giant stars, proposing that the stellar winds are driven by the mechanism of radiation pressure on grains. His 1975 paper is still widely cited to this date. In the same thesis, he showed that OH maser emission is a manifestation of the mass loss process and OH/IR stars are the most heavily mass-losing stars known.

He went back to Canada for postdoctoral studies, first at UBC and then at York University. While at York, he applied his knowledge of mass loss to the problem of formation of planetary nebulae, leading to now well-established interacting winds theory which revolutionized our understanding of the evolution of planetary nebulae. After he joined the University of Calgary, he collaborated with Bruce Hrivnak to discover many proto-planetary nebulae, the then missing link between asymptotic giant branch and planetary nebulae. Together with his former graduate student Kevin Volk, he analyzed the spectra obtained from the Low Resolution Spectrometer of the IRAS satellite and discovered the mysterious 21 micron emission feature, a feature that is still unidentified to this date.

Through his work on millimeter-wave and infrared spectroscopy, Sun promoted the idea that evolved stars are prolific molecular factories. In addition to molecules and minerals, he suggested that evolved stars are also producing complex organics. His paper in *Nature* in 2004 demonstrated that organic synthesis in the late stages of stellar evolution can contaminate the chemical content of the Galaxy and can even affect the chemical composition of the primordial solar system.

In addition to publishing in scientific journals, Sun has also written numerous articles in popular science magazines and given public talks to general audience all over the world. His two popular science books *Cosmic Butterflies* (Cambridge 2001) and *Stardust: the cosmic seeds of life* (Springer 2013) are extremely successful in the amateur astronomy communities.

Sun is a dedicated teacher. He invested a great deal of efforts in his teaching. From the notes he developed over 20 years, he wrote the textbook *Physics and Chemistry of the Interstellar Medium* (University Science Books 2006). This book is now widely used all over the world, including leading universities such as Caltech.

Sun has also been active in international service, having served as President of the IAU Working Group of Planetary Nebulae, President of IAU Commission 34 Interstellar Matter, and is the current President of IAU Commission F3 Astrobiology.

Sun returned to Hong Kong in 2006 to take up the position as the first appointed Dean of Science at the University of Hong Kong. As soon as he arrived, he began to reform the science curriculum at HKU, beginning with the introduction of a flexible major/minor system, Faculty common admission, experiential learning requirements, an academic advising system, creation of the interdisciplinary major Environmental Science, introduction of academic induction, undergraduate summer research fellowship and overseas research fellowship, and science service courses for engineering and medicine students. The most significant was his creation of a one-year core sequence of new science foundation courses, which introduced new ways of teaching mathematics and science at the undergraduate level. He is also committed to the internationalization of academic staff, having brought in new hires from many different countries and greatly raising the quality of the staff as a result.

In spite of his heavy administration duties, Sun continues to teach every year. He designed and taught the course *Our Place in the Universe* as part of the Common Core of HKU. The course discusses the impact of 5000 year of astronomical development on society, religion and society.

In 2016, Sun Kwok will have completed his 10-year tenure as Dean. I am glad to have the opportunity to dedicate the 11th Pacific Rim Conference on Stellar Astrophysics to celebrate the academic career of Sun Kwok. Many of us present here are well familiar with the influence Sun has effected over his career on the physics and chemistry of the late stages of evolution. I am very glad to see that so many experts in this field are gathered here in Hong Kong and I wish everyone a very successful and productive conference. I also wish Sun will continue his productive scientific research after he no longer has the administrative burdens and we look forward to more exciting science results from Sun in the years to come.

Kam Ching Leung

Journal of Physics: Conference Series **728** (2016) 011006 doi:10.1088/1742-6596/728/1/011006

Participant List

Name	Organization	Email
Stavros Akras	Valongo Observatory	akras@astro.ufrj.br
Rodolfo Angeloni	GEMINI Observatory	rangelon@gemini.edu
Bruce Balick	University of Washington	balick@uw.edu
Andrzej Baran	Pedagogical University of Cracow	andysbaran@gmail.com
Helen Barker	University of Manchester	helen.barker@postgrad.manchester.ac.uk
Ivan Bojicic	The University of Hong Kong	ibojicic@hku.hk
Valentin Bujarrabal	Observatorio Astronomico Nacional (OAN-IGN)	v.bujarrabal@oan.es
Franco Cataldo	Osservatorio Astrofisico di Catania and Actinium Chemical Research, Rome, Italy	franco.cataldo@fastwebnet.it
Jose Cernicharo	ICMM CSIC	jose.cernicharo@csic.es
Huan Cheng Chang	Institute of Atomic and Molecular Sciences, Academia Sinica	hchang@gate.sinica.edu.tw
Seok-Jun Chang	Sejong University	csj607@gmail.com
Wen-Ping Chen	National Central University	wchen@astro.ncu.edu.tw
You-Hua Chu	ASIAA	yhchu@asiaa.sinica.edu.tw; yhchu@illinois.edu
Romano Corradi	Instituto de Astrofísica de Canarias	romano.corradi@gtc.iac.es; rcorradi@iac.es
Nick Cox	IRAP/CNRS	nick.cox@irap.omp.eu
Kris Davidson	University of Minnesota	kd@astro.umn.edu
Elvire De Beck	Chalmers University of Technology	elvire.debeck@chalmers.se
Francesco Di Mille	Las Campanas Observatory	fdimille@lco.cl
José Jairo Díaz-Luis	Instituto de Astrofísica de Canarias	jdiaz@iac.es
Pham Ngoc Diep	Vietnam National Satellite Center	pndiep@vnsc.org.vn
Dinh Van Trung	Institute of Physics, Vietnam	dvtrung@iop.vast.ac.vn
Danica Draskovic	Macquarie University	danica.draskovic@mq.edu.au
David Frew	The University of Hong Kong	djfrew@hku.hk
Cezary Galan	Nicolaus Copernicus Astronomical Center	cgalan@camk.edu.pl

Name	Organization	Email
Domingo Aníbal García-Hernández	Instituto de Astrofisica de Canarias	agarcia@iac.es
Tom Geballe	GEMINI Observatory	tgeballe@gemini.edu
Stephan Geier	University of Warwick	geier@sternwarte.uni-erlangen.de
Marcin Gladkowski	Nicolaus Copernicus Astronomical Center	seyfert@ncac.torun.pl
Fiona Healy	University of Manchester	fionahealy17@gmail.com
Andrew Helton	USRA/SOFIA	ahelton@sofia.usra.edu
Jeong-Eun Heo	Sejong University	jeung6145@gmail.com
Kenneth Hinkle	NOAO	hinkle@noao.edu
Naomi Hirano	ASIAA	hirano@asiaa.sinica.edu.tw
Chae-Lin Hong	Sejong University	chealin93@gmail.com
Bruce Hrivnak	Valparaiso University	bruce.hrivnak@valpo.edu
Chih-Hao Hsia	The University of Hong Kong	xiazh@hku.hk
Po-Sheng Huang	ASIAA	posheng@asiaa.sinica.edu.tw
Roberta Humphreys	University of Minnesota	roberta@umn.edu
Kazunori Ishibashi	Nagoya University	bish@nagoya-u.jp
Christine Joblin	IRAP, Univ. Toulouse/CNRS	christine.joblin@irap.omp.eu
David Jones	Instituto de Astrofisica de Canarias	djones@iac.es
Devika Kamath	KU Leuven	devika@ster.kuleuven.be; devika.kamath@ster.kuleuven.be
Young-Woon Kang	Sejong University	kangyw@sejong.ac.kr
Denise Keller	Max Planck Institute for Radio Astronomy	dkeller@mpifr-bonn.mpg.de
Hyosun Kim	ASIAA	hkim@asiaa.sinica.edu.tw
Hanindyo Kuncarayakti	Millennium Institute for Astrophysics & Dept. Astronomia, Universidad de Chile	hanin@das.uchile.cl
Sun Kwok	The University of Hong Kong	sunkwok@hku.hk
John Lattanzio	Monash University	john.lattanzio@monash.edu
Chin-Fei Lee	ASIAA	cflee@asiaa.sinica.edu.tw
Hee-Won Lee	Sejong University	hwlee@sejong.ac.kr
Man Hoi Lee	The University of Hong Kong	mhlee@hku.hk
Kam-Ching Leung	University of Nebraska-Lincoln	kleung@unl.edu
Xiaohu Li	National Astronomical Observatories, Chinese Academy of Sciences	xiaohu@nao.cas.cn
Wenping Liao	Yunnan Observatories, Chinese Academy of Sciences	liaowp@ynao.ac.cn
Jeremy Lim	The University of Hong Kong	jjlim@hku.hk

Name	Organization	Email
Xiaowei Liu	Peking University	x.liu@pku.edu.cn; liuxw@bac.pku.edu.cn
Fred K.Y. Lo	National Radio Astronomy Observatory	flo@nrao.edu
Robin Lombaert	Onsala Space Observatory	lombaert@chalmers.se
Jose Alberto López	Instituto de Astronomia, Universidad Nacional Autonoma de Mexico	jal@astrosen.unam.mx
Matthias Maercker	Chalmers University of Technology	maercker@chalmers.se
Arturo Manchado	Instituto de Astrofísica de Canarias	amt@iac.es
Tom Millar	Queen's University Belfast	tom.millar@qub.ac.uk
Young Chol Minh	Korea Astronomy & Space Science Institute	minh@kasi.re.kr
Mark Morris	UCLA	morris@astro.ucla.edu
Chow-Choong Ngeow	National Central University	cngeow@astro.ncu.edu.tw; cngeow@gm.astro.ncu.edu.tw
Masatoshi Ohishi	National Astronomical Observatory of Japan	masatoshi.ohishi@nao.ac.jp
Hans Olofsson	Chalmers University of Technology	hans.olofsson@chalmers.se
Takashi Onaka	University of Tokyo	onaka@astron.s.u-tokyo.ac.jp
Gabor Orosz	Kagoshima University	gabor.orosz@gmail.com
Masaaki Otsuka	ASIAA	otsuka@asiaa.sinica.edu.tw
Miyako Oyadomari	Kagoshima University	k7219031@kadai.jp
Tali Palma	Pontificia Universidad Católica de Chile	tpalma@astro.puc.cl
Quentin Parker	The University of Hong Kong	quentinp@hku.hk
Mario Pasquato	Yonsei University	mario.pasquato@gmail.com
A. Beate C. Patzer	Zentrum fuer Astronomie und Astrophysik	patzer@astro.physik.tu-berlin.de
Ondrej Pejcha	Princeton University	pejcha@astro.princeton.edu
Qiuhe Peng	Nanjing University	qhpeng@nju.edu.cn
Víctor Pérez Mesa	Instituto de Astrofísica de Canarias	vperezme@iac.es
Shengbang Qian	Yunnan Observatories, Chinese Academy of Sciences	qsb@ynao.ac.cn
Guojun Qiao	Peking University	gjn@pku.edu.cn
Guillermo Quintana-Lacaci	ICMM-CSIC	g.quintana@icmm.csic.es
Nicole Reindl	Eberhard Karls University Tübingen	reindl@astro.uni-tuebingen.de
SeyedAbdolreza Sadjadi	The University of Hong Kong	ssadjadi@hku.hk
Raghvendra Sahai	Jet Propulsion Laboratory, Caltech	sahai@jpl.nasa.gov; raghvendra.sahai@jpl.nasa.gov

Journal of Physics: Conference Series **728** (2016) 011006 doi:10.1088/1742-6596/728/1/011006

Name	Organization	Email
Itsuki Sakon	University of Tokyo	isakon@astron.s.u-tokyo.ac.jp
Pablo Saz Parkinson	The University of Hong Kong	pablosp@hku.hk
Detlef Schönberner	Leibniz Institute for Astrophysics Potsdam	deschoenberner@aip.de
Wolfgang Steffen	Instituto de Astronomia, Universidad Nacional Autonoma de Mexico	wsteffen@astro.unam.mx
Ryszard Szczerba	Nicolaus Copernicus Astronomical Center	szczerba@ncac.torun.pl
Toshiya Ueta	University of Denver	tueta@du.edu
Shane Vickers	Macquarie University	shane.vickers@students.mq.edu.au
Robert Williams	Space Telescope Science Institute	wms@stsci.edu
Emily Wong	The University of Hong Kong	emily79@connect.hku.hk
Ka Tat Wong	Max Planck Institute for Radio Astronomy	ktwong@mpifr-bonn.mpg.de
Yuefang Wu	Peking University	ywu@pku.edu.cn
Andres Yague	Instituto de Astrofisica de Canarias	and.yague@gmail.com; andyague@gmail.com
Yuan-Gui Yang	Huaibei Normal University	yygcn@163.com
Hong Kiu Bosco Yung	Nicolaus Copernicus Astronomical Center	boscohky@gmail.com
Miloslav Zejda	Masaryk University	zejda@physics.muni.cz
Yong Zhang	The University of Hong Kong	zhangy96@hku.hk
Liying Zhu	Yunnan Observatories, Chinese Academy of Sciences	zhuly@ynao.ac.cn

Reception on Sunday evening

From left to right: Fiona Healy, Helen Barker, Stavros Akras, Alberto López

Bruno Steffen, expert in Rubik's cube.

Journal of Physics: Conference Series **728** (2016) 011007 doi:10.1088/1742-6596/728/1/011007

From left to right: Arturo Manchado, Mark Morris, Franco Cataldo

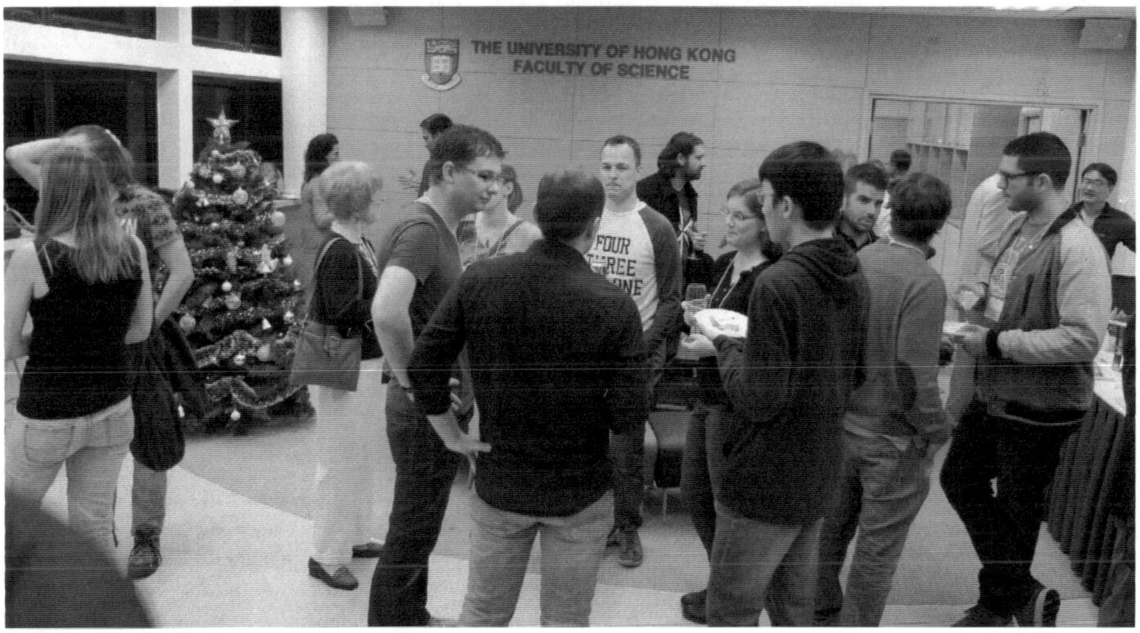

Journal of Physics: Conference Series **728** (2016) 011007 doi:10.1088/1742-6596/728/1/011007

Top from left to right: Bruce Hrivnak, John Lattanzio, Bruce Balick

Below from left to right: Wenping Liao, Yuan-Gui Yang, Yiling Zhu

From left to right: Danica Draskovic, Quentin Parker, Shane Vickers

The 11th Pacific Rim Conference
on Stellar Astrophysics
14-17 December, 2015
The University of Hong Kong

Sun Kwok giving the opening talk

Above: Ken Hinkle

Below: Detlef Schönberner

Left: Matthias Maercker

Journal of Physics: Conference Series **728** (2016) 011007 doi:10.1088/1742-6596/728/1/011007

Above: Hans Oloffson.

Right: Ramano Corradi.

Below: Hee Won Lee asking a question.

Journal of Physics: Conference Series **728** (2016) 011007 doi:10.1088/1742-6596/728/1/011007

Left: Xiaowei Liu.

Below: Fiona Healy.

Lower left: Ryszard Szczerba.

In Repulse Bay during the Wednesday afternoon tour.

Left: Xiaohu Li and Xiaowei Liu.

Below from left to right: Bruce and Della Balick, Elaine and Bob Williams.

Below: Ramano Corradi (left) and Bruce Hrivnak (right) in deep discussion on the beach.

Left Quillermo Quintana-Lacaci at Repulse Bay.

Below: Christine Joblin at the cold and misty Victoria Peak.

Below: Tom Millar at the Blake Pier in Stanley.

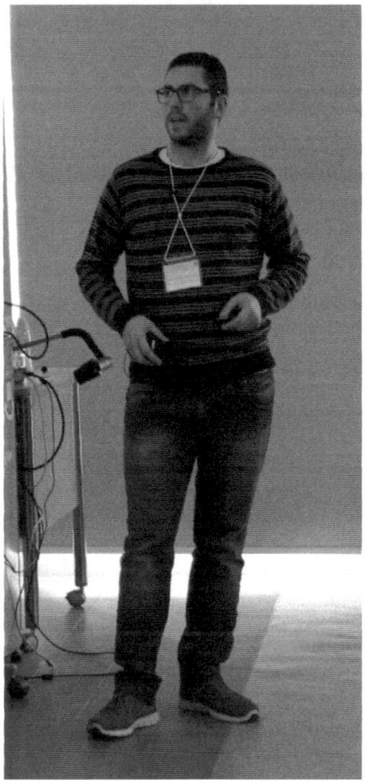

Top left: SeyedAbdolreza Sadjadi.

Above: Tom Millar.

Left: José Diaz-Luis

Journal of Physics: Conference Series **728** (2016) 011007 doi:10.1088/1742-6596/728/1/011007

Above: Jose Cernicharo asking a question. Below right: Itsuki Sakon answering questions from the audience. Below left: Christine Joblin giving her talk.

Journal of Physics: Conference Series **728** (2016) 011007 doi:10.1088/1742-6596/728/1/011007

Clockwise: Xiaowei Liu, Arturo Manchado, Detelf and Maria Schönberner, Sun and Emily Kwok, Bruce and Lucy Hrivnak, Elaine and Bob Williams, Della and Bruce Balick

Clockwise: Tom Millar, Shengbang Qian, Kam Ching Leung, Yvonne Leung, Young Chol Minh, Mrs. Minh, Tom Geballe, Franco Cataldo, Anibal Garcia-Hernandez, Venus Chu, Laura and Alberto López.

Face changing show

Below: Detlef Schönberner

Left: Bob Williams

Journal of Physics: Conference Series **728** (2016) 011007 doi:10.1088/1742-6596/728/1/011007

From left to right: Charlotte Hinkle, Kris Davidson, Roberta Humphreys, Masatoshi Ohishi, Takashi Onaka, John Lattanzio

Clockwise: Ryszard and Aldona Szczerba, Raghvendra Sahai, Nick Cox, Bruno Steffen, Alejandra Sánchez Vázquez, Wolfgang Steffen, Christine Joblin, Jose Cernicharo, Hans Olofsson, Mark Morris

Journal of Physics: Conference Series **728** (2016) 011007 doi:10.1088/1742-6596/728/1/011007

Clockwise: Yong Zhang, Chin Fei Lee, Wen Ping Chen, Toshiya Ueta, Itsuki Sakon, Dinh Van Trung, Jeremy Lim, Matthias Maercker, Hee-Won Lee, Anisia Tang, Chih Hao Hsih, Huan Cheng Chang

First course: suckling pig

Clockwise: David Frew, Guillermo Quintana-Lacaci, Devika Kamath, Stavros Akras, Rodolfo Angeloni, Cezary Galan, Laurence Demond, Quentin Parker, Robin Lombaert, Matthias Maercker, Elvire De Beck, Romano Corradi.

doi:10.1088/1742-6596/728/1/011007

Top: Kam Ching Leung. Above left: Bruce Balick. Above right: a table of happy faces. Right: Young Chol Minh.

Left: leaving Victoria Harbor on a bright December morning.

Above: Tom Geballe, Tom Millar, Bob Williams and Seyedabdolreza Sadjadi on a boat in Victoria Harbor.

Left: Fracno Cataldo, Sun Kwok, Ryszard Szczerba, and Arturo Manchado;

Journal of Physics: Conference Series **728** (2016) 011007 doi:10.1088/1742-6596/728/1/011007

Left: to Macau on a Cotai water jet.

Left: Franco Cataldo on the Macau Citadel.

Below: from let to right: Mrs. Qiao, Emily and Sun Kwok, Arturo Manchado, Franco Cataldo, Bruce and Lucy Hrivnak, Detlef and Maria Schönberner in the Clube Militar de Macau at the end of the Macau tour.

Journal of Physics: Conference Series **728** (2016) 011007 doi:10.1088/1742-6596/728/1/011007

Staff in the HKU Faculty of Science office that helped make this conference the success it was. From left to right: Linda Lee, Venus Chu, Anisia Tang, Jenny Cheng , and Y.L. Lee.

Journal of Physics: Conference Series **728** (2016) 011008 doi:10.1088/1742-6596/728/1/011008

Peer review statement

All papers published in this volume of *Journal of Physics: Conference Series* have been peer reviewed through processes administered by the proceedings Editors. Reviews were conducted by expert referees to the professional and scientific standards expected of a proceedings journal published by IOP Publishing.

Group photo taken on the second day of the conference at the Centennial campus of HKU.

Journal of Physics: Conference Series **728** (2016) 022001 doi:10.1088/1742-6596/728/2/022001

Physics and chemistry of the late stages of stellar evolution – an introduction

Sun Kwok[1]

[1]Laboratory for Space Research, Faculty of Science, The University of Hong Kong, Hong Kong, China

E-mail: [1] sunkwok@hku.hk

Abstract. The stellar evolution from the asymptotic giant branch (AGB) to planetary nebulae (PN) contains some of the most interesting physical and chemical processes in the Universe. Within a time period of one million years starting from the nucleosynthesis of carbon in the core, we witness the chemical synthesis of molecules in the atmosphere, followed by the condensation of minerals and organics in the stellar outflow. Different phases of supersonic stellar winds, both spherical symmetric and highly collimated, and their interactions lead to a series of dynamical processes and morphological transformation of the stellar ejecta.

Most interestingly, PN are now known to be major sources of complex organics in the Galaxy. Organic compounds of mixed aromatic and aliphatic structures have been observed to form in the post-AGB evolution over time scales as short as hundreds of years. There is likely that these stellar organics journeyed through the Galaxy and were embedded in early Solar System.

1. Introduction

The asymptotic giant branch (AGB) phase of stellar evolution is characterized by two major events: one is the nucleosynthesis of the chemical element carbon (C) by helium (He)-shell burning above the core, and the other is the initiation of a strong stellar wind at the surface. Through convection, C is dredged up from the core to the envelope. As the star ascends the AGB, the envelope expands, the luminosity increases, and the surface temperature drops. When the surface temperature falls to below 3000 K, simple molecules begin to form in the photosphere and solid-state materials condense in the extended atmosphere. Radiation pressure on the solid grains drags the gas along, resulting in large-scale mass loss of the stellar envelope. Such mass loss is capable of completely depleting the hydrogen (H) envelope. The exposure of the hot core and the subsequent initiation of another phase of high-speed mass loss results in dynamical wind interactions. The rapidly evolving hot core gradually photo-ionize the circumstellar medium, creating the spectacular phenomenon of planetary nebulae (PN)[1]. At the same time, chemical synthesis of increasingly complex molecules and solids occurs in the circumstellar environment, cumulating in the formation of complex organics. In this paper, we review the rich physical and chemical processes at play during the AGB-PN evolution.

2. Mass loss on the AGB

Large-scale mass loss from AGB stars was discovered as the result of the detection of infrared excesses in the spectra of AGB stars [2]. Analysis of the double-peaked OH maser and CO

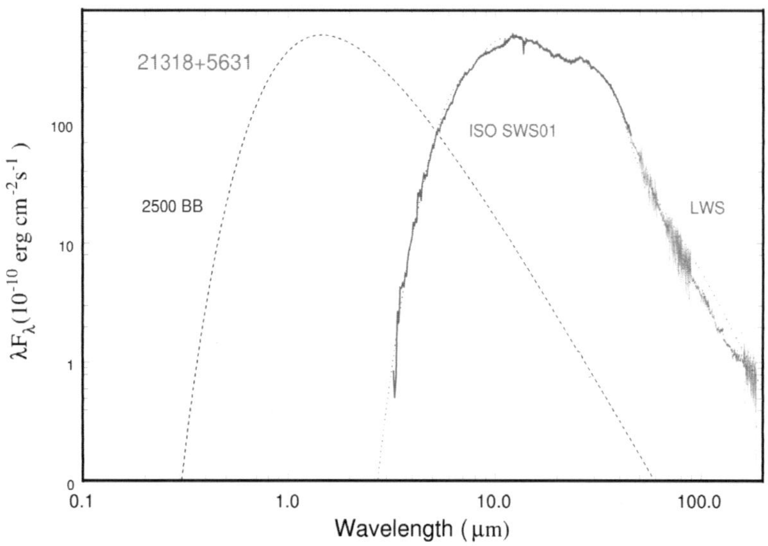

Figure 1. The extent of mass loss is illustrated in the spectral energy distribution of the extreme carbon star IRAS 21318+5631 where the stellar photosphere is completely obscured by its own ejected circumstellar dust envelope. The central star suffers from 360 magnitudes of extinction in the visible (A_V), and is undetectable in the optical region. Its infrared spectrum (solid green line) is due to dust emission and has a color temperature of 300 K. The dotted line represents the theoretical fit to the spectrum based on a 1-D radiation transfer model with a hidden 2500 K central star (dashed line) as the energy source. The absorption feature near the peak of the spectrum is the 13.7 μm band of acetylene.

thermal emissions suggests that these molecular phenomena are manifestations of a steady, low-velocity ($\sim$10 km s^{-1}) stellar wind [3, 4]. The mechanism of mass loss was identified as radiation pressure on solid-state dust grains and the subsequent transfer of momentum from dust to gas through dust-gas collisions [5, 6]. The rates of mass loss can be estimated by both molecular-line and infrared observations. The mass loss rate steadily increases as the star ascends the AGB, from $\sim 10^{-7}$ M$_\odot$ yr^{-1} during the early AGB to as high as 10^{-4} M$_\odot$ yr^{-1} at the tip of the AGB. At the end of the AGB, the mass loss rate can be so high that the photosphere of the star is completely obscured by circumstellar dust and the star appears only as an infrared object [7] (Fig. 1). Such mass loss has significant effect on stellar evolution as it allows stars with initial mass as high as 8 M$_\odot$ to evolve through the PN phase to white dwarfs without undergoing supernova explosions [8].

3. Effects of mass loss on the formation of planetary nebulae

While PN were commonly thought to be the result of sudden ejection of the H envelope of their parent stars, the discovery of mass loss on the AGB let us to the idea that the remnant of circumstellar envelopes ejected during the AGB phase must have an effect on the formation of PN. When the H envelope is completely depleted by mass loss and the core is exposed, the star will become hotter and bluer and the AGB wind ceases. A subsequent phase of less massive but faster mass loss, probably driven by radiation pressure on ionized gas, can compress and accelerate the remnant of the AGB wind to form the shell structure observed in PN [9]. If the interaction of the two winds is energy-conserving, then the expansion of the PN shell is driven by thermal pressure from the shocked fast wind [10].

Soon after the proposal of the interacting winds theory, the hypothesized fast winds from the central stars of PN were discovered by the newly launched *International Ultraviolet Explorer (IUE)* satellite [11]. In the following decade, development of CCD imaging technology made possible the detection of faint outer haloes of PN, therefore confirming the presence of AGB envelopes in PN [12, 13]. The AGB envelopes were also detected by infrared continuum observations by the *Infrared Astronomical Satellite (IRAS)* [14] and millimeter-wave observations of molecular lines [15]. Evidence for the shocked fast wind was found by diffuse X-ray continuum emission by the *ROAST* and *Chandra* satellites.

The detection of neutral materials in PN suggests that PN contain more than just ionized gas, but also molecular and solid-state components as well. The amount of mass contained in the molecular component is much higher than in the ionized gas component and can play a role in the morphological transformation of PN (see section 5). The chemical constituents of the neutral matter also led to the use of PN as chemical laboratories (see section 8).

4. Morphological structure of planetary nebulae

High dynamic-range CCD imaging observations of PN show that PN have complex morphological structures. In addition to the high-density shell, there are also rims, crowns, and haloes [16]. At the same time, the central star is evolving rapidly to high temperatures, at a rate that is highly dependent on its mass [17]. The rapid change in temperature of the central star results in a hardening of radiation emitted by the central star and a changing ionization structure of the circumstellar material. The observed multiple-shell structures of PN can only be understood by a consistent treatment of dynamics of interacting stellar winds coupled with stellar evolution (see Schönberner, these proceedings).

5. Stellar metamorphosis

While the 1-D dynamical models can successfully explain the morphological structure of spherically symmetric PN, there exist many PN which have clear bipolar structures. Even some very well-known PN such as the Ring, the Helix, and the Dumbbell are likely to be bipolar systems viewed near pole-on. When one takes into account of projection effects, the fraction of bipolar nebulae is actually very high. The discovery and imaging of proto-planetary nebulae (PPN) shows that the transformation of the PN morphology occurs early, during the short (10^3 yr) transition from the end of the AGB to the PN stage [18]. The mechanism of shaping is still not understood, although time-dependent outflows collimated by an equatorial torus probably play a role.

PN are optically bright because of the strengths of the recombination lines of H and He and collisionally excited lines of metals. As a result, our view of PN are highly biased by their optical images. The amount of ionized mass in PN are typically $10^{-2} - 10^{-1}$ $M_{\odot}$, but the amount of molecular mass can be 10 to 100 times higher. It would be desirable to directly map the structure of the neutral matter. This can be done by molecular-line and far-infrared imaging observations. Submillimeter-wave imaging observations have resulted in the detection of the molecular torus [21] and imaging observations by the *Spitzer* satellite has found evidence of cold dust beyond the optical nebula [22]. Indirectly, the tight waists seen in bipolar nebulae such as NGC 6302 and NGC 2346 is evidence that these PN are confined by an unseen external medium. The optical lobes therefore do not represent where the majority of masses reside, but cavities cleared out by collimated outflows [19]. In the case of PPN, the bright optical lobes are the result of illumination effects [20].

High dynamic-range imaging observations have resulted in the detection of an increasing numbers of PN with multipolar structures [23, 24]. It is possible that multipolar structures are very common as many of the observed morphologies of PN can be explained by a universal multipolar model of PN [25]. While multipolar PN can have several pairs of bipolar lobes oriented at different directions, they are usually of similar physical lengths when projection effects are taken into account. These are therefore unlikely to represent masses ejected by separate events, but instead are evacuated cavities in the neutral envelope cleared out by high-speed outflows leaked out from holes in the torus.

Although we have made a lot of progress in understanding the physical origin of the morphological structure of PN through the interacting winds model, there are still many unsolved questions. The fast outflows can be directional, or even time dependent. The physical mechanism of collimation is not understood. Future observations of higher sensitivity and dynamic range (e.g., by *ALMA* and *SOFIA*) are needed to map out the distribution of neutral matter outside of the ionized nebulae for us to have a complete picture of the structure of PN.

6. Symbiotic stars, novae and binary evolution

When an AGB star is a member of a binary system, its stellar wind can be accreted onto the main sequence companion, resulting in the symbiotic phenomenon. If the companion has a higher initial mass and has already evolved through the PN stage to become a white dwarf (WD), then the wind accretion could result in the ignition of H burning on the WD, creating a symbiotic nova. The most famous examples of this class are V1016 Cyg and HM Sge. If the accretion rate is high, the WD can even evolve backwards to the red, leading to the born-again phenomenon [26]. For an AGB-WD system, the WD can photoionize the AGB wind, resulting in a PN-like emission-line system. The WD can even have a high-speed wind, causing a colliding winds situation between the winds of the two components [27]. Morphologically, these binary systems can resemble the appearance of PN because many similar physical processes are at work. PN and symbiotic stars have to be distinguished not so much by their observational properties but their evolutioanry status [28].

The declining light curve of classical novae is the result of shrinking photosphere and the re-ignited, H-shell-burning WD evolving to the blue at constant luminosity [29]. The radio and visible light curves of novae can be explained by a decreasing rate of mass loss after the outburst [30]. The fact that novae can change from a free-free dominated spectrum to a dust-emission dominated spectrum over a period of days is the best illustration that solids can form efficiently in an low-density environment.

7. Circumstellar synthesis of molecules and solids

Rotational transitions of over 80 molecules have been detected in the circumstellar envelopes of AGB stars and PN (see Cernicharo, these proceedings). The list of molecules include inorganics (CO, SiO, SiS, NH_3, AlCl), organics (C_2H_2, CH_4, H_2CO, CH_3CN), radicals (CN, C_2H, C_3, HCO^+), rings (C_3H_2), and chains (HC_9N). AGB stars are therefore prolific molecular factories. Infrared spectroscopic observations also found that solid minerals can condense in large quantities directly from the gas phase in the stellar wind. Over 4000 O-rich AGB stars are detected to show the 9.7 μm feature of amorphous silicates and 700 C-rich AGB stars to show the 11.3 μm SiC features [31]. Since the dynamical times scales of AGB envelopes, PPN, and PN are 10^4, 10^3, and 10^4 yr, respectively, we are actually witnessing chemical synthesis at work on these short time scales.

8. Circumstellar synthesis of complex organics

When the stars evolve from the AGB to PN, the chemical composition of the solid-state component changes significantly. While amorphous silicates, silicon carbide, and various refractory oxides dominate the spectra of AGB stars, a family of unidentified infrared emission (UIE) features emerges in PN [32]. These features were identified as vibrational modes of aromatic compounds [33]. Since the UIE features are not seen in the progenitor AGB stars, these organic compounds must have been synthesized in situ in the stellar outflow during the short ($\sim 10^3$ yr) PPN phase. Indeed, infrared spectroscopic observations of PPN show that the UIE features are already present in PPN [34]. In addition to the aromatic features, aliphatic features due to C–H stretches of CH_3 and CH_2 around 3.4 μm are also detected in PPN [35, 36]. Accompanying the UIE features are strong and broad emission plateau features around 8 and 12 μm, which are most likely due to superpositions of in-plane and out-of-plane bending modes of a variety of aliphatic chains attached to the peripherals of aromatic rings [37].

9. Origin of the UIE features

The UIE phenomenon consists of aromatic features at 3.3, 6.2, 7.7, 8.6, and 11.3 μm, aliphatic features at 3.4 and 6.9 μm, weaker unidentified features at 15.8, 16.4, 17.4, 17.8, and 18.9 μm, and broad plateau features at 8, 12, and 17 μm. What is the chemical structure of the carrier of all these features? The most popular model is the polycyclic aromatic hydrocarbon (PAH) hypothesis, where the UIE features are the result of infrared fluorescence from small ($\sim$50 C atoms) gas-phase PAH molecules being pumped by far-ultraviolet photons [38]. The strongest argument for the PAH hypothesis is that single-photon excitation of PAH molecules can account for the 12 μm excess emission observed in cirrus clouds in the diffuse interstellar medium by IRAS [39, 40].

However, the PAH hypothesis faces a number or problems. PAH molecules have well-defined sharp lines but the observed UIE features are broad. PAH molecules are primarily excited by UV with little absorption in the visible, but UIE features are seen in PPN and reflection nebulae with no UV radiation. Actually, the shapes and peak wavelengths of UIE features are independent of temperatures of exciting stars [41]. The strong and narrow expected electronic transitions of PAH in the UV are not seen in interstellar extinction curves [42]. No specific PAH molecules have been detected in spite of the fact that the vibrational and rotational frequencies are well known [43].

Chemists are also uncomfortable with the PAH hypothesis as "no PAH emission spectrum has been able to reproduce the UIE spectrum w.r.t. either band positions or relative intensities" [44, 45]. In order to fit the astronomical observations, the PAH model has to appeal to a mixture of PAH of different sizes, structures (compact, linear, branched) and ionization states, as well as artificial broad intrinsic line profiles. The PAH model is so flexible that it has been shown to be able to fit the observed spectra of silicates, hydrogenated aromatic carbon (HAC), coal, and even artificial spectra [46].

What are the alternatives to the PAH model for the carrier of the UIE features? By introducing H into graphite (sp^2) and diamond (sp^3), a variety of amorphous C-H alloys can be created. By varying the aromatic to aliphatic ratio, geometric structures of mixed hybridization states can be artifically produced. A variety of laboratory techniques involving injection of different forms of energy into a mixture of gas-phase hydrocarbons followed by condensations on substrates has resulted in many different forms of amorphous hydrocarbons [47, 48, 49, 50]. These artificial substances share common properties with natural substances such as coal [51], kerogen [52], petroleum [53], soot and other products of combustion [54] in that they consist of islands of aromatic rings linked by aliphatic chains of different lengths of orientations. The amorphous nature of these particles is demonstrated by X-ray diffraction studies [55]. Figure 2 shows a comparison between the laboratory spectrum of HAC to the astronomical UIE spectrum of a PN.

10. The MAON model

Spectral analysis of the spectra of PN shows that a significant fraction of the energy emitted in UIE features above the infrared continuum arise from the aliphatic component. This led us to propose that the carrier of the UIE features are best represented by mixed aromatic/aliphatic organic nanoparticles (MAON for short). These are complex organic solids with disorganized

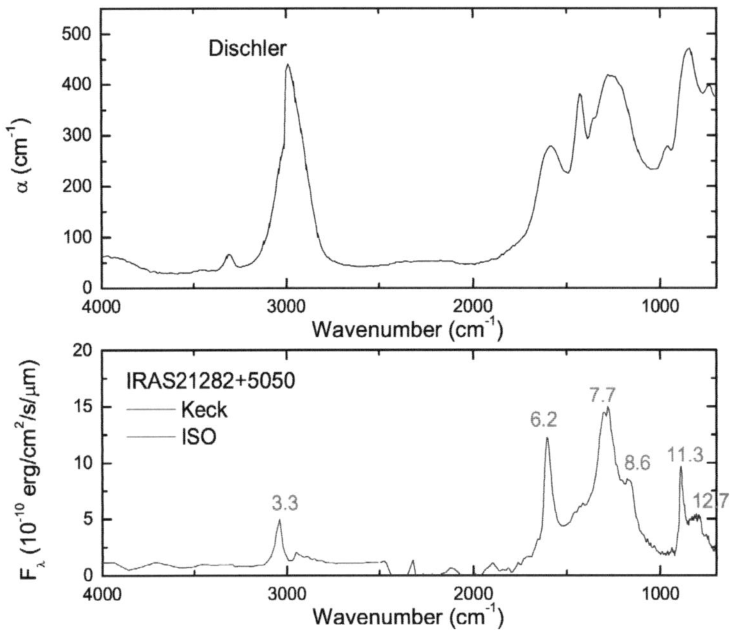

Figure 2. Laboratory infrared spectra of hydrogenated amorphous carbon (top panel) compared to the astronomical spectrum of the planetary nebula IRAS 21282+5050 (bottom panel). The UIE bands are labeled by their wavelengths in μm.

structures consisting of small units of aromatic rings linked by aliphatic chains, including impurities of O, N, S. A typical nanoparticle may contain multiple of such structures with several hundred C atoms [56, 57]. MAON are different from PAH in that they are amorphous with no fixed structure, contains a mix of rings and chains, contain impurities, and are three dimensional in structure.

How do such complex organics form in the circumstellar environment? Some find it very difficult to accept that organic solids can condense directly from gas-phase molecules under near vacuum conditions. However, observationally we see aliphatics and aromatics form in PPN on time scales as short as hundreds of years. In novae, they are actually observed to form on a time scale of days or weeks [58].

11. Star and Solar System connection

The traditional picture of objects in the Solar System is that they are made of minerals, metals, and ices. In the past 30 years, there has been increasing evidence that organic molecules and solids are common in the Solar System, and they can be found in planets and their satellites, asteroids, comets, meteorites, and minor bodies in the outer Solar System. Interplanetary dust particles (IDP) show the 3.4 μm aliphatic feature similar to that seen in PPN [59]. Almost all biologically relevant organic compounds are present in the soluble component of carbonaceous meteorites [60]. Seventy percent of the organic matter in carbonaceous meteorites is in the form of insoluble organic matter (IOM), which chemical structure is similar to MAON [61]. Complex organics containing N, commonly referred to as tholins, are found in the surface of Titan [62]. The 3.4 μm features observed in the Titan atmosphere and comets have very similar profiles as those observed in PPN [63].

This raises the question on the origin of these organics: were they made in the solar nebula or were they inherited from interstellar space? Since 95% of stars evolve through the AGB stage, a large quantity of complex organics is synthesized by stars and distributed throughout the Galaxy. The discovery of pre-solar grains in meteorites suggests that minerals produced by AGB stars can travel through the interstellar medium and end on the surface of the Earth. If stellar organics can also survive their journeys through the interstellar medium, it is quite possible that the early Solar System inherited some of these stellar organics [64].

12. Conclusions

The circumstellar envelopes of AGB stars and PN have provided us with excellent laboratories to study dynamical processes at work in the Galaxy. The lessons that we have learned from PN have been applied to Wolf-Rayet stars, supernovae, star-forming regions, and active galactic nuclei. Since PN are bright in all wavelengths across the electromagnetic spectrum from radio to X-ray, they can be studied with all the modern observational techniques in astronomy. Theoretical models of PN can be stringently tested by these multi-wavelength imaging and spectroscopic observations. The presence of all states of matter (ionized, atomic, molecular and solid-state) allows us to study the radiative interactions between all these components.

The circumstellar envelopes of AGB stars and PN also provide us with chemical laboratory to study chemical synthesis of molecules and solids in space. With the changing appearance and abundance of different species as stars evolve from AGB through PPN to PN, we have a direct view of chemical synthesis at work under a very low-density environment. The degree of complexity of the end products were totally unexpected and the observations creates great challenges to conventional theories. With the efficient synthesis of organics by stars, it is quite possible that stellar organics have enriched the entire Galaxy, and possibly even the early Solar System and the primordial Earth.

Recent research has demonstrated that the late stages of stellar evolution from the AGB to PN are important areas of study both from the physical and chemical points of view. It is even conceivable that the products of evolved stars could have implications on the biological evolution on Earth [65].

Acknowledgments

I would like to thank my collaborators Bruce Hrivnak, Kevn Volk, Yong Zhang for their contributions to the work reported here. The Laboratory for Space Research was established by a special grant from the University Development Fund of the University of Hong Kong. This work is also in part supported by a grant from the HKRGC (HKU 7027/11P).

References

[1] Kwok S 2000 *Origin and Evolution of Planetary Nebulae* (CUP)
[2] Woolf N J and Ney E P 1969 *Astrophys. J.* **155** L181
[3] Kwok S 1976 *J. R. Astron. Soc. Canada* **70** 49-66
[4] Kuiper T B H, Knapp G R, Knapp S L and Brown R L 1976 *Astrophys. J.* **204** 408-414
[5] Gilman R C 1972 *Astrophys. J.* **178** 423-426
[6] Kwok S 1975 *Astrophys. J.* **198** 583
[7] Volk K, Kwok S and Langill P P 1992 *Astrophys. J.* **391** 285-294
[8] Kwok S 1987 *Phys. Rep.* **156** 111-146
[9] Kwok S, Purton C R and FitzGerald M P 1978 *Astrophys. J. Lett.*, **219** L125
[10] Kwok S 1982 *Astrophys. J.* **258** 280-288
[11] Heap S R *et al* 1978 *Nature* **275** 385-388
[12] Jewitt D C, Danielson G E and Kupferman P N 1986 *Astrophys. J.* **302** 727-736
[13] Chu Y-H, Jacoby G H and Arendt R 1987 *Astrophys. J. Suppl. Series* **64** 529-544
[14] Zhang C Y Kwok S 1991 *Astron. Astrophys.* **250** 179-211
[15] Mufson S L, Lyon J and Marionni P A 1975 *Astrophys. J. Lett.* **201** L85-L89
[16] Frank A, Balick B and Riley J 1990 *Astron. J.* **100** 1903-1914

[17] Schönberner D 1981 *Astron. Astrophys.* **103** 119-130
[18] Kwok S 1993 *Annu. Rev. Astron. Astrophys.* **31** 63-92
[19] Kwok S 2010 *Publ. Astron. Soc. Australia* **27** 174-179
[20] Koning N, Kwok S and Steffen W 2013 *Astrophys. J.* **765** 92
[21] Dinh-V-Trung, Bujarrabal V, Castro-Carrizo A, Lim J and Kwok S 2008 *Astrophys. J.* **673** 934-941
[22] Su K Y L *et al* 2004 *Astrophys. J. Suppl. Series* **154** 302-308
[23] López J A, Vázquez R and Rodríguez L F 1995 *Astrophys. J. Lett.* **455** L63
[24] Sahai R, Morris M R and Villar G G 2011 *Astron. J.* **141** 134
[25] Chong S-N, Kwok S, Imai H, Tafoya D and Chibueze J 2012 *Astrophys. J.* **760** 115
[26] Paczynski B and Rudak B 1980 *Astron. Astrophys.* **82** 349-351
[27] Kwok S 1988 *The Symbiotic Phenomenon* (Proc. IAU Colloq. 103) eds Mikolajewska J *et al* p 129
[28] Kwok S 2003 *Symbiotic stars probing stellar evolution* (ASP Conf. Series vol 303) eds Corradi L M *et al* p 428
[29] Gallagher J S, III and Code A D 1974 *Astrophys. J.* **189** 303-314
[30] Kwok S 1983 *Mon. Not. R. Astron. Soc.* **202** 1149-1157
[31] Kwok S, Volk K and Bidelman WP 1997 *Astrophys. J. Suppl. Series* **112** 557
[32] Russell R W, Soifer B T and Willner S P 1977 *Astrophys. J. Lett.* **217** L149-L153
[33] Duley W W and Williams D A 1981 *Mon. Not. R. Astron. Soc.* **196** 269-274
[34] Kwok S, Volk K and Hrivnak B J 1999 *Astron. Astrophys.* **350** L35-L38
[35] Geballe T R, Tielens A G G M, Kwok S and Hrivnak B J 1992 *Astrophys. J.* **387** L89-L91
[36] Hrivnak B J, Geballe T R and Kwok S 2007 *Astrophys. J.* **662** 1059-1066
[37] Kwok S, Volk K and Bernath P 2001 *Astrophys. J.* **554** L87-L90
[38] Tielens A G G M 2008 *Annu. Rev. Astron. Astrophys.* **46** 289-337
[39] Sellgren K 1984 *Astrophys. J.* **277** 623-633
[40] Sellgren K 2001 *Spectrochimica Acta* **57** 627-642
[41] Uchida K I, Sellgren K, Werner M W and Houdashelt M L 2000 *Astrophys. J.* **530** 817-833
[42] Gredel R, Carpentier Y, Rouillé G, Steglich M, Huisken F and Henning T 2011 *Astron. Astrophys.* **530** A26
[43] Ali-Haïmoud Y, Pérez L M, Maddalena R J and Roshi D A 2015 *Mon. Not. R. Astron. Soc.* **447** 315-324
[44] Cook D J and Saykally R J 1998 *Astrophys. J.* **493** 793
[45] Wagner D R, Kim H and Saykally R J 2000 *Astrophys. J.* **545** 854-860
[46] Zhang Y and Kwok S 2015 *Astrophys. J.* **798** 37
[47] Sakata A, Wada S, Onaka T and Tokunaga A T 1987 *Astrophys. J.* **320** L63-L67
[48] Scott A and Duley W W 1996 *Astrophys. J.* **472** L123
[49] Herlin N, Bohn I, Reynaud C, Cauchetier M, Galvez A and Rouzaud J-N 1998 *Astron. Astrophys.* **330** 1127-1135
[50] Jäger C, Huisken F, Mutschke H, Jansa I L and Henning T H 2009 *Astrophys. J.* **696** 706-712
[51] Guillois O, Nenner I, Papoular R and Reynaud C 1996 *Astrophys. J.* **464** 810
[52] Papoular R 2001 *Astron. Astrophys.* **378** 597-607
[53] Cataldo F, Keheyan Y and Heymann D 2002 *Int. J. Astrobiology* **1** 79-86
[54] Pino T, Dartois E, Cao A-T, Carpentier Y, Chamaillé T, Vasquez R, Jones A P, D'Hendecourt L and Bréchignac P 2008 *Astron. Astrophys.* **490** 665-672
[55] Biennier L, Georges R, Chandrasekaran V, Rowe B, Bataille T, Jayaram V, Reddy K P J and Arunan E 2009 *Carbon* **47** 3295-3305
[56] Kwok S and Zhang Y. 2011 *Nature* **479** 80-83
[57] Kwok S and Zhang Y 2013 *Astrophys. J.* **771** 5
[58] Sakon I *et al* 2016 *Astrophys. J.* **817** 145
[59] Flynn G J, Keller L P, Feser M, Wirick S and Jacobsen C 2003 *Geochimica et Cosmochimica Acta* **67** 4791-4806
[60] Schmitt-Kopplin P, Gabelica Z, Gougeon R D, Fekete A, Kanawati B, Harir M, Gebefuegi I, Eckel G and Hertkorn N 2010 *PNAS* **107** 2763-2768
[61] Pizzarello S and Shock E 2010 *Cold Spring Harbor Perspectives in Biology* **2** a002105
[62] Lorenz R D *et al* 2008 *Geophysical Research Letters* **35** 2206
[63] Kim S J, Jung A, Sim C K, Courtin R, Bellucci A, Sicardy B, Song I O and Minh Y C 2011 *Planetary and Space Science* **59** 699-704
[64] Kwok S 2004 *Nature* **430** 985-991
[65] Kwok S 2013 *Stardust: the cosmic seeds of life* (Springer)

Journal of Physics: Conference Series **728** (2016) 022002 doi:10.1088/1742-6596/728/2/022002

Uncertainties in AGB evolution and nucleosynthesis

John Lattanzio[1] and Amanda Karakas[1]

[1]Monash Centre for Astrophysics, School of Phyics and Astronomy, Monash University, 3800, Australia

E-mail: [1]john.lattanzio@monash.edu

Abstract. We summarise the evolution and nucleosynthesis in AGB and Super-AGB stars. We then examine the major sources of uncertainty, especially mass-loss.

1. Introduction

AGB stars are those that return to the giant branch after core helium exhaustion, and show thermal instabilities (or pulses) of the helium shell while ascending the second giant branch. They begin their lives with masses between about 0.8 and 8 $M_\odot$. The Super-AGB stars (hereafter SAGB) are the higher-mass cohort, from about 8 to 12 $M_\odot$, which burn carbon in their cores, then proceed to show thermal pulses. Their mass range is bounded above by the massive stars, which go on to more advanced nuclear burning. The shape of the initial mass function ensures that most stars above 0.8 $M_\odot$ experience an AGB phase. They are bright, and hence they dominate the light of many galaxies. Their copious mass-loss, and advanced nucleosynthesis, ensures that they are major producers of dust, possibly producing as much as 90% of the dust in our galaxy [1]. Their nucleosynthesis is now understood to be crucial for understanding the chemical evolution of the Universe. AGB stars are significant producers of carbon, nitrogen, fluorine and about half of the elements heavier than iron.

2. Summary of AGB evolution

In what follows we restrict our discussion to single stars. We will discuss pre-AGB evolution below, when we consider the uncertainties. The structure of an AGB star is well known. In the centre is the degenerate core composed mostly of C and O, surrounded by a He-burning shell. The next layer consists of the ashes of the H-burning shell, and is topped by said shell. There is then usually a small radiative buffer zone, and finally a deep convective envelope. The AGB phase follows immediately after the star exhausts its core He supply. While the He shell is becoming established, as a result of the core contracting and bringing He at the edge of the core to higher temperatures, there is a substantial energy output and the outer layers expand and cool. This is called the *early AGB* and the cooling outer layers cause convection to extend inwards in mass. For stars exceeding about 3 $M_\odot$ the H shell is extinguished and the convection may reach into the He-rich region, composed of the ashes of H burning. This is called the *second dredge-up* and it can produce substantial increases in the surface He content, up to about 38% by mass [2]. It is this event that produces the classic core-envelope structure required for stars to ascend the giant branch, and we will see later that the details of the timing and depth of the second dredge-up are vital in determining the final fate for SAGB stars [3].

Thermally pulsing evolution has been reviewed previously in the literature, and we refer the reader to [2,4,5] for details. Here we give only a summary, due to space limitations. The He shell is thermally unstable and experiences flashes or pulses every 10^3 to 10^5 years, depending on the core mass. During this "on" phase the He shell drives a convective region that extends from the He shell almost to the H shell, and is known as the *intershell* or *pulse-driven* convective zone. This distributes the products of He burning, mostly C, from the He shell almost to the bottom of the H shell. This region now comprises about 75% He and 25% C (by mass). After about 200 years the convection ends as the shell energy input decreases. The regions outside the He shell expand due to the energy input, causing the H shell to be pushed out to lower temperatures where it is extinguished (or nearly so). The opacity in these cooler regions increases and the bottom of the convective envelope extends inwards in mass. If the envelope reaches into the (top of the) region that was previously the intershell convective zone, then newly synthesised C in this region is mixed to the surface. This is known as the third dredge-up, and it can happen repeatedly, following (nearly) each thermal pulse. Note that a low-mass star may not experience a second dredge-up, but it may still experience many third dredge-up events. The nomenclature is well established, but not perfect. After the energy from the thermal pulse has diffused to the surface, the expanded envelope contracts and heats, and the H discontinuity becomes the new H-burning shell. The star begins the longest phase of the pulse cycle, the inter-pulse phase. At this stage the He shell is essentially inert and the star is powered by the H shell. This phase ends with the next thermal pulse of the He shell, some 10^3 to 10^5 years later, depending on the (core) mass of the star.

3. Summary of AGB nucleosynthesis

The above explains perhaps the most obvious consequence of pulses and dredge-up, namely the increase in the surface C content. But just as important is the synthesis of *s*-process elements, lithium, and the occurrence of hot bottom burning (hereafter HBB).

3.1. Hot bottom burning

For the more massive stars, exceeding about 4 $M_\odot$ (for solar composition, and decreasing as [Fe/H] decreases) the bottom of the convective envelope reaches into the top of the H-burning shell during the interpulse phase. Proton captures can occur at these high temperatures, and this is called HBB. To follow it accurately requires simultaneous calculation of mixing and burning, usually done with a diffusive approximation although it is important to remember that convection is *advective* rather than *diffusive*. The major reactions in such a case are CNO cycles, which burn ^{12}C (and at extreme temperatues ^{16}O) into ^{13}C and ^{14}N. Often the ^{12}C is the same ^{12}C that was dredged-up following the previous thermal pulse. The result is that HBB can prevent the formation of a C-star, or even reduce the C/O ratio from above unity to below unity at the cost of increasing the (primary) ^{14}N content of the envelope. For higher temperatures, which means for more massive AGB stars or even SAGB stars, we find the activation of the Ne-Na chain and possibly the Mg-Al chain [6,7] or even beyond, for SAGB stars.

3.2. Lithium production

Lithium continues to cause headaches for astrophysics. It is very fragile and burns through pp chains at temperatures as low as 2.5 MK. The AGB stars contribute to the Li problems through being production sites for Li, but with an uncertain effect on Galactic Li content.

If the AGB star experiences HBB then it will produce Li through the Cameron-Fowler mechanism [8,9]. This seems to be verified by observations [10,11], which show that Li is present in AGB stars that are not C-stars, as expected when HBB operates. The problem is that the overall production or destruction of Li in these stars is unclear because of the uncertainties associated with mass loss [12,13]. Although Li can be efficiently produced, it is also destroyed by proton captures in the hot bottom of the convective envelope. The production and destruction is a delicate balance, with production driven by the initial supply of ^{3}He. When that is used up then the destruction will dominate. Hence the Li-rich phase is only temporary. Whether the overall yield of Li from such a star is positive or negative depends on when the majority of the mass-loss occurs. If it is after the destruction, as most models predict, then the stars are negligible producers of Li for the Galaxy. If, however, the mass-loss rate is high when Li is abundant then the stars can produce significant amounts of Li. Li is also a crucial guide to extra-mixing processes on the RGB, as we discuss below.

3.3. Producing s-process elements

The slow neutron capture process, or *s*-process, is responsible for producing about half of the elements heavier than Fe, and AGB stars are the main producers [14]. There are two ways for this to occur. The first is active in more massive AGB stars, and occurs in the intershell convective region during a thermal pulse. The H-shell transmutes essentially all CNO species into ^{14}N. During the flash these ^{14}N nuclei can capture two α particles (He nuclei) to produce ^{22}Ne. If the temperature exceeds about 300 MK then one more α capture can occur, producing free neutrons via ^{22}Ne(α,n)^{25}Mg and these neutrons are now available for capture on Fe and other heavy elements and the production follows the *s*-process path. The elements produced in this way are then mixed to the envelope at the next dredge-up episode.

The other way to produce neutrons is through the ^{13}C source. This requires some form of partial mixing of protons below the formal bottom of the convective envelope at the end of dredge-up. There is much debate about the mechanism and how exactly this occurs (see discussion in [2]). But let us suppose it happens, as seems to be required by observations. Then these protons can be captured by the abundant ^{12}C to produce ^{13}C which can then capture an α particle to produce a free neutron (and a ^{16}O nucleus). These neutrons produce *s*-processing as before, but at a lower neutron density than the ^{22}Ne source, albeit for a longer time. We note that the formation and properties of the ^{13}C pocket are major uncertainties in our understanding of the *s*-process.

4. Super-AGB stars

Recent years have seen a rise in the number of studies of SAGB stars. These are very demanding calculations and it is only recently that we have had the computer power to throw at this problem. SAGB stars ignite C in their cores, whereas the normal AGB stars do not. Subsequent evolution depends critically on the mass (and the details of how convection is calculated [15]). The C ignites off-centre in a small convective shell. This shell may burn all the C present in the convective region into Ne, and then a second shell ignites C, and so forth. These shells can be located further toward the centre or further outward, with the result that one eventually burns all of the core C in some cases, and in others the C burning may not reach the centre so we have a CO core and an ONe outer region. The crucial thing is that these stars then proceed to experience thermal pulses on the (S)AGB. They are quite separate beasts to "massive" stars, which go on to further nuclear burning stages after C burning. SAGB stars do not.

SAGB evolution is qualitatively the same as AGB evolution, but the quantitative differences are important. Firstly, the HBB occurs at high temperature [6,7,16] due to the deep convective envelopes. However, the intershell convective zone is very small in mass, typically 10^{-3} to 10^{-5} M$_\odot$ (compared to something like 0.01 M$_\odot$ for AGB stars). This means that the region undergoing neutron captures (from the ^{22}Ne source) is tiny and when this region is diluted in the much deeper envelope (the core mass is perhaps 1 M$_\odot$, leaving an envelope mass of a few M$_\odot$) then the enhancements of *s*-process elements are not expected to be large. This may change near the end of the evolution, when the envelope mass has been dramatically reduced due to mass loss. Further, SAGB stars have very small interpulse periods, more like 30–1000 yr as opposed to 1-100 kyr for AGB stars. Hence hundreds to thousands of pulses must be calculated.

5. Main uncertainties

We have indicated above some of the uncertainties in AGB evolution. We discuss these in more detail below. However, the AGB is the last phase of evolution for these stars, and hence the models begin with the uncertainties already accumulated over all of the earlier phases. Thus we need to briefly review these if we are to provide a realistic estimate of the confidence we should place in the models.

5.1. Extra-mixing on the first giant branch

It is now well established that the predictions for abundance changes resulting from first dredge-up are largely in agreement with observations. It is also well established that there is a second mixing event that changes these compositions, and it seems to begin at the position of the bump in the giant branch luminosity function (see [2] for a recent review, and [17] for a beautiful illustration using Li). The exact mechanism for this extra-mixing is not known, and early investigations focussed on the obvious candidate of meridional circulation in rotating stars [18], but modern models suggests that this does not match the observations [19,20]. Recent interest focusses on thermohaline mixing following the discovery of a molecular weight inversion that appears in RGB stars when they reach the bump [21]. Calculations show that this seems to match the observations reasonably well [22,23]. Debate exists concerning how to model this process, with 2D and 3D hydro calculations disagreeing with the typical 1D models used; see [24,25] and the extensive discussion in [2]. More work needs to be done before we can be confident in how to model this process, let alone verifying its role in the observed abundance patterns.

5.2. Extra-mixing on the asymptotic giant branch

If some process causes extra-mixing on the RGB, does it also operate on the AGB? This has been postulated by various authors as a possible explanation for O and Al isotope measurements in pre-solar grains [e.g. 26]. But there are also discrepancies with the C isotope ratios predicted for C stars. By the time the star has dredged-up sufficient ^{12}C to produce C/O $>$ 1 the ratio of ^{12}C/^{13}C greatly exceeds the observed values [27]. There are also other problems that may be alleviated by some extra-mixing on the AGB. However, [28] showed that, at least for solar metallicities, the inclusion of the effects of extra-mixing on the RGB (usually ignored in the models) removed the problems on the AGB. The final word is yet to be written.

5.3. Core helium burning

The core helium-burning phase has a history of challenging our modelling skills. This is where semiconvection was first recognised in the 70s [29,30] and later the core-breathing pulses added more unwelcome complications [e.g. 31]. It is now well documented that small variations in numerical details of the determination of the convective boundaries can produce enormous differences in the size of the convective core, not to mention the semiconvection region, as discussed recently in [32]. The reality or otherwise of the core breathing pulses remains in debate [e.g. 33] and recent work [34] has tried to use asteroseismology as a probe of mixing in the cores of these stars. The problem is very difficult, with a core opacity source that is higher (in the regions rich in C and O) than in the outer He-rich region. This drives overshooting at the core edge. But the stellar conditions contrive to produce a local minimum in the ratio of the adiabatic to radiative temperature gradients. When this minimum reaches unity, the traditional value for convection according to the Schwarzschild criterion, then the correct way to calculate the behaviour of the outer edge of the core is far from clear. Further, when the central He content

Journal of Physics: Conference Series **728** (2016) 022002 doi:10.1088/1742-6596/728/2/022002

drops below about 0.1 [35] then the cubic dependence of the triple-α energy production on the He content means that small perturbations on the He mass fraction can produce large changes in the energy output and hence drive larger convective cores – these are breathing pulses.

Clearly the only way forward is multi-dimensional hydrodynamical simulations as discussed in [32]. Such calculations are demanding because the simulation must be performed for many turnover times. Indeed, determining the behaviour and timescale for semiconvection is one of the hoped-for outcomes of such simulations.

5.4. Convective boundaries

Thankfully the timescale for convective mixing is usually much smaller than the evolutionary timescale for the star. Usually one uses a diffusion equation to approximate mixing. Because mixing is usually very rapid, there is not a lot of dependence on the diffusion coefficient, as long as it is sufficiently large to produce rapid mixing, although there are some notable exceptions, such as HBB. As you can see from the previous discussion, the calculation of convective borders remains a serious problem for stellar models. Some overshoot must carry material beyond the naive Schwarzschild border. This is simply the result of conservation of momentum. The Schwarzschild criterion considers the buoyancy force, and places the border where that goes to zero. However material arrives at the neutral border with a finite momentum so it must penetrate the border – but by how much? It has become common to modify the diffusive mixing implementation by including an overshoot region where the diffusion coefficient is chosen to reproduce an exponential decay in velocity beyond the formal border [36]. This is qualitatively fine, but as always there is a parameter (determining the decay length) that is usually fixed by appealing to some observations. This procedure produces the partial mixing required to produce a ^{13}C pocket for *s*-process nucleosynthesis, although of course the details depend on the overshoot procedure.

If one applies such a scheme to all convective borders, then interesting things happen. Of course, one must calibrate, or somehow choose, the decay length for each border, and there is no reason to think that overshoot inwards (to higher density) has a similar decay length to overshoot outwards into less dense material. Nevertheless, when applied to the intershell convective region, we find overshoot into the CO core with the result that the intershell becomes enriched in C and O. Models including this effect give better fits to observed abundances in H-deficient post-AGB stars [37] and possibly AGB stars as well [28].

5.5. Convection theory

Of course, the situation is even worse than described above. The Mixing-Length Theory (MLT) of convection has such a hold on stellar modelling that we forget that it is only one possible formulation. Another that has made substantial contributions is the Full Spectrum of Turbulence theory [38,39]. These two theories produce significantly different results [40] and it seems that this uncertainty is largely ignored in the literature. We note that researchers do continue to develop potential new convection models [e.g. 41] but these must be presented in a format that is easily implemented in an evolution code if they are to overcome the dominance of the MLT.

5.6. Opacities for varying envelope compositions

This is one area that has received a lot of attention recently, with the result that the current situation is very satisfactory. Thermal pulses increase the C (and possibly O, via overshoot, see above) content of the stellar envelope, and HBB can burn this C (and O) into N. After H and He, these can be the next most abundant species, and they are a very significant source of opacity. This is doubly so for stars with very low [Fe/H]. Such variations in the opacity have been ignored until recently, as tables for varying compositions were not available. This is no longer the case, with the AESOPUS tool now providing opacities for appropriate mixtures

[42,43]. The increase in opacity when the C content increases has also been shown to have a dramatic effect on the evolution of the AGB stars, increasing mass loss and hence terminating the evolution much sooner than in calculations that ignore this effect [44,45].

5.7. Envelope ejection?
In 1986 Wood and Faulkner [46] found convergence problems in a late AGB model which they described as due to the disappearance of hydrostastic solutions to the stellar structure for large cores. Further work on this problem was performed by [47] who confirmed that a super-Eddington luminosity developed at the bottom of the convective envelope. This was identified as being due to an opacity bump produced by Fe. An understanding of the subsequent behaviour of the star will require a hydrodynamical study. Of course this may be important for understanding the formation of planetary nebulae.

6. Final fate of AGB stars
An AGB (or SAGB) star ends its life when mass loss removes the envelope. Usually this produces a CO white dwarf. However the SAGB stars ignite core C and a number of outcomes become possible [3]. The C burning can ignite in a shell in the outer part of the CO core, and in many cases does not proceed further. This produces a CO(Ne) hybrid white dwarf. If the C burning proceeds to the core then we find the formation of an ONe white dwarf. The most massive SAGB stars experience "dredge-out" [48,49] where the convective C burning region meets with the convective envelope. This is a computationally demanding phase of the evolution and subject to all the uncertainties associated with time-dependent mixing, which are amplified by simultaneous rapid nuclear energy generation.

Following core He burning most SAGB stars have a core mass that easily exceeds the Chandrasekhar mass. We expect such a star to proceed through various nuclear burning stages and end life as a supernova. But the occurrence of second dredge-up in SAGB stars reduces the core mass below the critical value. The fate of such a star depends on the competition between core growth and mass loss. If the former dominates and the core reaches the Chandrasekhar mass then an electron-capture supernova will result. If mass loss terminates the evolution with the core mass below the critical value then the star ends as an ONe white dwarf [3].

7. Mass-loss
From the viewpoint of stellar models, what is required is a formula that specifies the mass-loss rate (MLR) in terms of known quantities. This mass loss is of course assumed to be steady and spherically symmetric, which we know is not always the case in reality. The mechanism believed to drive mass loss in AGB stars is the pulsation enhanced dust-driven wind scenario, where grains are driven outward by the photon wind. These are collisionally linked to the gas, and hence the gas is also removed. For stars with $C/O > 1$ we believe that amorphous C grains are involved. For O-rich stars it is presumably Mg and Fe silicate grains that are implicated. The difficulty is that these latter grains do not couple well with the gas unless they are very large [50]. It is only recently that such large grains were indeed observed [51].

There are many MLR expressions in the literature. We discuss here only the most commonly used formulae. The Reimers formula [52] was derived for giants and supergiants. Vassilliadis & Wood [53] instead fit the MLR to the pulsation period for red giants and AGB stars. The MLR was bounded by the radiation limit, and for massive stars the superwind phase was delayed to ensure that periods exceeding 500 d were obtained. The Blöcker formula [54] is one of many modifications to the Reimers rate, in this case motivated by dynamical pulsation models of Miras. The more recent (2005) Schröder & Cuntz formula [55] is a physically motivated, semi-empirical modification of the Reimers formula.

As one may expect, the MLR has a potentially enormous effect on the star's evolution [6,40] and nucleosynthesis [6,56]. Increasing the MLR removes the envelope more rapidly, thus terminating the evolution and also terminating thermal pulses and all nucleosynthesis. The reultant yields are very dependent on the MLR used (and the free parameters chosen for those formulae with such parameters).

Various authors have performed tests of the MLRs. Mostly these are crude sanity checks, but some quantitative tests have also been performed. Schröder & Cuntz [57] looked at detailed models for some of the best studied galactic giants and supergiants, with considerable success. Another nice test was a critical examination of AGB luminosity functions [58] which again was very favourable to the Schröder & Cuntz MLR.

However, the Vassilliadis and Wood [53] formula also has been carefully tested. Detailed evolution and pulsation models for thermaly pulsing stars in NGC419 and NGC1978 were compared with infrared data by [59] and an excellent agreement was found. The MLR in [53] also accurately predicted the magnitude of the tip of the AGB in these clusters. Another quantitative test of this rate with AGB stars in the SMC was performed by [60] and again the MLR produced a successful quantitative comparison.

In conclusion, one should carefully choose the MLR to be used, depending on the phase of evolution being investigated. The Schröder & Cuntz formula [55] seems suitable for giants and AGB stars, while the Vassilliadis and Wood [53] formula, tailored for AGB stars, does a very good job in that regime.

8. Conclusions

There remain many uncertainties in trying to model AGB and SAGB stars. As usual, these mostly centre on convectioon and its various manifestations. Recent work has led to substantial improvements in our understanding of mass loss and we now have formulae that seem to be quantitatively reliable, at least in a global sense. The use of the Reimers formula for AGB stars is not recommended.

Acknowledgments

JL thanks the organisers, and especially Sun Kwok and Linda Lee, for their hospitality and practical assistance. He also thanks Sun Kwok for 30 years of valued friendship.

References

[1] Sloan G C, Kramer K E, Wood P R, Ziljstra A A, Bernard-Salas J, Devost D and Houck J R 2008 *Astrophys. J.* **686** 1056
[2] Karakas A I and Lattanzio J C 2014 *Publ. Astron. Soc. Australia* **31** 30
[3] Doherty C L, Gil-Pons P, Siess L, Lattanzio J C and Lau H B 2015 *Mon. Not. R. Astron. Soc.* **446** 2599
[4] Busso M, Gallino R and Wasserburg G J 1999 *Annu. Rev. Astron. Astrophys.* **37** 239
[5] Herwig F 2005 *Annu. Rev. Astron. Astrophys.* **43** 435
[6] Doherty C L, Gil-Pons P, Lau H H B, Lattanzio J C and Siess L 2014a *Mon. Not. R. Astron. Soc.* **437** 195
[7] Doherty C L, Gil-Pons P, Lau H H B, Lattanzio J C, Siess L and Campbell S W 2014b *Mon. Not. R. Astron. Soc.* **441** 582
[8] Cameron A G W and Fowler W A 1971 *Astrophys. J.* **164** 111
[9] Sackmann I-J and Boothroyd A I 1992 *Astrophys. J. Lett.* **392** L71
[10] Smith V V and Lambert D L 1989 *Astrophys. J. Lett.* **361** L69
[11] Mazzitelli I, D'Antona F and Ventura P 1999 *Astron. Astrophys.* **348** 846
[12] Ventura P, D'Antona F and Mazzitelli I 2000 *Astron. Astrophys.* **363** 605
[13] Travaglio C, Randich S, Galli D, Lattanzio J C, Elliott L M, Forestini M and Ferrini F 2001 *Astrophys. J.* **559** 909
[14] Bisterzo S, Travaglio C, Gallino R, Wiescher M and Käppeler F 2014 *Astrophys. J.* **787** 10
[15] Doherty C L, Siess L, Lattanzio J C and Gil-Pons P 2010 *Mon. Not. R. Astron. Soc.* **401** 1453
[16] Gil-Pons P, Doherty C L, Lau H H B, Campbell S W, Suda T, Guilani S, Gutiérrez J and Lattanzio J C 2013 *Astron. Astrophys.* **557** A106

[17] Lind K, Primas F, Charbonnel C, Grundahl F and Asplund M 2009 *Astron. Astrophys.* **503** 545
[18] Sweigart A V and Mengel J G 1979 *Astrophys. J.* **229** 624
[19] Palacios A, Talon S, Charbonnel C and Forestini M 2003 *Astron. Astrophys.* **399** 603
[20] Palacios A, Charbonnel C, Talon S and Siess L 2006 *Astron. Astrophys.* **453** 261
[21] Eggleton P P, Dearborn D S P and Lattanzio J C 2006 *Science* **314** 1580
[22] Charbonnel C and Zahn J-P 2007 *Astron. Astrophys.* **467** L15
[23] Eggleton P P, Dearborn D S P and Lattanzio J C 2008 *Astrophys. J.* **677** 581
[24] Denissenkov P A 2010 *Astrophys. J.* **723** 563
[25] Traxler A, Garaud P and Stellmach S 2011 *Astrophys. J. Lett.* **728** L29
[26] Busso M, Palmerini S, Maiorca E, Cristallo S, Straniero O, Abia C, Gallino R and La Cognata M 2010 *Astrophys. J. Lett.* **717** L47
[27] Karakas A I 2010 *Mon. Not. R. Astron. Soc.* **403** 1413
[28] Karakas A I, Campbell S W, and Stancliffe R J 2010 *Astrophys. J.* **713** 374
[29] Castellani V, Giannone P and Renzini A 1971 *Ap Space Sci* **10** 340
[30] Castellani V, Giannone P and Renzini A 1971 *Ap Space Sci* **10** 355
[31] Castellani V, Chieffi A, Tornambe A and Pulone L 1985 *Astrophys. J.* **296** 204
[32] Campbell S W, Constantino T N, D'Orazi V, Meakin C, Stello D, Christensen-Dalsgaard J, Kuehn C, De Silva G M, Arnett W D, Lattanzio J C and MacLean B T 2015 Towards 21st Century Stellar Models: Star Clusters, Supercomputing, and Asteroseismology *Preprint* arXiv 1512.04774
[33] Constantino T N, Campbell S W, Lattanzio J C and van Duijneveldt A 2016 *Mon. Not. R. Astron. Soc.* **456** 3866
[34] Constantino T N, Campbell S W, Christensen-Dalsgaard J, Lattanzio J C and Stello D 2015 *Mon. Not. R. Astron. Soc.* **452** 123
[35] Sweigart A V and Demarque P 1973 *Variable Stars in Globular Clusters and Related Systems* **36** 221
[36] Herwig F, Blöcker T, Schönberner D and El Eid M 1997 *Astron. Astrophys.* **324** L81
[37] Herwig F Blöcker T, Langer N and Driebe T 1999 *Astron. Astrophys.* **349** L5
[38] Ventura P, Zeppieri A, Mazzitelli I and D'Antona F 1998 *Astron. Astrophys.* **334** 953
[39] Mazzitelli I, D'Antona F and Ventura P 1999 *Astron. Astrophys.* **348** 846
[40] Ventura P and D'Antona F 2005 *Astron. Astrophys.* **431** 279
[41] Arnett W D, Meakin C, Viallet M, Campbell S W, Lattanzio J C and Mocák M 2015 *Astrophys. J.* **809** 30
[42] Lederer M T and Aringer B 2009 *Astron. Astrophys.* **494** 403
[43] Marigo P and Aringer B 2009 *Astron. Astrophys.* **508** 1539
[44] Marigo P and Ventura P 2010 *Mon. Not. R. Astron. Soc.* **408** 2476
[45] Constantino T N, Campbell S W, Gil-Pons P and Lattanzio J C 2014 *Astrophys. J.* **784** 56
[46] Wood P R and Faulkner D J 1986 *Astrophys. J.* **307** 658
[47] Lau H B B, Gil-Pons P, Doherty C L and Lattanzio J C 2012 *Astron. Astrophys.* **542** A1
[48] Ritossa C, Garcia-Berro E and Iben I Jr 1999 *Astrophys. J.* **515** 381
[49] Siess L 2007 *Astron. Astrophys.* **476** 893
[50] Höffner S 2008 *Astron. Astrophys.* **491** L1
[51] Norris B R M, Tuthill P G, Ireland M J, Lacour S, Zijlstra A A, Lykou F, Evans T M, Stewart P and Bedding T R 2015 *Nature* **484** 220
[52] Reimers D 1975 *Problems in Stellar Atmospheres and Envelopes* ed Baschek B *et al* (Berlin Heidelberg: Springer-Verlag) p 229
[53] Vassiliadis E and Wood P R 1993 *Astrophys. J.* **413** 641
[54] Blöcker T 1995 *Astron. Astrophys.* **297** 727
[55] Schröder K-P and Cuntz M 2005 *Astrophys. J. Lett.* **630** L73
[56] Stancliffe R J and Jeffrey C S 2007 *Mon. Not. R. Astron. Soc.* **375** 1280
[57] Schröder K-P and Cuntz M 2007 *Astron. Astrophys.* **465** 593
[58] Rosenfield P, Marigo P, Girardi L, Dalcanton J J, Bressan A, Gullieuszik M, Weisz D, Williams B F, Dolphin A and Aringer B 2014 *Astrophys. J.* **790** 22
[59] Kamath D, Wood P R, Soszyński I and Lebzelter T 2010 *Mon. Not. R. Astron. Soc.* **408** 522
[60] Groenewegen M A T, Sloan G C, Soszyński I and Petersen E A 2009 *Astron. Astrophys.* **506** 1277

Journal of Physics: Conference Series **728** (2016) 022003 doi:10.1088/1742-6596/728/2/022003

Atmospheric structure and mass loss in Miras

Kenneth H Hinkle[1], Thomas Lebzelter[2] and Oscar Straniero[3]

[1] NOAO, PO Box 26732, Tucson, AZ 85726 USA

[2] Univ. of Vienna, Depart. of Astrophysics, Türkenschanzstrasse 17, A-1180 Vienna, Austria

[3] INAF, Osservatorio Astronomico di Collurania, I-64100 Teramo, Italy, and INFN-LNGS, Assergi (AQ), Italy

E-mail: [1] hinkle@noao.edu

Abstract.

The cool molecular region of late-type stellar atmospheres, the 'MOLsphere' located between the photosphere and expanding circumstellar shell, is explored by using time series of near-IR spectra of one large amplitude AGB (Mira) variable. The ~1400 K MOLsphere of the Mira R Cas is shown to undergo large amplitude velocity changes that are aperiodic with time scales many times longer than the photospheric pulsation. The mass of the MOLsphere is $\sim 3\times 10^{-5}$ $M_\odot$. Initial analysis suggests that MOLsphere is thin compared to the stellar radius.

1. Introduction

For low mass stars the majority of the mass loss occurs during the AGB evolutionary phase. Near the tip of the AGB mass rates are typically $\sim 10^{-6}$ $M_\odot$ yr^{-1}. Initial-to-final mass relations show that solar mass stars loose about 0.45 $M_\odot$ before becoming a white dwarf. The total mass loss exceeds 2 $M_\odot$ for a 3 $M_\odot$ star [1]. There is broad agreement on the basic mass loss mechanisms. Mechanical energy from the stellar pulsation enlarges the star and results in a cool, extended outer atmosphere. This allows dust to form and mass is lost due to radiation pressure on the grains [2].

Arguably the least understood subprocess in mass loss is the transfer of matter from the photosphere to the expanding circumstellar region. A number of theoretical studies show that dissipation of shocks from the stellar pulsation plays a critical role [3]. It is known that SiO masers are formed in the non-expanding circumstellar region and mid-IR and microwave interferometry techniques are also starting to probe the region. Here we discuss spectroscopy of near-IR lines that originate in the photospheric-circumstellar boundary.

2. The MOLsphere

Hinkle [4] discovered that the low excitation 2-0 CO lines of Miras are asymmetric. The asymmetric line profiles result from blending between photospheric lines and lines formed in a cooler, ~1000 K, atmospheric component. Tsuji [5] found similar lines in non-variable cool stars and attributed the origin of these lines to a cool, molecular region he called the MOLsphere. Initial observations [4, 6] of the bright, unobscured Miras R Leo and χ Cyg suggested that the Mira MOLsphere was stationary. The photospheres of all Miras, observable in many features in the near-IR spectra, undergo large amplitude velocity pulsations at the period of the optical light curve. There are a number of papers describing the photospheric pulsation as observed

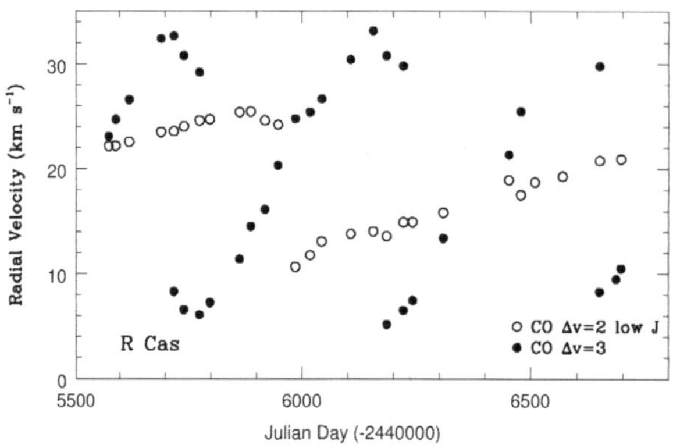

Figure 1. A 3 year time series of velocities is shown CO lines in the Mira R Cas. The filled symbols are CO second overtone lines formed in the photosphere. The velocities of these lines follow the 431 day Mira pulsation of R Cas [7]. The open symbols are low-excitation CO first overtone (2.3 μm) lines formed in the MOLsphere. The MOLsphere velocities are decoupled from the stellar pulsation.

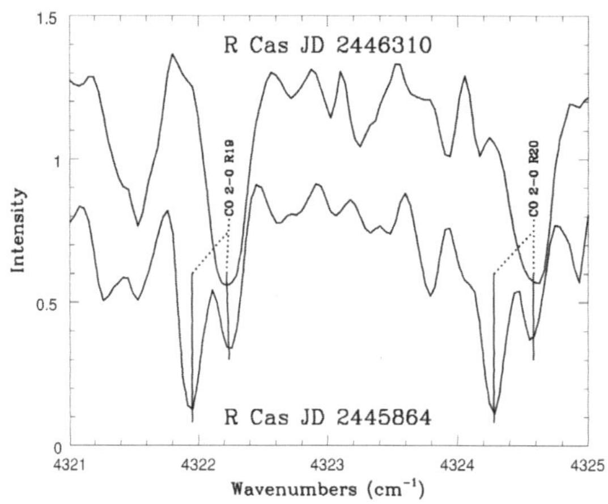

Figure 2. Two spectra of the low excitation CO first overtone lines in R Cas at different dates. Upper spectrum has blended photospheric and MOLsphere components but these are widely separated on the lower spectrum. JD 2445864=1984 Jun 12.

in CO second overtone lines [8]. A stationary MOLsphere was contrary to expectations from models that did not produce any stationary layers [3] .

Additional observations of a small sample of typical Miras show that the early observations suffered from unfortunate sampling. The MOLsphere lines can have brief periods of near constant velocity but in fact undergo velocity changes over much longer periods than the stellar pulsation period. This is seen in Figure 1 for R Cas. The aperiodic MOLsphere velocity changes

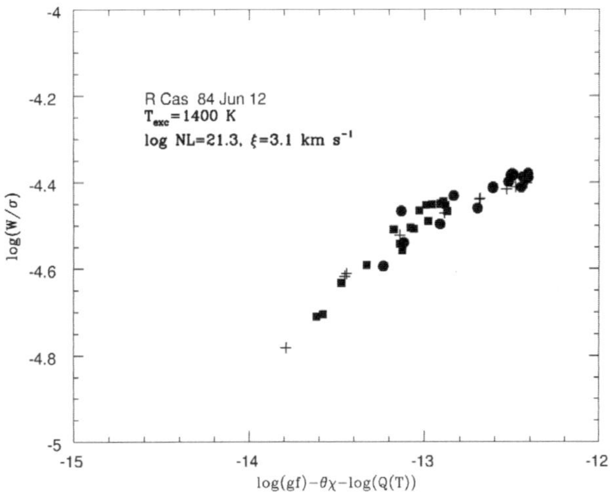

Figure 3. Curve of growth for the MOLsphere lines. The excitation temperature was selected by forcing high and low excitation lines onto a single curve of growth. Data filled symbols, simple model fit pluses.

observed in R Cas are similar to model predictions. Comparison with models suggests that the lines originate at several stellar radii.

As a consequence of changing velocities for both the MOLsphere and photospheric lines (Figure 1) there are some times when the spectral lines originating in these components are unblended (Figure 2). This greatly simplifies the measurement of the MOLsphere excitation temperature and permits curve of growth determination of the MOLsphere column density (Figure 3). For R Cas on 1984 Jun 12 T_{exc}=1400 K. The CO curve of growth is well fit by a plane-parallel isothermal model with log NL=21.3 and a microturbulance of 3.1 km s^{-1}. The fit implies that the MOLsphere region is thin compared to the stellar radius. Dynamic model atmospheres place gas at this temperature at ~2 R_* with dust condensation at similar or slightly larger radii (~3 R_* [9]) . An ~2 R_* radius is in excellent agreement with mid-IR interferometry of a cool molecular region [10] . A simple model of the MOLsphere shell assuming a typical Mira radius of 400 R_* gives a total MOLsphere mass of $\sim3\times10^{-5}$ $M_\odot$.

3. Mira Masses

In a recent paper [11] we discuss the determination of main sequence masses of Mira progenitors from isotopic abundances. The atmospheric carbon and oxygen isotopic ratios are altered by the CNO cycle, He burning, and mixing processes that take place in the stellar interior over the lifetime of the star [12]. For instance, in a region where an incomplete CNO cycle takes place ^{13}C and ^{17}O increase whereas ^{12}C and ^{18}O are depleted. In the heart of an H-burning zone, where the temperature is higher and the CNO cycle attains the equilibrium, ^{13}C and ^{17}O are also consumed. On the other hand, the ^{12}C is the main product of He burning through the well-known 3α reaction. The products of these fundamental nucleosynthesis processes appear at the stellar surface as a result of deep mixing events that produce substantial modification of the atmospheric abundances of carbon and oxygen. The values of $^{12}C/^{13}C$ and $^{16}O/^{17}O$ are both very dependent on the main-sequence mass of the progenitor stars.

By measuring carbon and oxygen isotopic ratios we demonstrate that most unobscured Miras in the solar neighborhood had progenitor masses ≤ 2 $M_\odot$ [11]. While we were not able to determine the oxygen isotopic ratios for R Cas, the low $^{12}C/^{13}C$ strongly suggests a low progenitor mass. In future studies it will be interesting to see how the structure of the MOLsphere is related to various inputs such as the surface abundances, mass loss rate, luminosity, etc. ultimately stemming from the nature of the progenitor.

Acknowledgments
TL acknowledges support by FWF project P23737.

4. References
[1] Weidemann V 2000 *Astron. Astrophys.* **363** 647
[2] Gail H P and Sedlmayr E 2014 *Physics and Chemistry of Circumstellar Dust Shell* (Cambridge University Press) Chapter 16
[3] Nowotny W, Aringer B, Höfner S, Gautschy-Loidl R and Windsteig W 2005 *Astron. Astrophys.* **437** 273
[4] Hinkle K H 1978 *Astrophys. J.* **220** 210
[5] Tsuji T 2009 *Astron. Astrophys.* **504** 543 and references therein
[6] Hinkle K H, Hall D N B and Ridgway S T 1982 *Astrophys. J.* **252** 697
[7] Hinkle K H, Scharlach W W G and Hall D N B 1984 *Astrophys. J. Suppl. Series* **56** 1
[8] Lebzelter T, Hinkle K H, Wood P R, Joyce R R and Fekel F C 2005 *Astron. Astrophys.* **431** 623
[9] Reid M J and Menten K M 1997 *Astrophys. J.* **476** 327
[10] Perrin G *et al* 2004 *Astron. Astrophys.* **426** 279
[11] Hinkle K H, Lebzelter T L and Straniero O 2016 *Astrophys. J.* submitted
[12] Dearborn D S P 1992 *Phys. Rep.* **210** 367

Journal of Physics: Conference Series **728** (2016) 022004 doi:10.1088/1742-6596/728/2/022004

Mineralogy of an OH/IR superwind

R. Lombaert[1], B. L. de Vries[2] and M. Hillen[3]

[1] Chalmers University of Technology, Department of Earth and Space Sciences, Onsala Space Observatory, 439 92 Onsala, Sweden

[2] ESA, Scientific Support Office, Directorate of Science and Robotic Exploration, European Space Research and Technology Centre (ESTEC/SRE-S), Keplerlaan 1, 2201 AZ Noordwijk, The Netherlands

[3] KU Leuven, Instituut voor Sterrenkunde, Celestijnenlaan 200D B-2401, 3001 Leuven, Belgium

E-mail: [1] `robin.lombaert@chalmers.se`

Abstract. The mineralogy of the dense, dusty superwind of OH/IR stars can provide important constraints for understanding the dust formation process as well as the acceleration of the wind. We aim to model the ISO-SWS spectrum of the OH/IR star OH 30.1-0.7 in detail by reproducing the slope of the spectral energy distribution in the continuum regions between 3 and 8 μm and 12 and 15 μm. In addition to metallic Fe, we find that larger amorphous olivine grains of size on the order of 1 μm may also contribute to the continuum emission. We report here on the preliminary model results and discuss their validity given the important assumption of a 1D spherically symmetric stellar wind.

1. Introduction

When cool stars with intermediate main-sequence masses (5-8 $M_\odot$) ascend the asymptotic giant branch (AGB) they form an O-rich outflow with both a dusty and a gaseous component. These objects can lose mass at rates as high as ($\dot{M} \sim 10^{-4}$ $M_\odot$/yr) causing them to be enshrouded by a dense, dusty superwind [1, 2]. Bright hydroxyl (OH) maser emission and strong infrared (IR) dust emission allow them to be classified as OH/IR stars, many of which are long-period (up to 2000 days) Mira pulsators with a luminosity of $\sim$1-5$\times 10^4$ $L_\odot$.

The high mass-loss rate and cool stellar atmosphere lead to high-density environments in the inner wind where large amounts of dust can form. Constraints on the dust formation process can be inferred by modelling the mineralogy of the stellar wind. The general shape of the spectral energy distribution (SED) is relatively well understood. Two strong absorption features at 10 and 18 μm indicate the presence of amorphous oxygen-rich silicates, including olivine [3, 4]. Other oxygen-rich minerals have been suggested to contribute in the mid-IR as well, such as alumina [5]. However, an additional source of *continuum* opacity was still needed to explain the lack of visual and near-IR flux from these sources. Kemper et al. (2002) suggested that a small abundance of metallic Fe, a solid-state form of Fe atoms, can readily provide this source of continuum opacity in oxygen-rich environments [6].

Metallic Fe can provide the missing continuum opacity, and can help to reproduce the overall shape of the SEDs of OH/IR stars. However, it remains difficult to reconcile the near-IR slope between 3 and 8 μm with the mid-IR flux between 10 and 30 μm.A second contributor to the continuum opacity that allows to change the near-IR slope can alleviate this issue. Here, we

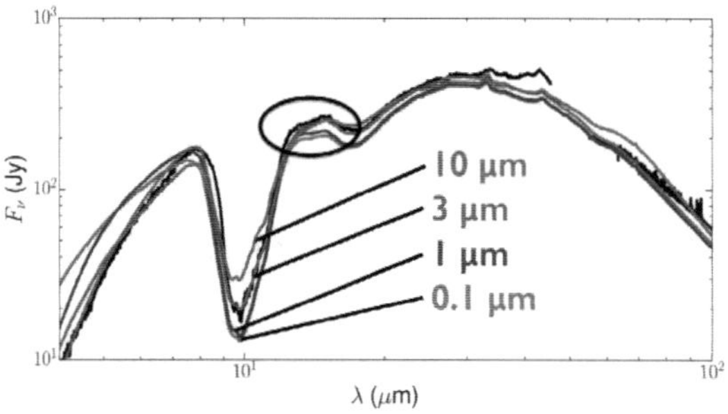

Figure 1. The SED of OH 30.1-0.7 is shown in black (ISO-SWS at $\lambda < 45$ μm and Herschel-PACS at $\lambda > 45$ μm), overlayed with four models that include olivines with a grain size distribution up to different maximum grain sizes as indicated in the figure. The 12-15 μm region is highlighted.

report on a work in progress to show that micron-sized dust grains can provide a solution. We apply this idea to the dusty stellar wind of the OH/IR star OH 30.1-0.7. The results will be published in an upcoming paper and we refer to that paper for further details.

2. Data and Modelling

We use the *Short Wavelength Spectrometer* (SWS) observed with the *Infrared Space Observatory* (ISO) and the *Photodetecting Array Camera and Spectrometer* spectra observed with the *Herschel* telescope, shown in Fig. 1. The relative flux calibration uncertainty between short and long wavelengths in the ISO-SWS spectrum is estimated to be 10%.

We calculate models with the radiative-transfer code *MCMax*, which self-consistently determines the dust temperature profile and resulting SED [7]. We assume a spherically symmetric expanding stellar wind with a terminal velocity of $v_{\infty,\mathrm{g}} = 18$ km s^{-1}, as constrained by the low-J CO lines observed for this source, and find a mass-loss rate of $\dot{M}_\mathrm{d} = 2.5 \times 10^{-6}$ M$_\odot$/yr. The outer radius is taken to be 7600 AU to reproduce the relatively low flux in the far-IR. This is in agreement with the short-superwind scenario reported by de Vries et al. 2014 [8], although the size we use is a factor of three larger. The tenuous wind at radii larger than the outer radius of 7600 AU does not contribute significantly to the IR flux. We calculate the pressure-dependent dust condensation temperature for each species [9] and derive an inner radius of the wind of $R_{\mathrm{i,d}}$ = 22-45 AU. The dust is composed of the amorphous olivine $(Mg_xFe_{1-x})_2SiO_4$ with $x = 0.5$ and $x = 1$, the amorphous pyroxene $MgSiO_3$, the alumina Al_2O_3, and metallic Fe. For completeness we also include crystalline silicates, i.e. forsterite and enstatite [10], and water ice [11], but these species do not have a significant effect on the overall shape of the SED. We assume an MRN grain size distribution [12] given by

$$\frac{n_\mathrm{d}}{da} = A n_\mathrm{H} a^{-3.5}, a_\mathrm{min} < a < a_\mathrm{max} \tag{1}$$

where A is a constant factor, n_H is the hydrogen number density and a is the grain size varying between minimum and maximum values. The minimum grain size is always 0.01 μm and the maximum grain size is varied between 0.1 and 10 μm. We use a dust grain shape model given by the distribution of hollow spheres (DHS [13]). The resulting mass extinction opacities are shown in Fig. 2 for metallic Fe and for amorphous olivine with different maximum grain sizes.

3. Results and discussion

Preliminary results indicate that metallic Fe alone cannot account for the near-IR slope below 8 μm while remaining consistent with the mid-IR flux between 10 and 40 μm. In a first step, we

Journal of Physics: Conference Series **728** (2016) 022004 doi:10.1088/1742-6596/728/2/022004

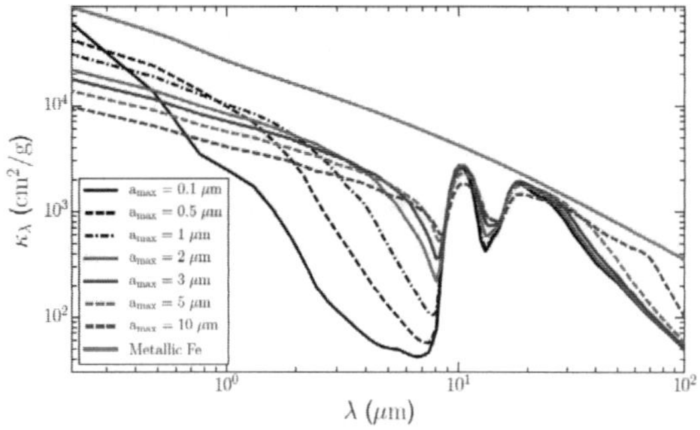

Figure 2. Mass extinction opacities as a function of wavelength for metallic Fe (in red) and the amorphous olivine $(Fe_{0.5}Mg_{0.5})_2SiO_4$ in grain size distributions with several maximum grain sizes (in black, blue and green). The distribution is given by Eq. 1 with a minimum grain size of 0.01 μm.

distinguish between increasing the metallic Fe abundance and increasing the overall dust content through the dust mass-loss rate. For a mass-loss rate of 2.5×10^{-6} $M_\odot$/yr we derive a metallic Fe abundance of $(7 \pm 2)\%$ by mass. While this predicts the overall emission reasonably well in the mid-IR and far-IR wavelength region, the near-IR slope is not reproduced. If we increase the maximum size of the amorphous olivine grains to 1–5 μm, however, we can reproduce the slope much better. Fig. 1 shows the resulting model SEDs for four different maximum grain sizes. These models were not optimised to fit the observations. They illustrate instead the effect of varying the maximum grain size of amorphous olivine.

As highlighted in Fig. 1, the spectral slope in the 12-15 μm region is also affected by the maximum grain size. The slope in the ISO-SWS observations compares well to that measured by the *Infrared Astronomical Satellite*. Moreover, the ISO-SWS spectrum between 12 and 15 μm is measured within a single subband of the spectrometer. The slope in this short wavelength region is thus very reliable. Two model properties affect this slope. First, the slope at a certain wavelength is significantly affected by the maximum grain size only when the Rayleigh limit is not applicable anymore, i.e. when the maximum grain size becomes similar to or larger than that wavelength. While for smaller grains the 12-15 μm region is not affected, the slope does change for maximum grain sizes of 1 μm or larger. This translates to a slope change in the SED. Second, the inclusion of Al_2O_3 in the dust composition changes this slope. However, when including an Al_2O_3 abundance in line with the solar abundance, the amorphous silicate still dominates the emission at 12-15 μm. As a result, the only way to reproduce the 12-15 μm slope is to include a grain size distribution with a maximum grain size of ~ 1 μm.

This result is in agreement with recent theoretical findings regarding wind-driving models of oxygen-rich winds. Grain sizes of at least 0.2 μm are needed to drive even a low mass-loss rate oxygen-rich wind due to the importance of scattering on otherwise nearly transparent silicate grains [14, 15]. Grains of similar size were detected in the extended atmosphere of the low mass-loss rate oxygen-rich AGB star W Hya [16]. Hence, it is possible that grains of 0.1–0.5 μm in size are common in the stellar winds of oxygen-rich AGB stars. Given the much higher mass-loss rates in OH/IR stars, it may be possible that grains can grow to even larger sizes. Interestingly, μm-sized grains may also exist in the interstellar medium, even though there is a preference for carbon-based grains rather than silicate grains [17].

Finally, we note that some of our assumptions may affect these results. We model the source with a spherically symmetric stellar wind, which may not be justified for OH 30.1-0.7. Data taken with *MID-infrared Interferometric instrument* at the *Very Large Telescope Interferometer* array indicate that the source appears smaller in the plane of the sky than predicted by our radiative-transfer models. Further analysis is required to interpret these spatially resolved observations in the context of our models.

4. Conclusions

We make a number of preliminary conclusions regarding the requirements of the dust model to reproduce the ISO-SWS SED of OH 30.1-0.7. These are made under the assumption of one-dimensional spherical symmetry of the wind geometry, the validity of which is being verified in ongoing work.

(i) Metallic Fe contributes to the *continuum* opacity, but it, alone, cannot account for the slope between 3 and 8 μm and between 12 and 15 μm.

(ii) A contribution from micron-sized dust grains to the *continuum* opacity is required to reproduce the slope.

(iii) The silicates are made up of both olivines and pyroxenes. The olivines contain Fe and Mg; only a small abundance of pure-Mg olivines can improve the fit to the 8-μm bump.

(iv) Assuming solar abundance, a near 100% condensation of Fe, Mg and Si from the gas phase into the solid phase is needed.

(v) No alumina or other signs of early dust formation are reliably detected.

References

[1] Renzini A 1981 *Physical Processes in Red Giants* (*Astrophysics and Space Science Library* vol 88) ed I Iben Jr & A Renzini pp 431–446
[2] Iben Jr I and Renzini A 1983 *Annu. Rev. Astron. Astrophys.* **21** 271–342
[3] Sylvester R J, Kemper F, Barlow M J, de Jong T, Waters L B F M, Tielens A G G M and Omont A 1999 *Astron. Astrophys.* **352** 587–599
[4] Suh K and Kim H 2002 *Astron. Astrophys.* **391** 665–674
[5] Maldoni M M, Ireland T R, Smith R G and Robinson G 2005 *Mon. Not. R. Astron. Soc.* **362** 872–878
[6] Kemper F, de Koter A, Waters L B F M, Bouwman J and Tielens A G G M 2002 *Astron. Astrophys.* **384** 585–593
[7] Min M, Dullemond C P, Dominik C, de Koter A and Hovenier J W 2009 *Astron. Astrophys.* **497** 155–166
[8] de Vries B L, Blommaert J A D L, Waters L B F M, Waelkens C, Min M, Lombaert R and Van Winckel H 2014 *Astron. Astrophys.* **561** A75
[9] Kama M, Min M and Dominik C 2009 *Astron. Astrophys.* **506** 1199–1213
[10] de Vries B L, Min M, Waters L B F M, Blommaert J A D L and Kemper F 2010 *Astron. Astrophys.* **516** A86
[11] Dijkstra C, Dominik C, Hoogzaad S N, de Koter A and Min M 2003 *Astron. Astrophys.* **401** 599–611
[12] Mathis J S, Rumpl W and Nordsieck K H 1977 *Astrophys. J.* **217** 425–433
[13] Min M, Hovenier J W and de Koter A 2003 *Astron. Astrophys.* **404** 35–46
[14] Höfner S and Andersen A C 2007 *Astron. Astrophys.* **465** L39–L42
[15] Höfner S 2012 *Nature* **484** 172–173
[16] Norris B R M, Tuthill P G, Ireland M J, Lacour S, Zijlstra A A, Lykou F, Evans T M, Stewart P and Bedding T R 2012 *Nature* **484** 220–222
[17] Wang S, Li A and Jiang B W 2015 *Astrophys. J.* **811** 38

Journal of Physics: Conference Series **728** (2016) 022005 doi:10.1088/1742-6596/728/2/022005

A rotating spiral structure in the innermost regions around IRC +10216

G. Quintana-Lacaci[1], J. Cernicharo[1], M. Agúndez[1], L. Velilla Prieto[1,2], A. Castro-Carrizo[3], N. Marcelino[4], C. Cabezas[5], I. Peña[5], J.L. Alonso[5], J. Zuñiga[6], A. Requena[6], A. Bastida[6], Y. Kalugina[7,8], F. Lique[7] and M. Guélin[3,9]

[1] Group of Molecular Astrophysics. ICMM, CSIC. C/ Sor Juana Inés de la Cruz 3, 28049 Cantoblanco, Madrid, Spain

[2] Centro de Astrobiología, INTA-CSIC. E-28691 Villanueva de la Cañada, Madrid, Spain

[3] Institut de Radioastronomie Millimétrique. 300 rue de la la Piscine, F-38406, Saint Martin d'Hères, France

[4] INAF, Istituto di Radioastronomia, via P. Gobetti 101, 40129, Bologna, Italy

[5] Grupo de Espectroscopía Molecular, Edificio Quifima, Laboratorios de Espectroscopía y Bioespectroscopía. Unidad asociada CSIC, Parque científico Uva, Universidad de Valladolid, Paseo de Belén 5, E-47011, Valladolid, Spain.

[6] Universidad de Murcia. Faculdad de Química. Dpto. de Química-Física. Campus Espinardo E-30100, Murcia, Spain.

[7] LOMC-UMR 6294, CNRS-Université du Havre, 25 rue Philippe Lebon, BP. 1123, 76063 Le Havre cedex, France

[8] Department of Optics and Spectroscopy, Tomsk State University, 36 Lenin av., Tomsk 634050, Russia

[9] LERMA, Observatoire de Paris, PSL Research University, CNRS, UMR 8112, F-75014, Paris, France

E-mail: [1] `g.quintana@icmm.csic.es`

Abstract. ALMA interferometer is providing us molecular maps with unprecedent precision and sensitivity, Key processes in the ejection of matter and dust from these objects occur in their inner zones. We have obtained sub-arcsecond interferometric maps of rotational transitions of metal-bearing molecules towards the prototypical C-rich evolved star IRC+ 10216. While Al-bearing molecular emission presents a roughly spherical shell, the molecular emission from NaCl and KCl presents an elongation in the inner regions, with a central minimum. The presence of the observed features only in KCl and NaCl might be a direct result of their comparatively high dipole moment with respect to the Al-bearing species. The most plausible interpretation for the spatial distribution of the salts is a spiral with a NaCl mass of 0.08 $M_\odot$ with an inner radius of 27 AU. The gas kinematics suggests that it is slowly expanding and rotating. Alternative gas distributions which could result in the presence of the elongation are also explored.

1. Introduction

IRC +10216 is the best studied AGB star. Due to its proximity (~230 pc) and chemical richness it has been the perfect testbench to study the ejection of material in the AGB phase and the circumstellar chemistry around this type of objects. Indeed, around half of all the molecular

Journal of Physics: Conference Series **728** (2016) 022005 doi:10.1088/1742-6596/728/2/022005

species detected in the interstellar medium had been observed for the first time in the ejecta around this object.

The molecular emission observed towards these object has been found, in general, to be roughly spherical, presenting arcs. These arcs have been suggested to be consequence of an spiral structure [1, 2, 3]), which is also a consequence of the presence of a companion star. Infrared observations have shown that the structures present in the innermost region of this star are more complicated than the presence of a spiral. In particular, several authors have shown the presence of a dark-lane with a P.A. of 120° (e.g. [4]) and the presence of an outflow orthogonal to this lane (e.g. [5]). However, while the optical an infrared images could provide better spatial resolution than those reached in the millimeter domain, these former observations lack of kinematical information which is in contrast provided by the molecular emission.

Only recently, the spatial resolution achieved with ALMA is allowing us to probe the structures and, more important, their kinematics of the innermost regions of the circumestellar envelope (CSE).

2. Metal emission as seen with ALMA

We obtained, during ALMA Cycle 0, interferometric maps of the emission from several metal species, such as AlCl, AlF, NaCl and KCl, as well as some of their isotopologues. These maps presented two different spatial distributions. In particular, while the Al-bearing species showed more or less spherical brightness distributions, the emission from the salts KCl and NaCl showed an elongation with a P.A. of ~77° with a central minimum (see Fig. 1). While this elongation seems compatible with the dark-lane already mentioned, there is a clear difference in the P.A. of both structures.

Since it has been shown that the structures of arcs observed towards IRC +10216 might be consequence of the presence of a spiral distribution of dust and gas we simulated the expected result of an ALMA observation of an spiral brightness distribution with the very same characteristics of uv-plane coverage and integration time as the actual observations (see Fig. 2). We found that this spiral distribution is compatible with the structures observed for the salts.

The reason why this structure is not observed in the Al-bearing emission maps is the difference is the dipole moment of the different species. In particular, as the dipole moment of AlCl and AlF (1.00 D and 1.53 D respectively) are much lower than those of the NaCl and KCl (9.0 D and 10.27 D), the excitation of the former molecules is much easier than that of the salts. Due to this, the salts are good tracers of the high density regions. In this particular case, the low density inter-arm regions of the spiral do emit in the Al-bearing species while do not in the salt emission. To reproduce this effect, we created another simulation, this time adding a diffuse inter-arm emission in the brightness distributions. The result showed that for such brightness distribution neither the elongation nor the central minimum are detected, resembling a simulated interferometric map similar to those of the Al-bearing molecules (see Quintana-Lacaci et al. 2016 [6] for more details).

3. Modeling kinematics: rotation

We prepared a simple kinematic model of an spiral structure with both an expansion and rotation velocity field to understand the kinematics of the gas. The expansion velocity field is well known (e.g. [7]). In contrast, no information was available on the possible inclination of the spiral with respect to the plane of the sky, neither on the rotation velocity field. We found that adding a rotation to the velocity field of the spiral resulted in a rotation of the P.A. of the elongation. Therefore, the P.A. of the dark lane and that of the elongation are compatibles if we add rotation.

In order to constrain these parameters, we assumed that the real position of the elongation was that of the dark lane observed in the IR images (i.e images not affect by kinematics). We

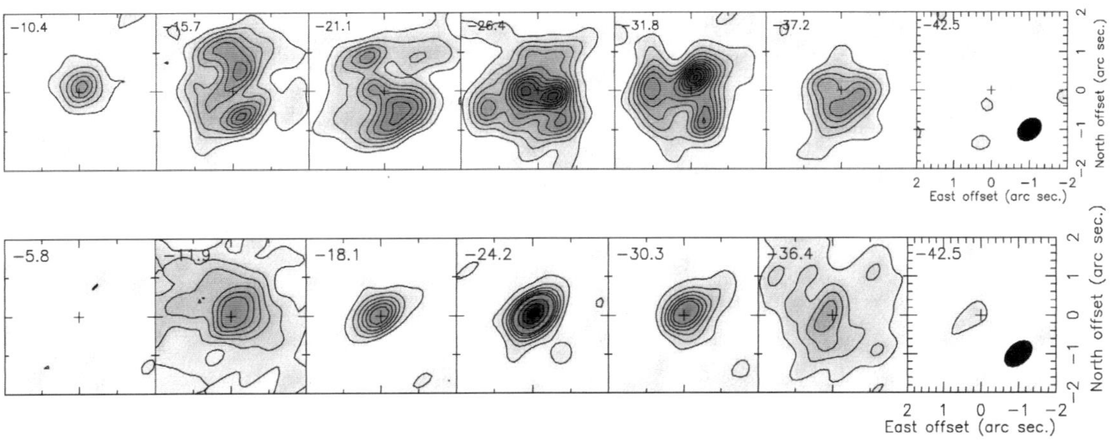

Figure 1. *Top:* Interferometric map of the NaCl $J = 21 - 20$ transition. The HPBW of the synthesized beam is $0.''704 \times 0.''516$ with a P.A. of -52.00°. The lowest contour corresponds to a value of 3σ and the rest are equally spaced in jumps of 5σ with respect to the first contour. *Bottom:* Interferometric map of the AlF $J = 8 - 7$ transition. The HPBW of the synthesized beam is $0.''858 \times 0.''557$ with a P.A. of -228.6°. The lowest contour corresponds to a value of 3σ and the rest of contours until 18σ are equally spaced in jumps of 3σ with respect to the first contour. From 18σ the intensity jump in the contours is 10σ. The rms of the map is $\sigma = 6.4\,\mathrm{mJy/beam}$. On the **upper-left** corner of each panel is written the v_{LSR} of the channel. The beam size is drawn in the last panel. Flux density scale is in Jy/beam.

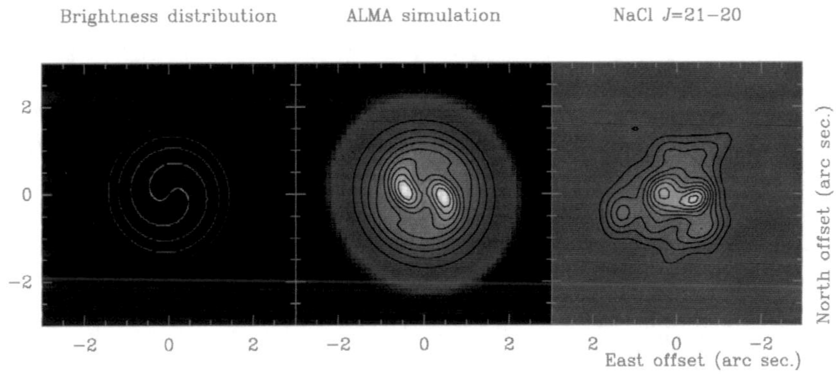

Figure 2. *Left:* Brightness distribution of a face-on Fermat Spiral with $a = 0.2$ at the systemic velocity. *Center:* Result of the ALMA Cycle 0 simulation of the brightness distribution *Right:* NaCl $J = 21 - 20$ emission map for the systemic velocity. The contours of the maps correspond to 10%, 20%, 30%, 40%, 50%, 60% , 70%, 80% and 90% of the peak flux.

found that for a fixed value of the expansion velocity field, the other pair of parameters to by determined, inclination and rotation velocity, where coupled. Therefore, for an expansion velocity of $\sim$11 km/s, as found for the gas at the radius of the NaCl emission, the found an inclination with respect to the plane of the sky of $i \sim 15 \pm 10°$ and a rotation velocity of 8 km/s. This low inclination with respect to the plane of the sky support the arguments of Cernicharo et al. (2015) [2] in favor of an spiral almost in the plane of the sky, in contrast with the conclusions of Decin et al. (2015) [3] suggesting a spiral almost edge-on.

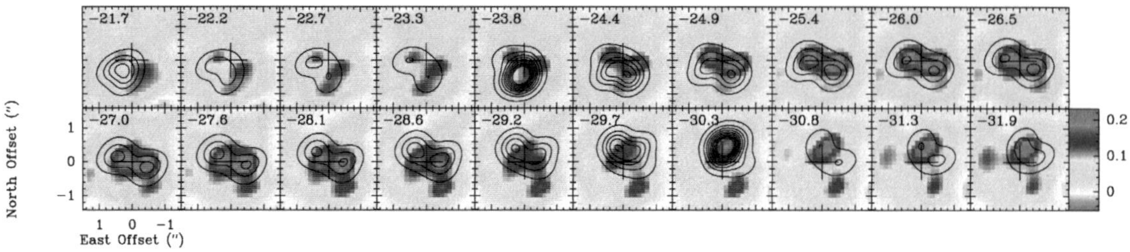

Figure 3. Simulation of an expanding and rotating spiral overplotted to the NaCl $J = 21 - 20$ emission map.

4. Results

Thanks to the capabilities of ALMA and to the high dipole moment of the NaCl and KCl species we have been able to infer a rotation an a tilt in the spiral structure observed around IRC +10216. In fact, this rotation allows to build up a complete picture of the structures around these object. In particular, in the large scale we have an almost spherical structure which responsible is an spiral distribution of gas and dust almost in the plane of the sky. In addition, if we move towards inner regions of the CSE around these object we found that this spiral is rotation, and orthogonal to this rotation plane, an outflow of collimated gas.

Might it be the case that this star is undergoing the shaping process observed for many post-AGB objects, i.e. the appearance of fast bipolar outflows which will carve through the AGB shell, powered by the rotation of a binary star?

Acknowledgments

The research leading to these results has received funding from the European Research Council under the European Union's Seventh Framework Programme (FP/2007-2013) / ERC Grant Agreement n. 610256 (NANOCOSMOS). We would also like to thank the Spanish MINECO for funding support from grants CSD2009-00038, AYA2009-07304 & AYA2012-32032. YK acknowledges the use of HPC resources of SKIF-Cyberia (Tomsk State University).

References

[1] Mauron N and Huggins P J 1999 *Astron. Astrophys.* **349** 203
[2] Cernicharo J, Marcelino N, Agúndez M and Guélin M 2015 *Astron. Astrophys.* **575** A91
[3] Decin L, Richards A M S, Neufeld D, Steffen W, Melnick G and Lombaert R 2015 *Astron. Astrophys.* **574** A5
[4] Jeffers S V, Min M, Waters L B F M, Canovas H, Pols O R, Rodenhuis M, de Juan Ovelar M, Keller C U and Decin L. 2014 *Astron. Astrophys.* **572** A3
[5] Tuthill P G, Monnier J D, Danchi W C and Lopez B. 2000 *Astrophys. J.* **543** 284
[6] Quintana-Lacaci G *et al* 2016 *Astrophys. J.* **818** 192
[7] Agúndez M, Fonfrá J P, Cernicharo J, Kahane C, Daniel F and Guelin M 2012 *Astron. Astrophys.* **543** A48

ALMA observations of the not-so detached shell around the carbon AGB star R Sculptoris

Matthias Maercker[1]

[1] Department of Earth and Space Sciences, Chalmers University of Technology, Onsala Space Observatory, 43992 Onsala, Sweden

E-mail: [1] matthias.maercker@chalmers.se

Abstract. I present our ALMA observations of the CO emission around the carbon AGB star R Sculptoris. The data reveal the known detached shell and a previously unknown, binary induced, spiral shape. The observations confirm a formation of the shell during a thermal pulse about 2300 years ago. The full analysis of the ALMA data shows that the shell around R Scl in fact is entirely filled with molecular gas, and hence not as detached as previously thought. This has implications for the mass-loss rate evolution immediately after the pulse, indicating a much higher mass-loss rate than previously assumed. Comparing the ALMA images to our optical observations of polarised, dust scattered light, we further show that the distributions of the dust and gas coincide almost perfectly, implying a common evolution of the dust and gas, and constraining the wind-driving mechanism. The mass-loss process and amount of mass lost during the thermal pulse cycle affect the chemical evolution of the star, its lifetime on the AGB, and the return of heavy elements to the ISM. New high-resolution ALMA observations constrain the parameters of the binary system and the inner spiral, and will allow for a detailed hydrodynamical modelling of the gas and dust during and after the last thermal pulse. Our results present the only direct measurements of the thermal pulse evolution currently available. They greatly increase our understanding of this fundamental period of stellar evolution, and the implications it has for the chemical evolution of evolved stars, the ISM, and galaxies.

1. Introduction

All stars with a mass between $\approx 0.5 - 8\,M_{\odot}$ evolve along the asymptotic giant branch (AGB). AGB stars are effective producers of heavy elements that are returned to the ISM through mass loss from the stellar surface. The photosphere and expanding CSE show a rich molecular chemistry, and are the formation sites of microscopic dust particles. AGB stars contribute to the chemical enrichment of galaxies and provide up to 70% of all interstellar dust particles [1]. The expelled dust and gas are incorporated into molecular clouds, where new stars and planets form. Understanding the chemistry in AGB stars, the mass loss, and the mixing of material into the ISM is therefore crucial to understanding early star formation and the evolution of galaxies.

During its AGB evolution, a star periodically undergoes rapid helium burning in a shell around the core. This phenomenon is known as a thermal pulse (TP), and lasts for only a few hundred years every $10^4 - 10^5$ years. The release of the extra energy into the stellar envelope causes the star to restructure, leading to the formation of elements inside the star (mainly carbon and s-process elements). Most of the stellar mass is lost between pulses. The new elements are mixed to the stellar surface and incorporated into the stellar wind, leading to the chemical evolution of the CSE [2]. In particular, the mixing-up of extra carbon leads to the evolution

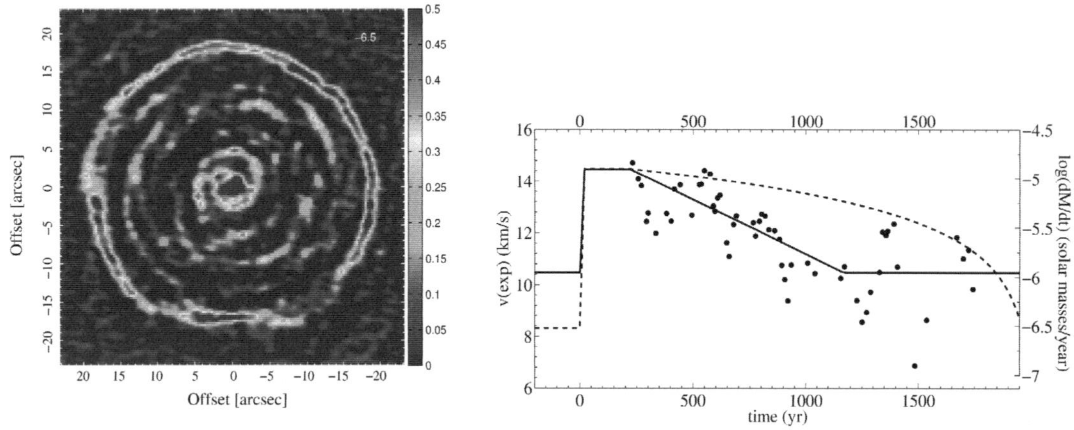

Figure 1. *Left:* ALMA Cycle 0 observations in ^{12}CO($J = 3 - 2$) of the detached shell around R Scl at the systemic velocity. The detached shell can be seen at 18.5", as well as the spiral structure due the shaping of the wind by a binary companion. The flux scale is given in Jy/beam. *Right:* The evolution of the mass-loss rate and expansion velocity as derived from hydrodynamical models [10].

from oxygen-rich (or M-type) AGB stars (with an atmospheric abundance ratio C/O<1), to carbon-rich AGB stars (with C/O>1).

The amount of new elements created depends critically on the physical parameters of the star during subsequent TPs. Owing to their short duration, and the long timescales between pulses, it is extremely unlikely to observe an AGB star during a TP directly. As a consequence models of TPs have been essentially independent of observations [3].

The increase in mass-loss rate and wind velocity during a TP cycle leads to the creation of a shell that may sweep up previously ejected material, leading to the appearance of a detached shell [4, 5, 6]. One of the few ways to study the thermal-pulse phenomenon are the observations of detached shells that have been found around ≈10 carbon stars in molecular line emission, as well as in stellar dust-scattered light at optical wavelengths [7, 8, 9, 10, 11].

2. The shell around R Sculptoris

We observed the circumstellar environment around the carbon AGB star R Scl using the compact configuration of ALMA in cycle 0. R Scl was already known to have a detached shell, observed in highest detail in dust scattered, stellar light with the Hubble Space Telescope [8]. The ALMA observations provided the highest-spatial resolution images in ^{12}CO of the CSE around R Scl, and clearly show the detached shell and a previously undetected spiral structure, shaped through interaction of the wind with a so far unknown binary companion (Fig. 1, left; [10]). Hydrodynamical and radiative transfer models of the observed ^{12}CO(3–2) emission from the spiral shape and shell, allowed us to constrain the evolution of the mass-loss rate and expansion velocity during and after the thermal pulse. The shell has an age of 2300 years. We can model the observed structure by assuming a thermal-pulse mass-loss rate of $\approx 2 \times 10^{-5}\,M_\odot\,yr^{-1}$, followed by a slow decline of the mass-loss rate to the present-day value of $3 \times 10^{-7}\,M_\odot\,yr^{-1}$ (Fig. 1, right; [10]). Theoretical models predict a much faster decline in mass-loss rate.

A full analysis of the ALMA Cycle 0 data, including the ^{12}CO(1–0) and ^{12}CO(2–1) lines, allowed us to model the CO emission from the shell and the extended CSE *separately* [12]. Radiative transfer models of the three CO line emissions from the shell constrain the shell mass and temperature. We derive a shell mass of $4.5 \times 10^{-3}\,M_\odot$ and a temperature of 50 K. For typical

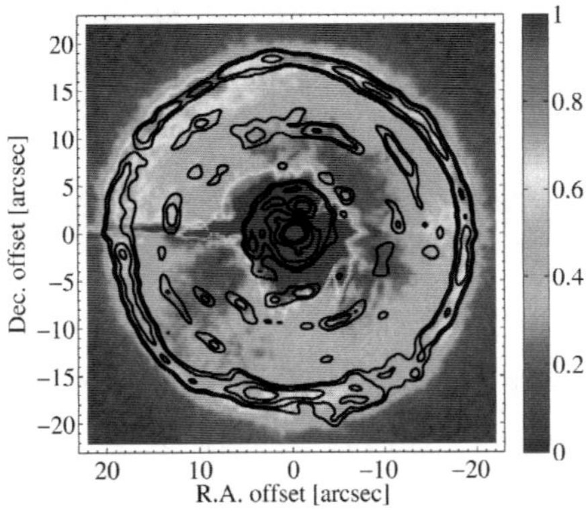

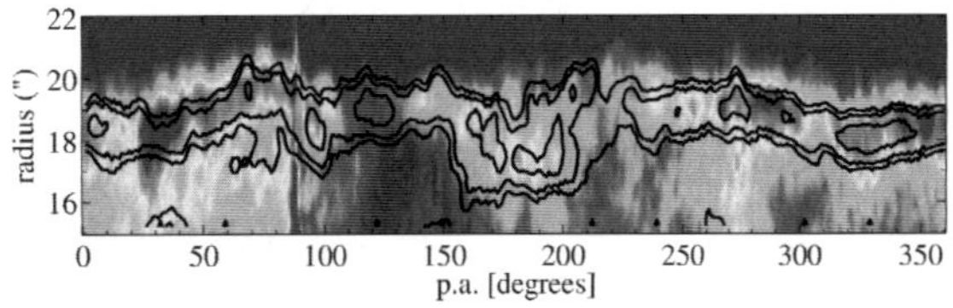

Figure 2. *Left:* Comparison of the PolCor polarised intensity image for R Scl (color scale) with the ALMA Cycle 0 $^{12}CO(3\text{–}2)$ data at the systemic velocity. *Right:* Position angle vs. radius for the same image, but of the detached shell only, showing the close correlation between the dust and gas shells [11].

thermal pulse timescales this implies a thermal pulse mass-loss rate of $2.3 \times 10^{-5}\,M_\odot\,yr^{-1}$. We then use the model of the shell and subtract the emission from single-dish observations, providing us with spectral lines from the extended CSE inside the shell. The analysis of the *overall* density distribution inside the shell (i.e. not modelling the spiral explicitly) shows that the mass-loss rate remains high even after the pulse (on average $1.6 \times 10^{-5}\,M_\odot\,yr^{-1}$ during the last 2100 years), confirming our results from the hydrodynamical models. As a consequence of the slow decline in mass-loss rate, the shell around R Scl does not appear detached, but is instead filled by the extended CSE [12]. The high post-pulse mass-loss rate causes the star to lose significantly more mass throughout the cycle ($0.03\,M_\odot$ vs. $0.007\,M_\odot$; [12]) than previously predicted [6, 3].

Comparison to our observations in dust-scattered optical light with PolCor additionally show an essentially identical distribution of the dust and gas in the detached shell around the star (Fig. 2), implying a common evolution [11]. This is in contrast to the detached shells observed around the carbon AGB star U Ant, where the dust and gas are clearly separated [9]. Detailed observations of the dust and gas in detached-shell sources hence describe the evolution of the mass loss, and the interaction between the dust and gas.

3. New high-spatial resolution ALMA observations

In ALMA Cycle 1 we observed the inner part of the spiral around R Scl at a spatial resolution of 0.12" (Fig. 3). The images resolve the beginning of the spiral, clearly revealing the effects of binary interaction on the wind, and indicating the binary separation. By measuring the deviations from a perfect spiral in the Cycle 1 observations, as well as the contrast between the spiral windings and the inter-winding material [13], we will be able to determine the eccentricity, inclination, and period of the orbit, as well as the companion mass. By constraining these parameters we will be able to fully exploit the Cycle 0 observations to understand the evolution of the mass-loss rate and wind expansion velocity during, and after, the last thermal pulse. Resolving the binary pair will additionally allow us to determine the present-day mass of R Scl.

4. Conclusions

Thermal pulses dominate the chemical evolution of the most numerous stars in the Galaxy. Determining the evolution of the mass-loss rate during this critical phase of stellar evolution is essential in understanding the chemical enrichment of the ISM and galaxies. Our observations combine images of dust and gas at multiple wavelengths and at high-angular resolution to

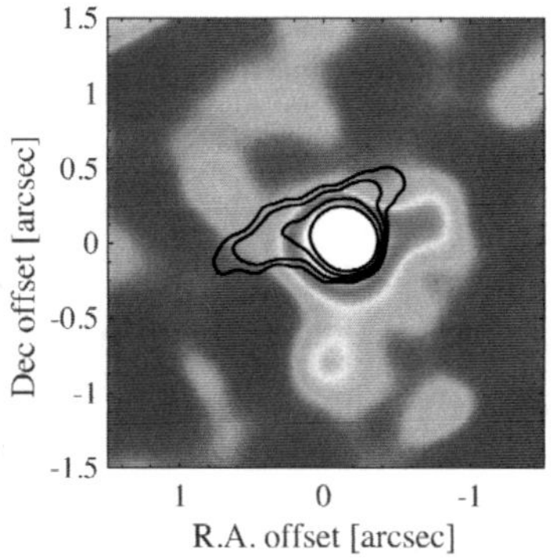

Figure 3. ALMA cycle 1 observations in $^{12}CO(3\text{–}2)$ of the inner part of the CSE around R Scl at the systemic velocity. The images show the beginning of the spiral (color scale), and an extension of the continuum emission (black contours). The observed structure show clear signs of the interaction of the binary companion with the AGB wind, and indicate the location of the companion.

describe the circumstellar structure. We set constraints on the thermal pulse cycle, one of the main drivers of AGB evolution. We find that the mass-loss rate remains significantly higher immediately after the pulse than predicted by stellar evolution models. As a consequence, much more mass may be lost during the thermal pulse cycle than previously expected, and hence limit the life-time of the star on the TP-AGB. This effects the amount of new material returned to the ISM, and hence the contribution of AGB stars to the chemical evolution of galaxies.

References

[1] Schneider R, Valiante R, Ventura P, dell'Agli F, Di Criscienzo M, Hirashita H and Kemper F 2014 *Mon. Not. R. Astron. Soc.* **442** 1440–1450
[2] Busso M, Gallino R and Wasserburg G J 1999 *Annu. Rev. Astron. Astrophys.* **37** 239–309
[3] Karakas A and Lattanzio J C 2007 *Publ. Astron. Soc. Australia* **24** 103–117
[4] Steffen M and Schönberner D 2000 *Astron. Astrophys.* **357** 180–196
[5] Schöier F L, Lindqvist M and Olofsson H 2005 *Astron. Astrophys.* **436** 633–646
[6] Mattsson L, Höfner S and Herwig F 2007 *Astron. Astrophys.* **470** 339–352
[7] Olofsson H, Bergman P, Eriksson K and Gustafsson B 1996 *Astron. Astrophys.* **311** 587–615
[8] Olofsson H, Maercker M, Eriksson K, Gustafsson B and Schöier F 2010 *Astron. Astrophys.* **515** A27
[9] Maercker M, Olofsson H, Eriksson K, Gustafsson B and Schöier F L 2010 *Astron. Astrophys.* **511** A37+
[10] Maercker M *et al* 2012 *Nature* **490** 232–234
[11] Maercker M, Ramstedt S, Leal-Ferreira M L, Olofsson G and Floren H G 2014 *Astron. Astrophys.* **570** A101
[12] Maercker M, Vlemmings W H T, Brunner M, De Beck E, Humphreys E M, Kerschbaum F, Lindqvist M, Olofsson H and Ramstedt S 2016 *Astron. Astrophys.* **586** A5
[13] Kim H and Taam R E 2012 *Astrophys. J.* **744** 136

Journal of Physics: Conference Series **728** (2016) 022007 doi:10.1088/1742-6596/728/2/022007

LBVs, hypergiants and impostors — the evidence for high mass loss events

Roberta M Humphreys[1]

[1] Minnesota Institute for Astrophysics, University of Minnesota, Minneapolis, MN, USA

E-mail: [1] roberta@umn.edu

Abstract. Mass loss from massive stars is common. It plays an important role in the evolution of stars above about 20 $M_{\odot}$. In the massive hot stars the winds and mass loss are driven by radiation pressure on the lines. The mass loss mechanism in post-main sequence red supergiants is still debated but pulsation and convection play a role. In this short talk, I am emphasizing the evidence for high mass loss episodes in evolved massive stars with specific examples such as VY CMa, IRC +10420 and the giant eruptions of LBVs, the possible origin of these episodes, and their importance in the final stages of massive star evolution. By analogy with the less massive AGB stars, I suggest that VY CMa is a candidate for a second red supergiant stage.

1. Introduction

We observe evidence for episodic high mass loss events across the upper HR Diagram ranging from the non-terminal giant eruptions of very massive stars such as eta Car, enhanced mass loss in the LBV stage, and in the evolved warm and cool hypergiants near the upper luminosity boundary. I'll begin this brief overview with the cool side of the HR diagram with highlights of recent work on the mass loss histories of the extreme red supergiant VY CMa and the post-red supergiant IRC +10420.

2. The warm and cool hypergiants

A few highly unstable, very massive stars lie on or near the empirical upper luminosity boundary in the HR diagram. In this paper I use the term hypergiant for the evolved stars that lie just below this upper envelope with spectral types ranging from late A to M. They represent a very short-lived evolutionary stage, characterized by high mass loss and eruptive events. Many of them are strong infrared sources and powerful OH masers.

2.1. The recent mass loss history of VY CMa

The powerful infrared source and OH maser VY CMa is one of the most luminous and largest evolved cool stars known. With its very visible asymmetric nebula combined with its high mass loss rate, VY CMa is a special case even among the cool hypergiants that define the upper luminosity boundary in the HR Diagram. Multi-wavelength HST/WFPC2 images of VY CMa [1] reveal a complex circumstellar environment (Figure 1) dominated by three prominent arcs plus bright clumps of dusty knots near the star, all of which are evidence for multiple and asymmetric mass loss episodes. The random orientations of the arcs suggested that they were produced by localized ejections, not aligned with an axis of symmetry.

Journal of Physics: Conference Series **728** (2016) 022007 doi:10.1088/1742-6596/728/2/022007

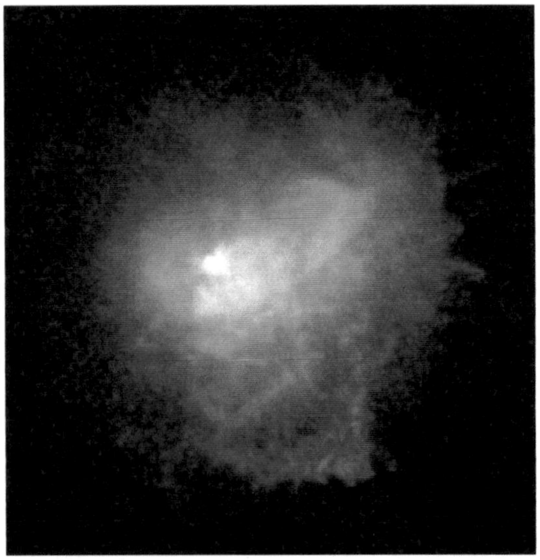

Figure 1. The multi-color HST visual image of VY CMa [1]

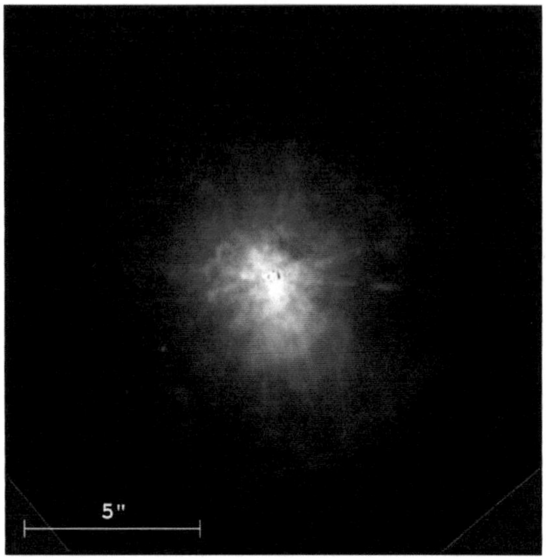

Figure 2. The multi-color HST visual image of IRC +10420 [10]

Second epoch HST/WFPC2 images to measure the transverse motions, which when combined with the Doppler velocities [2], provide a complete picture of the kinematics of the ejecta including the total space motions and directions of the outflows [3]. The arcs and clumps of knots are moving at different velocities, in different directions, and at different angles relative to the plane of the sky and to the star, confirming their origin from eruptions at different times and from physically separate regions on the star. They were ejected in separate events over the past 800 years. The clumps and arcs are massive, with masses on the order of few $\times 10^{-3}$ $M_{\odot}$ [4]. This activity could be due to magnetic/convective regions and events analogous to solar activity, i.e. "starspots". Starspots and large surface "asymmetries" have been observed on several stars including Betelgeuse and other red giants, AGB stars and supergiants. The magnetic field strength has now been measured in the ejecta of many of these stars from their OH, H_2O, and SiO masers[5]. For VY CMa they imply surface field strengths of the order of 200 – 400 G. Large-scale convective activity may thus be a cause of high mass loss episodes in evolved, luminous cool stars leading to the formation of intricate arcs and loops as the shocks move through the ejecta from the star's more quiescent outflow.

Long-wavelength (11–37μm) imaging with SOFIA reveals a cooler component with the same asymmetric nebula seen in the visible (Figure 3) but the lack of cold dust at much greater distances indicates that its high mass loss is limited to the last 1200 years [6].

VY CMa also has a unique chemistry compared to other red supergiants. Twenty different molecules including carbon compounds have been identified in its ejecta [7,8]. Many of these molecules are found in the massive arcs and may be produced by shocks in the outflows. In addition, VY CMa has an unusually high C^{12}/C^{13} ratio for an O-rich supergiant[9]. Its chemistry plus it unique circumstellar environment leaves us with some interesting questions about VY CMa's evolutionary state.

2.2. IRC +10420 – two high mass loss periods

IRC +10420 is a powerful infrared source, the warmest maser source known and in the past 30 years or so its apparent spectral type has gone from late F-type to a mid-A. HST/WFPC2

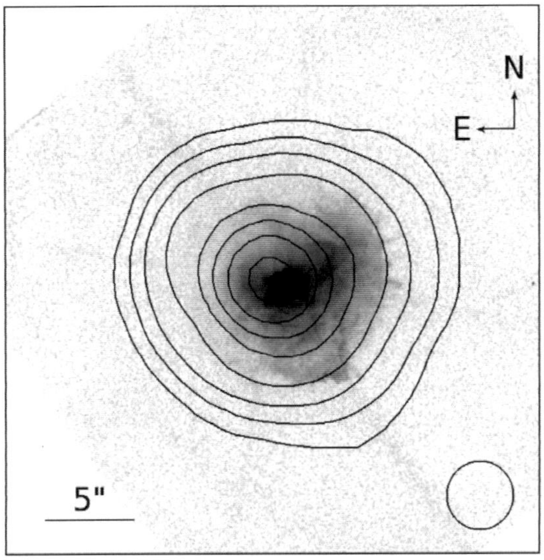

Figure 3. The SOFIA/FORCAST 37μm contours overlaid on the HST visual image, in Figure 1. The circle is the 37μm beam size. The extensions to the NW and SW correspond to the direction of ejection of the NW Arc and Arc 1 [2,3].

images [10] reveal a complex circumstellar environment (Figure 2), with a variety of structures including condensations or knots, ray-like features, and several small, semi–circular arcs or loops within 2" of the star, plus one or more distant reflection shells. These features are all evidence for high mass loss episodes during the past few hundred years. Like VY CMa, the circular polarization of its OH masers imply the presence of magnetic fields in its ejecta with a surface field of 300 – 400G at the star.

Following the example from VY CMa, Tiffany et al.[11], measured the transverse motions of numerous knots, arcs, and condensations in its inner ejecta from second epoch *HST/WFPC2* images. When combined with the radial motions for several of the features, the total space motion and direction of the outflows show that they were ejected at different times, in different directions, and presumably from separate regions on the surface of the star. These discrete structures in the ejecta are kinematically distinct from the general expansion of the nebula and their motions are dominated by their transverse velocities. We find that they are all moving within a few degrees of the plane of the sky. We are thus viewing IRC +10420 nearly *pole-on* and looking nearly directly down onto its equatorial plane. This result is confirmed by independent interferometry [12] and the polarimetry [13].

Recently Shenoy et al. [6] found that IRC +10420's spectral energy distribution at the long wavelengths (11 - 37 μm) from SOFIA/FORCAST imaging cannot be explained by a single mass loss rate. A substantial change in the rate of mass loss that occurred in the past several thousand years is required. Although the transition was very likely gradual, at least two high mass loss periods are required. The first, from about 6000 years ago ago with a very high mass loss rate of 2×10^{-3} $M_\odot$ per year, dropped to about 10^{-4} $M_\odot$ per year beginning about 2000 years ago. We suggest that this change in the mass loss rate is due its post-red supergiant evolution to warmer temperatures.

2.3. The mass loss mechanism and the evolutionary state

Mass loss from red giants and red supergiants (RSGs) has been known since the 1960s, but the mass loss mechanism for red supergiants is still not understood. The leading processes have included radiation pressure on grains, pulsation and convection. Pulsation and dust-driven winds have been successful at explaining the mass loss of the Miras and AGB stars which are fundamental mode pulsators, but are not adequate for the less variable RSGs with their very extended low density atmospheres.

The defining signature of mass loss in RSGs is the presence of circumstellar dust usually revealed as excess radiation in their spectral energy distributions (SEDs) from the silicate emission features at 10 and 20 μm. To a first order, the strength of the silicate emission feature appears to be correlated with the luminosity and apparent temperature as revealed by the spectral type; i.e, the higher the luminosity and later the type (or cooler the star) the stronger the silicate emission and the larger the infrared excess. But the measured mass loss rates at a given luminosity have a large scatter of 10 to 100 times [14], suggesting that other factors are important such as the mass loss mechanism or evolutionary state.

Ground and space-based high resolution imaging and interferometry of evolved massive and luminous stars are transforming our view of circumstellar ejecta, mass loss and the mass loss mechanism in evolved stars. The discovery of large-scale surface asymmetries or hot spots on the surfaces of red supergiants, which vary on timescales of months or years, lends support for convection as an important mechanism for the RSGs. The massive arcs and clumps of knots in the ejecta of VY CMa and IRC +10420, plus the presence of magnetic fields in their ejecta and in other RSGs and AGB stars, suggest that *enhanced convective activity together with magnetic activity may be important for these high mass ejections.*

So, why not more VY CMa's?

There is an observed correlation for increased mass loss with increasing luminosity and cooler temperatures among the red supergiants. Do RSGs evolve through the red supergiant stage getting apparently cooler with more extended envelopes and higher mass loss rates? Or like lower mass stars, could there be more than one RSG state? For example, a post-RSG warm hypergiant could evolve back to the red supergiant stage a second time becoming an extreme RSG, a VY CMa. With its massive arcs and clumps and evidence for extreme activity plus its peculiar chemistry with carbon compounds, VY CMa is a candidate for a second RSG stage.

3. Luminous Blue Variables (LBVs)

The term LBV has been rather loosely used in the astronomical literature during the past few years, so I begin with a description of the LBV/S Dor variability and the distinction with giant eruptions.

3.1. What is an LBV/S Dor variable?

An LBV or S Dor variable is distinguished by its rather unique spectroscopic and photometric variablilty. In quiescence, an LBV or S Doradus variable is a moderately evolved hot star, with a B-type supergiant or Of-type/late-WN classification. In its maximum light stage or "LBV eruption", enhanced mass loss causes its wind to become dense and opaque, with a large pseudo-photosphere at $T \sim$ 7000-8000K resembling the spectrum of an F-type supergiant. On an HR diagram the object thus appears to move toward the right. Since this alters the bolometric correction, the visual brightness increases by $\sim$ 1 to 2 magnitudes while the total luminosity remains nearly constant [15,16] or may decrease [17]. Such an event can last for several years or even decades.

There is no consensus on the origin of the LBV instability, but most explanations invoke their proximity to their Eddington limit, and include the opacity-modified Eddington limit, rotation, super-Eddington winds, or gravity-mode instabilities.

One of the distinguishing characteristics of LBV/S Dor variability is that during quiescence or minimum light they lie on the S Dor instability strip first introduced by Wolf [18], see Figure 3. The more luminous, classical LBVs above the upper luminosity boundary have very likely not been red supergiants, while those below may be post-red supergiant candidates. Thus LBVs with very different initial masses and different evolutionary histories occupy the same locus in the HR Diagram. This empirical relation remains unexplained.

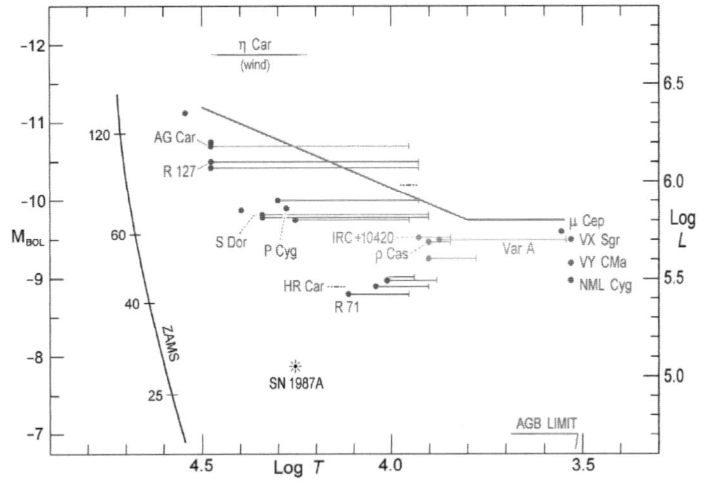

Figure 4. A schematic HR Diagram. A sample of known LBV/S Dor variables are shown in blue. The straight blue lines illustrate their apparent transits in the HRD during the LBV optically dense wind state. The dark green line is the upper luminosity boundary. Several cool (red) and warm hypergiants (green) are also shown.

3.2. The supernova impostors

In rare cases, however, the luminosity substantially increases during outburst; these have been called giant eruption LBVs [16] or "supernova impostors"[18] because they are often initially mistaken for true supernovae and receive a supernova designation. Their luminous energies often rival that of true supernovae. Historical examples include the "great eruption" of eta Car in 1843, P Cygni in 1600, SN 1961V, and V12 in NGC 2403 (SN 1954J). The distinction between giant eruptions and the more common LBV or S Dor-type variability is often overlooked in the literature. They may be related and originate from similar types of stars, perhaps in the same evolutionary stage, but the physical cause of the eruption or instability is very likely different. Certainly the energetics of the eruptions and what we observe are very different. There are numerous questions about the origin of the giant eruptions, their relation to normal LBV outbursts, and perhaps even to SNe. These eruptions are important. They may account for considerable mass loss and they indicate that some instability has been overlooked in stellar theory. *But it is important to distinguish them from the normal LBV/S Dor variability.*

References

[1] Smith N, Humphreys R M, Davidson K, Gehrz R D, Schuster M T and Krautter J 2001 *Astron. J.* **121** 1111
[2] Humphreys R M, Davidson K, Ruch G and Wallerstein G 2005 *Astron. J.* **129** 492
[3] Humphreys R M, Helton L A and Jones T J 2007 *Astron. J.* **133** 2716
[4] Shenoy D P *et al* 2013 *Astron. J.* **146** 90
[5] Vlemmings W H T, van Langevelde H J and Diamond P J 2005 *Astron. Astrophys.* **434** 1029
[6] Shenoy D P, Humphreys R M, Jones T J *et al* 2016 *Astron. J.* **151** 51
[7] Ziurys L M, Milam S N, Apponi A J and Woolf N J 2007 *Nature* **447** 1094
[8] Ziurys L M, Tenenbaum E D, Pulliam R L, Woolf N J and Milam S N 2009 *Astrophys. J.* **695** 1604
[9] Milam S N, Woolf N J and Ziurys L M 2009 *Astrophys. J.* **690** 837
[10] Humphreys R M, Smith N, Davidson K, Jones T J, Gehrz R T, Mason C G, Hayward T L, Houck J R and Krautter J 1997 *Astron. J.* **114** 2778
[11] Tiffany C, Humphreys R M, Jones T J and Davidson K 2010 *Astron. J.* **140** 339
[12] de Wit W J *et al* 2008 *Astrophys. J.* **685** L75
[13] Shenoy D P, Jones T J, Packham C and Lopez-Rodriguez E 2015 *Astron. J.* **150** 15
[14] Mauron N and Josselin E 2011 *Astron. Astrophys.* **526** A156
[15] Wolf B 1989 *Astron. Astrophys.* **217** 87
[16] Humphreys R M and Davidson K 1994 *Publ. Astron Soc. Pac.* **106** 1025
[17] Groh J H, Hillier D J, Daminelli A, Whitelock P A, Marang F and Rossi C 2009 *Astrophys. J.* **698** 1698

[18] van Dyk S D 2005 *The Fate of the Most Massive Stars* (ASP Conference Series vol 332) ed Humphreys R M and Stanek K Z p 47

Roberta Humphreys giving her talk.

Giant eruptions of very massive stars

Kris Davidson[1]

[1] School of Physics and Astronomy, University of Minnesota, Minneapolis MN 55455, U.S.A.

E-mail: [1] kd@astro.umn.edu

Abstract. Giant eruptions or supernova-impostor events are far more mysterious than true supernovae. An extreme example can release as much radiative energy as a SN, ejecting several $M_\odot$ of material. These events involve continuous radiation-driven outflows rather than blast waves. They constitute one of the main unsolved problems in stellar astrophysics, but have received little theoretical attention. The most notorious giant-eruption survivor, η Carinae, is amazingly close to us for such a rare event. It offers a wealth of observational clues, many of them quite unexpected in terms of simple theory.

1. Introduction

Let me steal a metaphor from Tom Wolfe. Some of us think that a Demon lives near the Eddington Limit. If you (as a massive star) try to approach that limit, he intervenes before you get very close to it. He shakes you so violently that you lose mass and energy, and he hurls you back away from the edge. After watching this happen to a number of stars, we have never seen the Demon's face. In hindsight his behavior almost makes sense in terms of physics, and it dramatically alters the evolution of very massive stars. But the only certain factor is that no theorist predicted it.

This idea grew from several disparate topics. Three decades ago, Luminous Blue Variable stars (LBVs) attracted attention because their sporadic mass-loss events could explain why there are no yellow and red supergiants with $L > 10^6 L_\odot$ [1]. LBVs have $L/M > 0.4\,(L/M)_{\rm Edd}$. After 1980 the nature of η Carinae's giant eruption in 1830–1860 also became apparent; as usual, one really good example provided better clues than dozens of less extreme ones. That event expelled 10 to 30 $M_\odot$ of material and the same amount of light as a typical supernova, in a timescale roughly 50 times as long as a SN event, and the star survived. By 1990 the role of eruptive mass loss in the most massive stars was widely recognized [2]. But then a sort of collective amnesia occurred after 1995; astronomers almost seemed to forget the earlier work. When mass loss in ordinary hot-star winds was reassessed downward around 2005, invalidating the published evolutionary tracks [3], experts again proposed eruptive loss. In fact its role had been familiar to many people 15 years earlier (see many refs. in [2]).

Meanwhile, SN surveys revealed abnormal explosions in other galaxies. Instead of obvious blast waves, they produced slower, more continuous outflows resembling η Car's great eruption [4]. Some were labeled "Type IIn supernovae," which implies pre-existing circumstellar ejecta. And occasionally the star survived; we call such cases Supernova Impostors, with η Car as the obvious prototype. Equally embarrassing, a few SN events had precursor outbursts, which seemed paradoxical in the textbook view of supernovae. One supernova in 2012, for instance, already had the name SN 2009ip!

Few researchers believe that any one instability mechanism produced all the eruptions that I've mentioned. Several types probably co-exist: (1) Core-collapse SNae that had unusual extended envelopes when they exploded; (2) other core phenomena, mostly but not entirely related to nuclear processes; (3) the Demon mentioned earlier, a hypothetical set of radiative/fluid instabilities that can arise when $L/M > 0.5\,(L/M)_{\rm Edd}$; (4) binary interactions, mass transfer, and/or mergers; and (5) whatever we haven't thought of. Incidentally, number 4 is frequently offered as a panacea, but the statistics make that very unlikely. Here let me emphasize a generality that hasn't been widely publicized: Since all these conjectural types of eruption share the same radiative outflow physics, *they look alike* when viewed from outside.

As you might guess, the situation has become too confused for a newcomer to learn easily. Some basic fallacies have propagated in the literature. For instance, most supernova specialists assume that progenitors of Type IIn must be LBVs, based on faulty logic. (That idea *may* be true, but the evidence cited for it has bad premises.) Many recent authors have applied the term LBV to stars that might belong to that class but are unproven, and to other stars that *don't* belong. Big evolutionary differences between luminous and less-luminous LBVs are seldom acknowledged [5]. Various explain-all "models" consist mainly of words, cartoon sketches, and/or computer runs with many unadvertised assumptions. Reciting their defects would fill many pages. Thus, I earnestly advise everyone to be wary of groupthink – and equally wary of claims that some paper has revolutionized the topic.

Since this is an account of some concepts, and not of the literature, few papers will be cited here. For general background see articles by many authors in [6], and older references in [2] – a review which, amazingly, has not been superseded nor seriously disproven after 22 years (except that it said too little about rotation).

2. Radiative physics in a giant eruption

Let's begin with a clear definition. A stellar "giant eruption" is a *super-Eddington mass outflow,* driven by continuum radiation pressure. It is not driven by a shock wave, though shocks may propagate through the flow. It's opaque, so the photosphere is located at a fairly large radius in the outflow. The eruption persists for months or years, much longer than any relevant dynamical timescale. It is quite likely to be non-spherical, e.g. the famously bipolar case of η Car. The word "eruption" is especially apt in some models that behave like geysers, with instability propagating inward while expelling mass outward [2]. Standard large LBV outbursts are not giant eruptions, but they have physical similarities.

We often mention the Eddington parameter $\Gamma = L/L_{\rm Edd}$, where $L_{\rm Edd} \approx 4\pi cGM/\kappa$. Most of the opacity κ is due to Thomson scattering, but the small absorption opacity determines the size of the photosphere (see below). Eta Car's great eruption had $\Gamma \sim 2$ to 10, but SN 2011ht and SN 2009ip had $\Gamma > 50$. Occasionally it is claimed that such large values cast doubt on the entire concept. In fact, however, the basic outflow math for $\Gamma \rightarrow \infty$ is not much different from $\Gamma \sim 4$. The star's mass M then has little effect and everything depends on L and on the sonic point where the flow originates; since higher eruption luminosities generally have larger size scales, the outflow speeds usually remain below 1000 km s^{-1} even if $\Gamma > 50$.

In principle a giant eruption can originate in more than one way. The Demon instability mentioned above, for example, might be an "opacity-modified Eddington limit" affair not far below the star's photosphere, or (more likely) it may involve strange-mode instabilities in the notorious high-opacity layers where $T \sim 3 \times 10^5$ K [2]. Either way, a lot of extra radiation cannot escape without pushing the outer layers. But a core-collapse SN can also become a giant eruption. Initially, of course, a SN blast wave occurs. Normally it produces a bright display when it reaches the star's surface. But suppose the star is surrounded by a large opaque envelope 1000 times as dense as an ordinary stellar wind – something resembling the wind that η Car had 70 years ago. It's easy to show that photons then diffuse outward ahead of the blast wave. (See, e.g., [10]. Supernova enthusiasts call this phenomenon "shock breakout," but it's really photon breakout.) The diffusing radiation accelerates a giant eruption that precedes the blast wave. The shock doesn't reach the outflow's photosphere until a time well after the peak brightness.

Logically, this account of a giant-eruption SN transfers the problem to *why* that dense circumstellar stuff was there. It requires a precursor outflow, a less luminous giant eruption in the last few years before the main explosion. But this seems counter-intuitive, because the tiny pre-SN core with its rapid nuclear timescale is not supposed to know about the star's outer layers, and vice-versa. This looks like evidence that the Demon lives in the core of the star, rather than the outer layers as some of us have usually supposed. But if that's true, then why is there an LBV instability strip in the HR diagram, representing only the outer layers? And why does it explain the HRD's upper limit so nicely? (See [2]). Are there two different Demons? Or more? These are among the biggest questions in stellar astrophysics, partly because no one has a credible answer yet.

Unfortunately a super-Eddington flow is very difficult to calculate, because 3-dimensional effects may be crucial. The ejected mass and velocities in η Car's giant eruption don't match simple 1-D outflow calculations [7,8]. This shouldn't surprise us, since it's conceptually "a light fluid pushing a heavy one" à la Rayleigh-Taylor. R-T instabilities within a star imply convection, but a giant eruption is a supersonic outflow. Likely result: local mass concentrations form and photons escape preferentially along the easiest paths between blobs – thus reducing the effective κ so far as radiative acceleration is concerned. Indeed the ejecta around η Car show obvious granulation with reasonable size scales. This phenomenon in a super-Eddington flow has been called "porosity" [8], though "granulation" may be a better term depending on the topology of the mass condensations. In order to avoid fresh 3-D calculations for each individual eruption, *we need a general, albeit rough, empirical prescription* based on many numerical simulations – in more or less the same spirit as mixing length theory for convection. (Some authors have recently asserted that 1-D models work better than I said above, see, e.g., [9]; but if this is true, it needs to be confirmed by 3-D investigations.)

Next let me say something about observed continuum slopes and emission lines. Giant eruptions, LBV eruptions, and other mass outflows typically have apparent temperatures between 7000 and 10000 K at maximum brightness [2]. This fact is a consequence of opacity physics, and does not imply that the outbursts had similar causes. The average temperature of escaping radiation represents the "thermalization depth" where $\surd(3\,\tau_{\rm tot}\tau_{\rm abs}) \sim 1$. A crucial fact is that opacity decreases rapidly below $T \sim 7500$ K, and *very* rapidly below 6500 K. This leads to some interesting generalities for a wind or outflow with a given luminosity [11]. First, a moderate mass-loss rate can produce apparent temperatures around 8000 K, defined in a particular way. But reducing that to 7000 K requires a large density and mass-flow rate; 6000 K implies a relatively huge rate; and lower photosphere temperatures are unlikely in almost all cases. Since temperatures in that range also have bolometric corrections near zero, it is entirely natural that opaque outflows of all kinds often look like $T \sim 7500$ K at maximum brightness.

Here's an elegant consequence for emission lines. With realistic opacities, a wind's continuum photosphere (a.k.a. thermalization depth) usually occurs at Thomson scattering depth $\tau_{\rm sc} \sim$ 2 to 4. Emission lines are formed mostly in the diffuse outflow outside that radius, i.e. in regions where $\tau_{\rm sc} \sim 1$ to 3. This is the range where Thomson-scattered line wings are apparent, recognizable, and moderate – just like the spectra of the best-observed giant eruptions (e.g. [12,13]). *When we see moderate Thomson-scattering wings on the Balmer emission lines, with* *a visual continuum slope like $T \sim$ 7000 to 15000 K, then we're probably looking at a super-Eddington flow.*

At this point I feel bound to warn against a particular spectroscopic fallacy: stellar spectral types applied to opaque winds. Casual analogies such as "reminiscent of an F-type supergiant" can be justified, but formally *one should not assign a stellar spectral type to the absorption features of an opaque outflow.* There are good reasons for this caveat. Imagine, for instance, a star with $T_{\rm eff} = 6500$ K, compared to an opaque wind with the same photosphere temperature. The star's atmosphere has practically no material with $T < 5200$ K, but outer parts of the wind can have $T < 4500$ K. Hence the wind can form absorption lines much cooler than anything in the star's spectrum. See the dispute in refs. [14] vs. [15]; contrary to published claims, the light-echo spectrum of η Car's great eruption agreed reasonably well with expectations.

We now have a sizable fund of excellent data on LBVs and related stars, η Car and other supernova impostors, and on giant-eruption supernovae; observers have done their job well. But theorists have given this topic far less attention than it deserves. As I implied earlier, *this subject is relatively unexplored territory* for theory.

References

[1] Humphreys R M and Davidson K 1979 *Astrophys. J.* **232** 409-420
[2] Humphreys R M and Davidson K 1994 *Publ. Astron. Soc. Pac.* **106** 1025-1051
[3] Fullerton A W, Massa D L and Prinja R K 2006 *Astrophys. J.* **637** 1025-1039
[4] van Dyk S D and Matheson T 2012 *Eta Carinae and the Supernova Impostors* **384** 249-274
[5] Humphreys R M, Weis K, Davidson K and Gordon M S 2016 arXiv:1603.01278
[6] Davidson K and Humphreys R M (eds) 2012 *ASSL 384, Eta Carinae and the Supernova Impostors* (New York: Springer)
[7] Shaviv N 2000 *Astrophys. J.* **532** L137-L140
[8] Owocki S and Shaviv N 2012 *Eta Carinae and the Supernova Impostors* **384** 275
[9] Quataert E, Fernandez R, Kasen D, Klion H and Paxton B 2015 arXiv:1509.06370
[10] Margutti R *et al* 2014 *Astrophys. J.* **780** 21
[11] Davidson K 1987 *Astrophys. J.* **317** 760-764
[12] Dessart L, Hillier D J, Gezari S, Basa S and Matheson T 2009 *Mon. Not. R. Astron. Soc.* **394** 21-37
[13] Humphreys R M, Davidson K, Jones T J, Pogge R W, Grammer S H, Prieto J L and Pritchard T A 2012 *Astrophys. J.* **760** 93
[14] Davidson K and Humphreys R M 2012 *Nature* **486** E1
[15] Rest A *et al* 2012 *Nature* **482** 375-378

Journal of Physics: Conference Series **728** (2016) 022009 doi:10.1088/1742-6596/728/2/022009

ALMA observations of TiO_2 around VY CMa

Elvire De Beck[1], Wouter Vlemmings[1], Sébastien Muller[1], John H. Black[1], Eamon O'Gorman[1], Anita M. S. Richards[2], Alain Baudry[3,4], Matthias Maercker[1], Leen Decin[5,6] and Elizabeth M. Humphreys[7]

[1] Department of Earth and Space Sciences, Chalmers University of Technology, Onsala Space Observatory, 43992 Onsala, Sweden
[2] Jodrell Bank Centre for Astrophysics, School of Physics and Astronomy, University of Manchester, Manchester M13 9PL, UK
[3] Université de Bordeaux, LAB, UMR 5804, 33270 Floirac, France
[4] CNRS, LAB, UMR 5804, 33270 Floirac, France
[5] Instituut voor Sterrenkunde, Katholieke Universiteit Leuven, Celestijnenlaan 200D, 3001 Leuven, Belgium
[6] Sterrenkundig Instituut Anton Pannekoek, University of Amsterdam, Science Park 904, 1098 Amsterdam, The Netherlands
[7] European Southern Observatory, Karl-Schwarzschild-Strae 2, 85748 Garching, Germany

E-mail: [1] elvire.debeck@chalmers.se

Abstract. Titanium dioxide, TiO_2, is a refractory species that could play a crucial role in the dust-condensation sequence around oxygen-rich evolved stars. We present and discuss the detections of 15 emission lines of TiO_2 with ALMA in the complex environment of the red supergiant VY CMa. The observations reveal a highly clumpy, anisotropic outflow in which the TiO_2 emission likely traces gas exposed to the stellar radiation field. We find evidence for a roughly east-west oriented, accelerating bipolar-like structure, of which the blue component runs into and breaks up around a solid continuum component. We see a distinct tail to the south-west for some transitions, consistent with features seen in the optical and near-infrared. We find that a significant fraction of TiO_2 remains in the gas phase outside the dust-formation zone and suggest that this species might play only a minor role in the dust-condensation process around extreme oxygen-rich evolved stars like VY CMa.

1. Introduction

The driving mechanism of the winds of oxygen-rich evolved stars is not well understood. Radiation pressure on dust through absorption and scattering of stellar light is thought to play a crucial role. The size and chemical composition of dust grains are critical to the efficiency of these processes. It is hence essential to address which gas-phase species provide the primary seeds for the dust-formation sequence in oxygen-rich outflows. TiO_2 is considered an important seed refractory species with possibly higher nucleation rates than SiO [1, 2]. Detections of presolar TiO_2 grains [3] support the importance of this species in the dust-condensation sequence.

Emission from gas phase TiO_2 has to date only been detected towards VY CMa [4, 5, 6], an extreme red supergiant at 1.2 kpc distance [7]. VY CMa's circumstellar environment exhibits a high degree of morphological complexity, from optical to radio wavelengths and on spatial scales from a few to several thousands of AU [6, 8, 9, 10, 11, 12, 13]. ALMA observations spatially resolved the H_2O maser emission and the submillimeter continuum emission, leading to a most accurate determination of the stellar position [14] and the discovery of a bright continuum component south-east of the star [15].

2. Observations

We retrieved ALMA science verification data on VY CMa from the ALMA archives. The observations, data calibration and reduction are described by [14]. An overview of the spectral coverage and representative rms noise values for the six spectral windows in ALMA's band 7 ($\sim$0.9 mm; $\sim$320 GHz) and one in band 9 ($\sim$0.45 mm; $\sim$660 GHz) is given by [4]. With projected baselines of 14 m up to 2.7 km, the spatial resolution at $\sim$320 GHz and 658 GHz is $\sim 0.2''$ and $\sim 0.1''$, respectively, and the maximum recoverable scales are $8.3''$ and $4.0''$.

3. Results

We analysed spectra extracted for a $1''$ diameter region around the stellar position, since no TiO_2 emission is detected beyond this aperture. We detect 15 lines with upper-level energies $E_{\rm up}/k$ in the range $48 - 676$ K and signal-to-noise ratios $S/N \approx 5 - 17$ at velocity resolutions $0.9 - 7.6\,{\rm km\,s^{-1}}$ (table 1, figure 1). Of the 15 lines of TiO_2 detected with ALMA, only 2 were detected with SMA, owing to a $\sim$10 times lower noise level in the ALMA data.

The TiO_2 emission is for the first time spatially resolved in the ALMA observations and shows a high level of morphological complexity, described in detail by [4]. The TiO_2 emission traces multiple wind components, as seen in figure 2. We find a red-shifted outflow to the west and a blue-shifted outflow to the east. At increasingly blue-shifted velocities the emission moves towards the bright clump C south-east of the star and then appears to curve and break up around it with the northern peak brighter than the southern one. With the clear exception of this interaction of the TiO_2 gas with C, the west and east outflows seem roughly symmetric around the star and aligned with the axis connecting the star and C. We rule out an equatorially enhanced environment such as a disk or ring, based on the spatial distribution of the TiO_2 emission at different velocities. We rather suggest an accelerating bipolar-like outflow at lower densities. We also find a predominantly blue-shifted south-west outflow, approximately perpendicular to the axis connecting the star and C, connecting the star and the south-west clump, a feature detected in the mid-infrared [11].

Whereas the H_2O maser emission in the "valley" between the star and C implies that C is

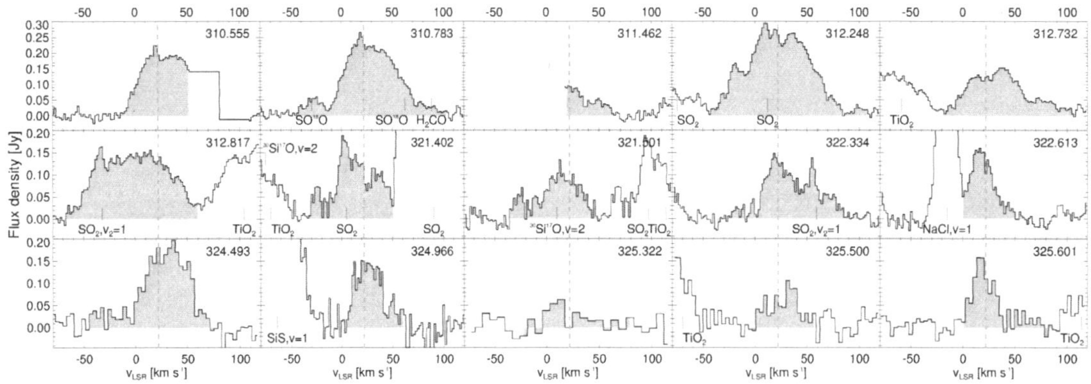

Figure 1. TiO_2 spectra extracted for a $1''$ diameter aperture around the stellar position. The vertical dashed lines indicate the stellar $v_{\rm LSR}$ of $22\,{\rm km\,s^{-1}}$, the shaded areas the $v_{\rm LSR}$-ranges from table 1. We indicated (tentative) identifications of species other than TiO_2 in the panels.

Table 1. Overview of detected TiO_2 lines as detected in the spectrum extracted in a 1″ diameter aperture centred on the star. Columns list transition, rest frequency, upper-level energy, minimum and maximum v_{LSR} reached, velocity resolution and rms noise at which the identification was made, and peak and integrated flux.

Transition $J'_{K'_a,K'_c} - J_{K_a,K_c}$	ν_{lab} (MHz)	E_{up}/k (K)	v_{min} (km s^{-1})	v_{max} (km s^{-1})	Δv (km s^{-1})	Rms (mJy)	Peak (Jy)	I (Jy km s^{-1})
22(1, 21) − 21(2, 20)	310554.735	180.5	−15.2	51.3	1.0	16	0.22	9.57
23(1, 23) − 22(0, 22)	310782.713	182.4	−43.9	102.3	1.0	16	0.27	15.28
40(8, 32) − 39(9, 31)	311462.082	675.9	19.4	66.7	1.9	10	0.10	2.31
7(5, 3) − 6(4, 2)	312248.341	47.9	−43.7	80.96	1.9	16	0.29	20.32
10(4, 6) − 9(3, 7)	312732.066	57.7	−19.4	105.1	1.9	16	0.15	9.43
30(3, 27) − 30(2, 28)	312816.809	354.4	−68.2	60.0	1.9	16	0.16	12.98
11(4, 8) − 10(3, 7)	321401.936	65.7	−33.9	50.5	0.9	24	0.19	7.24
28(2, 26) − 28(1, 27)	321501.043	298.8	−35.0	51.2	0.9	24	0.13	4.76
35(8, 28) − 35(7, 29)	322333.594	532.4	−36.4	108.1	1.0	19	0.15	7.03
33(8, 26) − 33(7, 27)	322612.696	481.4	−0.1	63.9	1.0	18	0.16	4.18
23(2, 22) − 22(1, 21)	324492.930	196.1	−47.9	79.4	1.9	29	0.21	9.64
37(8, 30) − 37(7, 31)	324965.693	586.5	3.6	59.8	1.9	30	0.15	4.92
26(1, 25) − 26(0, 26)	325322.035	246.5	−25.2	81.8	7.6	21	0.11	1.85
26(4, 22) − 25(5, 21)	325500.415	281.5	−7.7	55.7	3.8	28	0.13	2.75
28(8, 20) − 28(7, 21)	325600.792	367.0	−0.4	90.2	3.8	28	0.18	3.89

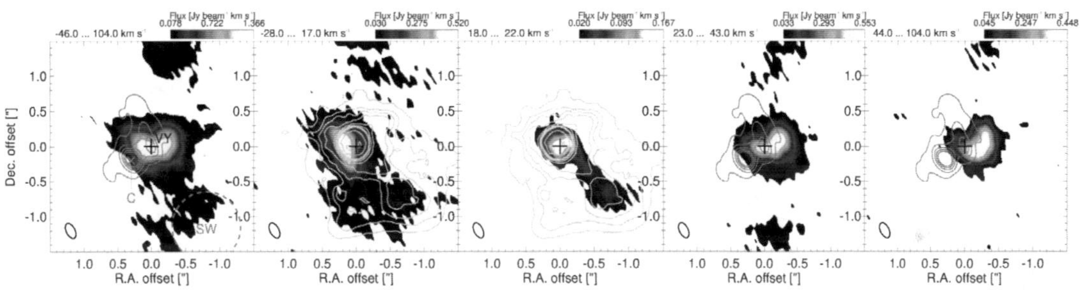

Figure 2. TiO_2 morphology. Colour maps of emission at 310.78 GHz integrated over the v_{LSR}-ranges indicated at the top left of each panel, cut off at 3σ. Red contours (1st, 4th, and 5th panel) show the 321 GHz continuum at $[3, 20, 40, 60, 80]\sigma$; green contours (2nd and 3rd panel) show HST emission at $[3, 5, 7, 10, 20, 30, 40, 50, 100, 200]\sigma$ [12]. In the 1st panel we marked the position of the star (+, VY; black) and of the continuum component (x, C; red) to the south-east [14, 15], and the position and approximate extent of the south-west clump (SW, dashed 1″ diameter circle; red) [11]. The apparent north-south emission is thought to arise from dynamic-range limitations in the peak channels.

close to or in the plane of the sky [14], the observations suggest that the TiO_2 gas breaks up around C while moving towards the observer, placing C "at least partially" in front of the plane of the sky. It is therefore likely that the H_2O masers and the TiO_2 emission probe parts of the outflow east of the star with different physical properties. Whereas the masers are probably excited through shocks at high densities [14], TiO_2 is more likely excited through radiation, at lower outflow densities (see below). In the denser regions, TiO_2 might not be excited and/or it might be efficiently depleted from the gas phase. The latter is, however, less likely (see below). We therefore suggest that TiO_2 traces the blue-shifted wind to the east with lower densities which runs into and curves around C.

From a rotational-diagram analysis we derive a column density $N_{\rm col} = 5.65 \pm 1.33 \times 10^{15}\,{\rm cm}^{-2}$ and a rotational temperature $T_{\rm rot} = 198.0 \pm 28.5\,{\rm K}$, in agreement with [5], although the kinetic temperature in the excitation region of TiO_2 varies from more than 1000 K down to ~100 K [16]. Assuming an average mass-loss rate of $2 \times 10^{-4}\,M_\odot\,{\rm yr}^{-1}$, an average velocity of $20\,{\rm km\,s}^{-1}$, and an $0.9''$ diametric extent, at 1.2 kpc, we find an abundance $TiO_2/H_2 \approx 3.8 \pm 0.9 \times 10^{-8}$. We however urge careful interpretation of these numbers, since the high dipole moment of TiO_2 supports efficient radiative excitation and possibly electron–TiO_2 collision rates large enough to exceed H_2–TiO_2 collision rates in case of a sufficient ionisation degree close to the star [4]. These effects could invalidate the LTE approach used in this simple analysis.

Taking into consideration the high dipole moment of TiO_2, the ratio between the collisional and radiative transition rates of TiO_2, and the strong resemblance between the TiO_2 emission measured with ALMA and the earlier scattered-light observations at $1\,\mu$m we put forth that the TiO_2 gas is radiatively excited. As such, the measured emission traces parts of the outflow where the stellar radiation field is least attenuated. From the comparison to the continuum at ~321 GHz, we derive that TiO_2 is excited in the directions with lower dust densities. The absence of detected TiO_2 emission north of the star could then point at efficient obscuration of this part of the outflow, in line with the observations of e.g. [12]. Attenuation of the stellar radiation field to the east and west of the star is limited. From this, we do not expect an equatorial enhancement of the mass-loss rate, since this would likely have induced more efficient dust formation, which is not seen in the continuum.

Since the TiO_2 emission is seen at radial distances from the star far beyond the dust-condensation region, and along directions where dust continuum is observed, we claim that TiO_2 is not a tracer for overall low grain-formation efficiency, but that much of it remains in the gas phase. The strong correspondence between the TiO_2 emission and the south-west tail of dust-scattered stellar light supports this. We also consider the high derived column density and suggest that TiO_2 might play only a minor role as a primary dust seed around VY CMa.

References

[1] Jeong K S, Winters J M, Le Bertre T and Sedlmayr E 2003 *Astron. Astrophys.* **407** 191
[2] Lee G, Helling C, Giles H and Bromley S T 2015 *Astron. Astrophys.* **575** A11
[3] Nittler L R and Alexander C M O 1999 *30th Annual Lunar and Planetary Science Conference* (Houston) p 2041
[4] De Beck E, Vlemmings W, Muller S, Black J H, O'Gorman E, Richards A M S, Baudry A, Maercker M, Decin L and Humphreys E M 2015 *Astron. Astrophys.* **580** A36
[5] Kamiński T, Gottlieb C A, Menten K M, Patel N A, Young K H, Brunken S, Muller H S P, McCarthy M C, Winters J M and Decin L 2013a *Astron. Astrophys.* **551** A113
[6] Kamiński T, Gottlieb C A, Young K H, Menten K M and Patel N A 2013b *Astrophys. J. Suppl. Series* **209** 38
[7] Zhang B, Reid M J, Menten K M and Zheng X W 2012 *Astrophys. J.* **744** 23
[8] Humphreys R M, Helton L A and Jones T J 2007 *Astron. J.* **133** 2716
[9] Monnier J D *et al* 2014 *Proc. SPIE* **9146** 91461Q
[10] Muller S, Dinh-V-Trung, Lim J, Hirano N, Muthu C and Kwok S 2007 *Astrophys. J.* **656** 1109
[11] Shenoy D P *et al* 2013 *Astron. J.* **146** 90
[12] Smith N, Humphreys R M, Davidson K, Gehrz R D, Schuster M T and Krautter J 2001 *Astron. J.* **121** 1111
[13] Ziurys L M, Milam S N, Apponi A J and Woolf N J 2007 *Nature* **447** 1094
[14] Richards A M S *et al* 2014 *Astron. Astrophys.* **572** L9
[15] O'Gorman E *et al* 2015 *Astron. Astrophys.* **573** L1
[16] Decin L, Hony S, de Koter A, Justtanont K, Tielens A G G M and Waters L B F M 2006 *Astron. Astrophys.* **456** 549

Journal of Physics: Conference Series **728** (2016) 032001 doi:10.1088/1742-6596/728/3/032001

The dynamical evolution of planetary nebulae

Detlef Schönberner

Leibniz-Institut für Astrophysik Potsdam (AIP), An der Sternwarte 16, 14482 Potsdam, Germany

E-mail: deschoenberner@aip.de

Abstract. Based on modern 1D-radiation-hydrodynamics simulations of formation and evolution of planetary nebulae, I discuss in detail the basic dynamical processes responsible for the "grand design" of most planetary nebulae, i.e. their double-shell morphology and their typical expansion properties. Special emphasis is given for a proper definition of a nebula's true expansion rate and its relation to spectroscopically measurable Doppler velocities of the expanding material. It is found that the typical nebular expansion is about twice as fast as hitherto assumed, viz. $\simeq$45 km s^{-1}.

1. The physical system and its modelling

Modern view. The process of formation and evolution of planetary nebulae (PNe) is intimately connected with mass-loss processes along the asymptotic giant branch (AGB) and during the final contraction and heating of the stellar remnant until the white-dwarf stage at the hot side of the Hertzsprung-Russell diagram (HRD) is reached. Thereby, the slow but very dense AGB wind is heated by photo-ionisation due to the intense radiation field of the hot central star which sets up an expanding shock wave and which becomes later compressed from within by the action of the tenuous but very fast central-star wind.

There is, however, no direct interaction of the fast wind with the former AGB matter: The wind is thermalised by a strong shock, and the system's steady attempt to achieve pressure balance between the ionised shell and this shocked wind material on one side, and between the shell and the still undisturbed AGB wind on the other side, is responsible for the formation of what we call a planetary nebula and its evolution with time. In this view, a PN is not only the evidence of matter once ejected from the stellar surface but can instead be described by a thermally driven shock wave through the ambient AGB wind envelope, starting at the inner edge of this envelope and powered by heating due to photo-ionisation. Because of the ever changing boundary conditions, a PN cannot be adequately described by static models or similarity solution for expanding shells.

The concept of expanding shock waves ensures that the system and its morphology remains stable for long times, at least as long as the boundary conditions do not change much. An example of a typical PN structure is rendered in figure 1. One clearly sees a system of two nested shells/shock waves: an inner bright "rim" and a larger but fainter (attached) "shell", both of which consist now of ionised and dynamical "reshaped" former AGB material which expands into the huge, round, even fainter "halo", the just ionised AGB wind. The rim encloses a "cavity", filled with the very hot, shocked wind gas which emit diffuse X-ray radiation (e.g. [2]). The PN proper consist of rim and shell, but note that most of the nebular mass (up to

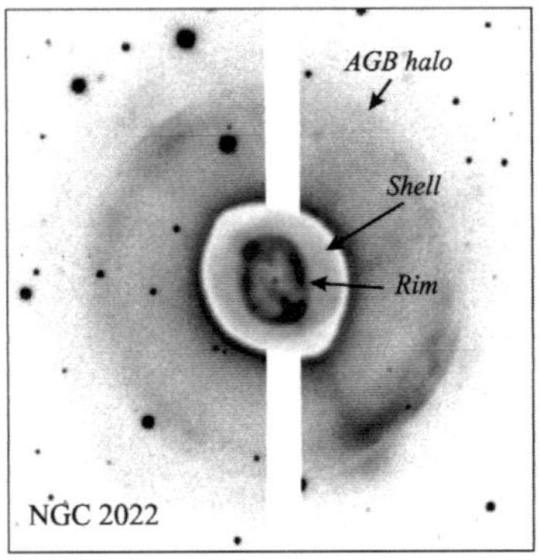

Figure 1. Combination of two [O III] images of the double-shell planetary nebula NGC 2022 with different (logarithmic) intensity scales for inset and main image (see [1] for details). The inset picture shows the PN proper, consisting of rim and shell. The nebula is expanding into the rather large, only slowly expanding round halo, consisting of ionised AGB-wind matter. The central star is about 100 000 K hot and has about 3600 $L_{\odot}$. The PN has a radius of about 0.2 pc and an estimated kinematic age of about 5000 yr; the halo is much older: $\simeq$ 40 000 yr (cf. [1]).

$\approx$ 80 %) is contained in the fainter shell! These nebular structures are obviously dynamical stable for expansion timescales of 10 000 years or even longer.

Based on the simple spherical models used in general, we can discuss only the shaping of round nebular structures around single central stars. Though we will show that the physics involved is of basic nature, other shaping mechanism responsible for distortions from sphericity, e.g. rotation, magnetic fields, or binarity, must be studied as well.

Historical review. The modelling of formation and evolution of a planetary nebula by means of hydrodynamics has already a long history. Early attempts ([3, 4]) were, as seen now, not really successful since the boundary conditions were not appropriate: no stellar wind for nebular "support", and expansion of an ad hoc assumed shell into a vacuum. Therefore, the shells of initially constant densities dispersed rather quickly, without showing any similarity to the observed long-lasting double-shell structures as seen in figure 1.

A breakthrough in our understanding of PN formation and evolution was the birth of the famous "interacting stellar winds" (ISW) theory introduced by [5]. These authors were the first who realised that stellar winds which change their properties strongly while the AGB remnant evolves from the tip of the AGB across the HRD are responsible for the formation of PNe: The tenuous but very fast (>1000 km s^{-1}) wind of the hot AGB-remnant (or central star) sweeps-up and engulfs the dense, slow ($\simeq$ 10–15 km s^{-1}) relic of the former AGB wind into a well-bounded dense nebular shell which increases permanently its mass and velocity. Follow-up papers presented refinements in terms of time-dependent central-star winds and radiation fields ([6, 7]). With suited chosen wind parameters, reasonable shell expansion velocities ($\simeq$ 20 km s^{-1}) could easily be achieved. The weakness of this approach comes from the fact that the hydrodynamics was treated by simple equations that could be solved analytically. The dynamic effects of ionisation could not be treated at all. Also, deep PNe images taken by [8] proved that the double-shell structure of the PN proper as seen in figure 1 is rather the rule, not the exception, whereas the ISW theory is only able to produce *one* expanding shell.

Modern approaches. Relief came from the first numerical 1D radiation-hydrodynamics simulations with realistic initial and time-dependent boundary conditions in the late 80s of the last century ([9, 10]), with some improvements of the wind model introduced later by [11]. The time-dependence refers to stellar post-AGB evolution in terms of stellar and wind parameters and to ionisation/recombination of hydrogen and helium. These simulations showed for the first time that the combined action of ionisation and wind interaction leads automatically to a double shell structure like that seen in figure 1.

In a series of papers, Mellema (e.g. [12, 14, 15]) expanded the previous works in terms of parameter range and better numerics. Mellema ([13]) performed also the first 2D simulations. He computed observable quantities like surface brightness distributions and line profiles/velocity-position diagrams which are necessary for checking the simulations against observations. He emphasised the importance of ionisation as compared to wind interaction and coined the terms

"I-shell" (I = ionisation) for the shell and "W-shell" (W = wind) for the rim in order to account for the different physics at work.

A more comprehensive study was performed by Villaver *et al* [16, 17]. These authors used the evolutionary AGB and post-AGB models of Vassiliadis and Wood ([18, 19]) including their (assumed) mass-loss rates and computed in a consistent way the hydrodynamical evolution of the stellar envelopes together with the central stellar object along the AGB and post-AGB. The predictive power of this study is, however, limited for the following reasons: (1) ionisation is treated by Strömgren spheres with a constant temperature of 10^4 K, and (2) no observables like intensity distributions or emission-line profiles have been computed.

The most sophisticated simulations so far are those presented by [20] and later by [22]. The code (NEBEL) is fully time-dependent, i.e. the actual ionisation/recombination and heating/cooling is computed at each time step, based on the prevailing ensemble of existing species, the physical state of the plasma, and the radiation field. Nine relevant elements up to argon and their ionisation stages are considered (fore more details, see [21] and the references therein). The NEBEL code includes also heat conduction by electrons which is important for a realistic description of the temperature and density profiles inside the bubble of shocked wind gas. Additional IDL-routines allow the (a posteriori) computation of a number of observable entities like emission-line strengths and profiles, intensity distributions, and X-ray emission from the hot bubble. In a series of papers, issues like circumstellar environment and expansion ([23]), internal kinematics and expansion ([24]), individual luminosity functions ([25]), diffuse X-ray emission ([26]), the extremely metal-poor PN G135.9+55.9 ([27]), PNe in distant stellar systems ([28]), and true expansion rates and visibilities of PNe ([29]) have been addressed.

Very recently, Toála and Arthur ([31]) computed spherical models in a similar way as in

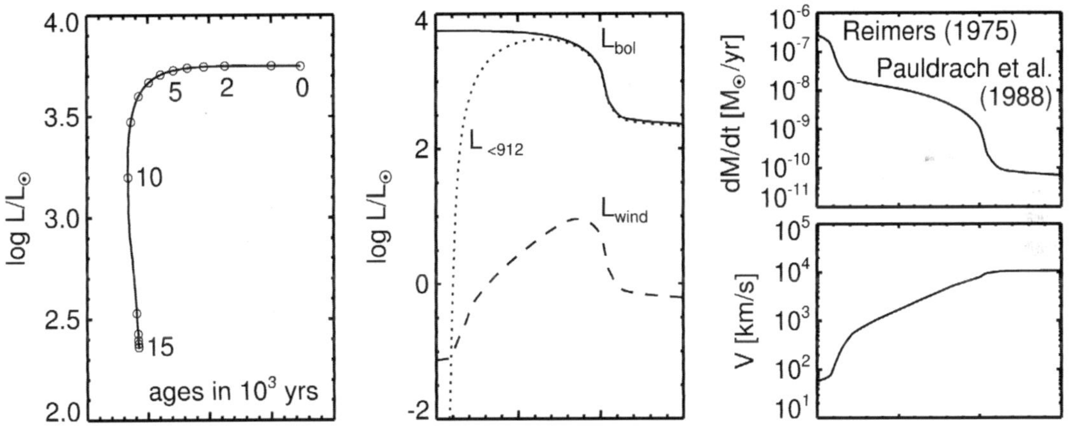

Figure 2. Example of stellar and wind properties used in our radiation-hydrodynamical simulations for the inner boundary conditions. *Left panel*: evolutionary path for a 0.595 $M_\odot$ post-AGB model with ages indicated; *middle panel*: the corresponding time evolution of the bolometric, L_{bol} (solid), Lyman continuum, $L_{<912}$ (dotted), and mechanical (wind) luminosity, $L_{wind} = \dot{M} V^2/2$ (dashed), all in solar units. *Right panels*: mass-loss rate $\dot{M}$ (*top*) and wind velocity V (*bottom*). The end of the strong, dense AGB wind sets the zero point of the post-AGB evolution and the beginning of the much weaker post-AGB wind, modelled first by the prescription of Reimers ([32]). Later, for $T_{eff} \geq 25\,000$ K, or post-AGB ages $\geq 2\times10^3$ years, the theory of radiation-driven winds as formulated by Pauldrach *et al* ([35]) is used for the rest of the evolution.

[16] but allowed for 2D effects like, e.g., Rayleigh-Taylor instabilities and mixing between "cool" nebular and "hot" shocked wind gas. To date it is too early to judge the quality of these models because detailed comparisons between the models and real objects are still pending.

Example of hydrodynamical PN modelling. In the following, I will thus refer mainly to results presented and discussed in [30], where an extensive comparison between 1D radiation-hydrodynamical PN models and the observations can be found. Thereby it is implicitly assumed that spherical models are able to give, at least on the average, a reasonable description of real objects, provided the distortion from sphericity of the latter is not too big.

The typical inner boundary conditions as they are given by the properties of an evolving central star are illustrated in figure 2. By comparing the wind power with that of the ionising radiation, one notices that the former increases only slowly and reaches a maximum value close before maximum stellar temperature is reached. The reason is the evolution of the wind velocity which is coupled to the stellar escape velocity and more than compensates for the slowly decreasing mass-loss rate until the white-dwarf stage with nearly constant stellar radius is reached (cf. right panels of figure 2).

The low wind power during the early PN evolution implies that its influence for shaping young objects is most likely rather small. This statement does not change even if one considers the uncertainty of the "early" post-AGB mass-loss rate which is assumed here to be of Reimers' type ([32]): The wind velocity is just too low. The mass-loss uncertainty translates, however, directly into the transition timescale of an AGB remnant from the tip of the AGB to, say, a position in the HR diagram with an effective temperature of about 25 000 K (cf. [33, 34]). We used in the simulations the older post-AGB mass-loss rates and wind velocities of Pauldrach *et al* ([35]) in order to be consistent with the evolutionary models of Blöcker ([34]). The more recent post-AGB mass-loss rates and wind velocities derived by Pauldrach *et al* ([36]) are somewhat different from the ones used here, but not by so much that existing hydrodynamical models are put into question, especially if one considers all the other uncertainties.

In order to cover a reasonable number of possible combinations between central star and initial envelope, a grid of PN simulations have been used in [30], encompassing central-star masses from $\simeq 0.57$ to $\simeq 0.70$ $M_\odot$ which are coupled to either envelopes with initial density profiles based on hydrodynamics of dusty envelopes on the AGB ([37]) or initial envelopes with simple radial power-law density distributions.

Figure 3 renders a typical middle-aged and fully ionised nebular structure taken from a simulation along the model track shown in figure 2 at a position in the HRD given in the figure caption. We see, next to the ionisation-generated shell which makes up for most of the nebula, also an already well-developed rim, and both are bounded by respective shocks whose radial positions are marked in the figure as well. The shell is somewhat diluted because it has already expanded into the former AGB wind with a rather steep density gradient, while the rim is of much higher density because of the bubble's comparatively high pressure, i.e. the rim's shock is catching up with the shell, thereby compressing and engulfing the inner shell matter. Consequently, both the rim and the shell show a very distinct behaviour in their radial density and velocity profiles, thus also their contribution to the intensity distributions and line profiles are distinct. Note especially the rather faint signatures of the shell in [O III] in both the brightnesses and line profiles, although the shell contains most of the nebular mass, i.e. 85 % of the $0.47\,M_\odot$ of ionised matter contained in the PN model shown in the left panel of figure 3. Note especially the rather low flow velocities within the rim, ≈ 12 km s^{-1}, as compared to the post-shock velocity of the shell, 30 km s^{-1}.

We emphasise here two facts: (1) Shell and rim are two distinct entities of a PN that are created by two different physical mechanisms, viz. ionisation and winds interaction, and (2) both have hence different expansion properties. Furthermore, the overall velocity field does not obey a simple $v(r) \propto r$ law, as generally assumed in spatio-kinematical modelling.

Journal of Physics: Conference Series **728** (2016) 032001 doi:10.1088/1742-6596/728/3/032001

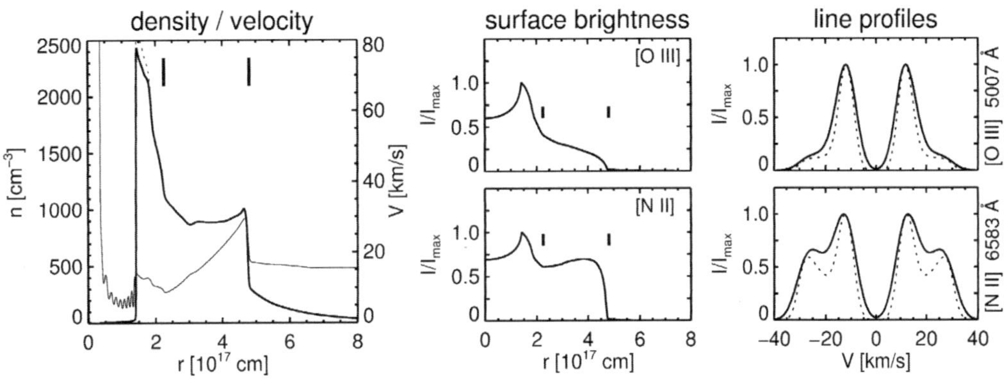

Figure 3. Snapshot of a typical middle-aged nebular model from a 1D-radiation-hydrodynamics simulation around a 0.595 $M_\odot$ central star whose properties are shown in figure 2. The stellar parameters are: post-AGB age $t = 6106$ yr, $T_{\text{eff}} = 80\,177$ K, and $L = 5.057{\times}10^3$ $L_\odot$. *Left panel*: heavy particle density (thick), electron density (dotted), and gas velocity (thin); *middle panels*: (normalised) surface brightnesses in [O III] λ5007 Å and [N II] λ6583 Å; *right panels*: the corresponding normalised line profiles computed for the central line-of-sight with infinite spectral resolution (dotted) and broadened with a Gaussian of 6 km s^{-1} FWHM (solid), both with a circular aperture of $1{\times}10^{16}$ cm. The line-profile peaks are the signature of the rim, while the fainter "shoulders" belong to the shell. The thick vertical marks (*left and middle*) indicate the positions of the leading shocks of the rim and shell, respectively. The nebular mass enclosed by the rim's shock is $M_{\text{rim}} = 0.07$ $M_\odot$, that enclosed by the shell's leading shock is $M_{\text{shell}} = 0.47$ $M_\odot$ (= total nebular mass). The shell's shock defines the outer edge of the PN.

Before continuing, we must emphasise that all the hydrodynamical simulations discussed here neglect magnetic fields. Whether they are responsible for shaping non-spherical PNe is still an open question. On the existence of magnetic fields in PNe, see also the recent work of Steffen *et al* ([38]).

Additionally, one should note also that both the stellar post-AGB models and the mass-loss modelling refer to objects which have normal stellar surface composition and which burn hydrogen while evolving across the HR diagram. Nebulae with hydrogen-deficient central stars are *not* considered because (1) the latter have much stronger winds with corresponding stronger dynamical impact on the nebula and (2) their formation and evolution is yet totally unknown.

2. Expansion properties of PNe

Although the typical "expansion velocity" of PNe is a quantity of utmost importance for a number of contexts, it is only poorly defined and cannot be specified without knowledge of the density structure and internal velocity field of the object in question. It is therefore important to use hydrodynamical models to get more insight into the expansion properties of PNe.

Guided by the model in figure 3, we can state that the velocity of the shell's leading shock ($\dot{R}_{\text{PN}}$ at $r = R_{\text{PN}} = 4.7{\times}10^{17}$ cm), whose position defines also the outer edge of the PN, is the *true* expansion velocity. Unfortunately, the propagation speed of this shock *cannot* be measured spectroscopically since it is a "pattern", not "matter" expansion! Several different definitions of the expansion velocity of PNe are in use, and their correspondence to the true expansion velocity $\dot{R}_{\text{PN}}$ is a priori not clear at all.

1. *Half-width of spatially/spectroscopically unresolved lines.* Very uncertain relation to $\dot{R}_{\text{PN}}$, and the latter can be up to three times higher (cf. figure 39 in [28]), depending on the phase of evolution. This velocity value should *not* be used at all since it is strongly biased towards the object's dense, low velocity structures.

2. Peak separation of resolved lines. A very frequently used method, but provides only the bulk velocity of the rim, $V_{\rm rim}$, which is usually well below $\dot{R}_{\rm PN}$ (see below). The canonical values of the PN expansion velocity used in the literature, 20–25 km s^{-1}, are mainly based on this method.

3. Post-shock velocity. A method introduced by Corradi *et al* ([39]): The (outer) inflection points of the line-profile shoulders as seen in figure 3 (right panels) are a good measure of the shell's shock post-shock gas velocity. This post-shock velocity is closely related to the shock propagation speed itself: $\dot{R}_{\rm PN} = V_{\rm post}\times F$, with $F > 1$. The value of F depends on the shock properties, and by means of PN models from different simulations, Schönberner *et al* ([30], figure 6 therein) found that a very reasonable value is $F = 1.25\pm0.05$ (optically-thin models only). Mellema ([40]) found a similar value, $F = 1.2$, by employing the jump conditions together with reasonable assumptions for the shell's shock properties.

The velocities measured by methods 1 and 2 may depend on the ion used and have unknown relations to the true expansion given by $\dot{R}_{\rm PN}$. Only method 3 leads to a physically sound velocity value, rather independently of the used species and close to the real expansion rate. Of course, the term "expansion velocity" as used here is only applicable for objects where we can be sure about the existence of a more or less regularily expanding shell.

In order to get more insight into the internal kinematics and expansion properties of PNe, Schönberner *et al* ([30]) analysed spatially resolved emission-line profiles of about 20 elliptical PNe. The bulk velocity of the rim, $V_{\rm rim}$, was measured from the line-peak separations, and the post-shock velocity, $V_{\rm post}$, from the line shoulders' inflection points, as discussed above. The results are rendered in figure 4, where the stellar effective temperature is used as proxy for the pace of evolution. One can conclude the following:

- The shell's edge expands considerably faster than the rim, but both accelerate during the evolution with $T_{\rm eff}$ (or time), albeit in a different way. The post-shock velocity of the shell increases with $\log T_{\rm eff}$ from about 15–30 km s^{-1} up to about 35–45 km s^{-1} at $\log T_{\rm eff} \geq 4.8$. In contrast, the rim is stalling at rather low velocities ($\simeq$ 5–10 km s^{-1}) for quite a while until it finally accelerates up to about 25–30 km s^{-1}.
- The initial shell expansion starts with post-shock flow speeds *exceeding* those at the tip of the AGB where outflow velocities of $\approx$ 10–15 km s^{-1} are typical. The rim gas, however, has initially velocities which are close or even well *below* the tip-AGB outflow velocities. Important is that for these younger PNe the post-shock velocity of the shell exceeds the rim expansion by factors up to five to six ($\sim$6 km s^{-1} vs. $\sim$34 km s^{-1} for NGC 6629)! In more advanced stages, $V_{\rm post}$ (and $\dot{R}_{\rm PN}$) still exceeds $V_{\rm rim}$ by factors well above unity.

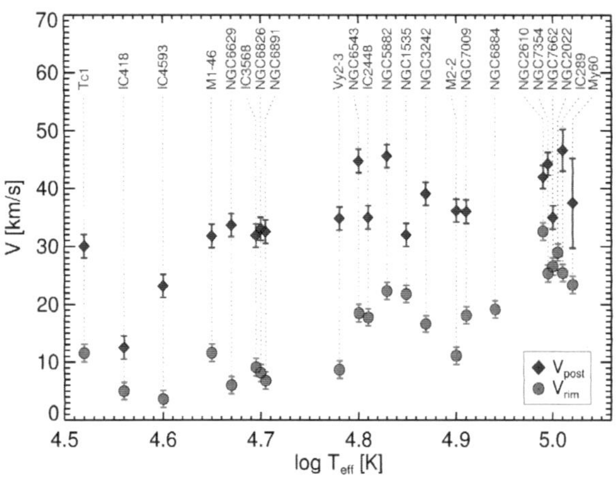

Figure 4. Shell post-shock velocities, $V_{\rm post}$ (blue diamonds), and rim bulk velocities, $V_{\rm rim}$ (half of line peak separation, red dots), of the objects investigated vs. $T_{\rm eff}$ of the respective central stars. There are generally two velocity entries for each object, except for those for which the post-shock velocity could not be measured (NGC 2022 and NGC 6894). Velocities are either from [N II] or [O III], or averaged if information from both ions is available. Vertical dotted lines connect the two velocities with the object's name. Some object entries are slightly shifted in temperature as to avoid overlapping.

These measurements fully confirm the basic predictions of the radiation-hydrodynamic simulations discussed above and illustrated in figure 3: It is the ionisation that sets up the expansion of a PN shell, not the stellar wind. Only in the more advanced stages of evolution plays the winds interaction a larger role. The high expansion speed of the shell's shock is obviously due to the steep density gradient of the upstream (= halo) matter, as predicted by analytical similarity solutions (cf. [41]) and confirmed by dedicated hydrodynamical simulations ([30]). In terms of power-law density distributions $\rho \propto r^{-\alpha}$, the observed post-shock gas velocities correspond to $\alpha \simeq 2.8$–3.4. Direct measurements of the halo density distributions from the observed intensity distributions give virtually the same answer (cf. [42, 30]).

We emphasise that the shock speed does *not* depend on the upstream density itself. Furthermore, the general steep density gradient means that the mass-loss rate increases immediately before the object departs from the tip of the AGB, thereby quenching the high mass-loss rate. Using the correction factor F to convert the observed post shock velocities into the respective shock speeds, one gets a reasonable value of the true PN expansion velocity, at least for the sample studied, of $\langle \dot{R}_{\rm PN} \rangle \simeq 45$ km s^{-1}, as compared to $\langle V_{\rm rim} \rangle \simeq 20$ km s^{-1}. A more comprehensive study concerning the true expansion rate of PNe and its relevance for the visibility time can be found in [29].

The ultimate test of the hydrodynamical models is provided by objects with lower metallicities. The weaker winds of metal-poor central stars lead to more slowly expanding rims, or no rims at all for the extreme metal-poor cases (cf. [28] for the details). In contrast, the high electron temperatures let the shell expand faster (cf. [41]), making the shell more diluted and fainter. Schönberner *et al* ([30]) succeeded in measuring $V_{\rm rim}$ *and* $V_{\rm post}$ for 4 metal-poor objects, and the results confirm the models (see figure 5): there is a tendency of $V_{\rm post}$ for being higher and for $V_{\rm rim}$ being lower than is typical for Milky Way disk objects at similar evolutionary stages, fully consistent with the predictions of the dynamical nebular models as presented here.

3. Summary

Modern radiation-hydrodynamics simulations, even in the 1D approximation, yield new insights into the formation and evolution of round/elliptical planetary nebulae:

- The often observed double-shell morphology is a natural consequence of ionisation and winds interaction, where shell and rim are driven by different physical mechanism: A shock wave is set up either by heating of circumstellar gas due to ionisation (shell) or by expansion of the hot, shocked wind gas (rim).

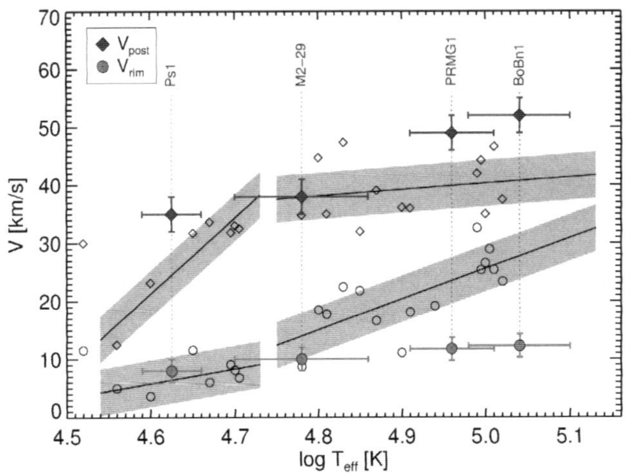

Figure 5. Same as figure 4, but with the positions of 4 metal-poor PNe, with horizontal bars indicating the range of temperature determinations. The general run of the velocities with effective temperatures for the PNe in the Milky Way disk are shown in a schematic way as well. The objects from figure 4 are underlain as small open diamonds and circles.

- The true mean expansion velocity of PNe isusually higher than hitherto assumed, $\simeq 45$ km s^{-1}, given by the expansion of the shell's leading shock which is exclusively ruled by electron temperature and upstream density gradient. The rim's expansion is much lower and does in general not exceed $\simeq 30$ km s^{-1}.
- The high shell expansion of PNe is caused by a rather steep halo density gradient ($\alpha \simeq 3$), which in turn means that the mass-loss rate during the past final AGB stage must have increased steadily.
- The shells of metal-poor PNe expand faster than their disk counterparts because of their high electron temperatures. On the other hand, their central cavity is smaller because of the weak stellar wind.

References

[1] Corradi R L M, Schönberner D, Steffen M and Perinotto M 2003 *Mon. Not. R. Astron. Soc.* **340** 417
[2] Kastner J H *et al* 2012 *Astron. J.* **144** 58
[3] Mathews W G 1966 *Astrophys. J.* **144** 206
[4] Ferch R L and Salpeter E E 1975 *Astrophys. J.* **202** 195
[5] Kwok S, Purton C R and Fitzgerald P M 1978 *Astrophys. J.* **219** L125
[6] Kwok S 1982 *Astrophys. J.* **258** 280
[7] Volk K and Kwok S 1985 *Astron. Astrophys.* **153** 79
[8] Chu Y-H, Jacoby G H and Arendt R 1987 *Astrophys. J. Suppl. Series* **64** 529
[9] Schmidt-Voigt M and Köppen J 1987 *Astron. Astrophys.* **174** 211
[10] Schmidt-Voigt M and Köppen J 1987 *Astron. Astrophys.* **174** 223
[11] Marten H and Schönberner D 1991 *Astron. Astrophys.* **248** 590
[12] Mellema G 1994 *Astron. Astrophys.* **290** 915
[13] Mellema G 1995 *Mon. Not. R. Astron. Soc.* **277** 173
[14] Mellema G 1997 *Astron. Astrophys.* **321** L29
[15] Mellema G and Frank A 1995 *Mon. Not. R. Astron. Soc.* **273** 401
[16] Villaver E, García-Segura G and Manchado A 2002 *Astrophys. J.* **571** 880
[17] Villaver E, Manchado A and García-Segura G 2002 *Astrophys. J.* 581 1204
[18] Vassiliadis E and Wood P 1993 *Astrophys. J.* **413** 64
[19] Vassiliadis E and Wood P 1993 *Astrophys. J. Suppl. Series* **92** 125
[20] Schönberner D, Steffen M, Stahlberg J, Kifonidis K and Blöcker T 1997 *Workshop Stellar Ecology: Advances in Stellar Evolution* ed R T Wood and A Renzini p 146
[21] Perinotto M, Kifonidis K, Schönberner D and Marten H 1998 *Astron. Astrophys.* **332** 1044
[22] Perinotto M, Schönberner D, Steffen M and Calonaci C 2004 *Astron. Astrophys.* **414** 993
[23] Schönberner D, Jacob R, Steffen M, Perinotto R L M and Acker A 2005 *Astron. Astrophys.* **431** 963
[24] Schönberner D, Jacob R and Steffen M 2005 *Astron. Astrophys.* **441** 573
[25] Schönberner D, Jacob R, Steffen M and Sandin C 2007 *Astron. Astrophys.* **473** 467
[26] Schönberner D, Steffen M and Warmuth A 2008 *Astron. Astrophys.* **489** 173
[27] Sandin C, Jacob R, Schönberner D, Steffen M and Roth M M 2010 *Astron. Astrophys.* **512** A18
[28] Schönberner D, Jacob R, Sandin C and Steffen M 2010 *Astron. Astrophys.* **523** A86
[29] Jacob R, Schönberner D and Steffen M 2013 *Astron. Astrophys.* **558** A78
[30] Schönberner D, Jacob R, Lehmann H, Hildebrandt G, Steffen M, Zwanzig A, Sandin C and Corradi R L M 2014 *Astronomische Nachrichten* **335** 378
[31] Toalá J A and Arthur S J 2014 *Mon. Not. R. Astron. Soc.* **443** 3486
[32] Reimers D 1975 *Problems in Stellar Atmospheres and Envelopes* ed B Baschek *et al* (Springer, Berlin) p 229
[33] Schönberner D and Blöcker T 1993 *Luminous High Latitude Stars* ed D Sasselov *ASP Conf. Series* **45** p 337
[34] Blöcker T 1995 *Astron. Astrophys.* **299** 755
[35] Pauldrach A W A, Puls J, Kudritzki R-P, Méndez R H and Heap S 1988 *Astron. Astrophys.* **207** 123
[36] Pauldrach A W A, Hoffmann T L and Méndez R H 2004 *Astron. Astrophys.* **419** 1111
[37] Steffen M, Szczerba R and Schönberner D 1998 *Astron. Astrophys.* **337** 149
[38] Steffen M, Hubrig S, Todt H, Schöller M, Hamann W-R, Sandin C and Schönberner D 2015 *Astron. Astrophys.* **570** A88
[39] Corradi R L M, Steffen M, Schönberner D and Jacob R 2007 *Astron. Astrophys.* **474** 529
[40] Mellema G 2004 *Astron. Astrophys.* **416** 623
[41] Chevalier R A 1997 *Astrophys. J.* **488** 263
[42] Plait P and Soker N 1990 *Astron. J.* **99** 1883

Journal of Physics: Conference Series **728** (2016) 032002 doi:10.1088/1742-6596/728/3/032002

The kinematic evolution of the ionized shells of planetary nebulae

J A López[1], M G Richer[1], M Pereyra[2] and M T García-Díaz[1]

[1] Instituto de Astronomía, Universidad Nacional Autónoma de México, Campus Ensenada, Ensenada, Baja California, C.P. 22860, México

[2] Schlumberger Foundation Fellow, School of Physics and Astronomy, University of Southampton, Southampton, SO17 1BJ, UK

E-mail: [1] `jal@astrosen.unam`

Abstract. In this contribution we present the results of measuring the bulk outflow, or global expansion velocities for a large number of planetary nebulae (PNe) that span a wide range of evolutionary stages and different stellar populations. The sample comprises 133 PNe from the galactic bulge, 100 mature and highly evolved PNe from the disk, 11 PNe from the galactic halo and 15 PNe with very low central star masses and low metallicities, for a total of 259 PNe. The long-slit, echelle data are drawn from the SPM Kinematic Catalogue of Galactic Planetary Nebulae. These results reveal from a statistical perspective the kinematic evolution of the expansion velocities of PNe in relation with the changing characteristics of the central star's wind and ionizing luminosity and as a function of the evolutionary rate determined by the central star (CS) mass. The large number of PNe utilized in this work for each group of PNe under study and the homogeneity of the data provide for the first time a solid benchmark from observations for models predictions. This project is still in progress and aims at measuring expansion velocities for about 300 more galactic PNe with the aim of providing a reliable parametric description of their expansion velocity in terms of different populations, morphologies, evolutionary stages and CS masses.

1. Introduction

Planetary Nebulae are expanding ionized shells evolving from the mass loss process in the previous AGB stage. The expanding nature of PNe was recognized nearly 70 years ago [1]. It was then inferred [2] that PNe should descend from fast-evolving, mass-losing red giant stars to eventually become white dwarfs. These ideas led to the concept of PNe originating in the ejected envelope of red giants [3] and shortly after the first basic post-AGB evolutionary model track was calculated [4] starting with an AGB star with an electron degenerate core and a H shell burning envelope. The times were ripe to realize then that the AGB envelope is the precursor of the PN [5]. However, the shells of AGB stars need additional pressure to keep an expanding motion after being ejected and various mechanisms involving ionization fronts and radiation pressure on gas and dust were explored with unsatisfactory results, until it was concluded [6] that mass loss from active stellar winds was needed to maintain the shell expansion. This problem was originally solved through the interacting stellar winds model (ISW) [7] [8] where after the initial AGB mass loss phase the exposed core of the now much smaller star becomes hotter and mass loss through radiation pressure develops a faster wind that interacts with the previous slower wind from the AGB producing a shocked interface between the two. The ISW model provides the mechanism to keep the ejected material expanding and the tools to understand the basic PN shapes produced by this slow wind - fast wind interaction. Once the central star evolves off the AGB stage and becomes even hotter the combined effects of the interacting winds and the ionizing luminosity over the nebula drive and dominate the shell expansion. In the energy-conserving case the ISW model predicts expansion velocities of the order of 30 km s^{-1} which agrees well with the average results of individual observations. The ISW model became the new paradigm to understand the formation and evolution of PNe, although we now know it does not explain many salient features of PNe discovered at later times with high spatial resolution (e.g. with the *HST*) and that clearly demand introducing additional elements to the original model. However, the ISW model remains a solid foundation of PN evolution and it is a fundamental input into all modern evolution models.

Although there are many works devoted to exploring the detailed kinematics of individual PNe, in contrast there have been few attempts to relate in a statistical way their global expansion velocities as a function of evolutionary status and compare the results with model predictions. Early attempts [9] tried to derive a relationship between the expansion velocity and the PN radius but using an heterogenous sample from different sources, obtaining a poor correlation. However, from a relatively small sample of observations with the IUE a positive correlation between the expansion velocity and the wind's terminal velocity was found [10]. Positive correlations were also found in the Magellanic Clouds [11] [12] between the expansion velocity and the excitation class. In general terms we know from observations and models that the mean expansion velocity of PNe lies between 15 - 40 km s^{-1}. We also know that the expansion velocities of PNe should evolve throughout their lifetimes as a function of the central star stellar wind, ionizing radiation and rate of CS evolution. However the available studies up to now do not provide enough information to draw a clear, parametric map of the kinematic evolution of diverse groups of planetary nebulae. This task requires measuring global expansion velocities, in a consistent way, for a large number of PNe divided by evolutionary stages, galactic populations, CS masses and metallicities. For this purpose we are using the SPM Kinematic Catalogue of Galactic PNe [13]. This catalogue contains long-slit, echelle, kinematic data for nearly 700 PNe. The spectral resolution in the data ranges from 6 km s^{-1} to 11 km s^{-1}, perfectly adequate to resolve the internal kinematics of PNe.

2. Kinematics and spatial resolution

For ground based observations it is relevant to distinguish between spatially resolved and unresolved cases. In the case of resolved cases, such as large PNe from the disk, the line profiles usually present either splitting of the emission line profile or an asymmetric profile, only very few cases with very low expansion velocities present symmetric gaussian-type, single profiles. The slit in these cases intersects only a small fraction of the object. For the unresolved cases, such as PNe in the halo or some cases in the bulge, the PNe are nearly point sources and the entire object, or most of it falls into the slit. In these cases the line profiles are either a single, symmetric profile or asymmetric, barely resolved profiles. For the cases with line splitting the expansion velocity of the nebular shell is derived as half the peak to peak difference between the main receding (redshifted) and approaching (blue-shifted) components of the line profile. When the emission line is a single symmetric profile we fit a gaussian profile and assign half the resulting FWHM after correcting for instrumental, thermal Doppler and fine structure broadening, if needed. Lastly, for asymmetric and barely resolved profiles we fit red and blue gaussian components and assign the expansion velocity as half the peak to peak difference.

From the description above it is clear that the global expansion velocity is defined here as the emission weighted expansion or bulk outflow velocity for the matter projected within the spectrograph slit. As such, this V_{exp} is an adequate parameter for characterizing the kinematic evolution of the ionized component of PNe. However, it should be noted that this V_{exp} may be

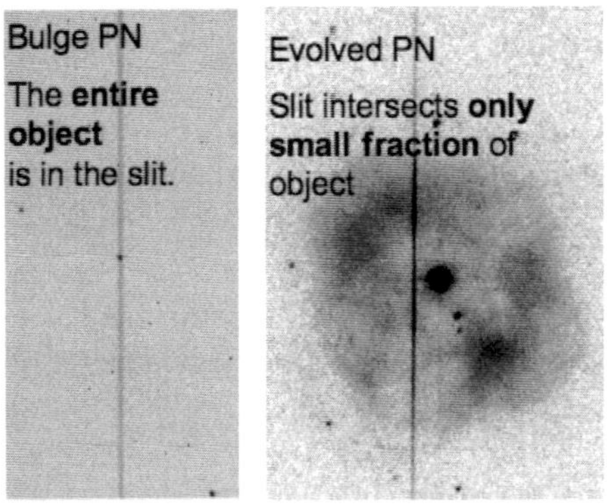

Figure 1. Examples that produce spatially unresolved (left) and resolved (right) long-slit spectra. The vertical line on both images is the projected image of the slit over the nebulae.

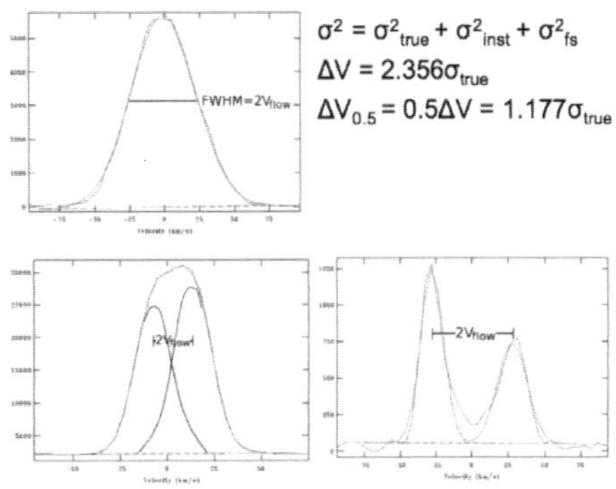

Figure 2. From top left to bottom right: spatially unresolved line profile with Gaussian fit. Correction for true Doppler width. Partially resolved line profile, two Gaussian components are fit. Resolved line profile is seen split.

unsuitable for determining kinematic ages or expansion parallax distances, for which the outer shock velocity, which cannot be measured directly, or other pattern of velocity may be a better option in some cases [14].

It should also be noted that our sample does not seem to contain PNe with a marked rim/shell structure and a hot X-ray emitting bubble, though some may be present in the bulge sample (spatially unresolved). In some models [15] the ISW mechanism causes that *all* PNe form and sustain throughout most of their lifetime a hot bubble produced by the fast, shocked stellar wind, contained by a rim and an outer shell. The mass loss is considered in these models always isotropic and resulting in a spherical and uniform envelope. The rim is dense and bright but moves slowly whereas the shell is fainter but expands faster. As a consequence the line profiles must contain four components, the two inner and brighter ones corresponding to the receding and approaching sections of the rim and the outer, fainter components corresponding to the shell. In this situation the global expansion velocity takes a non-ballistic structure and defining a mean, global expansion velocity requires additional assumptions. However our sample of large, evolved, spatially resolved nebulae has been chosen specifically to contain only PNe with no rims or filamentary shell structures. The sample of PNe from the halo, spatially unresolved, descend from very low mass progenitors and it is as yet unclear whether they form or preserve a hot bubble and a rim/shell structure for a long time due to their very slow evolution. As for the the bulge sample, also largely composed by unresolved objects with diverse properties, there is a possibility that in this case some rim/shell type PNe may be present here but their possible inclusion in the sample may only mean that we underestimate their relative weight in the region of high V_{exp} since for their characteristics, rim/shell PNe are of high excitation, but they would not alter in a significant way the general statistical conclusions of this work. Incidentally and contrary to expectations, rim/shell nebulae that host hot X-Ray emitting bubbles do not seem to be pervasive. From a 1.5 kpc volume limited *CHANDRA* survey in the solar neighborhood it has been found [16] that only 27% of the surveyed PNe show extended X-Ray emission indicative of the presence of a hot bubble.

Journal of Physics: Conference Series **728** (2016) 032002 doi:10.1088/1742-6596/728/3/032002

3. The bulge PNe sample

Here we aim at obtaining systematic observations of a single galactic population of PNe, spanning from the earliest evolutionary stages until the cessation of nuclear burning. The detailed individual results for the studies of this group have appeared in [17][18]. There are 133 PNe observed in this group. Line intensity ratios of $I(5007)/I(H_\beta)$ and H_β luminosities (LH_β) have been used to discriminate (broad) evolutionary stages in the sample by adopting $I(5007)/I(H_\beta) \leq 3$ and high (LH_β) for (24) young PNe; $I(5007)/I(H_\beta) \geq 6$ and high (LH_β) for (86) mature PNe, additionally subdivided by the presence or absence of He II 6560. And finally $I(4686) \geq 0.5\ I(H_\beta)$ for the (23) more evolved PNe in the bulge sample. In this way we identify 4 groups within the sample, see Figure 3, whose expansion velocities clearly increase with excitation conditions.

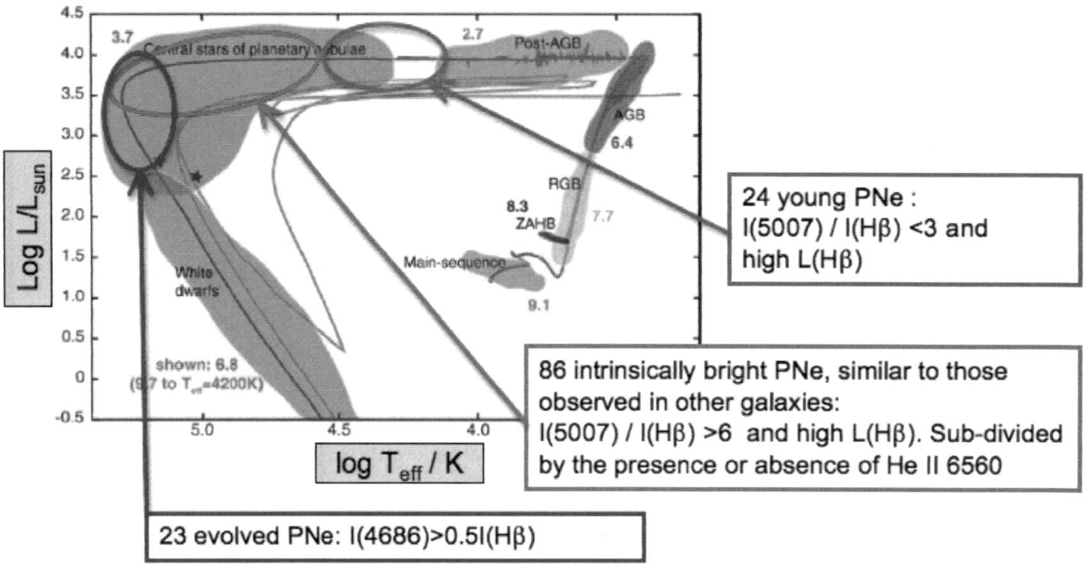

Figure 3. The approximate distribution of evolutionary stages of the PNe sample from the bulge as determined from excitation conditions are shown over a sketch version of the the H-R diagram [19]

For young PNe we find an average $V_{exp} = 10.7$ km s^{-1}, for warmer nebulae $V_{exp} = 15.7$ km s^{-1}, for hot nebulae $V_{exp} = 23.6$ km s^{-1} and for the nebulae with the highest excitation conditions in these groups, $V_{exp} = 26.5$ km s^{-1}. See Figure 4. There is a very good agreement between the CS evolutionary stage and the expansion velocity with a clear trend of increasing V_{exp} as the PN evolves.

4. The mature and highly evolved PNe sample

This sample aims at testing the global expansion velocity of PNe at late and very late evolutionary stages. These are relatively large, old nebulae located in the disk and composed of a single, closely spherical, smooth shell with no rims or filamentary structure. The sample is broadly divided into two groups, namely, mature PNe that still show some structure and bright outer edges but no rim and highly evolved PNe with low surface brightness and no inner structure. The detailed results of the study of these groups have appeared in [20]. There are 100 PNe in this group, 22 of them are considered mature and 78 highly evolved. The mature PNe concentrate around the "knee" in the evolutionary tracks, that is near the maximum effective

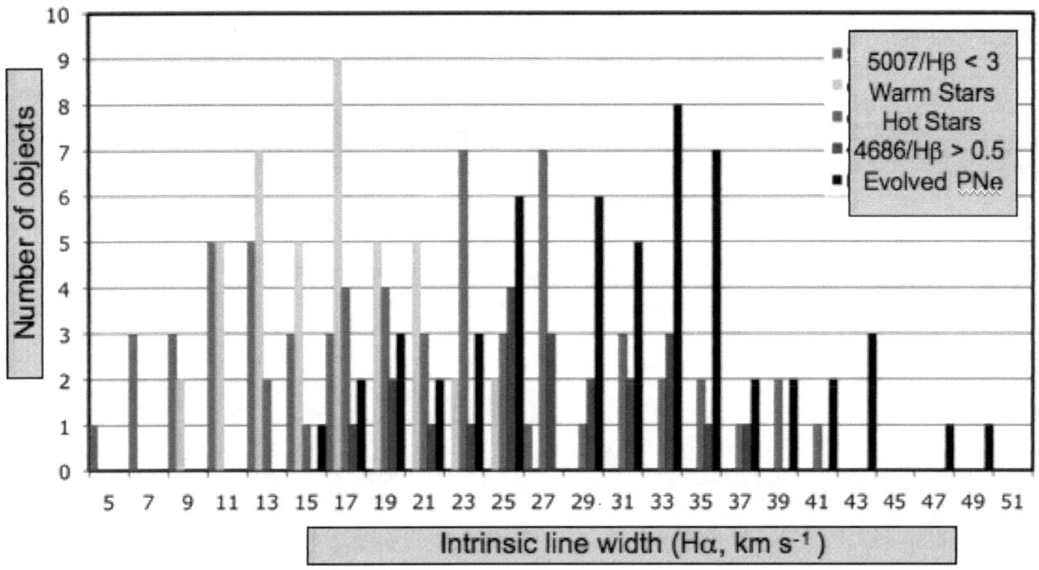

Figure 4. The expansion velocity distribution in the bulge sample is shown in colors for the different groups and in black for the mature and highly evolved sample. A continuos progression in V_{exp} with excitation degree or evolutionary stage is apparent.

temperature in the evolutionary tracks, and they show the highest expansion velocities of both groups with a mean $V_{exp} = 37 \pm 4$ km s^{-1}. The highly evolved group are distributed mostly along the low luminosity and cooling evolutionary tracks that lead to the white dwarfs zone. This group shows a decrease in expansion velocity with respect to the mature PNe, with a mean $V_{exp} = 28 \pm 10$ km s^{-1} which is in contrast with the previous trend. From the bulge sample and continuing with the group of mature PNe a clear positive correlation of increasing expansion velocity with increasing evolutionary stage, excitation conditions and central star effective temperature is apparent, see Figure 4. However, once nuclear reactions cease and the ionizing luminosity and presumably wind power decline rapidly, this trend no longer continues and the observed expansion velocity slows down. What appears as a deceleration of the expansion velocity in the highly evolved PNe near the end of their life may have additional contributions from effects such as ionization stratification, decreasing optical thickness, interaction with the surrounding medium and recombination of the outer shell regions in some cases. Overall we find that the more evolved and hotter the CS the faster the PN expansion, up to the point when the CS luminosity and stellar wind decline. At this stage this correlation breaks and V_{exp} seems to start to decline.

5. The halo and low metallicity sample

With this sample we aim to characterize the expansion velocities of PNe with very low mass progenitors ($\leq$1 $M_\odot$) and slow evolution rates. The sample contains all the PNe that have been previously classified as belonging to the galactic halo (11 objects) and PNe with log(O/H) + 12 $\leq$ 8.0 dex (15 objects) most of them located in the bulge. The selection of the second group is based on the assumption that PNe with very low metallicities also descend from very low mass progenitors. The detailed results of this work can be found in [21]. The main outcome of this study is that the expansion velocity distribution of both samples behave very similar indicating that although the groups may arise from different populations there is indeed a common link

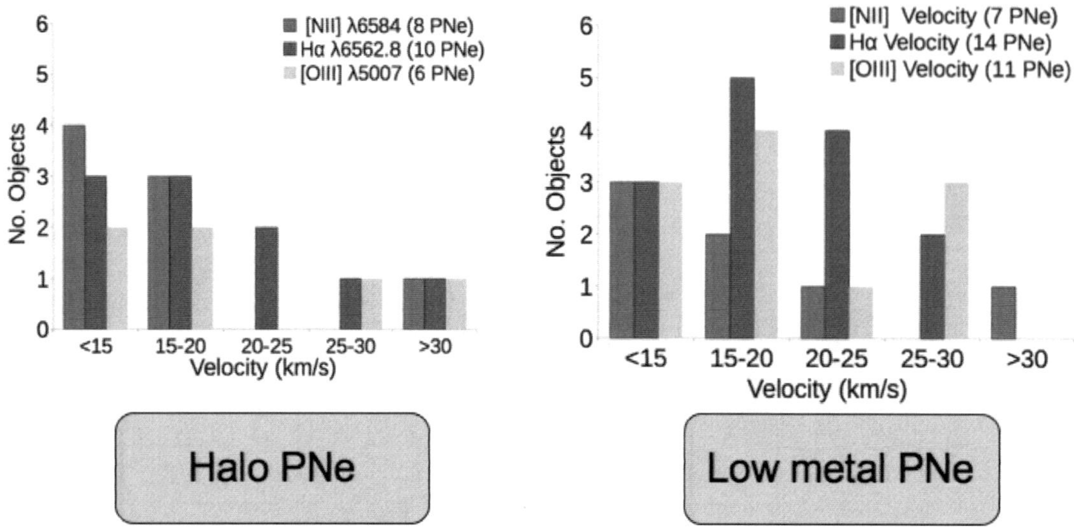

Figure 5. Distribution of the expansion velocities for the halo and the low metal groups. In both cases measurements obtained from each emission line are plotted. Both groups behave very similarly and regardless of the emission line considered most of the objects show low $V_{exp.}$

related to the stellar parameters of this kind of PNe that drives V_{exp} in these objects. In addition it is found that these groups show the slowest expansion velocities of all previous groups studied so far with mean $V_{exp} \simeq 19$ km s^{-1} for all the emission lines considered (H_α, [N II] and [O III]). Our kinematic observations are in excellent agreement with [22 -25] for the 4 halo PNe that we have in common. The objects with the highest expansion rates are also those located at high effective temperature, as has been the case in the others samples. However their average $V_{exp} \simeq 25$ km s^{-1} is substantially lower than what is found for the mature PNe ($V_{exp} \simeq 37$ km s^{-1}). We notice that according to some models [26] PNe with low metallicity have lower wind luminosities but also due to poor line cooling efficiency their electron temperature increases and tend to accumulate thermal pressure that drives a faster expanding shell than in PNe with normal abundances. Although the physics is sound behind this argument, this is contrary to our results for both groups in this sample where the opposite is observed, as mentioned above. Since the mass loss processes and rates of evolution from the AGB to the PN stage are poorly known for these groups further advances on theory and additional observations at higher spatial and spectral resolution are required to clarify this effect.

6. Conclusions

We have presented here the results of several works that cover a systematic and homogenous set of observations aimed at characterizing the expansion velocity of PNe as a function of evolutionary stage, considering diverse populations and CS masses. The global results from the samples considered here are shown in Figure 6. This figure shows for the first time a snapshot of the kinematic evolution of PNe as the CS evolves towards higher effective temperatures. For each group the number of members that have been measured and the average V_{exp} are indicated. Notice that even for the PNe with low CS masses, the halo and low metal groups, their lowest expansion velocities appear at the low T_{eff} boundary of the sample whereas the higher expansion velocities are found at higher T_{eff} just as in the other samples, though at different (lower) velocity ranges.

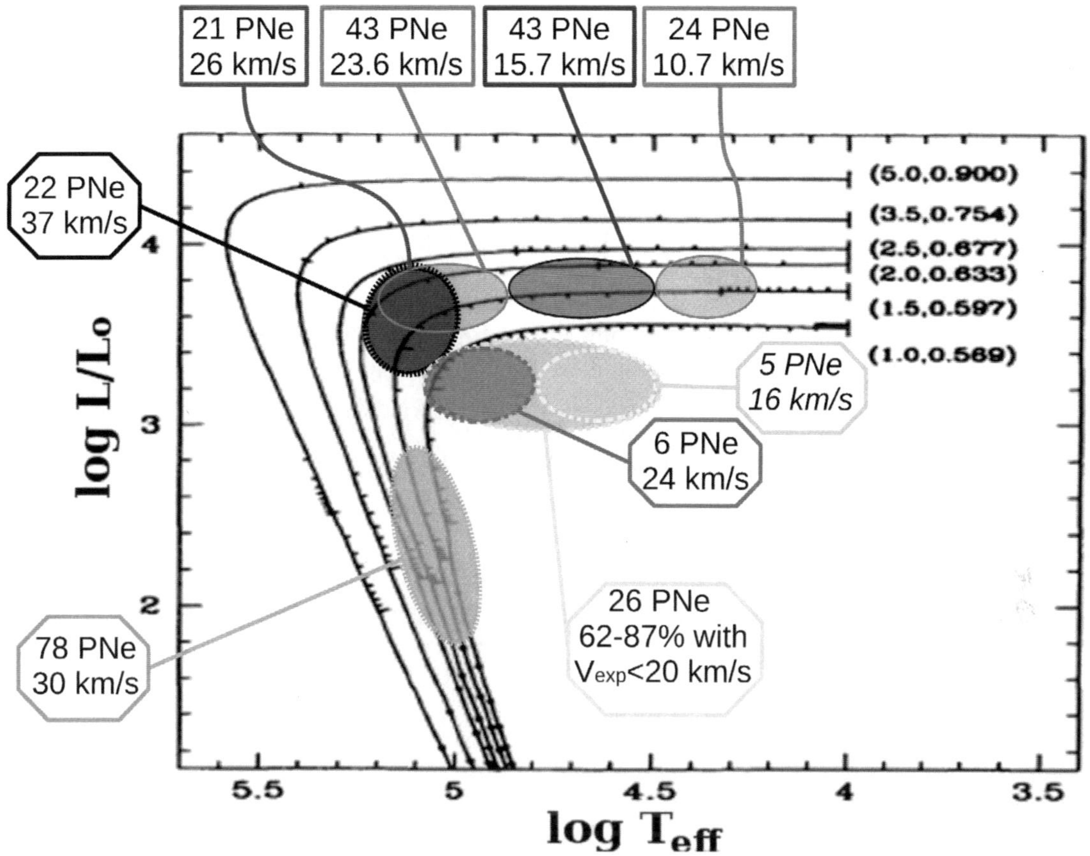

Figure 6. The approximate distribution in the H-R diagram of the groups presented in this work are shown over evolutionary tracks [27]. For each group are indicated the number of members measured and their average V_{exp}. The kinematic evolution of PNe with evolutionary stage is evident in this figure. V_{exp} increases wit T_{eff} in all cases, for the bulge and disk PNe and for the halo and low metal PNe, reaching maximum V_{exp} values at the point of maximum T_{eff} though within different velocity ranges that indicate a dependance on CS masses and possibly metallicity. Labels at the top of the diagram point to the groups from the bulge. Labels on the left of diagram point to the mature and highly evolved groups. Labels within the diagram point to the halo and low metal sample.

Our samples have not considered PNe that form rim/shell structures and contain a hot X-Ray emitting bubbles as a result of shocking winds predicted in the ISW theory. That type of PNe also show often fast collimated outflows known as FLIERS. In these cases the faint line components from the fast outer shell may escape detection [15] and cause to underestimate V_{exp} at high T_{eff} but otherwise it would not modify the trend and main conclusions of this work. In any case, the mature and highly evolved, spatially resolved PNe do not have these structures. There is the possibility that some PNe of the rim/shell class may be present in the sample of spatially unresolved objects from the bulge. We consider unlikely their presence in the halo and low metal sample due to the expected lower wind speeds and very slow evolution of these groups. From the 2 kpc volume limited CHANDRA survey of X-ray emission it is found that only 27% of the PNe in the sample show extended X-ray emission indicative of the presence of a hot bubble

[16]. Therefore it seems that these rim/shell PNe with hot bubbles are not ubiquitous. It may be that they always form, as expected from the theory, but are quickly disrupted by leakage of the post-shock, fast wind through a porous or partially open envelope, an effect that may be related with mass loss history and CS mass or pace of evolution. In addition, since the majority of cases of rim/shell + hot bubble PNe have extremely similar hydrodynamic structures and morphologies, this leads to suspect that their formation and preservation may be favored by certain conditions, possibly related to mass loss rate and a range of CS masses. This is an open subject that requires further attention. Lastly, the expansion velocities observed in the halo and low-met groups, both of low CS mass, do not agree with theoretical expectations. We find that these groups have on average lower V_{exp} than their counterparts with higher CS masses and solar-type abundances. A better understanding of the mass loss processes and transitions times for these PNe is required together with additional observations with higher spatial resolution to better understand their structures. It is our aim to continue studying the kinematic evolution of PNe by incorporating in the future the missing classes in the present study.

Acknowledgments

The authors thankfully acknowledge financial support from DGAPA - UNAM through PAPIIT project No. IN108416 and to CONACYT, Mexico through project 178253. M.P. is grateful to the Schlumberger Foundation for a postdoctoral fellowship for women from developing countries. JAL thanks Prof Sun Kwok for partial financial support to attend this meeting and for his gracious hospitality during the event.

References

[1] Wilson O C 1948 *Astrophys. J.* **108** 201
[2] Shklovsky I 1956 *Astr. Zh.* **33** 222
[3] Abell O B and Goldreich P 1966 *Publ. Astron. Soc. Pac.* **78** 232
[4] Paczyński B 1971a *Acta Astr.* **21** 417
[5] Paczyński B 1971b *Astrophys. Lett.* **9** 33
[6] Mathews W G 1966 *Astrophys. J.* **143** 173
[7] Kwok S, Purton C R and Fitzgerald M P 1978 *Astrophys. J. Lett.* **219** L125
[8] Kwok S 1982 *Astrophys. J.* **258** 280
[9] Chu Y H, Kwitter, K B, Kaler J B and Jacoby G H 1984 *Publ. Astron. Soc. Pac.* **96** 598
[10] Heap S 1993 *Proc of the IAU Symp 155 on Planetary Nebulae* ed Ronald Weinberger and Agnes Acker Innsbruck, Austria (Kluwer Academic Press; Dordrecht) p. 23
[11] Dopita M A, Lawrence C J, Ford H C and Webster B L 1985 *Astrophys. J.* **296** 390
[12] Dopita M A, Meatheringham S J, Webster B L and Ford H C 1988 *Astrophys. J.* **327** 639
[13] López J A, Richer M G, García-Díaz M T, Clark D M, Meaburn J, Riesgo H, Steffen W and Lloyd M 2012 *Rev. Mex. Astron. Astrof.* **48** 3
[14] Jacob R, Schönberener D and Steffen M 2013 *Astron. Astrophys.* **558** 78
[15] Schönberener D, Jacob R, Lehmnn H, Hildebrant G, Steffen, Zwansig A, Sandin C and Corradi R L M 2014 *Astron. Nachr.* **335** 378
[16] Freeman M *et al* 2014 *Astrophys. J.* **794** 99
[17] Richer M G, Lopez J A, Pereyra M, Riesgo H and García-Díaz M T 2008 *Astrophys. J.* **689** 203
[18] Richer M G, Lopez J A, García-Díaz M T, Clark D M, Pereyra M and Díaz-Méndez E 2010 *Astrophys. J.* **716** 857
[19] Herwig F 2005 *Annu. Rev. Astro. Astrophys.* **43** 435
[20] Pereyra M, Richer M G and Lopez J A 2013 *Astrophys. J.* **771** 114
[21] Pereyra M, Lopez J A and Richer M G 2016 *Astron. J.* **151** 53
[22] Otsuka M, Hyung S, Lee S-J, Izimura H and Tajitsu A 2009 *Astrophys. J.* **705** 509
[23] Otsuka M, Tajitsu A, Hyung S and Izimura H 2010 *Astrophys. J.* **723** 658
[24] Otsuka M, Hyung S and Tajitsu A 2015 *Astrophys. J. Suppl. Series* **217** 22
[25] Otsuka M, Tamura S, Yadoumaru Y and Tajitsu A 2003 *Publ. Astron. Soc. Pac.* **115** 67
[26] Schönberener D, Jacob R, Sandin C and Steffen M 2010 *Astron. Astrophys.* **523** A86
[27] Vassiliadis E and Wood P R 1994 *Astrophys. J. Suppl. Series* **92** 125

Journal of Physics: Conference Series **728** (2016) 032003 doi:10.1088/1742-6596/728/3/032003

Proof of shock-excited H_2 in low-ionization structure of PNe

Stavros Akras[1], Denise R. Gonçalves[1] and Gerardo Ramos-Larios[2]

[1] Observatório do Valongo, Universidade Federal do Rio de Janeiro, Ladeira Pedro Antonio 43, 20080-090, Rio de Janeiro, Brazil

[2] Instituto de Astronomía y Meteorología, Av. Vallarta No. 2602. Col. Arcos Vallarta, CP 44130, Guadalajara, Jalisco, Mexico

E-mail: [1] akras@astro.ufrj.br

Abstract. We report the detection of near-IR H_2 line emission from the low-ionization structures (LISs) in planetary nebulae. The deepest, high-angular resolution H_2 1-0 S(1) at 2.122 μm, and H_2 2-1 S(1) at 2.248 μm images of K 4-47 and NGC 7662, obtained using NIRI@Gemini-North, are presented here. K 4-47 reveals a remarkable high-collimated bipolar structure, with the H_2 emission emanating from the walls of the outflows and a pair of knots at the tips of these outflows. The H_2 1-0 S(1)/2-1 S(1) line ratio is ~7-8 which indicates shock interaction due to both the lateral expansion of the gas and the high-velocity knots. The strongest line, H_2 v=1-0 S(1), is also detected in several LISs located at the periphery of the outer shell of the elliptical PN NGC 7662, whereas only four knots are detected in the H_2 v=2-1 S(1) line. These knots have H_2 v=1-0 S(1)/v=2-1 S(1) values between 3 and 5. These data confirm the presence of molecular gas in both highly (K 4-47) and slowly moving LISs (NGC 7662). The H_2 emission in K 4-47 is powered by shocks, whereas in NGC 7662 is due to photo-ionization by the central star. Moreover, a likely correlation is found between the H_2 v=1-0 S(1)/H_2 v=2-1 S(1) and [N II]/Hα line ratios.

1. Introduction

Optical imaging surveys of planetary nebulae (PNe) have revealed that a fraction of PNe possess, besides the large scale structure such as rims, shells, haloes, some small-scale structures (e.g. [1], [2], [3]). These structures are prominent in the low-ionization emission lines such as [N II], [S II] and [O I] (hereafter LISs). They exhibit a variety of morphologies like knots, jets, and filaments, ([3]), whereas they cover a wide range of expansion velocities from 30 km s^{-1} up to 350 km s^{-1}. Akras & Gonçalves ([4]), studying a sample of Galactic PNe with LISs, demonstrate that the excitation mechanism of these structures is a combination of UV-photons and shocks. The contribution of each mechanism depends on paramteres like the distance of LISs to the central star, the stellar parameters (T_{eff}, $L\odot$) and LIS's expansion velocity.

The physical properties like T_e, N_e and chemical abundances of LISs have been studied by different groups (e.g. [4], [5], [6]), and some of the most important conclusions are: i) there is no difference in T_e between the LISs and the nebular components, ii) LISs have systematically lower N_e compared to the surrounding medium (e.g.[4], [5]) and iii) there is no difference in chemical abundances that could provide an explanation for the enhancement of the low-ionization emission lines.

The formations models of LISs predict that they are denser structures than the surrounding ionized medium. This result is found to be inconsistent with the observations. A possible explanation for this discrepancy may be that the formation models refer to the total density of gas (dust, atomic and molecular) and not only to the N_e. Gonçalves et al. ([6]) have proposed that LISs may also be made of molecular gas. H_2 emission has previously been detected from the cometary knots in the Helix PN ([7], [8]) and knots/filaments in NGC 2346 ([9]).

Generally, H_2 emission has been detected in several PNe (e.g. [10]). A comparison of H_2 and optical line (e.g. [N II] and [O I]) images has revealed similar morphologies, suggesting that both emissions emanate from the same regions. A recent theoretical work by Aleman & Gruenwald ([11]) has shown that the peak intensities of the optical low-ionization and H_2 lines occur in a narrow transition zone between the ionized and neutral (photo-dissociation) regions. Moreover, an empirical relation between the fluxes of the [O I] λ6300 and H_2 v=1-0 S(1) lines for Galactic PNe was reported by Reay et al. ([12]). Hence, the detection of strong low-ionization lines in LISs may also suggest the presence of H_2 gas.

2. Observations

The deepest, high-angular resolution H_2 v=1-0 S(1) at 2.122 μm and H_2 v=2-1 S(1) at 2.242 μm images were obtained for K 4-47 and NGC 7662 using the NIRI instrument on the Gemini-North 8 m telescope on Manua Kea in Hawaii. For these observations, the f/6 configuration (pixel scale=0.117 arcsec and field of view of 120 arcsec) were used. The exposure times were estimated using the empirical relation from Reay et al. ([12]). The final continuum-subtracted images are presented in Figures 1 and 2.

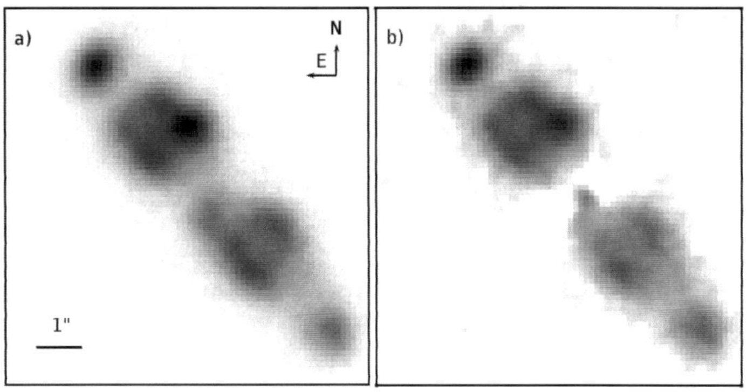

Figure 1. Line images of the bipolar planetary nebula K 4-47. (a) Continuum subtracted H_2 v=1-0 S(1), (b) H_2 v=2-1 S(1) emission lines.

3. Discussion

3.1. The highly collimated bipolar nebula K 4-47

The narrow-band near-IR H_2 images of K 4-47 (Fig. 1) reveal a remarkable high-collimated bipolar structure with the emission arising from the walls of the bipolar outflows. At the tips of these outflows lies a pair of highly moving, low-ionization knots (100 km s^{-1}; [13]).

These knots exhibit very strong [N II] λ5200 and [O I] λ6300 emission lines that usually are attributed to high-velocity shocks. Gonçalves et al. ([14]), running a number of shock models, came to the conclusion that the expansion velocity of these knots must be up to 250–300 km s^{-1}. Despite their very high velocities, their H_2 v=1-0 S(1)/H_2 v=2-1 S(1) line ratio is found to be 7.3, lower than the typical value of 10 for shock-excited regions ([15]).

Regarding the bipolar outflows, a comparison between our near-IR and optical images from [13] illustrates that the ionized gas (optical emission) is concentrated in a inner highly collimated structure, surrounded by the H_2 outflows. This structure of K 4-47 completely resembles that of the M 2-9 and CRL 618 PNe indicating a possible connection among these objects. The large

H_2 v=1-0 S(1)/H_2 v=2-1 S(1) line ratio, estimated in the bipolar shell, was quite unexpected. Although, the recent hydrodynamic models by Balick B. and collaborators ([16]) can adequately explain our findings. The H_2 emission from the bipolar outflows can be explained as the result of the interactions between a jet or bullet with the AGB material. As the jet/bullet moves through the AGB material forms an conical structure that expands laterally outward with a velocity that increases with the distance from the central star. At the same time, the jet/bullet continues moving outwards, with a velocity proportional to the distance from the central star, dissociating the AGB H_2 gas, which then is ionized by UV-photons (optical images). The surface brightnesses in the H_2 v=1-0 S(1) and H_2 v=2-1 S(1) emission lines are found to range from 0.2 to 1×10^{-15} erg cm^{-1} s^{-1} arc^{-2} and from 0.2 to 1 10^{-16} erg cm^{-1} s^{-1} arc^{-2}, respectively.

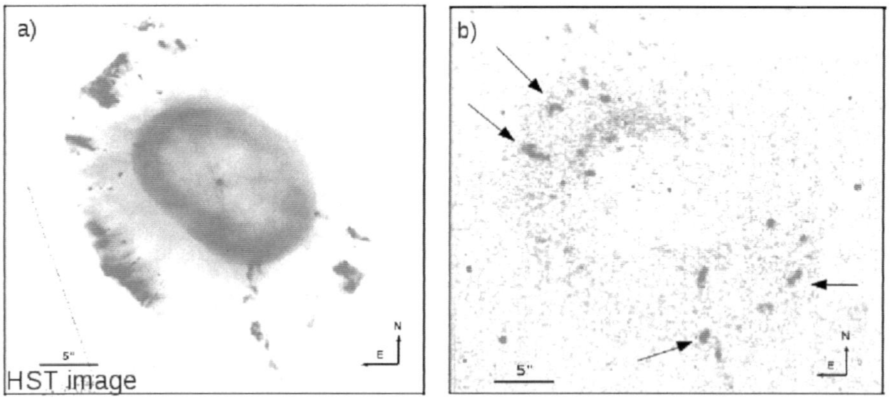

Figure 2. Continuum subtracted H_2 v=1-0 S(1) line image of the elliptical planetary nebula NGC 7662. The arrows indicate the four LISs in which the H_2 v=2-1 S(1) emission is detected.

3.2. The elliptical nebula NGC 7662
NGC 7662 is an elliptical PN that posses almost two dozens of LISs (knots and a jet-like structure) embedded in the outer shell with expansion velocities that vary from 30 to 70 km s^{-1} (see [17]). All these structures are easily discerned in the [N II] line image ([18]), whereas spectroscopic data have revealed that they also exhibit a strong [O I] λ6300 line ([17],[6]). Despite that the former line is a strong indicator of shock interactions, most of the LISs have low velocities except from the southern jet-like feature. (see [17]).

Here, we present the deepest near-IR images of this nebula. The H_2 v=1-0 S(1) emission line is detected in almost all optically identified LISs (Fig. 2, right panel.), whereas only four LISs are found to have H_2 v=2-1 S(1) emission. The former line is found to range from 1 to 4.8 $\times$ 10^{-16} erg cm^{-1} s^{-1} arc^{-2}, whereas the latter from 0.6 to 1 $\times$ 10^{-16} erg cm^{-1} s^{-1} arc^{-2}. No H_2 emission is found associated with the nebular shells. This confirmed that LISs are molecular condensations embedded in a high excitation nebula. The value of H_2 v=1-0 S(1)/H_2 v=2-1 S(1) line ratio is found to range between 3 and 5. These values implies that the LISs in this nebula are predominantly photo-ionized. Nevertheless, shocks cannot be discarded.

4. Conclusion

New, deep, high-angular resolution near-IR H_2 images confirmed the presence of molecular gas in highly (K 4-47) and slowly moving LISs (NGC 7662). H_2 emission was also detected at the dense walls of the bipolar outflows of K 4-47 which implies a lateral expansion in agreement with the predictions from hydrodynamic models. Important morphological similarities among K 4-47, M 2-9 and CRL 618 have been found.

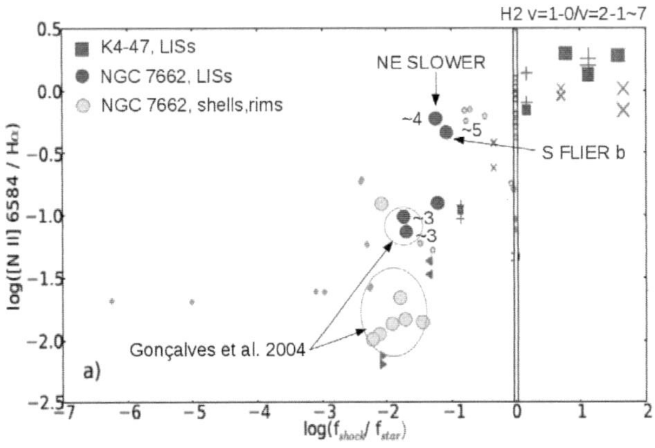

Figure 3. Plot of [N II]/Hα vs. log(f_{shocks} / f_{star}) for K 4-47 ([6]) and NGC 7662 ([17]) from the diagnostic diagram of Akras & Gonçalves ([4]) for photo-ionized and shock-excited regions. The numbers indicate the value of the H_2 v=1-0 S(1)/H_2 v=2-1 S(1) line ratio from six LISs (the pair of knots in K 4-47 and the four knots in NGC 7662).

In the NGC 7662 nebula, H_2 emission was detected in several LISs while it is totally absent in the nebular shells. Using the new diagnostic diagram from Akras & Gonçalves ([4]), and the spectroscopic data of the six LISs with measured H_2 v=1-0 S(1)/H_2 v=2-1 S(1) line ratio, we found that the near-IR line ratio increases with the [N II]/Hα optical line ratio (Fig. 3). This result may reflect a similar origin for the near-IR and optical line ratios. In conclusion, the H_2 emission in K 4-47 is powered by shocks, whereas in NGC 7662 is due to photo-ionization by the central star with a possible contribution of shocks.

Acknowledgments
Based on observations obtained at the Gemini Observatory, which is operated by the Association of Universities or Research in Astronomy, Inc., under a cooperative agreement with the NSF on behalf of the Gemini partnership. This work is supported by CAPES (program A35/2013) and FAPERJ (program E-26/111.817/2012).

References
[1] Balick B 1987 *Astron. J.* **94** 671
[2] Corradi R L M, Manso R, Mampaso A and Schwarz H E 1996 *Astron. Astrophys.* **313** 913
[3] Gonçalves D R, Corradi R L M and Mampaso A 2001 *Astrophys. J.* **547** 302
[4] Akras S and Gonçalves D R 2016 *Mon. Not. R. Astron. Soc.* **455** 930
[5] Balick B, Alexander J, Hajian A, Terzian Y, Perinotto M and Patriarchi P 1998 *Astron. J.* **116** 360
[6] Gonçalves D R, Mampaso A, Corradi R L M and Quireza C 2009 *Mon. Not. R. Astron. Soc.* **398** 2166
[7] Huggins P J, Forveille T, Bachiller R, Cox P, Ageorges N and Walsh, J R 2002 *Astrophys. J.* **573** 55
[8] Matsuura M, Speck A K, McHunu B M, Tanaka I, Wright N J, Smith M D, Zijlstra A A, Viti S and Wesson R 2009 *Astrophys. J.* **700** 1067
[9] Manchado A, Stanghellini L, Villaver E, García-Segura G, Shaw R A and Garcia-Hernandez D A 2015 *Astrophys. J.* **808** 115
[10] Marquez-Lugo R A, Guerrero M A, Ramos-Larios G and Miranda L F 2015 *Mon. Not. R. Astron. Soc.* **453** 1888
[11] Aleman I and Gruenwald R 2011 *Astron. Astrophys.* **528** 74
[12] Reay N K, Walton N A and Atherton P D 1988 *Mon. Not. R. Astron. Soc.* **232** 615
[13] Corradi R L M, Gonçalves D R, Villaver E, Mampaso A, Perinotto M, Schwarz H E and Zanin C 2000 *Astrophys. J.* **535** 823
[14] Gonçalves D R, Mampaso A, Corradi R L M, Perinotto M, Riera A and Lopez-Martin L 2004 *Mon. Not. R. Astron. Soc.* **355** 37
[15] Black J H and van Dishoeck E F 1987 *Astrophys. J.* **322** 412
[16] Balick B, Huarte-Espinosa M, Frank A, Gomez T, Alcolea J, Corradi R L M and Vinković D 2013 *Astrophys. J.* **772** 20
[17] Perinotto M, Patriarchi P, Balick B and Corradi R L M 2004 *Astron. Astrophys.* **422** 963
[18] Guerrero M A, Jaxon E G and Chu Y-H 2004 *Astron. J.* **128** 1705

Unraveling the histories of pre planetary nebulae

Bruce Balick[1]

[1] Astronomy Department, University of Washington, Seattle WA 98195-1580

E-mail: [1] `balick@uw.edu`

Abstract. HST images of pre planetary nebulae (pPNE) show a mesmerizing array of highly symmetric bi-lobed morphologies with pinched waists. I shall review the results of detailed Hydrodynamical model studies that nicely match the observed structures and reveal the outflow histories and the geometry of the nuclear outflows of two prePNe.

1. Introduction

The general shaping processes of mature planetary nebula have been very thoroughly investigated starting with [1] and continuing with the exquisite 1-D radiation hydro models of the Potsdam group and the two-dimensional simulations of Adam Frank, Vincent Icke, Garcia-Segura, van Marle, many other people and their many collaborators. This field of research has matured very nicely in recent years. With apologies to authors not mentioned, space limitations prevent me from a suitable literature review.

However, in the past decade it has been possible to probe the precursors or planetary nebulae, called pre planetary nebulae, or simply pPNe in this paper. These provide us with a powerful opportunity to trace the evolution of PNe in their infancy and prior to the disruptive effects of ionization, fast and sustained stellar winds, and the shocks and pressure waves that reverberate through the structures of more mature counterparts. This is vital since the underlying astrophysical question is how the stellar outflows of highly evolved stars are formed and collimated to shape the bi-polar pPNe.

Bipolar pPNe pinched waists and generally dark equatorial waistbands comprise the largest single segment, $\approx 50\%$, of the prePN population. Owing to their highly symmetric and articulated structures they are also the most opportune targets for understanding collimation processes in post-AGB stars, processes which have remained enigmatic despite a decade of talks, meetings, MHD simulations, and a new generation of very elaborate 3-D hydro simulations of outflows in close binary stars (Huarte-Espinosa et al., 2014[2], and refs within). That is, we are yet to find a flow paradigm that nicely encompasses the essential features of detailed observations of the structure and flow dynamics of pPNe. Indeed, we still dont know whether the equatorial disks and waistbands actively shape the polar outflows or are mere bystanders in the story of late stellar evolution and lobe formation.

Along with Adam Frank, Martin Huarte-Espinosa, Baowei Liu, and several students I have recently endeavored to match details HST images of some bipolar pPNe with the results of hydro simulations (without magnetic fields). A montage of the best-resolved bipolar pPNe is shown in Fig. 1. The subjects of this study are, for now, constrained by the images and by 2-D observations of their projected flow patterns (that is, images of the density as well as Doppler shifts). That is, the observational data give a reasonably complete picture of the essential hydro state variables, the density and momentum distributions. In a few cases localized radiation from shocks that lie between important parts of the lobes have been used to inform the models of radiative energy losses. The thermal pressure distribution within the flow is not directly measurable at this time since the most omnipresent line ratios that might reveal the temperatures arise in localized shocks or in the cold molecular gas that girds the lobes. However, the images are better fitted by models in which the thermal pressures are less than ram pressures, but not vice-versa.

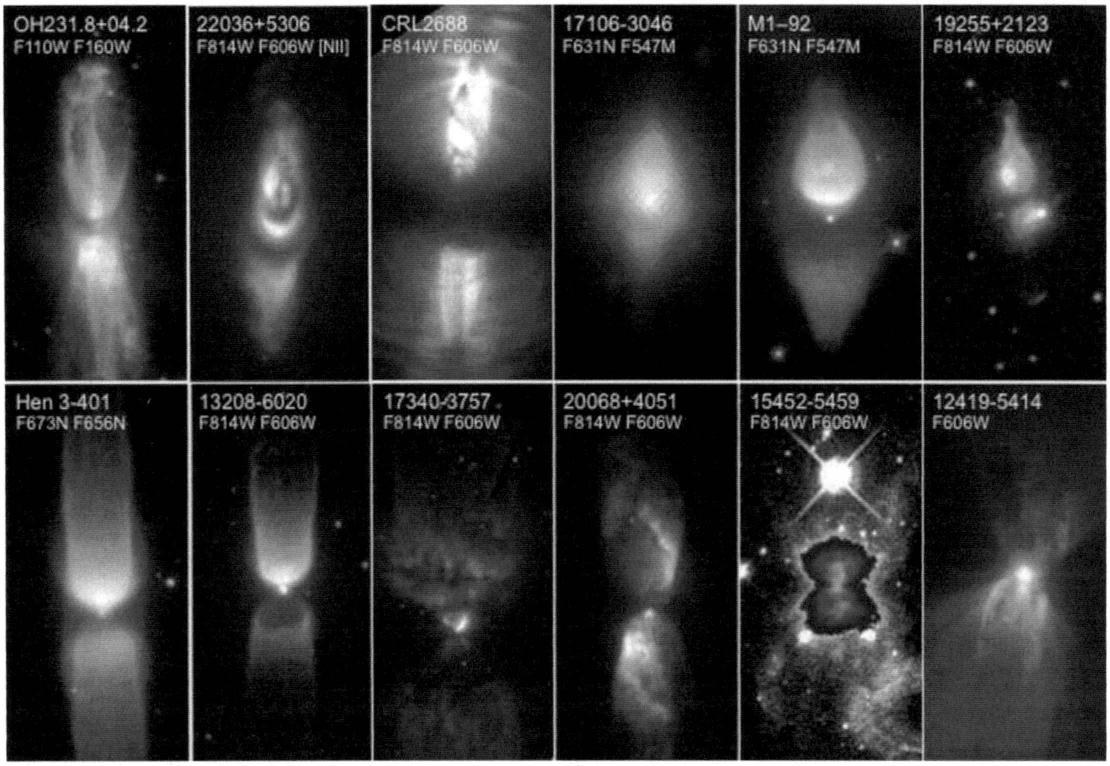

Figure 1. Examples of bright and well-resolved pPNe from HST images in scattered starlight. The colors arise from superimposing images of various filters (indicated) into image planes in Photoshop and adjusting the contrast until that stars are white.

2. General conclusions

There are some very important general conclusions to be made before describing the detailed models. For starters, most lobes that are seen in reflected starlight and molecular emission appear to be hollow (strongly edge-brightened), so that the vital medium that flows through them is not an observable. Rather we see their edges, generally in starlight that is reflected from their flow-compressed boundaries. It is these edges, their loci, and their Doppler shifts that we fit with models.

In addition the innermost portions of flows driven by sustained winds are also the youngest, most recently emerged gas and dust. Furthermore, the density structure of the environment is a vital part of the shaping story. That is, once the densities in the diverging flows drop to about the density of the ambient gas they begin to slow down as they accrete upstream gas of low specific momentum. Such lobes will have thin leading edges where the upstream gas is being accreted and "pushed" forward to "plugs". Shocks will lie on both sides of the plugs.

The forward shock has a speed of the plug speed (which declines with time). The speed of the reverse shock will be the difference of plug and wind speed and will increase with time. The fastest shock will appear along the leading (trailing) edge of the plug when the flows are young (old).

As a rule, we also find that the ambient medium falls off in density as about radius squared. Highly collimated (slowly diverging) flows can penetrate far into the ambient medium at steady speed, much more so than very divergent flows of the same initial momentum flux. In contrast, the edges of blunt flows are impeded and decelerated by the accumulation of the far slower ambient medium.

Finally, after a lengthy parameter study, we have learned that *diverging flows in which the low momenta are concentrated to the polar axis* (more specifically, fall off in zenith angle with a Gaussian form; e.g., [3]) *are the only way to match the flame-like shapes seen in Fig. 1.* We shall use the term "centrally concentrated sprays" for this type of flow to distinguish it from wide-angle flows of uniform density and speed that produce simple pie-wedge-shaped lobes. Pencil-jet flows never fit the HST images.

Another very important result for observers is the flow speed. Many observers have found that their Doppler images show a pattern of linearly increasing Doppler shift with distance from the flow origin. From this they (wishfully) conclude that the flow is ballistic (aka "Hubble-like"), as if the outflow is the result of a brief eruption and not the work of steady winds of roughly constant speed. (Appearances can be deceiving!) However, outbursts inject clumps, and the hydrodynamic evolution of clumps is quite different than that of steady flows.[1]

Our simulations in which the outflow is a centrally concentrated spray show something quite different. In such cases the hot, fast from the advancing tip of the candle-shaped lobe generates a backflow of steadily decelerating speed as it continually mixes with very slow ambient gas. The inevitable, robust result is a thermally cooling zone in the stagnation zone or pocket that trails the leading edge and where gas in the winds mix with slower ambient gas. Emission lines in this zone will display a steady speed gradient with distance behind the head of the lobe.

3. A case study

I presented three case studies in my talk. However I will only present one such study here owing to space limitations. The target is the pPN associated with intensively studied southern lobe of OH231.8+04.2, an OH-IR star. To save space I will list the references but not cite specific results. The optical and IR images are extracted from the Hubble Legacy Archive.[2] The maps of the kinematics that I will use here are ^{12}CO observations from Alcolea et al. 2001, hereafter A+01[4]. Note that the CO emission is restricted to the narrow spine of reflecting dust on the symmetry axis. Our first goal is to simulate the spatial and kinematic structure of the southern lobe. We will consider the bulb separately.

We have successfully simulated the structures of both lobes of OH231 by launching clumps of two types into a stationary ambient medium whose density declines as radius^{-2}. We will report the work for only the southern lobe (Fig. 3). The mass in the external medium is 0.1 $M_\odot$. The momentum flux and speed of the flow match the results of A+01. Since the tip of both lobe tips have a diameter of 7AU we ran models of a cold clump of that radius and a hot clump model in which the clump thermally expands to that size. Both clumps are launched with core densities of $\approx 10^{4.5}$ cm^{-3}.

Observationally the two sets of models give nearly indistinguishable results. Both leave "contrails" that match the observed shape of the lobe. Both have dense axial spines, and both

[1] While eruptions and clumps might be viable, as they are in η Carinae and OMC-1, theory and the examples of spiral flows favor less exotic modes of continuous mass loss for AGB stars. However, the jury is out.

[2] The HLA is a joint project of the Space Telescope Science Institute (STScI), the Space Telescope European Coordinating Facility (ST-ECF), and the Canadian Astronomy Data Centre (CADC).

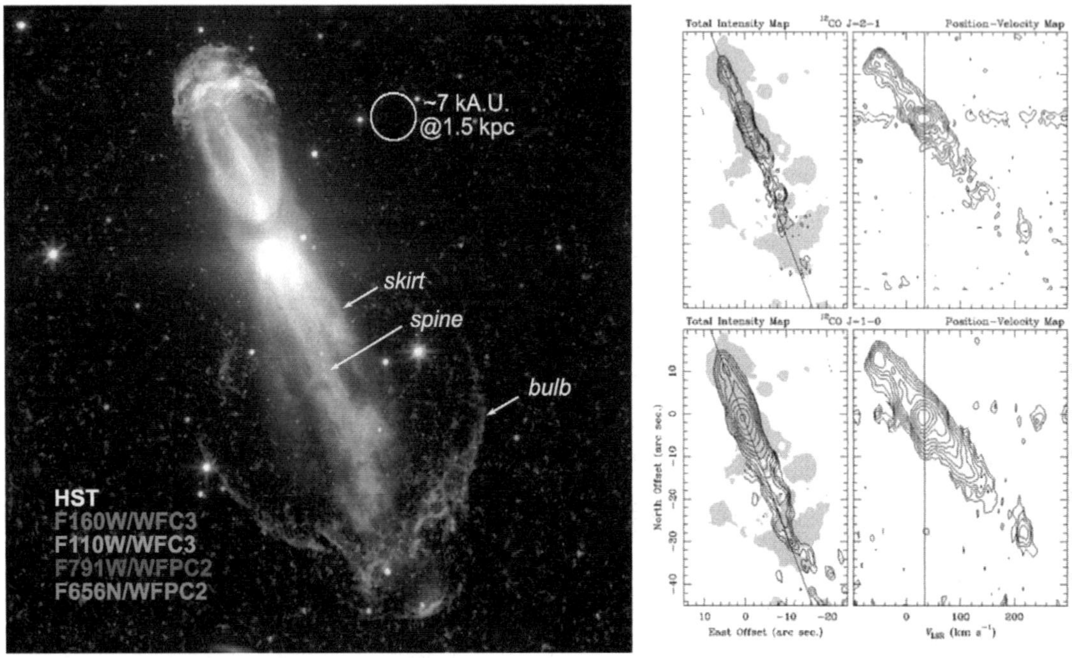

Figure 2. Left: Color overlay of images taken using the WFC3 camera on HST. The coor code is indicated in the legend. The lobes (shown in yellow) consist of dust-scattered starlight. The bulb is outlined only in shock-excited Hα. Right: spatial (left panels) and kinematic images (right panels) in ^{12}CO $J = 0 - 1$ (top) and $J = 2 - 1$ (bottom) from [4]. The linear kinematic pattern is not that of a jet (constant speed with radius). The kinematic age of the P-V slope is 770y at a distance of 1.5 kpc.

show linearly decreasing speed changes with offset that match the position-velocity pattern of ^{12}CO observations. One problem with the large cold clump is that its radius exceeds the width of the core seen in HST images. One nice feature of the hot clump is that the simulations match the bowl" at the base of the flow better than does the cold clump model. Both show hot leading tips ($\approx 10^6$ K) that expand laterally into the ambient gas as the tip advances onward. However, these hot backflows do not mimic the dimensions of the bowl. If we reduce the external density to "fatten" the bulb then other problems arise. In particular, the lobe also fattens unacceptably.

The bulb is easily simulated using the same ambient gas parameters as before, a spray of initial density of 10^5 cm^{-3}, and an initial outflow speed that matches that of the clump. These parameters are consistent with the CO observations of A+01. However, the lobes and the bulb really must be modeled as a single, interacting dynamical system. We are pursuing this now.

4. Conclusions

The characterization of outflows is our ultimate goal. Of course, case studies are of interest since they can demonstrate whether the physics of simulations are appropriate, provide feedback on the choice of outflow paradigm, and help to interpret detailed observations. However, we intend to simulate a variety of other well-studied pPNe in order to develop a general sense of the outcomes of the flow collimator, an important feature that is too small to be directly observed. So far we have successfully simulated CRL618 [5] and are close to finishing models for PN M2-9. We will report on these and other models in due course.

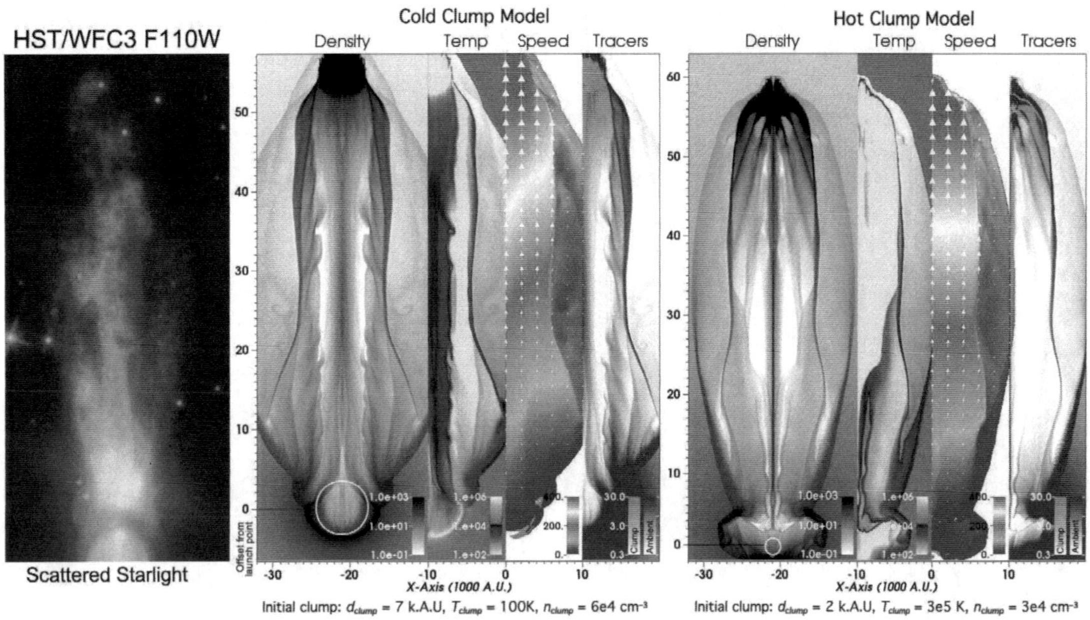

Figure 3. Left: F110W image of the southern lobe of OH231. Right: models of cold and hot clumps at the same spatial scale as the HST image (note the small vertical offsets). The four subpanels are, from left to right, log density, log temperature, flow speed, and tracers that show the location of the gas initially located in the ambient medium (blue) and clumps (orange).

References

[1] Kwok S, Purton C R and Fitzgerald P M 1978 *Astrophys. J. Lett.* **219** L125–L127

[2] Huarte-Espinosa M, Carroll-Nellenback J, Nordhaus J, Frank A and Blackman E 2014 *Asymmetrical Planetary Nebulae VI Conference* p 40

[3] Lee C F and Sahai R 2003 *Astrophys. J.* **586** 319–337

[4] Alcolea J, Bujarrabal V and Sanchez Contreras C 1996 *Astron. Astrophys.* **312** 560–564

[5] Balick B, Huarte-Espinosa M, Frank A, Gomez T, Alcolea J, Corradi R L M and Vinković D 2013 *Astrophys. J.* **772** 20

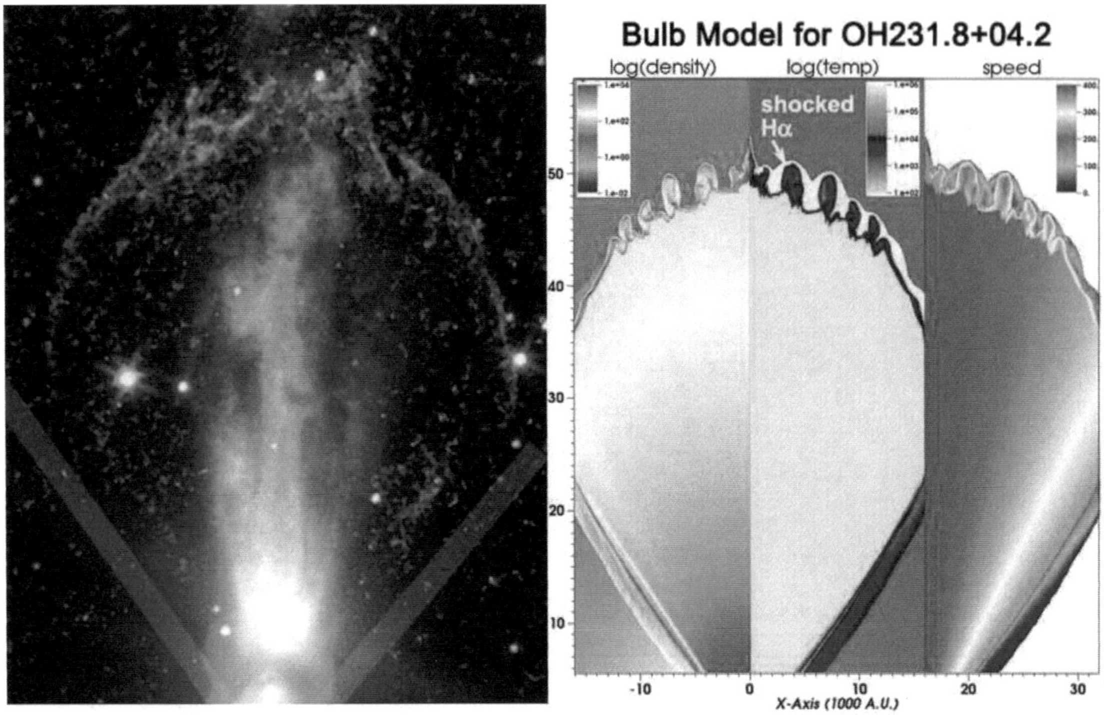

Figure 4. Left: Composite image of the southern bulb, much as in Fig. 2. Right: the hydro simulation that matches the shape of the bulb. The outflow is a centrally concentrated spray with a full opening angle of 40° within which the flow momentum is tapered by a Gaussian of $1/e$ width of 60°. This matches the straight bulb edges in the lower half of the image while providing the round, ruffled shape of the leading part of the bulb.

3-D structures of planetary nebulae

W Steffen[1]

[1] Instituto de Astronomía, Universidad Nacional Autónoma de México, Ensenada, B.C., México

E-mail: [1] wsteffen@astro.unam.mx

Abstract. Recent advances in the 3-D reconstruction of planetary nebulae are reviewed. We include not only results for 3-D reconstructions, but also the current techniques in terms of general methods and software. In order to obtain more accurate reconstructions, we suggest to extend the widely used assumption of homologous nebula expansion to map spectroscopically measured velocity to position along the line of sight.

1. Introduction

The main scientific motivation to generate 3-D representations of the structure of complex planetary nebula lies in the hope that they will contain information about the formation mechanism nebulae that is not obvious in the observations alone. A second reason for producing 3-D models is that of pure visualization, which helps to inspire new scientific insight or generate interest and better understanding by the general public.

Let me clarify what I mean by 3-D reconstruction by comparing the overall stages of 3-D research on planetary nebulae with that applied to study dinosaurs in palaeontology. The observations of a nebula correspond to extracting the remains of a dinosaur from the ground and the recording of the location and other properties of interest as they were found. More often than not, the relative location of the bones does not correspond to that in a living creature and many of the bones may actually be missing. Thus the observation and processing of the information from a palaeontological dig, corresponds to the observation of a nebula using a telescope and its instrumentation, followed by the processing of the data.

The 3-D reconstruction of the structure of a planetary nebulae based on imaging and spectro-kinematic information from the telescopic observations then corresponds to the interpretation of the data and the relative positioning of bones in the 3-D reconstruction of a skeleton. In both sciences, palaeontology and astronomy, the step from the collection of bones to the reconstructed skeleton and from processed observations to a 3-D reconstruction involve a significant amount of prior knowledge and assumptions on the type of creature and nebula that is being reconstructed.

Furthermore, the animal bones that were found or the radiation that was observed from a nebula, are only a relatively small part of the complete being or nebula, respectively. Organs, muscles, skin and brain, among others, are what make up most of the living being, but these have been lost to decomposition. While any ordinary matter in space is expected to generate some kind of radiation, when we observe a nebula over a given spectral range with a limited sensitivity, a large part of the gas flow is likely to go undetected. The 3-D reconstruction of a nebula therefore has to be "fleshed out" in order to convey a complete picture of its present structure and potential past evolution. Multi-wavelength observations can help constrain the structure further even if a complete 3-D reconstruction of the multi-wavelength data is rarely possible, for the lack of a mapping to position along the line of sight.

The fact that astronomical observation directly only yields angular position on the sky is the key problem in 3-D reconstruction, since the structure along the line of sight for each pixel has to be found indirectly using additional constraints. Such constraints may have quite a diverse nature. In the case of an animal skeleton, the constraints may be the shape of the connecting regions between different bones. A tail bone simply will not sensibly fit to a vertebra near the neck. In the case of expanding nebulae symmetry is often a key constraining property, even in the case of non-spherical complex nebulae.

Wenger *et al.* [1] [2] developed automated numerical algorithms to reconstruct the 3D structure of nebulae using only an image with the assumption of general axi-symmetry of the emission. While their algorithm can handle minor deviations from axi-symmetry and exactly reproduces the projected structure for the original viewing direction, the ring-like artifacts typical for axi-symmetric models remain present. The algorithm works well when the symmetry axis is sufficiently inclined, but fails for small angles to the line of sight.

The key constraint for the 3-D reconstruction of more asymmetrical *expanding* nebula is the correlation between position and velocity. It was originally noted by Wilson [3] as a correlation between the expansion velocity and the nebula size for different spectral lines. This observation led to the general assumption of a homologous expansion law in which the position and velocity vectors are proportional to each other. The spectroscopically observed Doppler-velocity can then be mapped directly to the position along the line of sight, if the constant that links them can be determined from other constraints or observations [4].

Physically a homologous expansion can be obtain from a ballistic expansion, i.e. when the gas parcels are sorted by their constant velocity vector. This requires that there is no significant interaction with the external medium or later ejections in the form of a wind, jets or "bullets". A rarefaction wave also generates a linearly increasing velocity field, but is not likely to be the origin of homologous expansion in planetary nebulae except in the quick ejection of dense massive stellar envelopes. A low-density wind interacting with dense clumps and filaments can also generate approximately homologous velocity fields for the clumps [5].

Some nebulae, such as NGC 6302, show no evidence from imaging, spectroscopy and internal proper motion measurements for significant deviations from a homologous expansion [6]. The ejection mechanism for most of the optically visible nebula must therefore have been very short when compared to the evolutionary timescale.

Hydrodynamic simulations in one or more dimensions do, however, show that a homologous expansion law, in general, has only a limited applicability and may yield reconstructions with first order accuracy only [7] [8] [9]. There are two main deviations. First, from the two-wind shock interaction, along a given direction which shows discontinuous changes in velocity are crossed. Second, in regions where the flow is oblique to the shock surface, the flow direction changes discontinuously and can deviate significantly from the radial direction from the central star.

For bipolar and multipolar nebulae, there are *characteristic* deviations [8], which have been confirmed by observations and their 3-D reconstructions in a study of an object that show an overall bipolar and axi-symmetric structure [10] [11]. Such deviations occur strongest at mid-latitudes from the equatorial plane and generate false point-symmetric structures and flaring of the position-velocity (P-V) diagrams in these regions. Naturally, when jets are involved, interacting with earlier wind structures, strong local disturbances are expected in the form of a bowshock. The kinematic signatures of bowshocks from jets or bullets have been studied in detail in the context of young stellar objects and proto-planetary nebulae and do not follow a homologous expansion pattern [12] [13].

In general, in a pressure driven bubble, due to wind interaction or a blast wave, that evolves into a non-spherical structure, oblique shocks will appear. These shocks produce non-radial velocity fields, and hence deviations from homologous expansion in the kinematics.

We have to distinguish between the velocity field of the gas flow and the pattern speed of the visible structures. They may be different. This difference has a strong impact on the interpretation of internal proper motion measurement of the expansion from images taken at different times. In addition, the local ionization conditions may change and induce motion of a pattern that has no correspondence in the flow of the gas [14].

All these deviations from the common assumption of homologous expansion for the reconstruction of the 3-D structure of planetary nebulae lead to the need of more detailed and systematic simulations, through (radiation-)hydrodynamical studies, with the specific aim of improving the methods for 3-D reconstruction of individual nebulae.

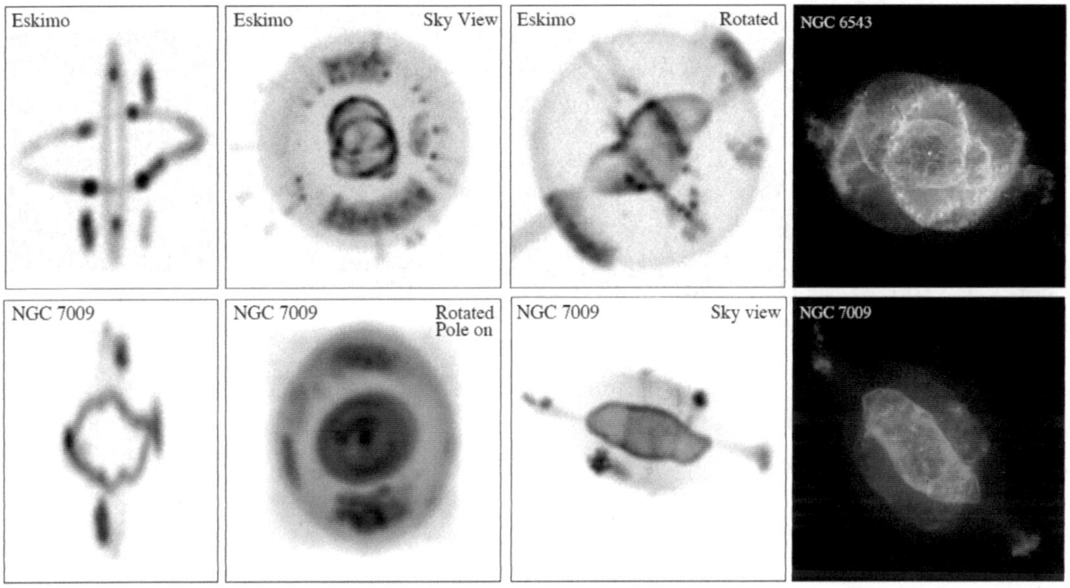

Figure 1. A comparison of images and P-V diagrams from three planetary nebulae seen from different real and synthetic viewing directions demonstrating their extreme similarities in structure and kinematics. Figure adopted from [28].

2. Methods

In recent years, in addition to the traditional custom programs, several publicly available software packages have been introduced to ease the 3-D modeling and reconstruction of astrophysical nebulae. Most of these packages specialize in particular aspects of the modeling, such as structure, photo-ionization, dust or molecular radiation transfer. One-dimensional photo-ionization has been successfully done for many years using CLOUDY [15] and its extension to pseudo-3D with pyCloudy by C. Morisset [16]. The dimensional limitations of CLOUDY led Ercolano *et al.* [17] to develop the fully three-dimensional Monte-Carlo radiation transfer code (MOCASSIN) that accepts any distribution of gas and dust density on a cartesian grid, including outputs from hydrodynamic simulations [18]. Recently, Hubber *et al.* [19] have introduced a Voronoi grid algorithm in MOCASSIN, which is meant to improve spatial resolution at the same or reduced computational cost. In their paper the authors compare the original cartesian grid based algorithm with the Voronoi mesh method and find that the new method performs better on complex density distributions.

While using the radiation transfer computations as a post-processing step can not take into account the dynamical effects from the photon heating, it is a reasonable first approximation in highly supersonic flows, where the structure is dominated by wind-wind interaction shocks. Frank & Mellema [20] introduced methods to efficiently include photo-ionization from a central source, photon-heating and cooling in numerical hydrodynamical calculations. An example for the incorporation of diffuse radiation in 3-D hydrodynamic simulations is given in [21] for the case of Herbig-Haro outflows.

In the realm of molecular radiation transfer there are new publicly available options, too. The ARTIST/LIME combination is promising for modeling data of high resolution observations from the *Atacama Large Millimeter Array* (ALMA). A European team of researchers has been working on the development of the components for this software package in recent years [22] [23]. This software was used, for instance, to model ALMA observations from a molecular 3-D spiral around the AGB star R Sculptoris, where the authors applied LIME radiation transfer to smoothed particle hydrodynamic simulations from the GADGET-2 code [24].

CO molecular line transfer computations can now also be done in SHAPE [25]. SHAPE is a more general purpose 3-D interactive astrophysical laboratory, which has grown out of a basic interactive morpho-kinematic modeling program. The key difference to other modeling software is that the user does not interact with the code of the software. The setup, modeling and analysis of the models is done in an interactive user interface. In addition to 3-D mesh based interactive modeling, it incorporates a eulerian hydrodynamics module. The hydrodynamic simulations are also set up interactively in the same fashion as other morpho-kinematic models are set up using 3-D meshes. The interactive mesh setup in SHAPE allows very complex emission distributions to be modeled, which is the reason why most recent morpho-kinematic modeling in the area of planetary nebulae has been done using this software package.

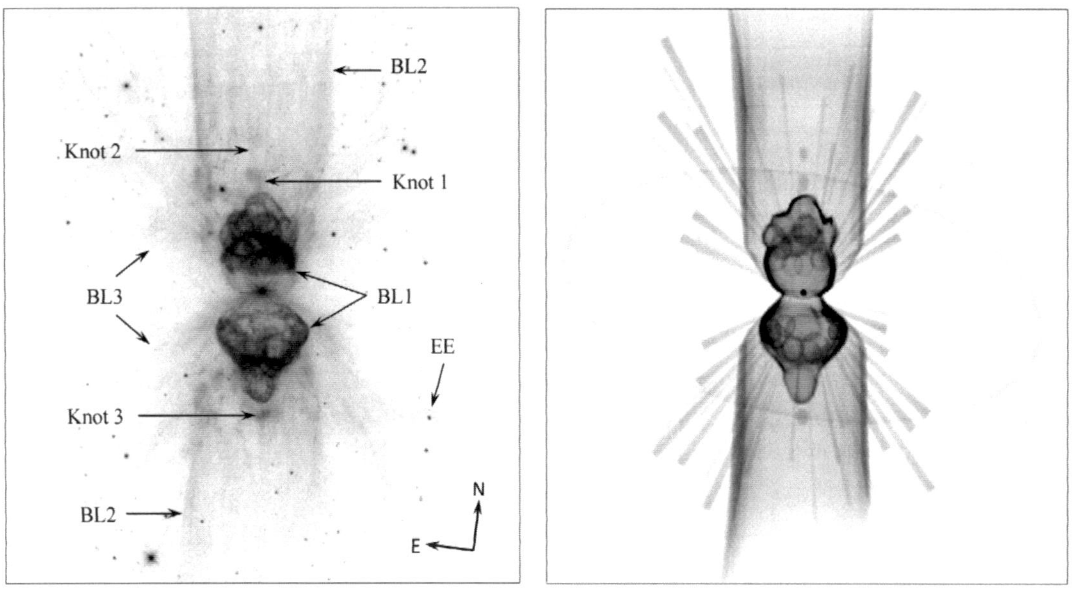

Figure 2. [36] presented some of the most complex 3-D morpho-kinematic reconstructions to date. On the left the observed HST-image of Mz 3 is shown, while the right panel shows the 3-D reconstruction. Figure adopted from [36].

3. Some recent results

One key property of planetary nebulae is their diversity in structure and ionization properties which manifest themselves in the variety that they show in their famous color images. The observed structures go from beautiful spherical bubbles, through highly collimated bipolar jet like nebulae to multipolar and even rather irregular shapes. Recent morpho-kinematic modeling has shown that, within each of the classes of nebulae, the diversity might not be as large as initially thought, which helps to reduce the number of formation mechanisms that have to be invoked for their explanation.

Chong *et al.* [26] and Hsia *et al.* [27] showed that many images of planetary nebulae can reasonably be approximated by variants of a multipolar structure seen from different directions. Hence, projection effects heavily way in when classifying objects. Ambiguities introduced by projection must be eliminated by additional observational or theoretical constraints. The main tool to disentangle the ambiguities is of course the kinematics as obtained from spectroscopic observations of the gas velocity along the line of sight, with the caveats that have been described in the Introduction.

Using spectroscopic observations with good coverage of the kinematics in the Saturn Nebula (NGC 7009) and the Eskimo Nebula (NGC 2392), García-Díaz *et al.* [28] showed that these two very complex bipolar nebulae have a 3-D emission structure that is very similar to that of the Cats Eye Nebula (NGC 6543)(Figure 1). Using their reconstructions they have been able to adjust the viewing angle of each of the nebulae to allow a direct comparison of the images and P-V diagrams. The striking similarities lead them to conclude that the formation and evolution of these nebulae must have been basically the same. Following the analogy of reconstructing the skeleton of a dinosaur, now that the "skeleton" of these nebulae has been obtained with the conclusion that they basically are of the same "species", dynamical studies can be pursued to find out how they were "born" and "grew up".

A detailed study of the Homunculus Nebula around Eta Carinae revealed detailed structures that may be linked to the wind-wind interaction of the central binary star [11] [29]. Two extrusions at a small angle from the equator, one on each of the bipolar lobes, are located in directions, with respect to the central binary, that correspond very accurately to the opening angle and direction of the cone carved out by the interaction of the stellar winds as derived by Madura *et al.* [30] using independent methods combining hydrodynamic simulations and spectroscopic observations of the central region of the nebula. Recent work by Teodoro *et al.* [31] find a similar orientation based on spectral line intensities and width variations over several orbital cycles of the central binary.

The work on the Homunculus Nebula was the beginning of 3-D printing of astrophysical nebulae. Steffen *et al.* [11] and Madura *et al.* [32] published 3-D mesh models of their results that have been 3-D printed. Such solid models are useful as a visual aid for scientific discussion or as a visualization tool for the general public. In particular, they help the visually impaired to literally getting a better "grip" on the shape of objects in the universe, which are inaccessible to them through flat images. Hopefully 3-D printing becomes an increasingly useful tool for visualizing astrophysical findings, thereby reaching out for new audiences.

Imaging observations of planetary nebulae with the Hubble Space Telescope over more than a decade produced the hope to obtain detailed internal proper motion studies that would reveal kinematic information that is independent of spectroscopic measurements and serve as additional constraint to the velocity field beyond global expansion measurements (e.g. [33]). Szyszka *et al.* [6] performed a detailed proper motion analysis of NGC 6302. In the region that was measured, the nebula showed no significant systematic deviations from a homologous expansion within the dispersion of the measurements. A proper motion analysis of hydrodynamic simulations meant to reproduce the structure of NGC 6302 by Uscanga *et al.* [34] showed rather different results, with hardly any proper motion in the dense inner regions and increasing velocities further out, but with very strong scattering. Furthermore, the direction of the proper motion vectors as determined by their algorithm was rather erratic and no save conclusion can be obtained from these results. In part, this is likely to be due to the currently employed algorithms not properly capturing the motion of extended filamentary structures. Similar observational studies have been done by García-Diaz *et al.* [35] for the Eskimo Nebula (NGC 2392), where the authors compare the proper motion obtained from two different methods. Research on the

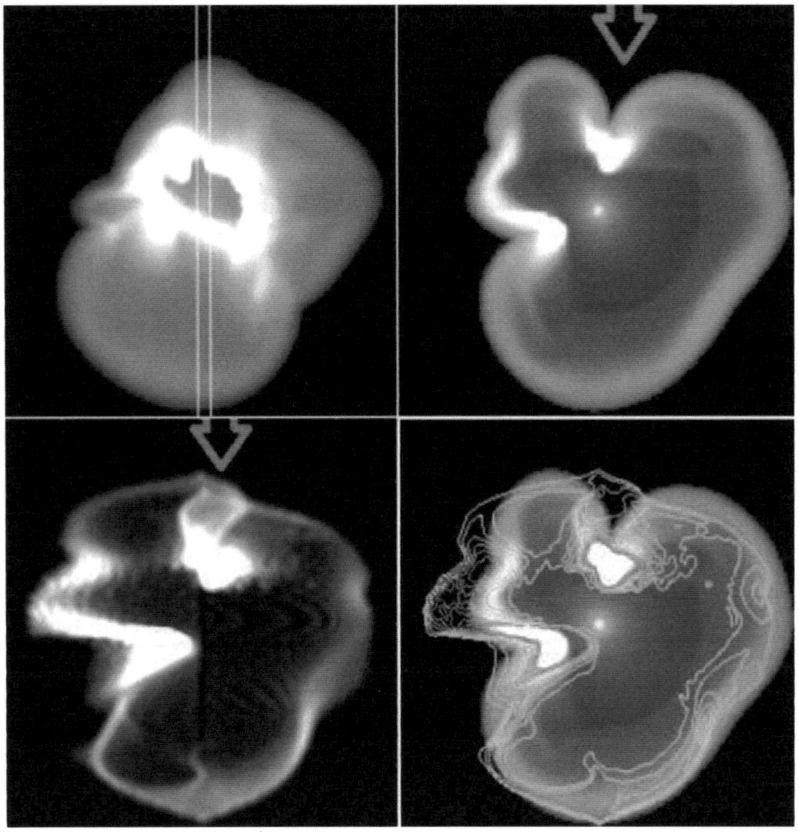

Figure 3. Hydrodynamic simulations show that converging shock structures can produce topological defects in 3-D reconstructions. Top left: Density image of a simulated multipolar planetary nebula with synthetic spectrograph marked. Top right: spatial density cut at position of synthetic slit. Lower left: P-V diagram with kinematic feature of converging shocks marked. Lower right: superposition of spatial cut and P-V diagram (contours). Figure adapted from [37].

quantification of proper motion in filamentary objects will be very helpful. Improvements along these lines could come from adapting pattern recognition methods to this problem such that they are capable of measuring the motion and distortions of knots, filaments or other recognizable features in easily quantifiable ways.

Most morpho-kinematic reconstruction work has concentrated on finding the 3-D structure of an object using an assumed homologous velocity field as a constraint. Observations and modeling techniques are now becoming sufficiently precise to allow, in some cases, to derive constraints on the velocity field itself, revealing deviations from the homologous expansion when assumptions on the symmetry can be made.

The observations and reconstruction of Hubble 5 by López *et al.* [10] showed that significant deviations from a homologous expansion velocity field had to be adopted to accommodate the apparent axi-symmetry of the double lobed planetary nebula. The assumption of homologous expansion would have produced point-symmetries and deviations from axi-symmetry in the reconstruction, which would have been unreasonably well aligned with the line of sight. The derived deviations of the velocity field from homologous expansion were in agreement with the expectations from hydrodynamic simulations of wind-blown bipolar nebula shells. The deviations consisted in a non-linear increase of the magnitude of the velocity as a function of distance from the central star and a directional deviation that peaks at intermediate latitudes.

Also the paper by Clyne *et al.* [36] showed that a more flexible velocity field and appropriate assumptions on the structure of an object can lead to relevant information on details of the structure itself or on the velocity field. Their model for M 2-9 required a non-homologous expansion in order to maintain the apparent axi-symmetry. It turned out an additional expansion component away from the symmetry axis did the job, which is consistent with what is found in hydrodynamical simulations of such collimated structures.

Clyne *et al.* [36] also presented a reconstruction of Hen 2-104 (the "Southern Crab"). This case is interesting since in addition to the global axi-symmetry, local assumptions on the "allowed" regions along a well defined ring of the knots and filaments in the highly fragmented outer hourglass structure, led to the identification of two opposite knots on the symmetry axis within the hourglass. These might be slower residuals of the axial ejecta that are prominent further out or they may have been later ejections. In the same paper a reconstruction of Mz-3 is presented, which includes the small-scale "bubbly" structures of the lobes (Figure 2).

Hydrodynamical simulations analyzed by Bermúdez [37] revealed that converging curved shocks can produce topological artifacts in 3-D reconstructions, since the kinematic signatures in the P-V diagrams can be superimposed, while in the actual spatial distribution the shocks do not even get into contact yet (Figure 3).

4. Summary

In recent years the observational material for precise morpho-kinematic 3-D reconstruction of planetary nebulae has increased considerably, in particular with the optical *San Pedro Mártir Kinematic Catalog of Planetary Nebulae* (López, this volume) and with ALMA for molecular outflows. In addition, software developments for modeling complex 3-D structures (SHAPE), 3-D photo-ionization (MOCASSIN) and 3-D molecular radiation transfer (LIME-ARTIST, SHAPEMOL) has made it possible to achieve new insight. Similarities between nebulae that appear very different in their images have been found, thereby reducing somewhat the intrinsic variety of shapes. 3-D hydrodynamic simulations have also lead to new extensions [38] to the already generalized interacting wind model originally devised by Kwok *et al.* [39]. Studies have shown that the common assumption of homologous expansion for morpho-kinematic reconstructions can lead to misinterpretations and artificial structures, thereby jeopardizing the search for the formation mechanism. The detailed observational investigation of the kinematics together with hydrodynamic simulations, with global and local constraints on the structure, have allowed to show that at least in some planetary nebulae the expansion is non-homologous and information on the structure of the velocity field itself can be derived. Future detailed radiation-hydrodynamic characterizations of the velocity fields for the diverse structures of planetary nebulae should help to improve morpho-kinematic 3-D reconstructions.

Acknowledgments

The author acknowledges support by grant UNAM-PAPIIT 101014 and thanks the organizers and their institutions for support to attend the conference.

References

[1] Wenger S, Ament M, Guthe S, Lorenz D, Tillmann A, Weiskopf D and Magnor M 2012 *IEEE Transactions on Visualization and Computer Graphics* **18** 2188-97
[2] Wenger S, Lorenz D and Magnor M 2013 *Computer Graphics Forum* **32** 93-100
[3] Wilson O. 1950 *Astrophys. J.* **111** 279-304
[4] Sabbadin F, Turatto M, Cappellaro E, Benetti S and Ragazzoni R 2004 *Astron. Astrophys.* **416** 955-81
[5] Steffen W and López J A 2004 *Astrophys. J.* **612** 319-31
[6] Szyszka C, Zijlstra A A and Walsh J R 2011 *Mon. Not. R. Astron. Soc.* **416** 715-26
[7] Schönberner D, Jacob, R and Steffen M 2005 *Astron. Astrophys.* **441** 573-88
[8] Steffen W, García-Segura G and Koning N 2009 *Astrophys. J.* **691** 696-704
[9] Steffen W 2011 *Proc. Int. Astron. Union* (S283) **7** 168-75
[10] López J A, García-Díaz Ma T, Steffen W, Riesgo H and Richer M G, 2012, *Astrophys. J.* **750** 131
[11] Steffen W, Teodoro M, Madura T I, Groh J H, Gull T R, Mehner A, Corcoran M F, Damineli A and Hamaguchi K 2014 *Mon. Not. R. Astron. Soc.* **442** 3316-28
[12] Raga A C and Bohm K H 1986 *Astrophys. J.* **308** 829-35
[13] Balick B, Huarte-Espinosa M, Frank A, Gomez T, Alcolea J, Corradi R L M and Vinković D 2013 *Astrophys. J.* **772** 20
[14] Mellema G 2004 *Astron. Astrophys.* **416** 623-29
[15] Ferland G J, Porter R L, van Hoof P A M, Williams R J R, Abel N P, Lykins M L, Shaw G, Henney W J and Stancil P C 2013 *RevMexAA* **49** 137-63
[16] Gesicki K, Zijlstra A A and Morisset C 2016 *Astron. Astrophys.* **585** 69
[17] Ercolano B, Barlow M J, Storey P J and Liu X-W 2003 *Mon. Not. R. Astron. Soc.* **340** 1136-52
[18] Monteiro H and Falceta-Gonçalves D 2011 *Astrophys. J.* **738** 174
[19] Hubber D A, Ercolano B and Dale J 2016 *Mon. Not. R. Astron. Soc.* **456** 756-66
[20] Frank A and Mellema G 1994 *Astron. Astrophys.* **289** 937-45
[21] Esquivel A and Raga A C 2013 *Astrophys. J.* **779** 111
[22] Brinch C and Hogerheijde M R 2010 *Astron. Astrophys.* **523** 25
[23] Padovani M *et al* 2012 *Astron. Astrophys.* **543** 16
[24] Maercker M *et al* 2012 *Nature* **490** 232-34
[25] Santander-García M, Bujarrabal V, Koning N and Steffen W 2015 *Astron. Astrophys.* **573** 56
[26] Chong S-N, Kwok S, Imai H, Tafoya D and Chibueze J 2012 *Astrophys. J.* **60** 115
[27] Hsia C-H, Chau W, Zhang Y and Kwok S 2014 *Astrophys. J.* **87** 25
[28] García-Díaz Ma T, López J A, Steffen W and Richer M G 2012 *Astrophys. J.* **761** 172
[29] Steffen W, Teodoro M, Madura T I, Groh J H, Gull T R, Mehner A, Corcoran M F, Damineli A and Hamaguchi K 2014B *The Messenger* **158** 26-29
[30] Madura T I Gull T R, Owocki S P, Groh J H, Okazaki A T and Russell C M P 2012 *Mon. Not. R. Astron. Soc.* **420** 2064-86
[31] Teodoro M *et al* 2016 *Astrophys. J.* 819 131
[32] Madura T I, Clementel N, Gull T R, Kruip C J H and Paardekooper J-P 2015 *Mon. Not. R. Astron. Soc.* **449** 3780-94
[33] Balick B, Gomez T, Vinković D, Alcolea J, Corradi R L M and Frank A 2012 *Astrophys. J.* **745** 188
[34] Uscanga L, Velázquez P F, Esquivel A, Raga A C, Boumis P and Cantó J 2014 *Mon. Not. R. Astron. Soc.* **442** 3162-65
[35] García-Díaz Ma T, Gutiérrez L, Steffen W, López J A and Beckman J 2015 *Astrophys. J.* **798** 129
[36] Clyne N, Akras S, Steffen W, Redman M P, Gonçalves D R and Harvey E 2015 *Astron. Astrophys.* **582** 60
[37] Bermúdez L C 2015 MSc-Thesis, Universidad Nacional Autónoma de México
[38] Steffen W, Koning N, Esquivel A, García-Segura G, García-Díaz Ma T, López J A and Magnor M 2013 *Mon. Not. R. Astron. Soc.* **436** 470-78
[39] Kwok S, Purton C R and Fitzgerald P M 1978 *Astrophys. J. Lett.* **219** L215-L217

Journal of Physics: Conference Series **728** (2016) 032006 doi:10.1088/1742-6596/728/3/032006

The rapid evolution of the central star of the Stingray Nebula – latest news from the HST

Nicole Reindl[1], Thomas Rauch[1], Marcelo M. Miller Bertolami[2] and Klaus Werner[1]

[1] Institute for Astronomy and Astrophysics, Kepler Center for Astro and Particle Physics, Eberhard Karls University, Sand 1, 72076 Tübingen, Germany

[2] Instituto de Astrofísica de La Plata, UNLP-CONICET, La Plata, Buenos Aires, 1900, Argentina

E-mail: reindl@astro.uni-tuebingen.de

Abstract. SAO 244567 is an unusually fast evolving star. Within twenty years only, it had turned from a B-type supergiant into the central star of the Stingray Nebula. Space- and ground-based observations obtained over the last decades have revealed that its spectrum changes noticeably over just a few years, showing stellar evolution in real time. The low mass of SAO 244567 is, however, in strong contradiction with canonical post-asymptotic giant branch evolution. Thus, its fast evolution has been a mystery for decades. We present preliminary results of the non-LTE spectral analyis of the recently obtained HST/COS observations, which finally allow us to shed light on the evolutionary history of this extraordinary object.

1. Introduction

Rapid changes of the observable properties of SAO 244567 were at first noticed by [1, 2]. Based on a spectral classification of the optical spectrum obtained in 1971 and the UBV colors, they concluded that SAO 244567 was a B-type supergiant with an effective temperature of $T_{\rm eff} \approx 21\,{\rm kK}$. However, the optical spectra from 1990 and 1992 as well as UV spectra from 1992 on display many nebular emission lines, indicating that SAO 244567 has ionized its surrounding nebula only two decades later. [3] and [4] presented the first spatially resolved images of the planetary nebula (PN) and named it Stingray Nebula.

In [5] the first quantitative spectral analysis of all available spectra from 1988 to 2006, that were taken with various space-based telescopes, was presented. It was found that the central star has steadily increased its $T_{\rm eff}$ from 38 kK in 1988 to a peak value of 60 kK in 2002. During the same time, the star was contracting, as concluded from an increase in surface gravity from $\log g = 4.8$ to 6.0 and a drop in luminosity. Simultaneously, the mass-loss rate declined from $\log(\dot{M}\,/\,M_\odot\,{\rm yr}^{-1}) = -9.0$ to -11.6 and the terminal wind velocity increased from $v_\infty = 1800\,{\rm km/s}$ to $2800\,{\rm km/s}$. The surface composition is largely solar with the exception of slightly subsolar C, P, and S, indicating that the possible asymptotic giant branch (AGB) phase of the star was terminated before the third dredge-up. We confirmed previous findings, that SAO 244567 must be a low mass star ($M < 0.55\,M_\odot$). This low mass, however, contradicts the rapid heating if a canonical post-AGB evolution is assumed [2, 4], because for such a rapid evolution the core mass should be around $0.7\,M_\odot$ [6]. [5] speculated that the star might had experienced a late thermal pulse (LTP) shortly after leaving the AGB, because the evolutionary speed of late He-shell flash objects is very high. Within the next decades an LTP scenario would predict for SAO 244567 an evolution back to the AGB, i.e., a cooling of the star. To follow the evolution of the surface properties of SAO 244567 and verify the LTP hypothesis, we applied for further observations with HST/COS.

2. Observations and spectral analysis

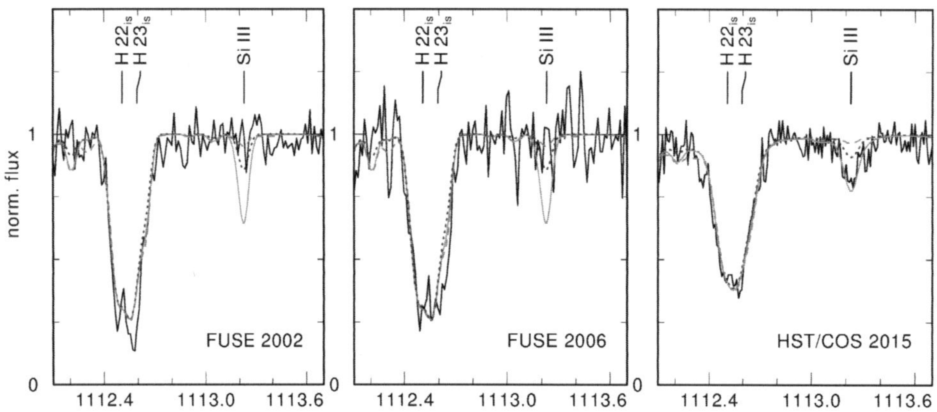

Figure 1. Emergence of Si III λ 1113.2 Å in the HST/COS observation taken in 2015 (right). This line was not visible in the FUSE observation in 2002/2006 (left/middle) when the star was about 10/5 kK hotter. The dotted red / dashed purple line corresponds to TMAP models with $T_{\mathrm{eff}} = 60/55$ kK (best fit for 2002/2006), while the solid blue line represents a model with $T_{\mathrm{eff}} = 50$ kK (best fit for 2015). Note that the differences in the line profiles are caused by the instruments' line spread functions.

The observations were successfully performed on August 9, 2015, more than nine years after the star was observed by an UV telescope the last time. We obtained FUV medium-resolution (grating G130M) and low-resolution FUV and NUV spectra (gratings G140L and G230L) spectra.

The new observations reveal that the flux level decreased by a factor of 1.55/1.4 compared to FUSE observations in 2002/2006. Thanks to the good signal-to-noise ratio (S/N), we could identify lines of Cr and Ni for the first time. Furthermore, we find that the N V resonance lines show blue shifted absorption wings, which might indicate the presence of a very weak stellar wind. Finally, we discovered the emergence the Si III λ 1113.2 Å line in the new HST/COS spectra (Fig. 1). This line was not visible in the previous FUSE observations, which cover this wavelength range as well and have about the same resolving power ($R \approx 20\,000$).

For the quantitative spectral analysis we used the Tübingen non-LTE Model-Atmosphere Package (TMAP, [7, 8, 9]), which allows the computation of fully metal-line blanketed model atmospheres in radiative and hydrostatic equilibrium. Our model grid included opacities of the elements H, He, C, N, O, Si, P, S, Cr, Fe, and Ni. The model atoms were taken from the Tübingen Model-Atom Database TMAD and calculated (Cr, Fe, Ni) via the Tübingen IRon Opacity interface TIRO [10]. In addition, we employed the program OWENS to model the ISM line-absorption spectrum.

We found, that Si III λ 1113.2 Å was not visible in the FUSE observation in 2002, because the star was about 10 kK hotter in 2002 compared to 2015 where we find the best fit at $T_{\mathrm{eff}} = 50$ kK. This value is also confirmed by evaluating additionally the ionization equilibria of C III / C IV, N III / N IV, O III / O IV / O V, and S IV / S V / S VI. The good S/N allows to reduce the error on T_{eff}

to ±2.5 kK. We conclude that the star has cooled significantly since 2002. The newly identified lines of Cr and Ni allowed to measure their abundances for the first time. They are solar. Moreover, we confirm the abundance values of all the other elements derived in our previous analysis, i.e., no hint of a change in the chemical abundances was found.

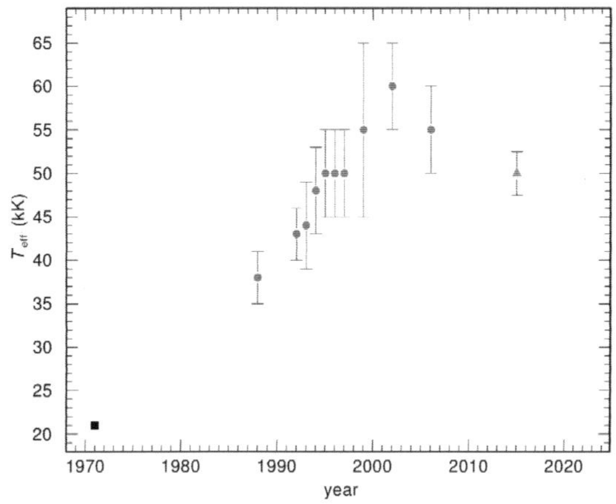

Figure 2. Temporal evolution of $T_{\rm eff}$. The square indicates the $T_{\rm eff}$ estimate of [2] when SAO 244567 was still a B-type supergiant. Dots correspond to $T_{\rm eff}$ values as derived by [5] and the triangle is the value measured with the new HST/COS observations performed in 2015. Note that the cooling rate of SAO 244567 (770 ± 580 K/yr) is similar to that of FG Sge (350 K/yr, [14]), the only other hitherto known LTP object.

3. Conclusions

SAO 244567 provides us with the rare opportunity to study stellar evolution in real time. Its primarily rapid heating (heating rate from 1988 to 2002: 1570 ± 570 K/yr, Fig. 2), followed by the rapid cooling (−770 ± 580 K/yr, Fig. 2) strongly supports the LTP evolutionary scenario (Fig. 3). LTPs are expected to occur to a significant fraction of the post-AGB stars while evolving with roughly constant luminosity from the AGB towards the white dwarf domain. Differently to very LTPs, which occur only on the early white dwarf cooling track and produce a hydrogen-free stellar surface already during the flash, a normal surface composition holds up after an LTP. Only when the star evolves back to the Hayashi limit on the AGB ($T_{\rm eff} \lesssim 7000$ K), envelope convection can set in again [11, 12, 13]. This was actually observed in case of FG Sge, the only other star known to date that certainly must have suffered an LTP. This star has been transformed over an interval of 120 years from a hot, hydrogen-rich post-AGB star into a very luminous cool supergiant (Fig. 3). [14] showed that, after its return to the AGB, the surface H fraction of FG Sge got diluted significantly (from 0.9 to 0.01, by number) while s-process abundances increased.

We note that SAO 244567 must be in an earlier evolutionary stage than FG Sge, as it was observed during its heating as well as its cooling phase. Moreover, the numerous UV observation taken over three decades as well as the use of sophisticated non-LTE model atmosphere codes allowed us to precisely record the evolution of this extraordinary object. Therefore the derived surface parameters will establish constraints for LTP evolutionary calculations. These calculations in turn may not only explain the nature of SAO 244567, but they could also provide a deeper insight in the formation process of hydrogen-deficient stars, which make up about a quarter of the post-AGB stars and white dwarfs.

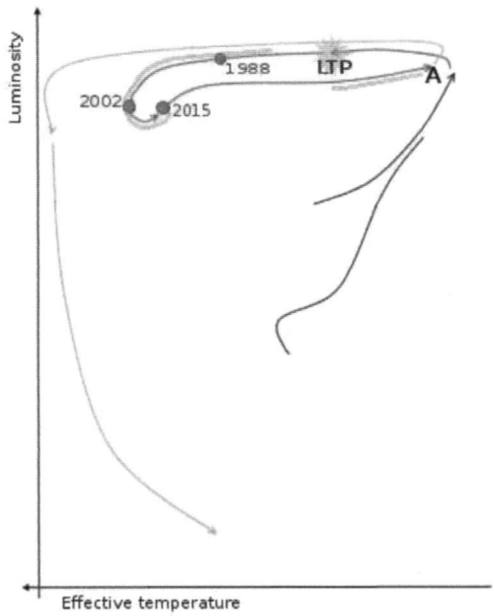

Figure 3. Sketch of an LTP evolution of SAO 244567 in the Hertzsprung-Russell diagram. The observed evolutionary path of SAO 244567 is indicated by the pink line, whereas the observed path of FG Sge is shown in gray. "A" corresponds to the point when the star reaches the Hayashi limit on the AGB and becomes H-deficient. "A" also corresponds to the recent position of FG Sge.

Acknowledgments
N.R. and T.R. are supported by the German Aerospace Center (DLR, grant 50 OR 1507).

References
[1] Parthasarathy M, Garcia-Lario P, Pottasch S R, Manchado A, Clavel J, de Martino D, van de Steene G C M, Sahu K C 1993 *Astron. Astrophys.* **267** L19
[2] Parthasarathy M, Garcia-Lario P, de Martino D, Pottasch S R, Kilkenny D, Martinez P, Sahu K C, Reddy B E, Sewell B T 1995 *Astron. Astrophys.* **300** L25
[3] Bobrowsky M 1994 *Astrophys. J. Lett.* **426** L47
[4] Bobrowsky M, Sahu K C, Parthasarathy M and García-Lario P 1998 *Nature* **392** 469
[5] Reindl N, Rauch T, Parthasarathy M, Werner K, Kruk J W, Hamann W-R, Sander A and Todt H 2014 *Astron. Astrophys.* **565** A40
[6] Miller Bertolami M M 2015 *Astron. Astrophys.* **588** A25
[7] Werner K, Dreizler S and Rauch T 2012 *Astrophysics Source Code Library* [record ascl:1212.015]
[8] Werner K, Deetjen J L, Dreizler S, Nagel T, Rauch T and Schuh S L 2003 *Stellar Atmosphere Modeling* eds Hubeny I, Mihalas D and Werner K (ASP Conf. Series vol 288) p 31
[9] Rauch T and Deetjen J L 2003 *Stellar Atmosphere Modeling* eds Hubeny I, Mihalas D and Werner K (ASP Conf. Series vol 288) p 103
[10] Müller-Ringat E 2013 Dissertation, University of Tübingen, Germany, `http://nbn-resolving.de/urn:nbn:de:bsz:21-opus-67747`
[11] Blöcker T and Schönberner D 1996 *Memorie della Societa Astronomia Italiana* **67** 665
[12] Blöcker T and Schönberner D 1997 *Astron. Astrophys.* **324** 991
[13] Schönberner D 2008 *Hydrogen-Deficient Stars* eds Werner W and Rauch T (ASP Conf. Series vol 391) p 139
[14] Jeffery C S and Schönberner D 2006 *Astron. Astrophys.* **459** 885

Journal of Physics: Conference Series **728** (2016) 032007 doi:10.1088/1742-6596/728/3/032007

Wolf-Rayet nebulae

You-Hua Chu[1]

[1] Institute of Astronomy and Astrophysics, Academia Sinica
AS/NTU, No.1, Sec. 4, Roosevelt Rd, Taipei 10617, Taiwan, R.O.C.

E-mail: [1] yhchu@asiaa.sinica.edu.tw

Abstract. Since the discovery of nebulae around Wolf-Rayet (WR) stars in the 1960s, it has been established that WR stars are massive stars at advanced evolutionary stages and that their surrounding nebulae result from the interactions between the stellar mass loss and the ambient interstellar medium. Surveys of WR nebulae have been made in the Galaxy, Magellanic Clouds, and other nearby galaxies in the Local Group. Some WR nebulae exhibit He II λ4686 line emission, indicating stellar effective temperatures of $90 - 100 \times 10^3$ K. The shocked fast stellar winds from WR nebulae have been detected in soft X-rays, but theoretical models have not been able to reproduce the observed X-ray spectral properties. Elemental abundances of WR nebulae consisting of synthesized stellar material can constrain stellar evolution models, but high-dispersion spectra are needed to kinematically separate the expanding shell of a WR nebula and the background interstellar medium for accurate abundance analyses.

1. Wolf-Rayet stars and their nebulae

The first three nebulae around Wolf-Rayet (WR) stars were reported in 1965 by Johnson and Hogg [1]: NGC2359, NGC6888, and S308. The ring-like morphology of these WR nebulae is indicative of a shell structure. It was immediately suggested that stellar outflows implied by the broad P Cygni line profiles were responsible for shaping the nebulae, although the evolutionary status of WR stars was not well known and the outflows were later called fast stellar winds.

Models of massive star evolution have been guided by observations, in particular, the distribution of massive stars in the Hertzsprung-Russel diagram (HRD) plotted in stellar effective temperature (T_{eff}) versus bolometric magnitude (M_{Bol}) by Humphreys and Davidson [2]. An absence of very luminous red stars in the HRD was observed, and this so-called Humphrey and Davidson Limit provided essential constraints to models of stellar evolution. It is now known that the most massive O stars, $\sim 60 M_\odot$ or higher, evolve off the main sequence to become luminous blue variables (LBVs), while the less massive O stars evolve into red supergiants (RSGs). During the LBV and RSG stages, stellar mass loss strips off the hydrogen-rich envelope to expose the helium-rich layer, and the star enters the WR phase.

Owing to its high luminosity, massive stars lose mass throughout their lifetime. During the main sequence, O stars have fast stellar winds; during the LBV or RSG phase, a dense slow wind develops; and during the WR phase, a dense fast stellar wind emerges. The velocities and mass loss rates of these winds are shown in the chart in Figure 1.

As a massive star evolves through different phases, its mass loss interacts with the ambient medium that has been modified by the previous mass loss. It is conceivable that a main-sequence O stars fast stellar wind will sweep up the ambient interstellar medium (ISM) into an interstellar bubble [3,4], the slow winds of LBVs and RSGs form circumstellar nebulae in the bubble interiors, and finally the WR wind sweeps up the circumstellar material to form a circumstellar bubble inside the interstellar bubble. An example of such nested bubbles, NGC6164-5, is shown in the left panel of Figure 2.

The formation of planetary nebulae (PNe) is very similar to the formation of circumstellar bubbles of WR stars. As a low- or intermediate-mass star evolves into a red giant or asymptotic giant branch (AGB) star, the copious mass loss forms a circumstellar nebula, and during the subsequent evolution toward white dwarf, a fast stellar wind develops and sweeps up the circumstellar nebula to form a PN [5,6]. The right panel of Figure 2 shows a PN, where the bright main shell consists of the swept-up AGB wind and the surrounding envelope unperturbed AGB wind.

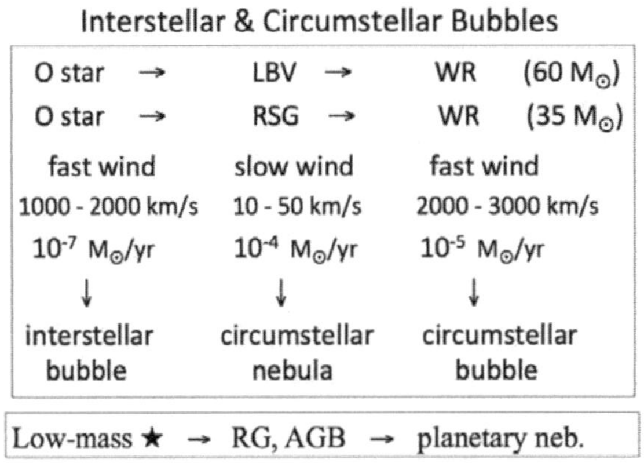

Figure 1. Stellar mass loss at different stages of stellar evolution. Stellar wind velocity and mass loss rates are listed.

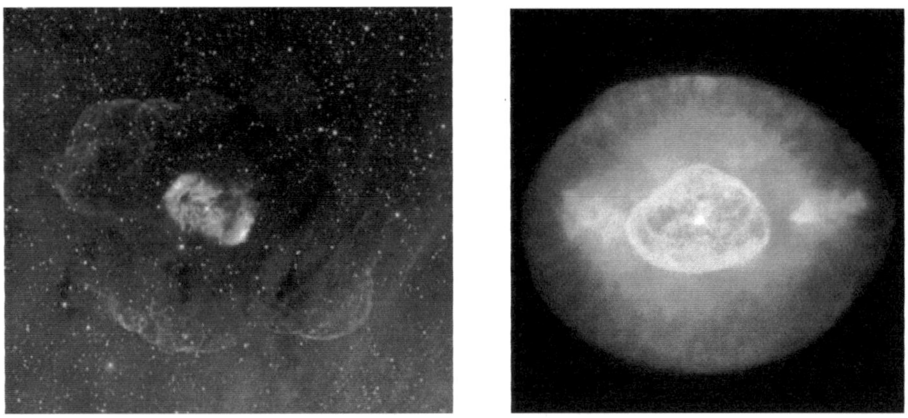

Figure 2. Left: Ground-based image of NGC 6164-5. The small (2 pc ×3 pc) nebula around the central star is a circumstellar bubble consisting of ejected stellar material, while the large (10 pc ×15 pc) shell is the interstellar bubble consisting of interstellar material. The central star is of spectral type O6.5f?p, not yet a WR star. Image credit: Don Goldman. Right: Hubble Space Telescope WFPC2 image of NGC 6826. The bright inner shell is a bubble consisting of swept-up AGB wind, while the envelope contains uninterrupted expanding AGB wind.

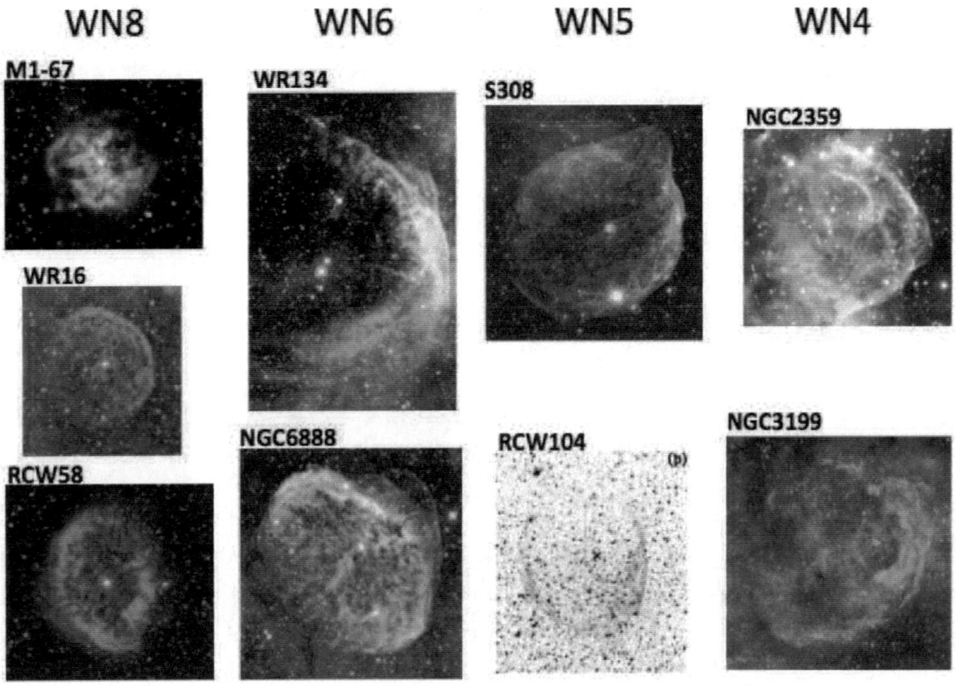

Figure 3. Images of WR nebulae. The spectral types of the central WR stars are marked above each column. Image credits: Hubble Space Telescope (M1-67), Don Goldman (WR16, RCW58), Bob Franke (NGC6888), Travis Rector (WR134), Robert Gruendl and You-Hua Chu (S308, RCW104), ESO/VLT (NGC2359), Velimir Popov and Emil Ivanov (NGC3199).

2. Observed morphologies of WR nebulae

Owing to the high qualities of affordable telescopes/CCD cameras/filters, amateur astronomers have been taking the best images of WR nebulae and made them available on the web. Figure 3 shows 9 Galactic WR nebulae associated with WR stars of different subtypes. There is an apparent correlation between the nebular morphology and the spectral type.

Most WR nebulae are associated with WN stars. The nebulae associate with late-type WN8 stars are dominated by stellar ejecta [7], and have the clumpiest morphology. Despite the large surface brightness fluctuations, the expansions of these nebulae are quite regular [8,9]. In contrast, the nebulae associated with early-type WN4 stars consist of mostly interstellar material [7,10]. The intermediate WN stars, WN5-6, show circumstellar bubble with various degrees of ablation by the fast WR winds.

The morphologies of WR nebulae have been satisfactorily reproduced by hydrodynamic models of nebular evolution from main sequence O to WR stages, taking into account of the mass loss history [11,12,33,34,35,36]. The clumpy morphology of circumstellar bubbles is caused by instability in the shell when the circumstellar material (the slow wind from the LBV or RSG phase) has been swept up.

An intriguing puzzle is that the interstellar bubble blown by a WR stars main-sequence progenitor is rarely observed. In fact, main-sequence O stars are hardly ever seen in interstellar bubbles, despite theoretical expectations [3,4]. To solve this puzzle, Hubble Space Telescope images of the H II region N11B around the OB association LH10 were obtained. LH10 still contains O3V stars, the most massive main sequence stars, indicating that this OB association has not had any supernova explosion yet, and the presence of dense ionized gas in N11B should allow O stars to blow interstellar bubbles. The Hubble images do not show any ring-like morphological features corresponding to bubbles; however, long-slit echelle spectrograms of the [N II] λ6583 line (Figure 4) do show line splitting indicating the existence of expanding shells of sizes $\sim$ 15 pc and expansion velocities 15 – 20 km s^{-1} [13]. For such low expansion velocities, line-split cannot be seen in the Hα line because of its large thermal width, $\sim$ 20 km s^{-1}, so they have escaped detection in the past. The absence of visible ring-like morphological features in Hα image is also caused by the low expansion velocity: for the 10 km s^{-1} sound velocity of a 10^4 K H II region, the expansion of the bubble is associated with very weak shocks, with Mach number of 1.5 – 2. Such weak shocks do not produce strong compression of the ISM to produce a density jump for a visible limb-brightened shell nebula. The puzzle is solved.

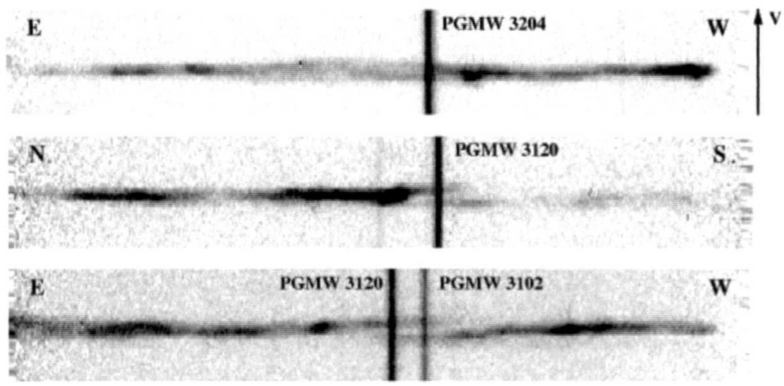

Figure 4. Three long-slit echelle spectrogram of [N II] λ6583 line centered on O stars in LH10/N11B. The horizontal axis is along the slit and the vertical axis is along the dispersion. The star names are marked next to the stellar continua.

3. Extragalactic WR nebulae

WR nebulae have been surveyed in the Large and Small Magellanic Clouds (LMC and SMC) [14,15]. Recent Hα survey of the LMC made with the CTIO Blanco 4m telescope and the MOSAIC camera has been used to study WR nebulae (in preparation). The examples displayed in Figure 5 show significantly more details than previous images taken with Curtis Schmidt Telescope. Some WR nebulae seem to be attached to the inner rims of superbubbles. Often the WR nebulae are best seen in the [O III] λ5007 line because they have higher excitation than the background H II region, for example, the WR bubble around Br12 in DEM L45 (Figure 5). From the surveys of the WR nebulae in the LMC and SMC, we have learned that the large shell structures around WR stars identified in low-resolution images are mostly superbubbles with sizes greater than $\sim$ 100 pc [14], and that high-resolution images are needed to reveal small circumstellar bubbles of WR stars.

Surveys of WR nebulae have been extended to M31 and M33 in the Local Group [16,17]. The ground-based images used in these surveys are of relatively low resolution. While many small WR nebulae (smaller than $\sim$ 40 pc in diameter) are identified as candidates of WR bubbles, their structures are not well resolved; furthermore, a significant fraction of the WR nebulae identified are actually superbubbles around OB associations.

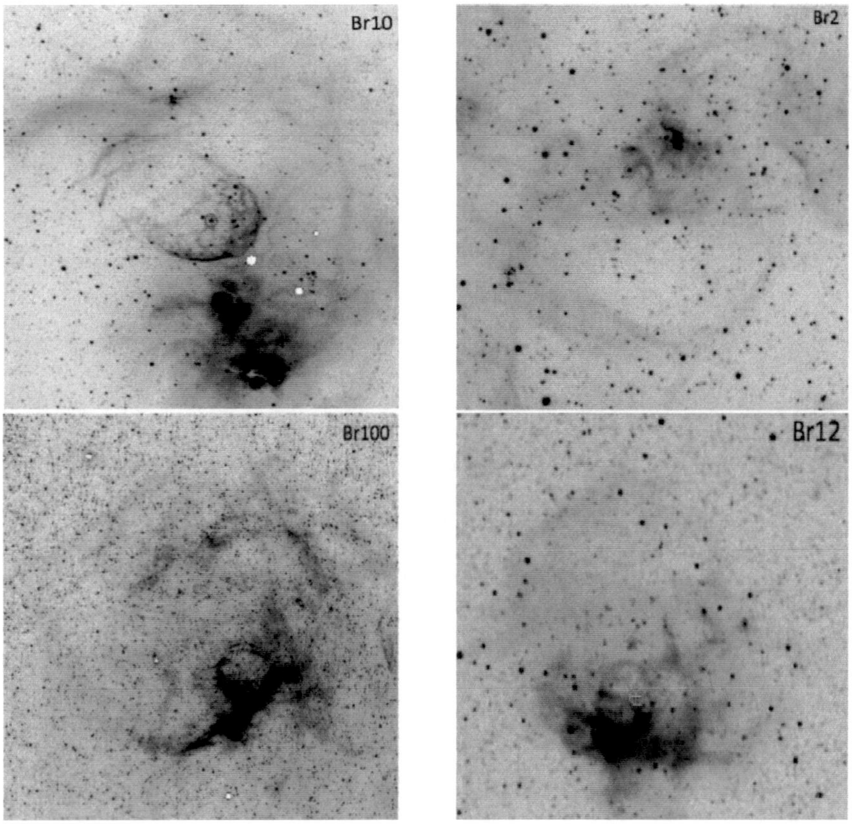

Figure 5. Images of WR nebulae in the LMC. The Hα images of Br2, Br10, and Br100 were taken with the CTIO 4m MOSAIC camera. The MCELS [O III] λ5007 image of Br12 was taken with the Curtis Schmidt Telescope.

4. He II-emitting WR nebulae

Unlike PNe, H II regions photoionized by main-sequence massive stars do not emit He II recombination lines because O stars do not have temperatures much higher than $\sim 50,000$ K; however, some He II regions ionized by WR stars exhibit He II λ4686 line emission, indicating that He is doubly ionized and that the stellar effective temperatures are as high as $85,000 - 100,000$ K. Examples include G2.4 + 1.4 around the WO star WR102 in the Galaxy [18], N79 around the WN2 star Br2 in the LMC, and N76 around the WN2 star AB7 in the SMC [19]. See Figure 6.

5. Hot gas in WR nebulae

Circumstellar bubbles of WR stars are ideal laboratories to study a number of astrophysical problems: hydrodynamic interactions between fast stellar winds and circumstellar material, stellar wind energy feedback, and thermal conduction between hot gas and cool gas. The bubble interior filled by shocked fast wind is expected to emit in X-rays, while the conduction layer between the hot interior gas and the cool bubble shell gas contains highly ionized species produced by thermal collisions, and is best studied in ultraviolet wavelengths. Multi-wavelength observations are needed to obtain thorough understandings of WR bubbles.

X-ray emission from the hot gas in bubble interiors is first detected in the WR nebula NGC

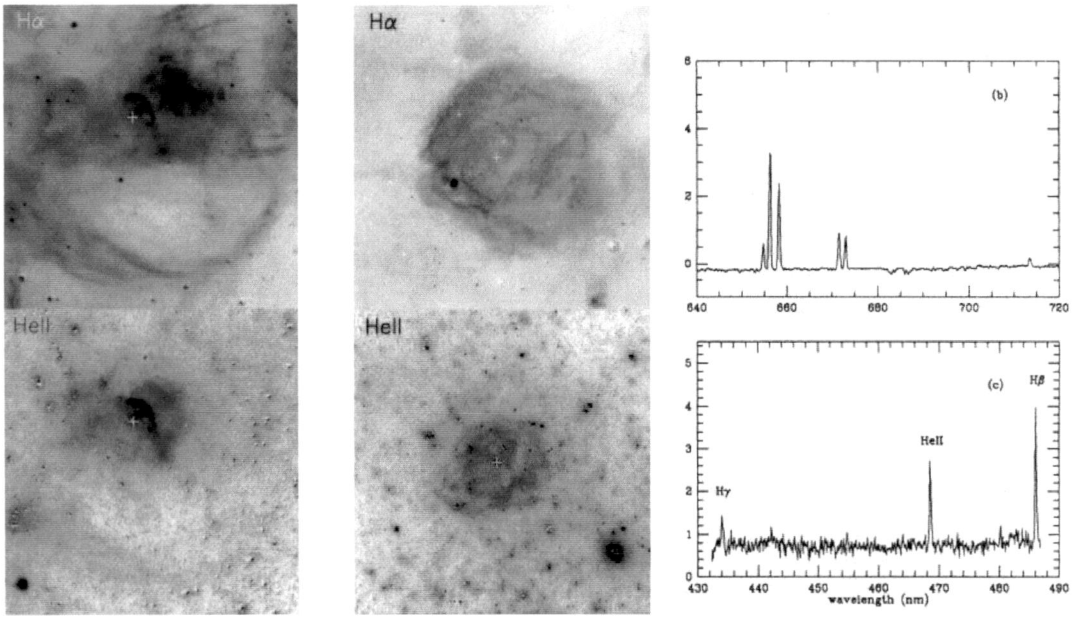

Figure 6. Left - Hα and He II λ4686 images of Br2 in the LMC [19]. Middle - Hα and He II λ4686 images of AB7 in the SMC [19]. Right - optical spectrum of G2.4 + 1.4 around WR102 in the Galaxy [18].

6888 using Einstein Observatory data [20]. ROSAT observations of the WR bubble S308 show hint of diffuse X-ray emission [21]. XMM-Newton observations of S308 confirmed the diffuse X-ray emission and produced its first useful X-ray spectrum; furthermore, XMM-Newton mapped S308 and presented a complete X-ray view of S308 [22,23]. Recently, XMM-Newton observations detected faint diffuse X-ray emission from the interior of WR bubble NGC 2359 [24,25]. The X-ray spectra of these three WR bubbles are very soft, and thermal plasma emission model fits indicate that the hot gas temperatures are only $1 - 2 \times 10^6$ K (Figure 7) [26]. As the emission measure is proportional to the density squared, the diffuse X-ray emission from a bubble interior should peak near the conduction layer where the density is high and temperature is low, qualitatively in agreement with the plasma temperatures implied by the X-ray spectra. However, hydrodynamic models of bubbles taking into account of thermal conduction, dynamic mixing, and radiative cooling have not been able to reproduce the very soft spectra of diffuse X-ray emission from WR bubbles [27].

Thermal conduction layers in bubbles or superbubbles had been studied with C IV and Si IV absorption lines, but the early investigations using International UV Explorer (IUE) observations of early-type stars yielded ambiguous results because C^{3+} and Si^{3+} can be produced by photoionization. The first unambiguous detection of conduction layer was provided by HST GHRS observations of the N V absorption line against the spectrum of HD50896, the central WR star of S308 [28]. Attempts to use HST STIS observations to detect N V λλ1238,1242 lines in emission from the interface layer of S308 has not been successful.

PNe have much higher density than WR bubbles, so their interface layers can be more easily detected. For example, O VI λλ1031,1037 lines and N V λλ1238,1242 lines have been detected both in emission (off the central star) and absorption (against the stellar spectra) in the Cat's Eye PN, NGC 6543 [29,30].

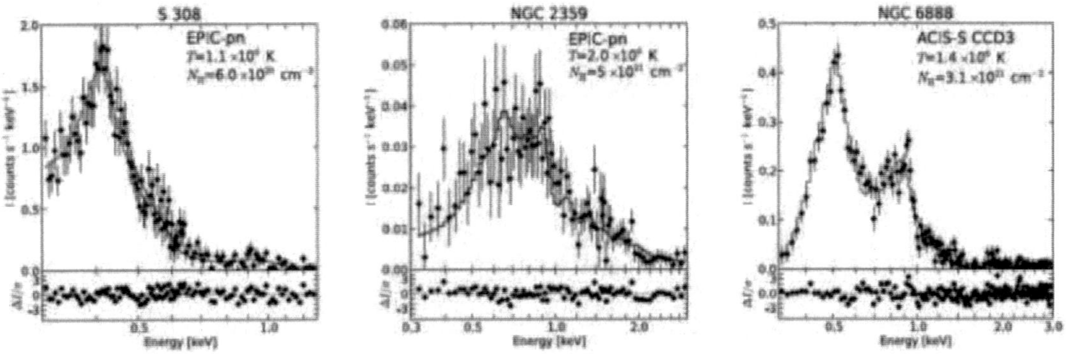

Figure 7. X-ray spectra of the hot gas in the WR bubbles S308, NGC2359, and NGC6888. The X-ray observations of S308 and NGC2359 were made with the XMM-Newton X-ray Observatory, and the X-ray observation of NGC6888 was made with the Chandra X-ray Observatory. This figure is taken from [26].

6. Stellar evolution constrained by WR nebular abundances

Circumstellar bubbles of WR stars contain material ejected by the progenitors. Depending on the previous evolutionary paths, whether through LBV or RSG, the stellar ejecta consist of processed stellar material whose elemental abundance reflects the nature of the WR stars' progenitors [7]. It is difficult to determine elemental abundances of WR bubbles because they are more often than not superposed on bright H II regions, and the background H II region contamination needs high-resolution spectra to identify and remove it. Figure 8 shows high-dispersion spectra of RCW58 and NGC6888; it can be seen that the background H II region has a nearly constant velocity throughout the slit length and the approaching and receding sides of the WR bubble appear blue- and red-shifted from the H II region component, respectively [9]. It is interesting to note that despite the large surface brightness variations, the expansion of the WR bubble is quite regular.

High-dispersion spectra covering all essential nebular lines have been obtained and analyzed for abundance only for the WR bubble NGC6888 [31,32]. These studies investigated trace of the CNO cycle and determined that the progenitor of the WR star in NGC6888 had an initial mass between 25 and $40 M_{\odot}$. Advances in modern instrumentation have made it possible to obtain high-dispersion spectra for large wavelength ranges (e.g. 3700 – 7400 Å, the High-Dispersion Spectrograph of the Subaru Telescope) so that bubble and background H II region components can be resolved and analyzed separately. Future observations should be made systematically for WR bubbles of different spectra types and physical properties, so that their abundances can be compared with those expected from models of stellar evolution. WR bubbles from galaxies of different metallicities should also be studied to see how metallicity affects stellar evolution.

References

[1] Johnson H M and Hogg D E 1965 *Astrophs. J.* **142** 1033
[2] Humphreys R M and Davidson K 1979 *Astrophys. J.* **232** 409
[3] Castor J, McCray R and Weaver R 1975 *Astrophys. J.* **200** L107
[4] Weaver R, McCray R, Castor J, Shapito P and Moore R 1977 *Astrophys. J.* **218** 377
[5] Kwok S 1983 *Planetary Nebulae* ed Flower D R (Proc. Int. Astron. Union S103) 293
[6] Frank A, Balick B and Riley J 1990 *Astron. J.* **100** 1903
[7] Esteban C, Vílchez J M, Smith L J and Clegg R E S 1992 *Astron. Astrophys.* **259** 629
[8] Solf J and Carsenty U 1982 *Astron. Astrophys.* **116** 54
[9] Chu Y-H 1988 *Pub. Astronom. Soc. Pac.* **100** 986
[10] Esteban C, Vílchez J M, Manchado A and Edmunds M G 1990 *Astron. Astrophys.* **227** 515

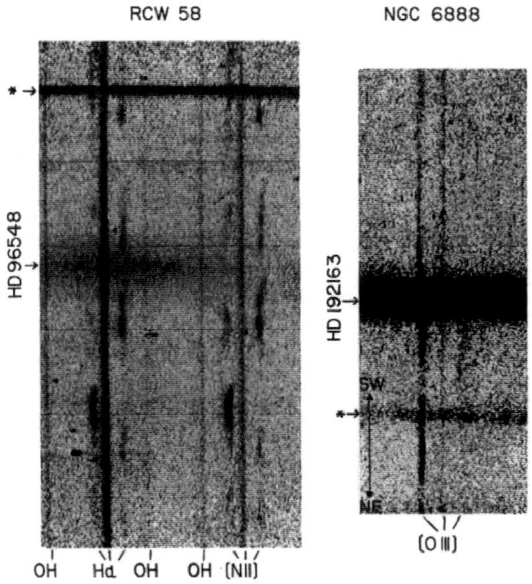

Figure 8. Long-slit high-dispersion echelle spectra of RCW58 and NGC6888. The spectral dispersion is along the X-axis and the spatial direction is along the Y-axis. This figure is reproduced from [9].

[11] García-Segura G, Langer N and Mac Low M-M 1996 *Astron. Astrophys.* **316** 133
[12] García-Segura G, Mac Low M-M and Langer N 1996 *Astron. Astrophys.* **305** 229
[13] Nazé Y, Chu Y-H, Points S D, Danforth C W, Rosado M and Chen C-H R 2001 *Astronom. J.* **122** 921
[14] Chu Y-H and Lasker B 1980 *Pub. Astron. Soc. Pac.* **92** 730
[15] Dopita M A, Bell J F, Chu Y-H and Lozinskaya T A 1994 *Astrophys. J. Suppl. Series* **93** 455
[16] Bransford M A, Thilker D A, Walterbos R A M and King N L 1999 *Astronom. J.* **118** 1635
[17] Drissen L, Shara, M M and Moffat A F J 1991 *Astronom. J.* **101** 1659
[18] Polcaro V F, Rossi C, Norci L and Viotti R 1995 *Astron. Astrophys.* **303** 211
[19] Nazé Y, Rauw G, Manfroid J, Chu Y-H and Vreux J-M 2003 *Astron. Astrophys.* **408** 171
[20] Bochkarev N G 1988 *Nature* **332** 518
[21] Wrigge M 1999 *Astron. Astrophys.* **343** 599
[22] Chu Y-H, Guerrero M A, Gruendl R A, Garcia-Segura G and Wendker H J 2003 *Astrophys. J.* **599** 1189
[23] Toalá J A, Guerrero M A, Chu Y-H, Gruendl R A, Arthur S J, Smith R C and Snowden S L 2012 *Astrophys. J.* **755** 77
[24] Zhekov S A 2014 *Mon. Not. R. Astronom. Soc.* **443** 12
[25] Toalá J A, Guerrero M A, Chu Y-H and Gruendl R A 2015 *Mon. Not. R. Astron. Soc.* **446** 1083
[26] Toalá J A, Guerrero M A, Chu Y-H, Arthur S J and Gruendl R A 2015 *Wolf-Rayet Stars: Proceedings of an International Workshop* eds Hamann W-R, Sander A and Todt H p 333
[27] Toalá J A and Arthur S J 2014 *Mon. Not. R. Astron. Soc.* **443** 3486
[28] Boroson B, McCray R, Clark O, Slavin J, Mac Low M-M, Chu Y-H and Van Buren D 1997 *Astrophys. J.* **478** 638
[29] Gruendl R A, Chu Y-H and Guerrero M A 2004 *Astrophys. J.* **617** L127
[30] Fang X, Guerrero M A, Toalá J A, Chu, Y-H and Gruendl R A 2016 *Astrophys. J.* **822** L1
[31] Esteban C and Vílchez J M 1992 *Astrophys. J.* **390** 536
[32] Mesa-Delgado A, Esteban C, García-Rojas J, Reyes-Pérez J, Morisset C and Bresolin F 2014 *Astrophys. J.* **785** 100
[33] Freyer T, Hensler G and Yorke H W 2003 *Astrophys. J.* **594** 888
[34] Freyer T, Hensler G and Yorke H W 2006 *Astrophys. J.* **638** 262
[35] Toalá J A and Arthur S J 2011 *Astrophys. J.* **737** 100
[36] van Marle A J and Keppens R 2012 *Astron. Astrophys.* **547** A3

Journal of Physics: Conference Series **728** (2016) 032008 doi:10.1088/1742-6596/728/3/032008

HASH: the Hong Kong/AAO/Strasbourg Hα planetary nebula database

Quentin A Parker[1,2], **Ivan S Bojičić**[1,2] **and David J Frew**[1,2]

[1] Department of Physics, The University of Hong Kong, Hong Kong SAR, China

[2] Laboratory for Space Research, The University of Hong Kong, Hong Kong SAR, China

E-mail: quentinp@hku.hk

Abstract. By incorporating our major recent discoveries with re-measured and verified contents of existing catalogues we provide, for the first time, an accessible, reliable, on-line SQL database for essential, up-to date information for all known Galactic planetary nebulae (PNe). We have attempted to: i) reliably remove PN mimics/false ID's that have biased previous studies and ii) provide accurate positions, sizes, morphologies, multi-wavelength imagery and spectroscopy. We also provide a link to CDS/Vizier for the archival history of each object and other valuable links to external data. With the HASH interface, users can sift, select, browse, collate, investigate, download and visualise the entire currently known Galactic PNe diversity. HASH provides the community with the most complete and reliable data with which to undertake new science.

1. Introduction

PNe, the ejected, ionised shrouds of dying stars, are a complex and brief (~25,000 year) phase of late stellar evolution. They offer rich science as vital probes of stellar nucleosynthesis processes in mid to low-mass stars. These stars make-up 90% of all stars above 1 solar mass. They provide a detectable, fossil record of stellar mass loss off the AGB/post-AGB and are powerful tracers of our Galaxy's star-forming history. Furthermore, they are useful kinematical probes visible to large Galactic distances due to their rich, strong, emission line spectra. These scientific levers and others encourage the search for and study of PN in our own and other Galaxies.

We are currently in a golden age of PN discovery. New high sensitivity, wide-field, narrow-band, Galactic plane surveys undertaken on the UK Schmidt Telescope in Australia [1][2], the Isaac Newton telescope on La Palma [3] and now the ESO/VST in Chile [4] have facilitated this. These Hα surveys have provided significant Galactic PNe discoveries that have more than doubled the totals accumulated by all telescopes over the previous 260 years, e.g. [5][6][7][8] and these proceedings, including posters by Kronberger et al. and Acker et al.

Most new PNe found are more redenned, evolved and of lower surface brightness than previous compilations such as [9][10] and [11] while others are faint but compact and more distant. The scope of any future large-scale PNe studies, particularly those of a statistical nature or undertaken to understand true PNe diversity and evolution, should now reflect this fresh PN population landscape. Studies should make us of the combined sample of ~3500 Galactic PNe now available in our HASH database. HASH takes into account recent major discoveries and the power invested in the wide-field, high sensitivity, high resolution, multi-wavelength imaging surveys now available across much of the electromagnetic spectrum.

Our group, now based at Hong Kong University, has played a lead role not only in the discovery of over 1300 new Galactic PNe but has also made some key contributions in understanding and quantifying the PNe phenomena. These include multi-wavelength identification, verification and testing techniques, e.g. [12][13], provision of accurate Hα and [O III] fluxes for large numbers of PNe, [14][15][16] and the establishment and verification of a new, statistical distance scale based on a robust incarnation of a surface-brightness radius relation [17]. We have also catalogued thousands of interesting non PNe and re-assigned significant numbers of supposed PNe into other object-types. All these activities, outputs and experience have been combined and incorporated into HASH in one form or another.

This paper can provide only a short HASH introduction. It briefly describes some of the power and broad functionality that HASH offers. A more comprehensive journal paper detailing the extensive capabilities of HASH is in preparation (Bojičić et al).

2. Rationale for establishing the HASH PN research platform

During our discovery and verification work for the MASH [5][6] and IPHAS [7] catalogues it was clear that previous PNe compilations were variable in quality and integrity. This is unsurprising as they contain heterogeneous assemblages of PNe identified, mis-identified and re-identified again over many decades by dozens of astronomers working with a wide variety of telescopes, detectors, resolutions, wave-bands and sensitivities. Furthermore, it became clear that not only does the availability of high-resolution, deep, optical, narrowband imaging surveys of high astrometric integrity provide the basis for significant new discoveries, but that these surveys also allow us to revisit the identity, morphologies and recorded positions for most PNe in existing catalogues. The advent of other high quality broad-band optical, near-IR, mid-IR and radio wide-field surveys provides strong, additional diagnostic and discriminatory power across the Galaxy. Taken together we can now construct a new type of PNe repository that can effectively federate all these new data and the extant spectroscopy into a single 'research platform' for investigating and evaluating all objects currently or previously identified as a PN in our Galaxy.

We have largely completed this task and the first HASH release is available in beta-form (contact authors for access). We have tried to resolve problems with mimics and dubious identifications for a purer, uncontaminated sample. We give more accurate positions, diameters, morphologies and designations. Vast amounts of disparate and scattered data have been consolidated and is presented in an intuitive, graphical MySQL-based interface for ease of use. Astronomers can use HASH on-line to interrogate, collate, evaluate, visualise and download chosen PN multi-wavelength images, spectra or tables from the entire currently known Galactic PNe population. HASH provides the community with the most complete, carefully vetted and reliable data to do this. HASH allows for highly efficient and effective examination and comparison of a variety of key observational PN properties for very large samples.

3. Input Catalogues

First we ingested data from the three largest catalogues of Galactic PNe: the Strasbourg-ESO catalogue of PNe (SEC [9]) and its supplement [10]; Version 2000 of the Catalogue of Galactic Planetary Nebulae [11] and the Macquarie/AAO/ Strasbourg Hα (MASH) catalogues [5][6], together with 159 new PNe from the related IPHAS survey in [7]. Our database contains all true, likely, possible, and misclassified PNe from these major catalogues. We then incorporated $\sim$400 true and candidate PNe from a large number of recent papers, that have been primarily discovered (or confirmed) optically since 2001. The most important are [8][18][19] and [20]. We also include a significant number of unpublished candidates found by us from on-going examination of Hα survey data, plus sundry other objects. Additional candidate PNe recently found in near/mid-IR surveys were also added, chiefly taken from [13][22][23] and [24] while another 320 PN candidates found in the mid-IR at 24μm were added from [25][26] and [27].

Using the Simbad Criteria Query Tool, we also searched for all objects called PNe, possible PNe, or proto-PNe (Simbad Object codes: PN, PN? and post-AGB, respectively) in the literature. All objects were cross-checked for duplication and identity (also using [28][29] and [30]), noting that the classifications of many nebulae have been fluid over time, e.g. [12] and that many current Simbad PN or possible PN identifications have turned out to be erroneous. For example [31] showed that in the GLIMPSE zone 45% of previously known, pre MASH PNe are in fact contaminants (mostly compact H II regions).

We reference objects to the main catalogues and compilation papers where we found a PN identification or candidate and whose given positional data and other standard identifiers are input to Simbad for initial assessment. These should not be confused with the original, discovery papers for individual or small groups of PNe which we did not consider in this work. For the majority of objects readers are referred to the main input catalogues above which contain an extensive discovery list for almost all associated PNe, and of course, to the Simbad database. The original data for each object is recorded in the main MySQL table and includes: a unique identifier (usual name), PNG identifier, coordinate (from the original source), a reference to the catalogue of origin and the status of the object (see section 7 for information of status flags).

Over 6000 Galactic objects are currently in our working database, including ~3500 true, likely, and possible PNe together with ~1500 mimics of various kinds (e.g.[35]) including about 50 transitional objects [32] but not ~500 post-AGB stars and related objects [33][34]. We are re-investigating mimics in HASH using multi-wavelength discrimination techniques we developed (e.g.[12]). There are many unclassified objects that need more investigation before they can be properly identified (see [36] and [37] for example case-studies).

4. Presentation of the data (online interface)

The HASH PN database and its online interface is currently hosted and maintained on our own server at HKU. In the near future we will mirror this server in Europe in order to secure faster service to users outside of Asia and better availability in case of technical problems. There is also an agreement to host a HASH front-end at the CDS in France. The backbone of the system is a relational database (MySQL) which provides data consistency and quick, efficient search and manipulation. HASH objects are characterised by a unique set of parameters (id, coordinates and PNG designation) and extensive observational data, collected from the literature and astronomical data repositories and mapped to corresponding objects.

The online interface is organised in *views*. The main view is the *table view* shown in Fig.1 which provides the tabulated data for all objects from the selected sample. The table presents the best available data for each observational parameter. The method for selecting the best available data differs from parameter to parameter and it is mostly based on the quality of observational parameters. We provide other available data on each PN's *individual page* - see Fig.2 below. In the *image views* one can select one or more pre-made total intensity or RGB thumbnail images to be presented for the currently selected sample.

The search can be on text (e.g. a PN name like NGC 6543) or on one or a combination of parameters including tabulated observational data, positions and association to pre-made samples. In HASH V1.0 we do not include derived data such as PN distances, radial velocities, metallicities, etc. These await a future release though much of the data is available to us.

The last sample selected is logged and is the current working sample unless a new sample is selected or the user chooses to restart sample selection. The working sample can be subjected to further manipulation (selection of a subsample), exported, saved for a future use and shared between users. For each HASH entry we provide links to the Simbad and Vizier entries based on name or position. Internet links are also provided to data pages hosted by the Hubble Legacy Archive, Planetary Nebula Image Catalogue (PNIC; [38]) and the SPM Kinematic Catalogue of Planetary Nebulae (SPMC; [39]). Links for these appear on the individual HASH pages via

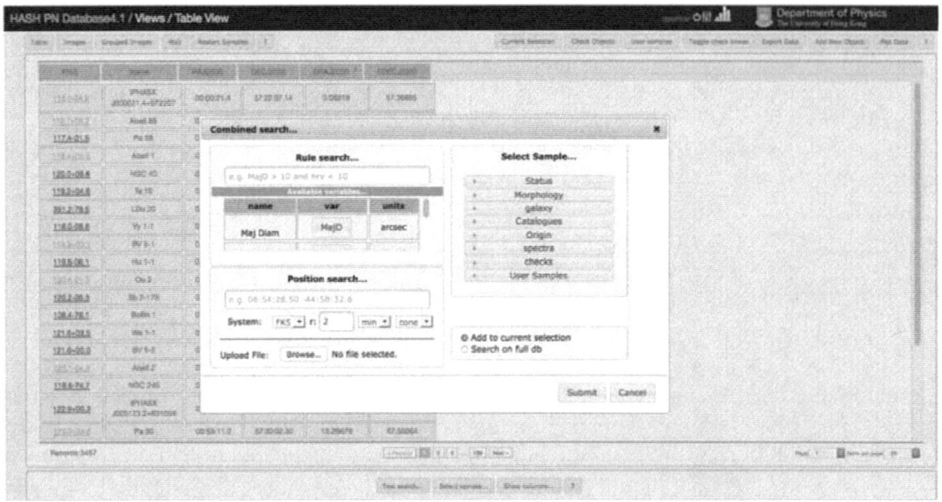

Figure 1. HASH PN basic table view of PN catalogue superimposed with the search box that appears when you want to search by any available parameter(s) to select a sub-sample.

clickable logos on the left hand side of the page under the summary data.

5. HASH multi-wavelength image data

We have collected and ingested imaging data around each object from over 30 large scale, multi-wavelength surveys from the UV to the radio (as WCS fits files). The complete list of surveys incorporated and the full details of all observational parameters available will be published separately (Bojičić et al. in preparation). Particular emphasis is placed on the narrowband Hα imaging surveys that have recently become available, e.g. [2][3] and [4] that have provided the discovery medium for the significant numbers of new MASH and IPHAS PNe in particular. These modern surveys also provide a pathway to the determination of accurate integrated Hα fluxes (e.g. [14][15]; see below). Combined RGB colour images for each object from combining various narrow and broad optical, near-IR and mid-IR bands are also provided which also link directly to an associated *RGB* fits cube that can be downloaded. The field of view, for each image, is calculated using the resolution of the survey and angular size of an object. We adopted a simple algorithm based on an intensity percentile to present as much information as possible on the object's brightness and structure together with its immediate environment.

Pre-made composite and/or total intensity images are produced using the APLpy (http://aplpy.github.io/) and Astropy (http://www.astropy.org) packages, the Montage (http://montage.ipac.caltech.edu/) toolkit and custom made python tools. Composite images were made where red (R), green (G) and blue (B) colours are for images ordered from the less to more energetic band. Each RGB image layer is scaled using the same (linear) scaling function in a way that provides a representative colour image combination that depends on ratios between fluxes in the used bands. If the nebular size was unknown or smaller than the survey's PSF we used the PSF for the aperture size. The measured distribution of data points is used for estimation of the maximum scaling value. The minimum scaling value is estimated from the data points outside of the aperture. Prior to RGB composition, constituent fits images were 're-gridded' to a common projection and resolution and aligned to have north straight up and east to the left before being combined into fits cubes.

Each HASH object has an *individual page* that contains basic data, links to web pages, image

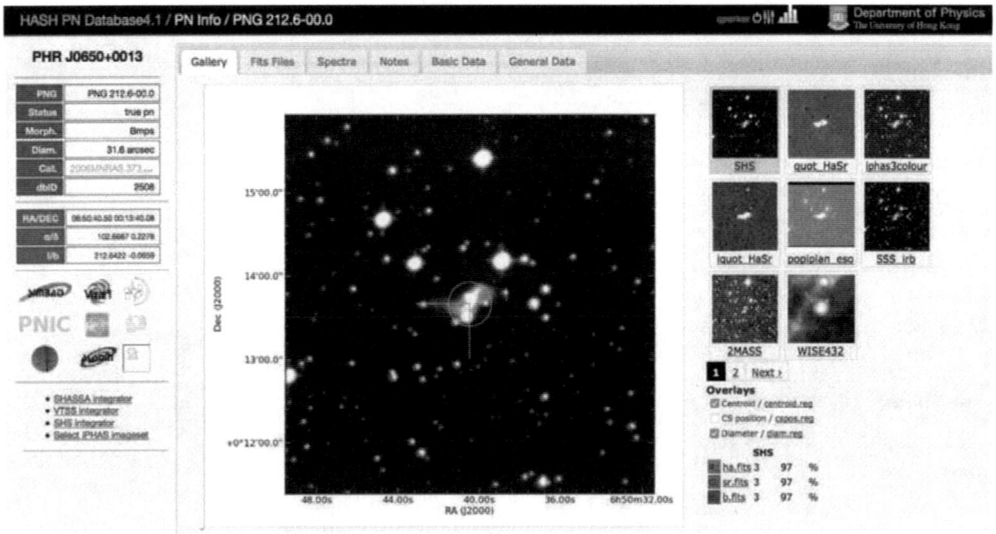

Figure 2. HASH PN image page for PHR J0650+0013 - a MASH PN. There is a series of clickable tabs at the top of the page which give access to the FITS images for each image, any 1-D spectra, notes pertaining to the object, basic data and more general details. All images in thumbnails at the right are clickable.The left hand side column provides summary information for the object and direct access via clickable logos to Simbad, Vizier, Aladin, PNIC, HST imagery and other useful sites and date-sets.

gallery, fits images repository, spectral gallery and repository, notes and observational data (refer to on-line documentation for further details). The gallery tab provides pre-made images from all available imaging datasets. Overlays of the catalogued position and diameter for each object are provided. An example HASH page for MASH PN PHR J0650+0013 is given in Fig.2.

6. Spectroscopic and other data

Where possible HASH gives 1-D optical spectra for all PNe. The primary source is the extensive MASH and IPHAS spectroscopic follow-up programs undertaken by our group and collaborators over the last 15 years. These mostly low to intermediate resolution spectra were taken mainly on a series of 2-m telescopes. Another major source is the 1050 low-resolution PN spectra from the Stenholm-Acker spectroscopic survey [40] that is now available to the community for the first time in HASH. Significant numbers (~1150) of reduced spectroscopic emission line data for PNe from [41]) have also been incorporated and are presented as pseudo 1-D spectral line intensity plots. Other true spectra are also included where available and an extensive search of the literature for additional spectra (as described in [14]) is on-going. HASH also includes reduced 1-D fits spectra from [18][19][32][42][43] and [44] kindly provided by these authors. Finally, 2-D and 1-D high resolution echelle spectra for accessible PNe from the San Pedro Martir Kinematic Catalogue of Galactic Planetary Nebulae [39] can be accessed from a dedicated clickable logo on the right hand side of the main page for each PN when available. Many PNe now have multiple spectra available of different resolutions and wavelength coverage and these are made available in HASH. Fig.3 presents a 1-D spectrum for MASH PN PHR J0650+0013 selected by clicking on the spectral tab at the top of the individual HASH PN page shown in Fig.2. Common PN emission lines can be selected/de-selected and these are shown as faint, vertical blue lines on the plot. If several spectra are available they can be over plotted or selected individually. The cursor can also be used to expand the spectra as desired around lines of interest. The spectral

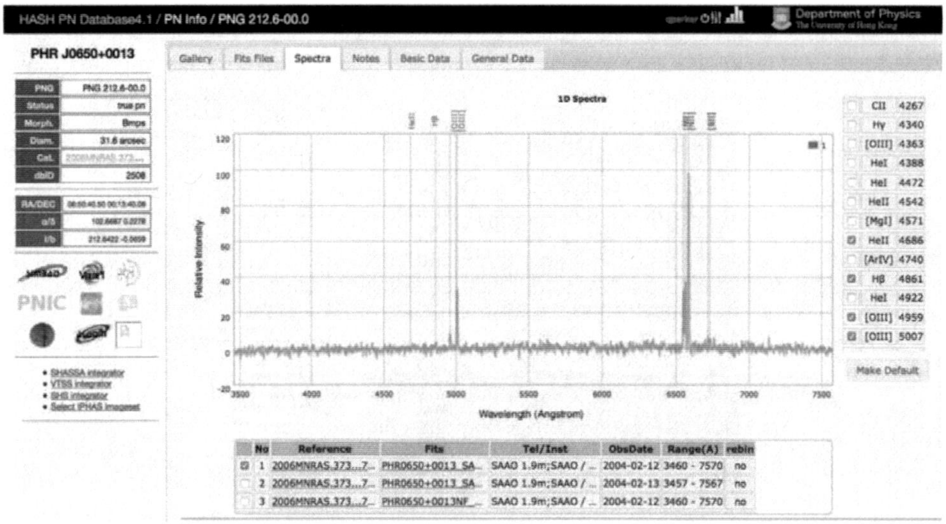

Figure 3. HASH basic spectral page for MASH PN PHR J0650+0013 selected by the spectral tab at the top of the individual HASH PN page in Fig.2. Common PN emission lines can be selected and are then indicated as faint, vertical blue lines on the plot. If several spectra are available they can be over plotted or selected individually. The cursor can also be used to expand the spectra as desired.

files are also available for download for more detailed evaluation and measurement.

Besides astrometric data and measurable properties like angular size and more subjective data like morphology we are adding other fundamental observational data for each object including integrated Hα, Hβ and [O III] fluxes (e.g. [14][15][16] and [45]), radio flux densities (e.g. [46]), and extinction estimates [9][17][47][48] and [49]. In future releases radial velocities, emission line ratios, plasma diagnostics, distances and other photometric measurements will be included. More extensive central star positions will also be provided as they become available.

7. The integrity of the HASH database: PN status

We define a PN following [12]. Identification is complicated by the wide variety of morphologies, ionization characteristics and brightness distribution exhibited by PNe. These reflect stages of nebular evolution, progenitor mass and chemistry and the possible influence of common envelope binaries, magnetic fields or even sub-solar planets. It is not surprising that contamination of previous catalogues has been a problem given such variables and the variable quality of the older data upon which identifications have been made. However, we have robustly tested a range of criteria to eliminate contaminants done by assessing morphology, emission-line intensities and widths, ionization structure, systemic velocity, ionized mass and the properties of the central star (where possible) as well as considering environment and multi-wavelength imaging properties [50]. This rigorous process is detailed in [12] and we utilised a range of diagnostic diagrams presented by [12][17][51][52] and Frew, Parker and Bojičić (these proceedings).

The HASH PN status flag is adopted from [5] and we classify objects as true (**T**), likely (**L**), possible (**P**) and not PN (**N**). As a preliminary step for non MASH/IPHAS PNe in HASH we initially adopt the most recent PN status from the literature. Using the available spectroscopic data and the extensive, modern, multi-wavelength imagery now conveniently available for each object in HASH we carefully re-examine each object against its current status. On this basis we assign objects as **T** that are confirmed PN with indicative multi-wavelength PN-type morphologies, PN spectral features and sometimes presence of an obvious CSPN. **L** indicates

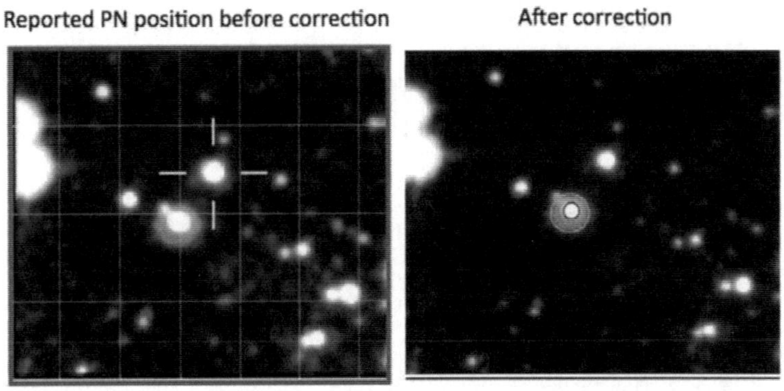

Figure 4. PN PBOZ 29 with position from Simbad based on [30] at left and the new HASH position for this PN on the right. The offset is 14.3 arcseconds and the image is 100×100 arcseconds with orientation NE to the top left. The HASH multi- wavelength combined RGB H-alpha, R-band and B-band images reveal the true PNe and not the star with which this compact PN had been confused in SIMBAD.

an object which is likely to be a PN but whose imagery or spectroscopy are not completely conclusive or unavailable. **P** indicates a possible PN where the morphology and spectroscopy are insufficiently conclusive, usually due to a combination of low S/N spectra, insufficient wavelength coverage, very low surface brightness or indistinct nebulosity. **N** indicates an object which has either been identified as an object of different nature and the current data does not provide any additional evidence of a possible PN status, or the newly available combined data that HASH provides strongly points to a non PN identification. We believe we have created the purest and most homogeneously assessed catalogue of PN as a result of this work. Once the status is established we compare the tabulated position against the accurate astrometric grid of the object's best resolution optical image. Despite recent papers of 'accurate' PN co-ordinates (e.g. [29], [30] and even MASH), it was surprising how poor some PN positions remained. Sometimes this is because only part of the PN extent was used to provide the position from broad band data whereas on-line surveys in various narrow and broad-bands with accurate astrometry now enables a more robust mechanism to check PNe positions. Sometimes the wrong object has been identified (see Fig.4 for PN PBOZ 29). Where differences are clear a new measurement of the object centroid is obtained that then replaces the old value. New estimates of the major and minor axes in arcseconds are also made at this time. Accurate, homogeneously derived PNe positions and sizes now form part of HASH and were obtained by aperture and centroid fitting to mainly Hα images at the 80% contour. Full details of this process will be in the main journal paper. However, 29% of PNe had positional offsets of 10 arcseconds or greater and many had smaller offsets down to 1 arcecond, all of which we have now corrected.

8. Conclusions

We are providing a fully integrated, functional and stable PN research platform that will be fully VO compliant and offers the community a one-stop shop for facilitating PN research. HASH is now available for beta-testing (contact the authors). Future releases will occur as new data and published results are ingested. These data and discoveries will be vetted and checked against our own standards and techniques to ensure, as far as is possible, the on-going integrity and homogeneity of the database. In future releases more functionality and parameters are planned. For example we intend to incorporate radial velocities and emission line ratios and associated plasma diagnostics for every PN where this is possible in a future release. HASH requires regular curation and maintenance of content, even on an individual PN basis, between major releases and we are committed to this task. Extension to the LMC and SMC is also in train.

References

[1] Parker Q A, Phillips S 1998 *Publ. Astron. Soc. Australia* **15** 28
[2] Parker Q A *et al* 2005 *Mon. Not. R. Astron. Soc.* **362** 689
[3] Drew J E *et al* 2005 *Mon. Not. R. Astron. Soc.* **362** 753
[4] Drew J E *et al* 2014 *Mon. Not. R. Astron. Soc.* **440** 2036
[5] Parker Q A *et al* 2006 *Mon. Not. R. Astron. Soc.* **373** 79
[6] Miszalski B, Parker Q A, Acker A, Birkby J L, Frew D J and Kovacevic A 2008 *Mon. Not. R. Astron. Soc.* **384** 525
[7] Sabin L *et al* 2014 *Mon. Not. R. Astron. Soc.* **443** 3388
[8] Jacoby G H *et al* 2010 *Pub. Astron. Soc. Aust.* **27** 156
[9] Acker A, Marcout J, Ochsenbein F, Stenholm B, Tylenda R and Schohn C 1992 *The Strasbourg-ESO Catalogue of Galactic Planetary Nebulae* (Garching: European Southern Observatory)
[10] Acker A, Marcout J and Ochsenbein F 1996 *First Supplement to the SECGPN* (Observatoire de Strasbourg)
[11] Kohoutek L 2001 *Astron. Astrophys.* **378** 843
[12] Frew D J and Parker Q A 2010 *Pub. Astron. Soc. Aust.* **27** 129
[13] Parker Q A, Cohen M, Stupar M, Frew D J, Green A J, Bojicic I, Guzman-Ramirez L, Sabin L and Vogt F 2012 *Mon. Not. R. Astron. Soc.* **427** 3016
[14] Frew D J, Bojičić I S and Parker Q A 2013 *Mon. Not. R. Astron. Soc.* **431** 2
[15] Frew D J, Bojičić I S, Parker Q A, Pierce M J, Gunawardhana M L and Reid W A 2014 *Mon. Not. R. Astron. Soc.* **440** 1080
[16] Kovacevic A V, Parker Q A, Jacoby G H, Sharp R, Miszalski B and Frew D J *Mon. Not. R. Astron. Soc.* **414** 860
[17] Frew D J, Parker Q A and Bojičić I S 2016 *Mon. Not. R. Astron. Soc.* **455** 1459
[18] Boumis P, Paleologou E V, Mavromatakis F and Papamastorakis J 2003 *Mon. Not. R. Astron. Soc.* **339** 735
[19] Boumis P, Akras S, Xilouris E M, Mavromatakis F, Kapakos E, Papamastorakis J and Goudis C D 2006 *Mon. Not. R. Astron. Soc.* **367** 1551
[20] Acker A 2015 *EAS Publications Series* **71-72** 309
[21] Kronberger M *et al* 2012 *Proc. Int. Astron. Union* (S283) **7** 414
[22] Phillips J P and Ramos-Larios G 2008 *Mon. Not. R. Astron. Soc.* **386** 995
[23] Kwok S, Zhang Y, Koning N, Huang H-H and Churchwell E 2008 *Astrophys. J. Suppl. Series* **174** 426
[24] Ramos-Larios G, Guerrero M A, Suárez O, Miranda L F and Gómez J F 2012 *Astron. Astrophys.* **545** A20
[25] Mizuno D R, Kraemer K E, Flagey N, Billot N, Shenoy S, Paladini R, Ryan E, Noriega-Crespo A and Carey S J 2010 *Astron. J.* **139** 1542
[26] Wachter S, Mauerhan J C, Van Dyk S D, Hoard D W, Kafka S and Morris P W 2010 *Astron. J.* **139** 2330
[27] Gvaramadze V V, Kniazev A Y and Fabrika S 2010 *Mon. Not. R. Astron. Soc.* **405** 1047
[28] Perek L and Kohoutek L 1967 *Catalogue of Galactic Planetary Nebulae* (Czechoslovak Acad. Sci.)
[29] Kimeswenger S 2001 *Revista Mexicana de Astronomia y Astrofisica* **37** 115
[30] Kerber F, Mignani R P, Guglielmetti F and Wicenec A 2003 *Astron. Astrophys.* **408** 1029
[31] Cohen M, Parker Q A, Green A J, Miszalski B, Frew D J and Murphy T 2011 *Mon. Not. R. Astron. Soc.* **413** 514
[32] Suárez O, García-Lario P, Manchado A, Manteiga M, Ulla A and Pottasch S R 2006 *Astron. Astrophys.* **458** 173
[33] Szczerba R, Siódmiak N, Stasińska G and Borkowski J 2007 *Astron. Astrophys.* **469** 799
[34] Vickers S B, Frew D J, Parker Q A and Bojičić I S 2015 *Mon. Not. R. Astron. Soc.* **447** 1673
[35] Boissay R, Parker Q A, Frew D J and Bojičić I S 2012 *Proc. Int. Astron. Union* (S283) **7** 316
[36] Frew D J, Madsen G J, O'Toole S J and Parker Q A 2010 *Pub. Astron. Soc. Aust.* **27** 203
[37] Frew D J, Bento J, Bojičić I S and Parker Q A 2014 *Mon. Not. R. Astron. Soc.* **445** 1605
[38] Balick B 2007 in *Proc. APN IV Conf.*, ed R Corradi, A Manchado, N Soker, `http://www.iac.es/proyect/apn4`
[39] López J A, Richer M G, Garcia-Diaz M T, Clark D M, Meaburn J, Riesgo H, Steffen W and Lloyd, M 2012 *Revista Mexicana de Astronomia y Astrofisica* **48** 3
[40] Stenholm B and Acker A 1987 *Astron. Astrophys. Suppl. Series* **68** 51
[41] Kaler J B Shaw, R A and Browning L 1997 *PASP* **109** 289
[42] Beaulieu S F, Dopita M A and Freeman K C 1999 *Astrophys. J.* **515** 610
[43] Hora J L, Latter W B and Deutsch L K 1999 *Astrophys. J. Suppl. Series* **124** 195
[44] Jacoby G H and Van de Steene G 2004 *Astron. Astrophys.* **419** 563
[45] Acker A, Raytchev B, Stenholm B and Tylenda R 1991 *Astron. Astrophys. Suppl. Series* **90** 89

[46] Bojičić I S, Parker Q A, Filipović M D and Frew D J 2011 *Mon. Not. R. Astron. Soc.* **412** 223
[47] Tylenda R, Acker A, Stenholm B and Köppen J 1992 *Astron. Astrophys. Suppl. Series* **95** 337
[48] Giammanco C *et al* 2011 *Astron. Astrophys.* **525** A58
[49] Schlafly E F and Finkbeiner D P 2011 *Astrophys. J.* **737** 103
[50] Cohen M *et al* 2007 *Astrophys. J.* **669** 343
[51] Sabin L, Parker Q A, Contreras M E, Olguín L, Frew D J, Stupar M, Vázquez R, Wright N J, Corradi R L M and Morris R A H 2013 *Mon. Not. R. Astron. Soc.* **431** 279
[52] Frew D J, Bojičić I S, Parker Q A, Stupar M, Wachter S, DePew K, Banehkar A, Fitzgerald M T and Douchin D 2014 *Mon. Not R. Astron. Soc.* **440** 1345

Quentin Parker.

David Frew giving his talk.

Journal of Physics: Conference Series **728** (2016) 032009 doi:10.1088/1742-6596/728/3/032009

On the formation of molecules and solid-state compounds from the AGB to the PN phases

D. A. García-Hernández[1,2] **and A. Manchado**[1,2,3]

[1] Instituto de Astrofísica de Canarias, C/ Via Láctea s/n, E-38205 La Laguna, Spain
[2] Departamento de Astrofísica, Universidad de La Laguna (ULL), E-38206 La Laguna, Spain
[3] Consejo Superior de Investigaciones Científicas, Spain

E-mail: agarcia@iac.es; amt@iac.es

Abstract. During the asymptoyic giant branch (AGB) phase, different elements are dredge-up to the stellar surface depending on progenitor mass and metallicity. When the mass loss increases at the end of the AGB, a circumstellar dust shell is formed, where different (C-rich or O-rich) molecules and solid-state compounds are formed. These are further processed in the transition phase between AGB stars and planetary nebulae (PNe) to create more complex organic molecules and inorganic solid-state compounds (e.g., polycyclic aromatic hydrocarbons, fullerenes, and graphene precursors in C-rich environments and oxides and crystalline silicates in O-rich ones). We present an observational review of the different molecules and solid-state materials that are formed from the AGB to the PN phases. We focus on the formation routes of complex fullerene (and fullerene-based) molecules as well as on the level of dust processing depending on metallicity.

1. Introduction

Low- and intermediate-mass ($1 < M < 8\ M_\odot$) stars suffer strong mass loss and a series of thermal pulses at the tip of the asymptotic giant branch (AGB) phase. The interstellar medium (ISM) is thus efficiently enriched with specific elements/isotopes and dust grains expelled by AGB stars. The principal nucleosynthetic processes occur on the thermal pulsing (TP) AGB phase, while the main molecular/dust processing takes place during the short transition phase between the end of the AGB phase and the planetary nebula (PN) stage. Carbon (^{12}C) and s-process neutron-rich elements (e.g., Rb, Zr, Sr, Nd, Ba, Tc) are synthesized and transported to the stellar surface during the TP-AGB phase. Solar metallicity low-mass ($M<1.5\ M_\odot$) AGB stars remain O-rich and they probably do not form an observable PN, while more massive ($1.5<M<4\ M_\odot$) AGBs are converted to C-rich (C/O>1) stars. The most massive ($M>4\ M_\odot$) AGBs remain also as O-rich stars because of the activation of the hot bottom burning (HBB) process. Also, the dominant neutron source (^{13}C vs. ^{22}Ne) at the s-process site depends on stellar mass, which translates into different s-process abundances patterns (e.g., the [Rb/Zr] ratio; see e.g., [1]). This basic AGB picture is strongly dependent on metallicity (e.g., [1,2,3] and references therein); e.g., the minimum progenitor mass to efficiently activate the HBB process decreases at lower metallicities. The higher mass AGB stars thus produce different elements/isotopes than the lower mass AGB ones, this being imprinted in the gas and circumstellar dust chemistry. At the end of the AGB, the more massive C-rich and O-rich stars are also heavily obscured by their thick circumstellar dust envelopes (CSE), experiencing a phase of total obscuration in their way to form PNe; they are thus only accesible in the infrared (IR) and millimeter spectral ranges.

AGB and post-AGB stars are also amazing molecular factories; i.e., more than 70 molecules have been detected in AGB CSE through their rotational transitions from the IR to millimeter wavelengths (e.g., [4]). These are mainly gas-phase molecules: i), inorganics like SiS and AlCl; ii) organics such as H_2CO and CH_3CN; iii) radicals (e.g., HCO^+); iv) rings (e.g., C_3H_2); and v) chains (e.g., HC_9N). We note that the formation of most (although not all) of these molecules may be explained by gas-phase reactions but solid-state chemistry has to be considered also in the models; i.e., molecules (e.g., H_2) may form on dust grains.

The CSE dust composition of sources evolving from the end of the AGB to the PN stage depends on the dominant star's chemistry (i.e., the C/O ratio). The C-rich (C/O>1) AGB stars show SiC and amorphous carbon dust, although other complex and disordered organic solids with mixed aromatic/aliphatic structures like kerogen and coal may provide also the strong dust continuum emission (e.g., [5]). The O-rich (C/O<1) AGBs, however, display amorphous silicates, weak crystalline silicates such as piroxenes, and refractory oxides (e.g., corundum and spinel) (e.g., [6]). On the other hand, young and evolved O-rich PNe (with a few exceptions that may still show amorphous silicates emission at $\sim$10 μm) mainly display strong crystalline silicate features, while the aromatic infrared bands (AIBs, generally attributed to polycyclic aromatic hydrocarbons; PAHs) and unidentified IR emission features (UIRs) are observed in the C-rich ones. Finally, double-chemistry PNe are characterized by both C-rich (e.g., PAHs) and O-rich (e.g., crystalline silicates) dust features (e.g., [7]).

Several dust types can be differentiated depending on the solid-state features observed in the *Spitzer Space Telescope* spectrocopic observations of Galactic PNe (see Figure 1, left panel), which otherwise are generally characterized by a strong IR dust continuum emission [8,9]. Carbon chemistry (CC) and oxygen chemistry (OC) PNe display respectively C-rich and O-rich dust features that may be aliphatic/amorphous and/or aromatic/crystalline. Double-chemistry (DC) PNe simultaneously show C-rich (mainly aromatic) and O-rich dust features (crystalline and/or amorphous). Finally, featureless (F) PNe show only nebular emission lines in their IR spectra with very weak dust continuum emission.

In general, there is a process of crystallization or aromatization during the short ($\sim 10^2$-10^4 years) transition phase between the end of the AGB and the PN stage (AGB–PN, hereafter) and the crystalline (O-rich) and aromatic (C-rich) structures are created (or more easily detected in the case of the crystalline silicates) during this short stage (e.g., [10,5,11]). Stars evolving from AGB stars to PNe are thus great laboratories for astrochemistry, providing us with strong observational constraints on the gas-phase and solid-state chemical models. The major challenge is to unveil the formation routes of these complex organic molecules and inorganic solid-state compounds. Here we present an observational review of the several molecular (and solid-state) materials that are formed in the AGB–PN phase, focussing on the formation pathways of complex fullerene (and fullerene-based) molecules as well as on the level of dust processing depending on metallicity.

2. Dust composition versus metallicity

The different *Spitzer* dust types observed in PNe are shown in Figure 1 (left panel). These *Spitzer* dust types show a strong dependence on metallicity; see Table 1 in [12] where the *Spitzer* dust type statistics for Galactic bulge (high metallicity) and disk (intermediate metallicity) PNe are compared with those observed in the lower metallicitiy PNe of the Magellanic Clouds (MCs). Dust features overimposed to the dust continuum are less frequent at the lower MC metallicities. Crystalline silicate features (and another O-rich dust features) are quite rare in the MCs, where carbon chemistry (CC) PNe are predominant [13]. In addition, double-chemistry (DC) PNe dominate towards the Galactic bulge [7], while they are not observed in the low-metallcity

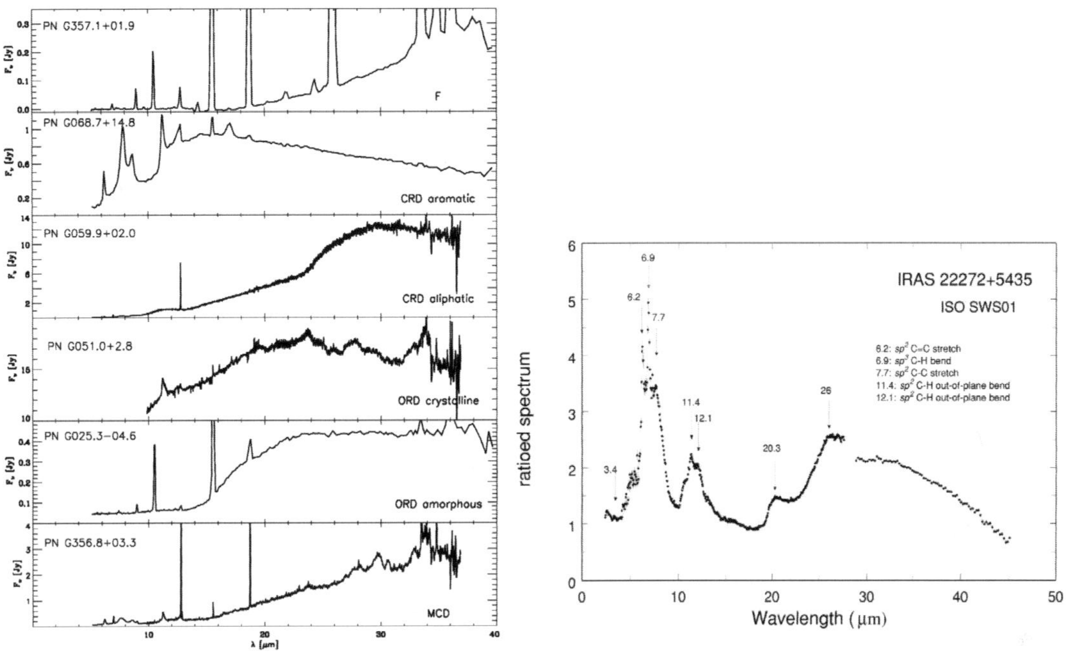

Figure 1. Left panel: Examples of the several *Spitzer* dust types found in compact (presumably young) Galactic PNe; from top to bottom: featureless (F); carbon-chemistry aromatic or aliphatic (CCar or CCal); oxygen-chemistry crystalline or amorphous (OCcr or OCam); and double-chemistry (DC) (updated from [8]). Right panel: Infrared Space Observatory (ISO) spectrum of the proto-PN IRAS 22272+5435 showing the presence of the still unidentified 21, 26, and 30 μm features (updated from [14]). The aliphatic discrete features (e.g., at 3.4 and 6.9 μm) and the 8 and 12 μm aliphatic plateaus together with AIBs (e.g., at 6.2 and 7.7 μm) are also seen.

environments of the MCs. The mixed-chemistry (DC), C-rich (CC), and O-rich (OC) PNe are equally distributed in the Galactic disk [8,9].

2.1. *Oxygen-rich dust PNe*

The O-rich dust PNe are less frequent at the lower MC metallicities [13] and the several *Spitzer* O-dust subtypes (amorphous or crystalline) vary with metallicity (e.g., [9]). In particular, the $\sim$9.7 μm broad emission band due to amorphous silicates is more common in the Galactic disk (intermediate metallicity), while the narrower crystalline silicate emission features (such as those of olivine and pyroxene at $\sim$23.5, 27.5, and 33.5 μm) completely dominate in the Galactic bulge (high metallicity), which otherwise are almost absent in the MCs. The O-rich dust features evolve from amorphous to crystalline silicates during the AGB−PN phase [10]. Two main scenarios may explain the silicates crystallization depending on temperature: a) at low temperature in long-lived circumbinary disks [15]; or b) at high temperature, because of the strong mass loss, at the tip of the AGB [6].

2.2. *Carbon-rich dust PNe*

A different situation is found for the C-rich dust PNe. They are more common at the low metallicity of the MCs than at higher metallicity (Galactic disk or bulge). In the MCs, the C-rich dust PNe generally display small dust grains (or less processed) [13]; the C-rich aliphatic dust features dominate their *Spitzer* spectra, while the PAH-like emission bands are very rare. In addition, the C-rich dust PNe dominate at the MC low-metallicity but they are not observed at the very high metallicity of the Galactic bulge. In particular, broad emission dust features of aliphatic character (e.g., at ~9−13, 15−20, and 25−35 μm; see Section 3) are usually detected in MC PNe, while Galactic disk C-rich dust PNe are usually compact (presumably young) PNe of slightly subsolar metallicity [9]. In the case of nearly solar metallicities, small hydrocarbon molecules (e.g., acetylene) are synthesized in the short transition phase AGB−PN, replacing the broad ~11.5μm feature due to SiC (as well as the strong dust continuum emission) seen in the previous AGB phase. The IR features rapidly change from aliphatic to PAH-like (aromatic) during the post-AGB phase and the nebular emission lines (sometimes the broad ~25−35 μm aliphatic emission feature is still detected) prevail in the PN stage [10]. The level of dust processing seems to vary with the metallicity of the environment. For example, the transition from aliphatic structures (the ~9−13, 15−20, and 25−35 μm features) to aromatic ones (the 6.2, 7.7, 8.6, and 11.3 μm PAH-like features) proceeds more slowly at the MCs metallicity [13]. Two models for the aromatization process from AGB stars to PNe have been proposed: a) the PAH precursors are acetylene (C_2H_2) and its radical derivatives [16]; or b) the PAHs are formed as a consequence of the photochemical processing of the dust grains, which transform aliphatics to aromatics (e.g., [14,10]).

3. Unidentified infrared emission features at 21, 26, and 30 μm

The still unidentified set of IR emission features at ~21, 26, and 30 μm is observed in C-rich sources evolving from AGB stars to PNe (see Figure 1, right panel).

The carrier of the 21 μm feature, generally observed in post-AGB stars, should be a fragile solid-state carbon compound (e.g., [17]). Several carriers such as hydrogenated fullerenes, TiC nanoclusters, nanodiamonds, HAC, and amides have been proposed in the literature but we still lack a firm identification.

The 30 μm feature (sometimes with substructure at ~26 μm), however, has been linked with magnesium sulfide [18], aliphatic chains [19] or HACs [20]. The carrier should be quite abundant in the CSE because it is detected from the AGB to the PN phases, carrying out a significant fraction of the total energy output. Remarkably, the HAC-like identification, some kind of solid with a mixed aromatic/aliphatic structure that may also explain the 21 and 26 μm features [20], could explain the formation of complex dehydrogenated organics like C_{60}, C_{70}, and planar C_{24} (a small piece of a graphene sheet) in some C-rich PNe (see below).

4. Detection of complex dehydrogenated molecules

Kroto et al. [21] discovered the tridimensional C_{60} and C_{70} fullerene molecules at laboratory and their high stability reinforced the hypothesis that they should be widespread in the ISM, being important species for the interstellar/circumstellar chemistry. These complex species may give an explanation to several phenomena in Astrophysics. For example, they may explain the diffuse interstellar bands (DIBs) (22; see e.g., [23] for a review on interstellar/circumstellar DIBs) and the UV bump (e.g., [24]). Fullerenes were indeed found on Earth and on meteorites and various unsuccessful fullerene searches (e.g., by using the Infrared Space Observatory) were made to find the four mid-IR features (at ~7.0, 8.5, 17.4, and 18.9 μm; see Figure 2, left panel) of the C_{60} fullerene towards proto-PNe (e.g., [25]) and R Coronae Borealis (RCB) stars (e.g., [26]).

Nowadays, C_{60} and C_{70} (first discovered in space by [27][1]) are known to be efficiently formed

[1] These authors interpreted the detection of C_{60} and C_{70} in the PN Tc 1 as due to the H-poor conditions in the inner core. This interpretation is in agreement with some laboratory experiments that show that fullerenes are efficiently produced under the absence of H. However, neither the central star, nor the inner core and the nebula are H-poor or H-deficient. Thus, current understanding of stellar astrophysics does not allow for Tc 1 being a H-poor (late TP) PN (see e.g., [28,29]).

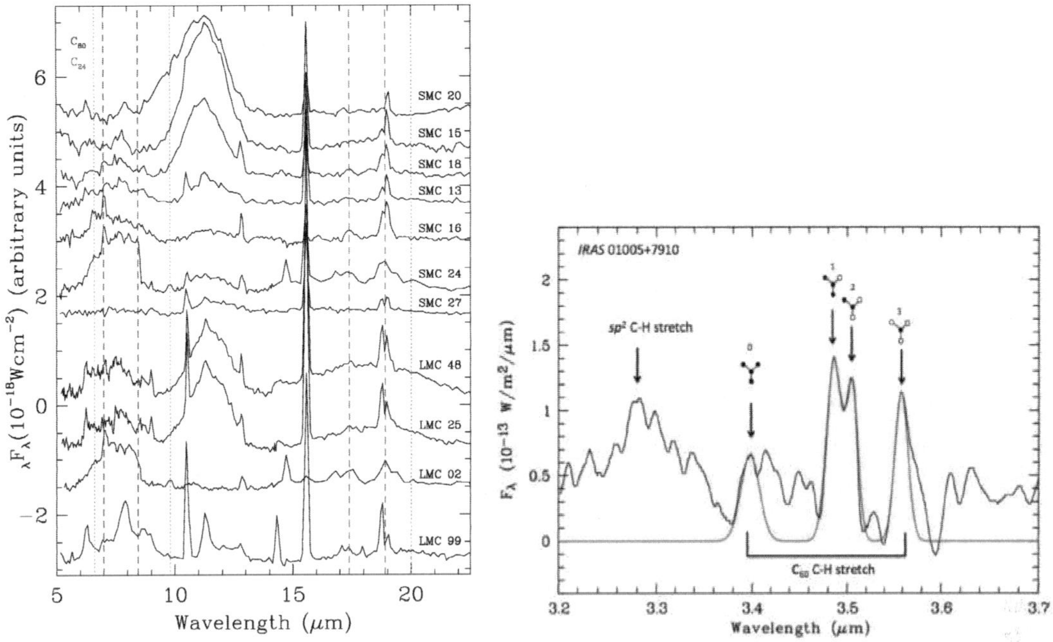

Figure 2. Left panel: *Spitzer* residual spectra of fullerene-containing PNe in the MCs. The band positions of C_{60} (dashed) and planar C_{24} (dotted) are marked (updated from [30]). Right panel: ISO spectrum of the PPN IRAS 01005+7910 in the 3-4 μm spectral region covering several fullerane features. Note the peaks for the 0, 1, 2, and 3 non-H-bonded neighboring carbon atoms that may be attributed to fulleranes (updated from [40]).

in H-rich circumstellar envelopes only [28,29]. Fullerenes are detected in low-mass C-rich PNe (both in the Milky Way and in the MCs) with normal H abundances [28,30] and in only those RCB stars with some H [29]; C_{60} is usually seen in conjunction with PAH-like features. Indeed, the C_{60} molecule has also been detected in other astronomical environments such as a PPN [31], reflection nebulae [32], and young stellar objects [33] and none of them is H-deficient.

Interestingly, unusual IR emission features at ~6.6, 9.8, and 20 μm, attributed to the planar C_{24} molecule (a small piece of a graphene sheet), have been also detected in MC and Galactic PNe [30,34] and in the ISM [35]. Figure 2 (left panel) shows the planar C_{24} and C_{60} emission features detected in some PNe. The detection of several complex dehydrogenated organic molecules like fullerenes and possible planar C_{24} (in conjunction with aromatic species like PAHs) suggests that other forms of carbon such as hydrogenated fullerenes (fulleranes like $C_{60}H_{36}$ and $C_{60}H_{18}$), fullerene-PAHs adducts, metallofullerenes or even buckyonions and carbon nanotubes, may be widespread in the Universe. For example: i) the recent studies of DIBs in C_{60}-rich PNe [36,37] suggests the possibility that some DIB carrier molecules may represent other fullerene-related molecules like buckyonions; ii) laboratory spectroscopy of adducts between fullerenes and PAHs (e.g.,C_{60}-anthracene/tetracene/pentacene) shows that such fullerene-PAHs species display the same mid-IR features of isolated C_{60} molecules, being non distinguishable through mid-IR spectroscopy only [38]; and iii) the recent non-detections of the most intense IR bands of several fulleranes in two fullerene PNe [39] together with the (tentative) fulleranes detection in the proto-PN IRAS 01005+7910 [40] (see also Figure 2, right panel) suggest that fulleranes may be formed in the short transition phase between AGB stars and PNe but they are quickly destroyed by the UV radiation field from the central star.

5. Formation routes of complex dehydrogenated organics

The formation process of fullerenic and graphenic nanostructures in space is still unclear to date. Several fullerene formation mechanisms have been proposed, the most notable ones being: i) the formation in H-poor environments [41]; ii) high-temperature formation in C-rich environments [42]; iii) photochemical processing of hydrogenated amorphous carbon grains (HACs; [28]); and iv) photochemical processing of large PAHs [43]. The first two mechanisms prevent the formation of hydrogenated species, something that seems to be difficult to reconcile with the astronomical observations. The alternative top-down approaches involve the photochemical processing of HACs and large PAHs as first proposed by [28] and [43], respectively. Both formation scenarios are based on top-down chemical models towards the most stable C_{60} and C_{70} fullerenes. These top-down fullerene formation scenarios (enumerated below) seem to be quite different to the fullerene production methods usually employed on Earth. The terrestrial vaporization or combustion synthesis routes to form fullerene-like species cannot work in space due to the low gas densities in the astrophysical environments; e.g., in the laboratory, the main formation channel for C_{60} is the build-up from atomic C, C_2, small C clusters, and rings [44].

5.1. Photochemical processing of HAC-like materials

Remarkably, fullerene PNe display broad HAC-like dust features at ~9–13 and 25–35 μm, suggesting that the most likely explanation for the simultaneous presence of fullerenes, (possibly) planar C_{24}, and PAH-like emission in the H-rich circumstellar envelopes of PNe is that these molecular species may be formed from the destruction (e.g., as a consequence of the UV radiation from the central star) of a carbonaceous compound with a mixture of aromatic and aliphatic structures - e.g., HACs - , which should a major constituent in their circumstellar envelopes [28,30,34]. It is to be noted here that there are other carbonaceous materials with a mixture of aromatic and aliphatic structures similar to HACs. For example, the mixed aromatic/aliphatic organic nanoparticles (MAONs) first introduced by Prof. Sun Kwok (e.g., [45,46]). The closest natural analogue of such structure is probably kerogen, which are random arrays of aromatic rings and aliphatic chains with functional groups made up of H, N, O, and S attached and that may contribute to the solid-state mid-IR features detected in proto-PNe (e.g., [47]). The coexistence of a large variety of molecular species such as HACs, PAH clusters, fullerenes, and small dehydrogenated carbon clusters (possibly planar C_{24} or graphene precursors) in PNe with fullerenes strongly supports the few laboratory experiments carried out by Scott and colleagues in the nineties, which showed that the decomposition of HACs is sequential with small dehydrogenated PAH molecules being released first, followed by fullerenes and large PAH clusters [48]. This fullerene formation scenario is also suggested by the strong - and unique - spectral variations (in a timescale of just a few years) seen in the IR spectrum of the fullerene-containing RCB star V854 Cen, which indicate that a significant fraction of the dust grains in the envelope have evolved from HACs to complex species such as PAHs and fullerene-like molecules [29]. The recent works by [49,50] also suggest that fullerene formation in PNe likely starts from HAC processing. Perhaps the most novel idea is introduced by [50] who propose that UV-photolysis produces structural changes in the HAC grains, forming 3D hollow structures or arophatic clusters (aromatic clusters linked by aliphatic bridging groups). These arophatic clusters end up with a cage-like, cup-like or tube-like aspect, being curled-up or folded-over graphene sheets. Further UV-induced dehydrogenation of these arophatic clusters may introduce pentagonal rings, permitting the curvature of the structure into large fullerenic cages that can shrink down to the most stable C_{60} and C_{70} fullerene configurations.

5.2. Photochemical processing of large PAHs

This fullerene formation route has been proposed to explain the formation of fullerenes in the ISM [43]. These authors propose an alternative top-down model where large PAHs (with ~70 C atoms) are converted into graphene, and subsequently fullerenes, under the action of UV photons from massive stars. A schematic view of this alternative top-down fullerene formation model is shown in Figure 3 of [43]. Upon UV irradiation, graphene formation takes place through PAH photolysis. The expected dehydrogenation, fragmentation, and isomerization can give rise to a rich organic chemistry, forming fullerenes, small cages, rings, and chains. Carbon loss followed by pentagonal defects formation that induces curvature of the graphene sheet may form fullerenes by migration of the pentagons. Remarkably, planar molecules such as C_{20} - C_{30} are expected to be formed by further graphene fragmentation. In the reflection nebula NGC 7023, emission at 6.6 μm (attributed to possible planar C_{24} molecules) is seen near the ionizing star [35], suggesting that the carrier is a photo-product of PAHs. In this context, the possible detection of the very stable planar C_{24} molecule in the ISM is also consistent with the idea of fullerene production from the photochemical processing of PAHs via graphene formation.

6. Complex dehydrogenated organics versus metallicity

Interestingly, the detection rate of complex dehydrogenated fullerene molecules in C-rich PNe decreases with increasing metallicity. In our own Galaxy, only ~5% of the PNe display the fullerene features, which otherwise are clearly detected in ~20% and ~44% of the PNe observed in the LMC and SMC, respectively [34]. This interesting finding suggests a higher presence of small dust grains (or a more limited dust processing) at the lower MCs metallicity. More recently, [51] demonstrate that all Galactic fullerene-containing PNe are low-mass objects of sub-solar metallicity, showing that low metallicity environments are more favourable to fullerene production and detection.

On the other hand, the still unidentified 21 μm feature is less frequent in the Galaxy than in the MCs [52] and the 21 μm carrier could be related with the formation of fullerenes. In particular, an anti-correlation between the 30 μm and other UIR features is seen among the 21 μm sources in the MCs [52] and such anti-correlation could result from the photochemical processing of HAC-like dust grains into fullerenes and PAHs. Under the HACs hypothesis for fullerene formation, the 21 μm feature should be related with the formation of fullerenes with the carrier being a fragile intermediate product from the decomposition of HAC or a similar material with mixed aromatic/aliphatic structures. However, [53], although based on a smaller sample of Galactic post-AGB stars, reported that the 30 μm feature and the other UIRs are not anti-correlated in Galactic 21 μm objects. Interestingly, the Galactic 21 μm emitters are spectroscopically different to the MC ones; i.e., the 21 μm proto-PNe in the MCs show more (typical) PAH-like aromatic features.

7. Concluding remarks

The IR dust characteristics (and evolution) observed from the AGB to the PN phases in different metallicity environments (Galactic disk, bulge, and MCs) agree well with the predictions of the AGB nucleosynthesis models (i.e., the expected effects of the third dredge-up and HBB processes). The coexistence of aromatic PAH-like species and complex dehydrogenated organic molecules such as fullerenes and planar C_{24} (a small piece of a graphene sheet) suggests a top-down model for the formation of fullerenes in space. Low metallicity environments like those of the MCs favour fullerene formation and detection as well as a more limited dust processing (or the general presence of small dust grains). Although more observational and laboratory efforts are needed, it seems clear that a diverse family of fullerene-related molecules (e.g., fulleranes, metallofullerenes, fullerene-PAH adducts, buckyonions, etc.) is very likely to be present in space.

Journal of Physics: Conference Series **728** (2016) 032009 doi:10.1088/1742-6596/728/3/032009

Acknowledgments

D.A.G.H. was funded by the Ramón y Cajal fellowship number RYC−2013−14182. D.A.G.H and A. M. acknowledge support provided by the MINECO grant AYA−2014−58082-P.

References

[1] García-Hernández D A, García-Lario P, Plez B, D'Antona F, Manchado A and Trigo-Rodríguez J M 2006 *Science* **314** 1751

[2] García-Hernández D A, García-Lario P, Plez B, Manchado A, D'Antona F, Lub J and Habing H 2007 *Astron. Astrophys.* **462** 711

[3] García-Hernández D A, Manchado A, Lambert D L, Plez B, García-Lario P, D'Antona F, Lugaro M, Karakas A I and van Raai M A 2009 *Astrophys. J.* **705** L31

[4] Herbst E and van DishoeckE F 2009 *Annu. Rev. Astron. Astrophys.* **47** 427

[5] Kwok S 2004 *Nature* **430** 985

[6] Waters L B F M *et al* 1996 *Astron. Astrophys.* **315** L361

[7] Perea-Calderón J V, García-Hernández D A, García-Lario P, Szczerba R and Bobrowsky M 2009 *Astron. Astrophys.* **495** L5

[8] Stanghellini L, García-Hernández D A, García-Lario P, Davies J E, Shaw R A, Villaver E, Manchado A and Perea-Calderón J V 2012 *Astrophys. J.* **753** 172

[9] García-Hernández D A and Górny S K 2014 *Astron. Astrophys.* **567** A12

[10] García-Lario P and Perea-Calderón J V 2003 *ESA Publication Series* **511** 97

[11] García-Hernández D A 2012 *Proc. Int. Astron. Union* (S283) **7** 148

[12] García-Hernández D A 2016 *in IAU GA XXIX, Cambridge Univ. Press* **(in press)** arXiv:1511.06165

[13] Stanghellini L, García-Lario P, García-Hernández D A, Perea-Calderón J V, Davies J E, Manchado A, Villaver E and Shaw R A 2007 *Astrophys. J.* **671** 1669

[14] Kwok S, Volk K and Bernath P 2001 *Astrophys. J.* **554** L87

[15] Molster F J *et al* 1999 *Nature* **401** 563

[16] Cernicharo J 2004 *Astrophys. J.* **608** L41

[17] Volk K, Kwok S and Hrivnak B J 1999 *Astrophys. J.* **516** L99

[18] Hony S, Waters L B F M and Tielens A G G M 2002 *Astron. Astrophys.* **390** 533

[19] Papoular R 2011 *Mon. Not. R. Astron. Soc.* **415** 494

[20] Grishko V I, Tereszchuk K, Duley W W and Bernath P 2001 *Astrophys. J.* **558** L129

[21] Kroto H W, Heath J R, O'Brien S C, Curl R F and Smalley R E 1985 *Nature* **318** 162

[22] Foing B H and Ehrenfreund P 1994 *Nature* **369** 296

[23] Luna R, Cox N L J, Satorre M A, García-Hernández D A, Suárez O and García-Lario P 2008 *Astron. Astrophys.* **480** 133

[24] Iglesias-Groth S 2004 *Astrophys. J.* **608** L37

[25] Kwok S, Volk K and Hrivnak B J 1999 *Astron. Astrophys.* **350** L35

[26] Lambert D L, Rao N K, Pandey G and Ivans I I 2001 *Astrophys. J.* **555** 925

[27] Cami J, Bernard-Salas J, Peeters E and Malek S E 2010 *Science* **329** 1180

[28] García-Hernández D A, Manchado A, García-Lario P, Stanghellini L, Villaver E, Shaw R A, Szczerba R and Perea-Calderón J V 2010 *Astrophys. J.* **724** L39

[29] García-Hernández D A, Rao N K and Lambert D L 2011 *Astrophys. J.* **729** 126

[30] García-Hernández D A, Iglesias-Groth S, Acosta-Pulido A, Manchado A, Larcía-Lario P, Stanghellini L, Villaver E, Shaw R A and Cataldo F 2011 *Astrophys. J.* **737** L30

[31] Zhang Y and Kwok S 2011 *Astrophys. J.* **730** 126

[32] Sellgren K 2010 *Astrophys. J.* **722** L54

[33] Roberts K R G, Smith K T and Sarre P J 2012 *Mon. Not. R. Astron. Soc.* **421** 3277

[34] García-Hernández D A, Villaver E, García-Lario P, Acosta-Pulido J A, Manchado A, Stanghellini L, Shaw R A and Cataldo F 2012 *Astrophys. J.* **760** 107

[35] Berné O, Mulas G and Joblin C 2013 *Astron. Astrophys.* **550** L4

[36] García-Hernández D A and Díaz-Luis J J 2013 *Astron. Astrophys.* **550** L6

[37] Díaz-Luis J J, García-Hernández, Rao N K, Manchado A and Cataldo F 2015 *Astron. Astrophys.* **573** A97

[38] García-Hernández D A, Cataldo F and Manchado A 2013 *Mon. Not. R. Astron. Soc.* **434** 415

[39] Díaz-Luis J J, García-Hernández D A, Manchado A and Cataldo F 2016 *Astron. Astrophys.* **589** 5

[40] Zhang Y and Kwok S 2013 *Earth, Planets and Space* **65** 1069

[41] Goeres A and Sedlmayr E 1992 *Astron. Astrophys.* **265** 216

[42] Jäger C, Huisken F, Mutschke H, Jansa I L and Henning Th 2009 *Astrophys. J.* **696** 706

[43] Berné O and Tielens A G G M 2012 *Proc. Nat. Academy Sci.* **109** 401

[44] Heath, J R 1992 *ACS Symposium Series* vol 481 1-23
[45] Kwok S and Zhang Y 2011 *Nature* **479** 80
[46] Zhang Y and Kwok S 2015 *Astrophys. J.* **798** 37
[47] Cataldo F, Keheyan Y and Heymann D 2002 *Int. J. Astrobiology* **1** 79
[48] Scott A, Duley W W and Pinho G P 1997 *Astrophys. J.* **489** L123
[49] Bernard-Salas J, Cami J, Peeters E, Jones A P, Micelotta E R and Groenewegen M A T 2012 *Astrophys. J.* **757** 41
[50] Micelotta E, Jones A P, Cami J, Peeters E, Bernard-Salas J and Fanchini G 2012 *Astrophys. J.* **761** 35
[51] Otsuka M, Kemper F, Cami J, Peeters E and Bernard-Salas J 2014 *Mon. Not. R. Astron. Soc.* **437** 2577
[52] Volk K *et al* 2011 *Astrophys. J.* **735** 127
[53] Mishra A, Li A and Jiang B W 2015 *Astrophys. J.* **802** 39

From left to right: Aníbal García-Hernández, Arthuro Manchado, SeyedAbdolreza Sadjadi, and Franco Cataldo at the Laboratory for Space Research, HKU.

Left: Nick Cox. Right: David Jones. Bottom: Valentin Bujarrabal.

Molecular line mapping of (young) planetary nebulae

Valentín Bujarrabal[1]
[1]Observatorio Astronómico Nacional (OAN, IGN), Ap. 112, 28803 Alcalá de Henares, Spain
E-mail: [1] v.bujarrabal@oan.es

Abstract. In this contribution, I will review recent results obtained from high-resolution observations of molecular emission of planetary nebulae in the millimeter and submillimeter waves, stressing the easy interpretation of the data and the great amount of quantitative results obtained from them. Radio interferometers have been shown to be very efficient in the observation of our objects and, particularly since the arrival of ALMA, the amount of results is becoming impressive. We will deal mainly with young planetary nebulae or protoplanetary nebulae, since, as we will see, molecular lines tend to be weak in evolved objects because of photodissociation. In relatively young nebulae, the molecular gas represents most of the nebular material and can be well observed in line emission in mm- and submm-waves. Those observations have yielded many quantitative and accurate results on the structure, dynamics, and physical conditions of this largely dominant nebular component. In more evolved sources, we can follow the evolution of the chemical composition, although the data become rare.

1. Introduction

This is a review talk intended, following suggestions of the organizers, to present the general properties of molecular line emission in planetary nebulae and to summarize recent results obtained from imaging of molecular lines in these objects.

Molecular lines in the mm- and submm-wave regimes are very useful tools to study planetary nebulae (PNe). In these objects, in which the central star is often hot, photodissociation by the stellar UV field becomes a key process. Molecules are very abundant in young PNe and protoplanetary nebulae (PPNe), and, as we will see, molecular gas is the dominant component, representing the bulk of the nebular material. However, in more evolved nebulae molecules are mostly destroyed because of the increasingly strong UV field and diffuse nebular shells (which allows photon penetration across the nebula), see e.g. [1]. At some intermediate point, when photodissociation is already efficient, radicals appear and can be relatively abundant, due to photon-induced chemistry [2], but with a trend to molecular destruction and weak molecular lines. Photodissociation from interstellar UV [3] is also important in PNe, limiting the outer extent of the molecule-rich shells, as already occurs in the AGB circumstellar envelopes. As a result, very extended outer halos are not detected in molecular line emission. However, due to presence of a very important increase of the mass-loss rate in the very late AGB phases (or very early post-AGB phases), the amount of material in the inner shells that are not affected by the interstellar UV field is much larger than that in the outer halos. The actual outer limit of the molecular shells is in fact given by a significant increase of density in those inner layers [4, 5].

Carbon monoxide is by far the most abundant molecule emitting at these frequencies and, in spite of its low dipole moment, that showing the strongest lines. CO is also known to be the molecule that better survives photodissociation, because its high abundance leads to a very important self-shielding [3]. As a result, most molecular observations of PNe and PPNe have been performed in CO lines, mostly in CO rotational lines, because of the low excitation requirements of these transitions. Other molecules have been rarely mapped with high resolution, because of there lower intensities (they are significantly less abundant) and small and irregular extent (they tend to occupy only regions with a peculiar chemistry). Most of the results I will show here come from observations of CO rotational lines.

1.1. Excitation of molecular lines: CO rotational transitions
CO rotational lines are characterized by their low excitation energy and Einstein coefficients. As for most observed molecules, the rotational levels of CO (in the ground vibrational and electronic states) are placed at a low energy over the ground. The excitation energy of the first excited level, J=1, is only of about 5 K; the energy of other levels is proportional to $J(J+1)$. The low-J transitions of CO (allowed only if ΔJ is equal to 1, i.e. J=1−0, J=2−1, ..., J=6−5, etc) therefore require very low excitation conditions, present even in the cool molecule-rich gas. These low level energies lead of course to low emitting frequencies, which can be efficiently observed with radio interferometers: for the abundant isotope ^{12}CO, λ(1–0) = 2.7 mm, ν(1–0) = 115 GHz, and ν(J–J-1) is proportional to J.

The Einstein A-coefficients of those lines, which give the spontaneous radiative deexcitation probability, are also particularly low, due to the anomalously low permanent dipole moment of CO. For instance, the value of the A-coefficient of J=2−1 is just 6.9 10^{-7} s^{-1}; for other transitions $J \rightarrow J$-1, A roughly varies with J^3. As a result the collisional transitions between the CO low-J levels easily dominate over radiative ones and the low-J lines are easily thermalized to the kinetic temperature; note that the probability in s^{-1} of the inelastic collisional transitions is typically $\sim 10^{-10}$ n, where n is the total density in cm^{-3}. Therefore, the excitation temperature of these transitions is often very similar to the kinetic temperature, $T_{\rm exc} \sim T_{\rm k}$, allowing the definition of a rotational temperature $T_{\rm rot}$ = $T_{\rm exc}$(low-J). Detailed excitation calculations for our sources confirm this result [6], the mm-wave transitions of CO, J=1−0 and J=2−1, are thermalized for $n \gtrsim 10^4$ cm^{-3}. Higher-J transitions, in the submm or FIR, require higher, but in any case moderate, excitation conditions [7]. In any case, because of the simple level structure and efficient collisional excitation of the carbon monoxide rotational ladder, detailed calculations of the level populations and line intensities are in our case simple and reliable.

The same conclusions apply for rare isotopic substitutions of CO, like ^{13}CO. Their observations, in combination with those of ^{12}CO, are particularly useful, because the ^{13}CO lines also present favorable excitation conditions but show a much lower optical depth, with often optically thin emission, allowing the analysis of the opacity effects.

Because of the low temperatures of molecule-rich gas in PNe and PPNe, only rotational lines in the ground vibrational state are observed for almost all molecules (only a few vibrationally excited lines are detected, including notably some lines of HC_3N and, of course, H_2 emission, which is not discussed in this contribution). For other molecules, the line excitation tends to be more difficult to describe, but calculations can be performed, using for instance the well known *LVG* approximation, which is very efficient and justified in our case (large macroscopic velocities are almost always present in PNe, but we will see some exceptions later). The line emission of rarer molecules depends more critically on the density and temperature, and observations often select condensations (in which, moreover, it is easier for those species to survive UV effects).

2. Molecular lines in protoplanetary nebulae
Very young PNe or PPNe represent a particularly important phase in the late stellar evolution, because in them the spectacular post-AGB dynamics is still active and the typical PN axisymmetric shapes incipiently appear. For the definition and identification of PPNe, the

pioneering work by Kwok and colleagues [8, 9] is often followed. As expected, the SEDs of PPNe shows already that the star is not loosing material at present, with two-component shapes due to the stellar and the nebular emissions, but does not show yet the high-excitation characteristics of PNe. Many of the PPNe studied in molecular lines (and by other means) have been identified from those properties.

2.1. M 1–92: an example of the power of molecular line mapping

M 1–92 is a protoplanetary nebula that can be very well studied using molecular lines. A lot of quantitative and accurate information on the nebula structure, velocity, and physical conditions has been derived. It was first mapped in CO lines in the 90's, using the Plateau de Bure interferometer (PdBI) [10]. The maps reveal a double bubble with a radial expansion velocity characterized by a clear linear gradient (a Hubble-like field). The velocities are high, ~ 70 km s^{-1} at the tips of the nebula. The total mass derived from CO emission is ~ 1 M$_\odot$, and about 0.5 M$_\odot$ are found to be in expansion at $\gtrsim 30$ km s^{-1}. This implies a very high linear momentum, much higher than what radiation pressure can provide, showing that other engine was necessary to power the nebular dynamics. Densities of about 10^5 cm^{-3} were found and the molecule-rich gas was shown to be cold, $T_k \lesssim 30$ K. Molecular gas represents by far the dominant nebular component and probably includes most initial stellar mass, now ejected to form the nebula. These properties are found to be characteristic of PPNe and young PNe and similar results are found in M 2–56, CRL 618, HD 101584, OH 231.8+4.2, the Boomerang Nebula, NGC 7027, etc.

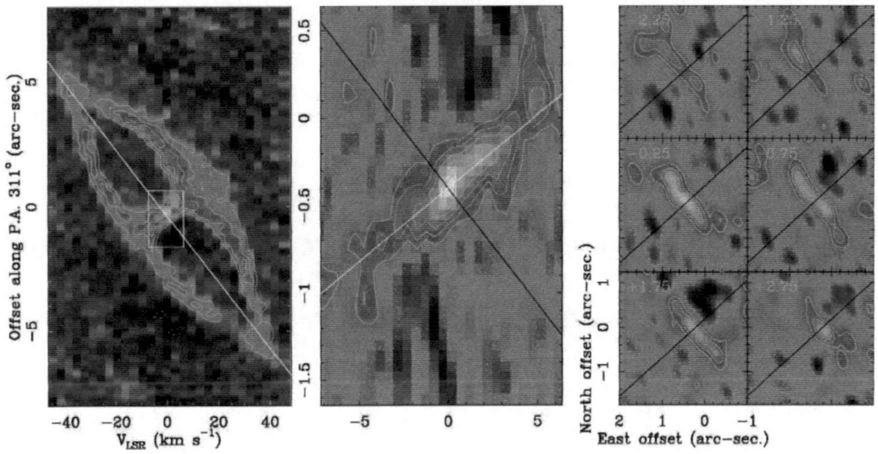

Figure 1. CO-line maps of M 1–92, adapted from [11]. Left: position-velocity diagram along the nebula axis. Center: the same for the central region of the nebula. Right: maps of the center of the nebula for selected velocities. The radial velocity gradient is remarkably constant and holds even very close to the star and for very low velocities.

Some years later, M 1–92 was mapped again with the PdBI [11], reaching a resolution higher than 0.″5. A very accurate description of the nebula was obtained, see Fig. 1. In particular, the linear velocity was found to apply down to distances $\lesssim$ 0.″3 (around 10^{16} cm) and expansion velocities of about 1 km s^{-1}. Surprisingly, in equatorial regions close to the central star, the expansion velocity of this source is significantly smaller than the expansion velocity in AGB circumstellar envelopes (in general ~ 10 km s^{-1}). It is not yet understood if this persistent velocity gradient, also found as we will see in other objects, is due to ejection of material starting not from the star but from a stable disk (probably in rotation, but not yet detected in this source) or to a peculiar ejection process that takes place in the very late AGB and could involve binary interaction [11].

2.2. Data of other PPNe and young PNe

Among other recent observations, we can mention those of IRAS 19475+3119 [12], a PPN identified from the properties of its SED. These data are part of COSAS, a big survey of observations of evolved nebulae performed with the PdBI. CO data confirmed its classification as PPNe. IRAS 19475+3119 shows a central empty region and a bipolar outflow sharing the main results found in M 1–92, including the linear velocity gradient and high mass and momentum.

A particularly well studied object is CRL 618. It was first observed with OVRO [13]. The general properties of the complex nebula where derived, CRL 618 was found to be composed of a dense halo extending more than 10^{17} cm and expanding at moderate velocity, a central equatorial region very dense and expanding a low velocity (which decreases towards the center down to less than 5 km s^{-1}), a double bubble, and a very fast outflow inside the bubbles with velocities that increase to more than 300 km s^{-1}. CRL 618 was later observed in CO J=3−2 with the SMA [14], with a resolution as high as 0.″3. The very fast jet was resolved and found to be composed of multiple outflows and to show a bow-like tip apparently due to shock acceleration. These authors [15] also mapped the source in lines of other molecules, namely HCN and HC_3N, which were found to be mainly confined to the dense central region of the slow condensation. For more details, see contribution by Lee et al. to this symposium

Data of similar sources have been recently obtained with ALMA. HD 101584 was found to show a high-mass double bubble with a clear linear velocity gradient [16]. Velocities are very high in this source, higher than 100 km s^{-1} in the prominent condensations placed at the tips of the bubbles. These results, together with maps of other molecules that show the presence of rich chemical processes, are discussed in the contribution by Olofsson et al. The Boomerang Nebula, a source that seems to show a very cold and fast outer halo, was also observed with ALMA [17]. The data mainly give information on the central part of the nebulae, again a double bubble expanding at high velocity. ALMA observations of other young PNe (NGC 6302, Mz–3, OH 231.8+4.2, ...) have also yielded impressive data, but they are still under analysis.

A source that presents a completely different distribution of molecule-rich gas is M 2–9. M 2–9 is thought to harbor in its center a binary star and shows a beautiful butterfly-like nebula in the optical that extend almost 1′. However, high-quality observations performed with the PdBI [18] have shown that molecules are only abundant in two equatorial rings. As the rest of the data we are presenting until now, the rings are expanding, in this case with quite low velocities, respectively at 4 and 8 km s^{-1}. They correspond apparently to two ejection events separated by about 500 yr. The mass of the rings is low, 0.05 $M_\odot$, about a 20% of this nebula, most of whose (relatively low) mass is placed in a PDR. A careful analysis of the M 2–9 maps has shown that the centroids of the rings both in position and velocity are slightly but noticeably different. The easiest explanation is to assume that the ejections of the rings took place in two different phases in the orbit of the post-AGB star, such that the systemic velocities of the rings are different, which also yields a detectable difference in the position of the center of the rings after several hundred years; the process is similar to that thought to yield the spiral-like structures found in many AGB and post-AGB sources (see contributions by Maercker, Quintana-Lacaci, and colleagues). Assuming this scenario, it is possible to derive the orbital parameters, it was in particular found that the mass of the secondary is very low, $\lesssim 0.2$ $M_\odot$.

I will finally mention the SMA maps of NGC 7027 [19], a nebula surrounding a very hot blue dwarf that is often catalogued as a young PN. The maps show a double elongated shell that surrounds the optical image of the nebula. Those maps were analyzed in combination with single-dish (i.e. low-resolution) observations of low- and high-J CO lines [20], including FIR lines up to J=16−15 observed with Herschel/HIFI. For the analysis, the code SHAPEMOL was used, which allows predicting brightness distributions and angle-integrated profiles of molecular lines from PNe and PPNe. The use of low- and high-excitation lines allowed the identification and study of warm layers, which in this source are particularly important. The molecular shell was divided for modelling purposes in 4 adjacent subshells (I, M1, M2, and O), in the center of which a PDR, an HII region, and even X-ray emitting gas are placed (Figs. 2, 3). In addition, two fast axial clumps are found to remain molecule-rich within the hot gas. The properties of the inner molecular subshell (labeled I) are remarkable: it is found to be faster and denser than outer regions. It is probably a thin layer compressed by the expansion of the HII region, an example of the *interacting wing paradigm*, expected to affect the evolution of many PNe [21].

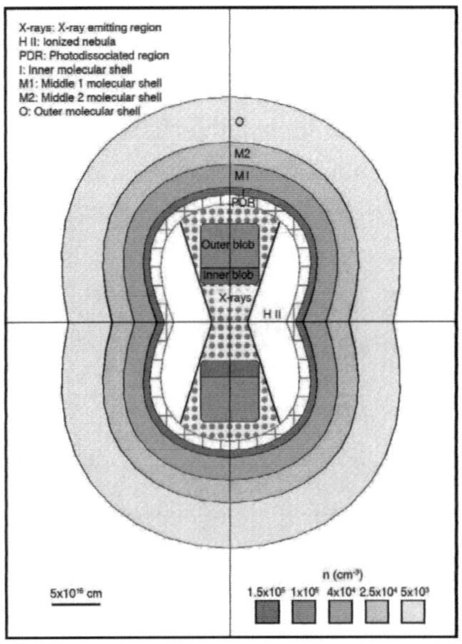

Figure 2. Sketch of the model deduced for NGC 7027 from maps and single-dish observations of molecular lines, adapted from [20]. The description of the nebula structure is particularly detailed for this source.

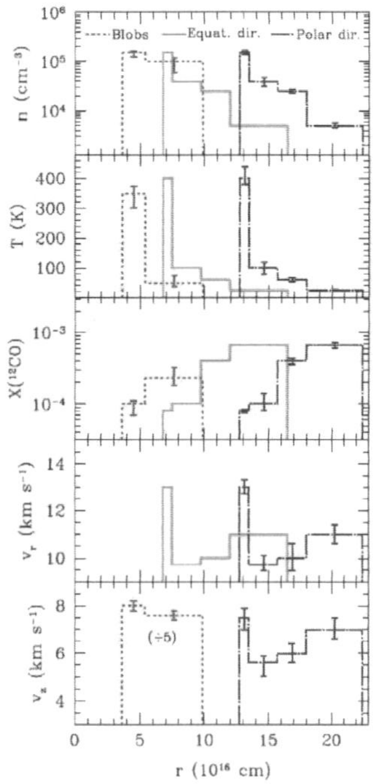

Figure 3. Details of the physical conditions deduced for the molecular shells of NGC7027, adapted from [20].

3. More evolved nebulae: rings, bullets and globules

As mentioned, the effects of photodissociation are very relevant to understand the properties of molecule-rich components in evolved planetary nebulae. Often, molecules only survive the stellar UV effects in dense regions in which shielding is particularly important.

In many PNe, the densest regions are equatorial rings or tori and molecules are often abundant only in such regions. This is the case of NGC 6720, the Ring Nebula [22]. The line emitting region is a ring similar to the optical image but slightly more extended. The probed mass is relatively low, 0.1 $M_\odot$, confirming the effects of photodissociation, but still larger than the mass detected by other means, particularly from ionized gas emission. The characteristic temperature of the molecule-rich gas is low, $\sim$ 30 K, and the (expansion) velocity is moderately high, $\sim$ 20 km s^{-1}, somewhat higher than in most AGB stars. Similar properties have been found in other PNe, in which molecular emission only comes from equatorial regions. This is the case of, for instance, M 1–16, which shows a ring like emitting region, in spite of the very extended axial extent [23], and of NGC 2346, KjPn 8, the Helix Nebula, etc.

A completely different case is that of BD +30 3639. This source show molecular line emission coming from two clumps flowing very fast along the nebula axis [24], at $\sim$ 50 km s^{-1}. In this source, the high density of the axial condensations, probably similar to the *fliers* (fast low-ionization emission regions) detected in other wavelengths, preventing photodissociation, while a ring-like equatorial feature is found to be ionized.

Another interesting case is the Helix, whose large angular size ($\sim 15'$) allowed high-quality maps years ago [25]. Complex structure and dynamics are found, the nebula apparently shows several superposed components with quite different velocities. As for the Ring Nebula, CO is mostly placed in a ring, similarly but slightly more extended than the well-known optical image. The high expansion velocities found are remarkable. Although the ring is placed almost in the plane of the sky, velocities projected in the line of sight as high as 25 $\mathrm{km\,s^{-1}}$ were measured. It was also possible to map one of the cometary globules detected in the optical [26]. The molecule-rich gas is placed *behind* the optically bright head, which is able to shield molecular gas from the stellar UV. Again, it was possible to derive physical conditions of the molecule-rich clump, which was found to be massive, $\gtrsim 10^{-5}$ $M_\odot$ (a lot for such small features), and cold, $\sim$ 25 K.

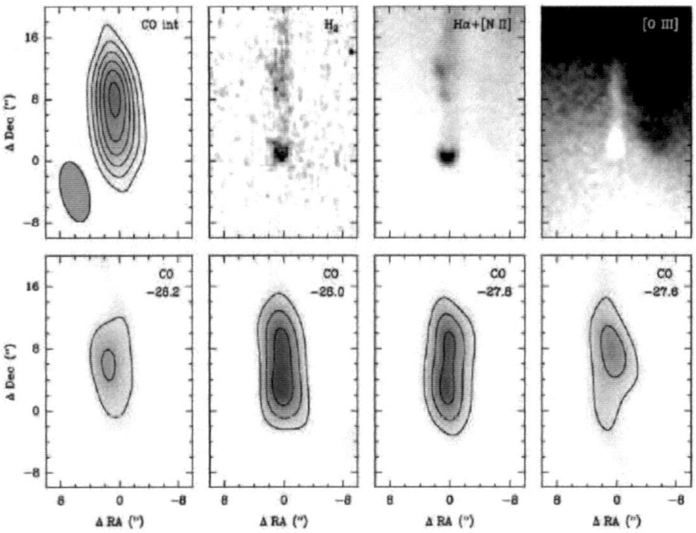

Figure 4. Observations of one of the globules detected in the Helix Nebula [26]. The integrated emission and several velocities are shown, together with images at other wavelengths. The head of the globule seem to be shielding the molecule-rich gas from the stellar UV (which comes from the south in this case).

4. Keplerian disks around post-AGB stars

To finish, I would like to present studies of a kind of objects in which I have been working in the last years: Keplerian disks orbiting post-AGB stars. Practically all the gas we detect in post-AGB nebulae, PPNe or PNe, is found to be in expansion. In particular, all the observations I have mentioned before reveal **only** the presence of expanding gas. However, rotating disks are particularly important to understand the reaccretion of circumstellar material by the post-AGB star (or a companion) and the launching of the very fast post-AGB jets. These post-AGB jets are thought to generate shocks when impinging the previously ejected gas, which contains most of the nebular material, accelerating it and so explaining the high axial velocities and bipolar shapes found in most PPNe and PNe, see e.g. [27] and the contribution by B. Balick to this symposium. Such rotating disks have been very elusive for our telescopes, particularly due to the need of very accurate spectroscopic mapping. Until two years ago, equatorial gas in rotation had been well detected only in one post-AGB object, the Red Rectangle, by means of PdBI maps of CO lines [28]. The Red Rectangle belongs to a relatively wide class of binary post-AGB stars which were suspected to be surrounded by rotating disks, see [29], [30], etc. These sources show some characteristics that suggests the presence of disks. In particular, their SEDs show, together with the usual two-maxima structure, a remarkable NIR excess that can be explained by the presence of hot dust kept in some way close to the star. This stable component should in principle be in rotation. A new argument supporting the presence of rotating disks in these sources was recently published [31]. Single-dish observations of CO emission systematically show peculiar line profiles, narrow and with a prominent central peak. Such profiles, found also in the Red Rectangle, are expected in Keplerian disks, both from previous observational data and theoretical predictions [31]. These disks detected in molecular line emission must be extended, around 10^{16} cm, much more than the tiny disks invoked to explain the NIR excess, which would just occupy $\lesssim 10^{15}$ cm. The CO profiles in these sources also show relatively intense wings, which were suggested to be due to the presence of low-velocity outflows (together with the disks).

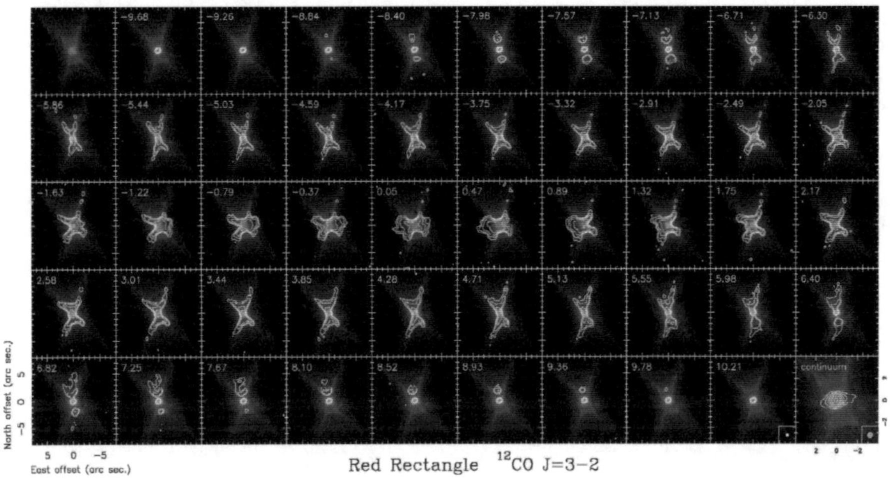

Figure 5. ALMA observations of the Red Rectangle in ^{12}CO J=3−2 emission, superposed on an HST image. Note the emissions of the equatorial rotating disk, in the central velocities, and of the outflowing gas, the X-shaped feature at more extreme velocities. Adapted from [32].

Recent ALMA observations of the Red Rectangle have spectacularly confirmed the presence of the rotating disk as well as of a faint outflow, probably composed of gas escaping the disk [32]. These data included high-quality maps of the J=3−2 and J=6−5 lines of ^{12}CO and rare isotopes; a resolution as high a $0''.25$ and a S/N ratio of several hundred were obtained, allowing the identification of the different components. The maps of different lines, with different excitation requirements and opacities, also help in the analysis, allowing a good study of the physical conditions, although sophisticated numerical treatments are necessary, since LTE is not satisfied by high-J lines and the LVG approximation does not apply in the presence of rotation.

Two other NIR-excess post-AGB stars, 89 Her and IRAS 19125+0343, have been mapped with high resolution, but in them only outflowing gas was found. The disks may be hidden in a central unresolved condensation. Only very recently, a new rotating disk has been well detected around one of these NIR-excess post-AGB stars, AC Her [33]. The Keplerian disk is very well identified from the position-velocity diagram, but no trace of expanding gas is found. New ALMA data of IW Car, which remain to be well analyzed and are not yet published, have found a third Keplerian disk. In this case, outflowing gas is also detected. With all these (more or less new) data, we can conclude that extended Keplerian disks are systematically present in this class of post-AGB objects with NIR excess. Their mass is relatively small, $\lesssim 10^{-2}$ $M_\odot$, but, in most cases, it represents most of the detected nebular material. The temperature of the disk gas is high, compared with molecule-rich gas in most PNe, reaching $\sim$ 100 K in central regions.

Molecule-rich outflows with moderate velocity, $\sim$ 5 km s^{-1}, also seem to be systematically present in these NIR-excess post-AGB stars. Their mass is also small, in general smaller than that of the disk, and their temperature is also high, in general over 100 K. Comparing the mass and velocities of the outflow with that of the disk, we can deduce the lifetime of the rotating structures, provided that the mass ejection rate is kept roughly constant. The estimates are uncertain, but we get from the different observations quite coherent lifetimes of $\sim$ 10000 yr. This time is short but not much shorter than the total life of detectable PNe, around 20000 yr. Therefore, it seems that the disk-phase is not a minor stage in the post-AGB evolution of these binary stars and that disks could survive around a significant fraction of PN central stars.

References

[1] Huggins P J, Bachiller R, Cox P and Forveille T 1996 *Astron. Astrophys.* **315** 284–302
[2] Bachiller R, Forveille T, Huggins P J and Cox P 1997 *Astron. Astrophys.* **324** 1123–1134
[3] Mamon G A, Glassgold A E and Huggins P J 1988 *Astrophys. J.* **328** 797–808
[4] Bujarrabal V, Gomez-Gonzalez J, Bachiller R and Martin-Pintado J 1988 *Astron. Astrophys.* **204** 242–252
[5] Bujarrabal V, Castro-Carrizo A, Alcolea J and Sánchez Contreras C 2001 *Astron. Astrophys.* **377** 868–897
[6] Bujarrabal V, Alcolea J, Neri R and Grewing M 1997 *Astron. Astrophys.* **320** 540–552
[7] Bujarrabal V *et al* 2012 *Astron. Astrophys.* **537** A8
[8] Hrivnak B J, Kwok S and Volk K M 1989 *Astrophys. J.* **346** 265–276
[9] Kwok S 1993 *Annu. Rev. Astron. Astrophys.* **31** 63–92
[10] Bujarrabal V, Alcolea J and Neri R 1998 *Astrophys. J.* **504** 915–920
[11] Alcolea J, Neri R and Bujarrabal V 2007 *Astron. Astrophys.* **468** L41–L44
[12] Castro-Carrizo A *et al* 2010 *Astron. Astrophys.* **523** A59
[13] Sánchez Contreras C, Bujarrabal V, Castro-Carrizo A, Alcolea J and Sargent A 2004 *Astrophys. J.* **617** 1142–1156
[14] Lee C F, Yang C H, Sahai R and Sánchez Contreras C 2013 *Astrophys. J.* **770** 153
[15] Lee C F, Sahai R, Sánchez Contreras C, Huang P S and Hao Tay J J 2013 *Astrophys. J.* **777** 37
[16] Olofsson H, Vlemmings W H T, Maercker M, Humphreys E M L, Lindqvist M, Nyman L and Ramstedt S 2015 *Astron. Astrophys.* **576** L15
[17] Sahai R, Vlemmings W H T, Huggins P J, Nyman L Å and Gonidakis I 2013 *Astrophys. J.* **777** 92
[18] Castro-Carrizo A, Neri R, Bujarrabal V, Chesneau O, Cox P and Bachiller R 2012 *Astron. Astrophys.* **545** A1
[19] Nakashima J, Kwok S, Zhang Y and Koning N 2010 *Astron. J.* **140** 490–499
[20] Santander-García M, Bujarrabal V and Alcolea J 2012 *Astron. Astrophys.* **545** A114
[21] Kwok S, Purton C R and Fitzgerald P M 1978 *Astrophys. J. Lett.* **219** L125–L127
[22] Bachiller R, Bujarrabal V, Martin-Pintado J and Gomez-Gonzalez J 1989 *Astron. Astrophys.* **218** 252–256
[23] Huggins P J, Forveille T, Bachiller R and Cox P 2000 *Astrophys. J.* **544** 889–894
[24] Bachiller R, Forveille T, Huggins P J, Cox P and Maillard J P 2000 *Astron. Astrophys.* **353** L5–L8
[25] Young K, Cox P, Huggins P J, Forveille T and Bachiller R 1999 *Astrophys. J.* **522** 387–396
[26] Huggins P J, Forveille T, Bachiller R, Cox P, Ageorges N and Walsh J R 2002 *Astrophys. J. Lett.* **573** L55–L58
[27] Soker N 2001 *Astrophys. J.* **558** 157–164
[28] Bujarrabal V, Castro-Carrizo A, Alcolea J and Neri R 2005 *Astron. Astrophys.* **441** 1031–1038
[29] van Winckel H 2003 *Annu. Rev. Astron. Astrophys.* **41** 391–427
[30] Hillen M, de Vries B L, Menu J, Van Winckel H, Min M and Mulders G D 2015 *Astron. Astrophys.* **578** A40
[31] Bujarrabal V, Alcolea J, Van Winckel H, Santander-García M and Castro-Carrizo A 2013 *Astron. Astrophys.* **557** A104
[32] Bujarrabal V, Castro-Carrizo A, Alcolea J, Van Winckel H, Sánchez Contreras C, Santander-García M, Neri R and Lucas R 2013 *Astron. Astrophys.* **557** L11
[33] Bujarrabal V, Castro-Carrizo A, Alcolea J and Van Winckel H 2015 *Astron. Astrophys.* **575** L7

PAH and H_2 emission in the Ring Nebula

N L J Cox[1,2], P Pilleri[1,2], O Berné[1,2], J Cernicharo[3], and C Joblin[1,2]

[1] Université de Toulouse, UPS-OMP, Institut de Recherche en Astrophysique et Planétologie, Toulouse, France
[2] CNRS, IRAP, 9 Av. Colonel Roche, BP 44346, 31028, Toulouse Cedex 4, France
[3] Group of Molecular Astrophysics, ICMM, CSIC, C/Sor Juana Inés de La Cruz N3, E-28049, Madrid, Spain

E-mail: nick.cox@irap.omp.eu; christine.joblin@irap.omp.eu

Abstract.
This paper presents the *Spitzer IRS* (Infra Red Spectrograph) detection of mid-infrared polycyclic aromatic hydrocarbon (PAH) emission features and H_2 associated with dense knots in the ring of the "oxygen-rich" (C/O∼0.6–0.8) planetary nebula (PN) NGC 6720 (Ring Nebula). **We explored a further three oxygen-rich extended PNe with similar dataset available. These turned out to be non-detection of PAHs, although two of these do contain H_2 emission knots.** The presence of PAHs is discussed in the context of a bottom-up formation mechanism, in which first small hydrocarbons, and later PAHs, are formed in warm dense knots inside the ionised regions of PNe.

1. Introduction

Polycyclic aromatic hydrocarbons (PAHs) are expected to be formed and processed in the circumstellar envelopes of evolved stars such as asymptotic giant branch (AGB) stars and (proto)-planetary nebulae (PNe). The processes related to PAH formation in oxygen-rich or carbon-rich environments are not known precisely, and both top-down or bottom-up formation scenarios have been proposed. Generally, silicate dust formation is related to oxygen-rich winds, while carbonaceous dust formation occurs primarily in carbon-rich winds. However, in several cases both oxygen-rich and carbon-rich dust features are seen in infrared spectra of PNe [1, 2]. For **currently carbon-rich PNe the presence of dual-dust chemistry within the system can be explained by turning from oxygen-rich to carbon-rich along the evolution. Silicate dust was formed earlier when the star was oxygen-rich, and eventually the star became carbon-rich, enabling to form carbon-rich dust [3].** For PAHs observed in oxygen-rich Galactic Bulge PNe [1, 4, 5] this scenario is less satisfactory, and it is has been proposed that PAHs in these objects are formed in-situ in an UV-irradiated dense torus [3, 6]. In this scenario, CO is dissociated by UV photons, producing free carbon which through chemical production of hydrocarbons leads to the formation of PAHs. We explore this scenario by revisiting mid-infrared spectral imagery of several oxygen-rich, extended PNe.

2. Spitzer IRS maps

We reinvestigate 4 oxygen-rich (C/O<1) PNe (Table 1) for which *IRS-SH Map* observations are available from the *Spitzer* Science Archive. The *IRS-SH* footprints are shown on Legacy

Journal of Physics: Conference Series **728** (2016) 032011 doi:10.1088/1742-6596/728/3/032011

Table 1. PNe observed with *Spitzer* in *IRS-SH Map* mode.

Target Name	Size (")	C/O	PAHs?	H_2?
NGC 2346	>55 x 55	0.35	no	yes
NGC 2392	44 x 44	0.41	no	yes
NGC 6720	68 x 98	0.62, 0.82	yes	yes
NGC 7009	24 x 32	0.32	no	no

Angular sizes are determined from HST images.

C/O references:	[9, 10, 11]
PAH references:	[9, 11, 8] (this work)
H_2 references:	[12, 13, 14, 15]

Archive HST images in Figure 1. The spectral cubes cover the range from 10–19.6 μm, with a spectral resolving power of ~600 [7]. For NGC 6720 a low-resolution IRS-SL spectrum was also retrieved from the archive. **PAH emission was detected in** NGC 6720 [8], but **not in any of the other three** PNe (c.f. Figure 2). In the next sections we focus on the 11.3 μm PAH and H_2 emission detected in the main dust ring of NGC 6720.

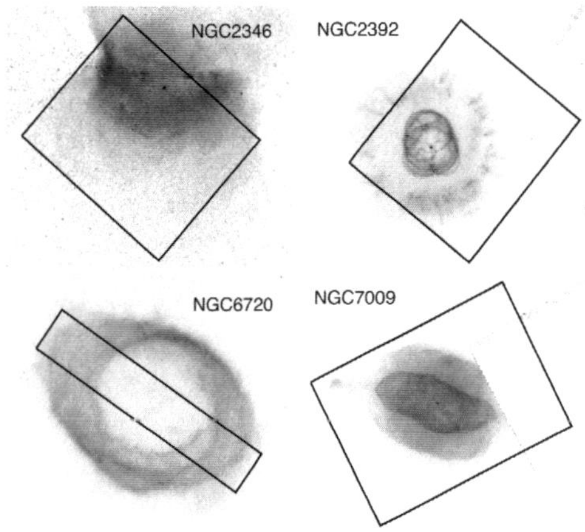

Figure 1. *Spitzer IRS-SH* footprints are shown on top of HST/WFPC2 Hα images of the 4 selected PNe (c.f. Table 1).

3. PAH emission in NGC 6720

The 10–13.5 μm mid-infrared spectra for two PNe, NGC 2346 and NGC 6720 are shown in Figure 2. The 11.3 μm PAH emission feature is clearly detected at the position of the main ring of NGC 6720 [8]. Note the presence of numerous strong forbidden emission lines, highlighting the necessity of sufficiently high spectral resolution observations. **Although we searched for** PAH bands at 6.2 and 7.7 μm in the Spitzer IRS-SL spectra, **there was no evidence for these bands.** They are however of insufficient quality to set constraining upper limits. A similarly peculiar PAH spectrum, with a relatively strong 11.3 μm band, but weak 6.2 and 7.7 μm bands, is observed for the Horsehead H II region [16]. The integrated line intensity of the 11.3 μm PAH band (normalised to the total integrated infrared emission) is consistent with that expected from the C/O ratio of 0.6 [9]. The 11.3 μm PAH emission band profile in NGC 6720 is compared with other environments, such as NGC7027, Horsehead H II region, and M2-9 (Figure 3). The 11.3 μm PAH band in NGC 6720 reveals a pronounced red wing, similar to that of the Horsehead.

4. H_2 knots in NGC 6270

Many PNe show evidence for 2.12 μm H_2 emission from **photo-dissociation regions** (PDRs) associated with dense knots in their ionised regions [12]. Examples include NGC 2346 [15],

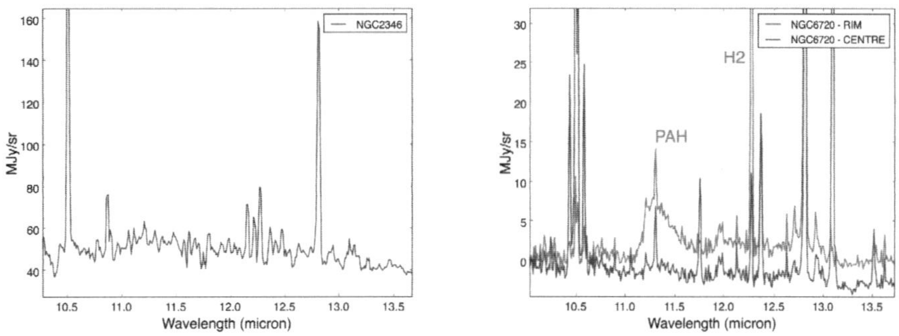

Figure 2. *Spitzer IRS-SH* spectra for NGC 2346 (centre) and NGC 6720 (rim and centre).

NGC 7293 [17], and NGC 6720 [13]. The *Spitzer IRS SL* and *SH* spectra (overlapping in the north-east part of the ring) provide line intensities for S(1) to S(4) H_2 rotationally excited lines. From the H_2 rotational diagram we obtain **a molecular hydrogen density**, $N(H_2) = 10^{18}$ cm^{-2} and **excitation temperature** $T_{H_2} = 620$ K (Figure 4).

From the increase of the H_2 intensity with radial distance (occurring at the PDR interface due to attenuation of UV photons) we derive a mean density $n_H > 10^4$ cm^{-3} [8] (which corresponds also roughly to the pressure equilibrium at the ionisation front). From both the total integrated infrared dust emission and geometrical dilution we derive an effective radiation field at the location of the ring corresponding to $G_0 \sim 200$–400 [8].

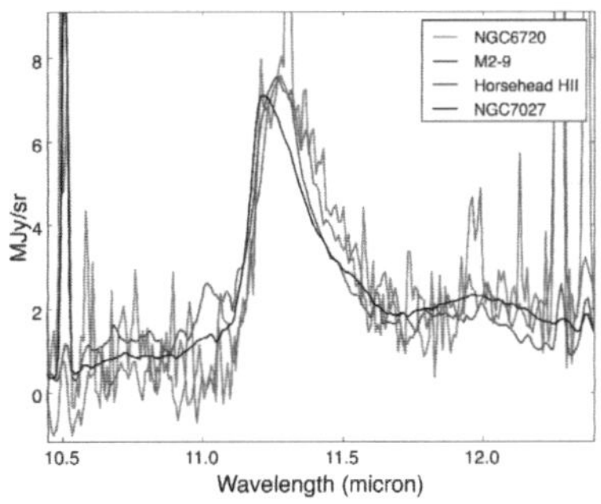

Figure 3. 11.3 μm PAH features in *Spitzer IRS-SH* spectra of NGC 6270, NGC 7027, M2-9, and Horsehead H II region.

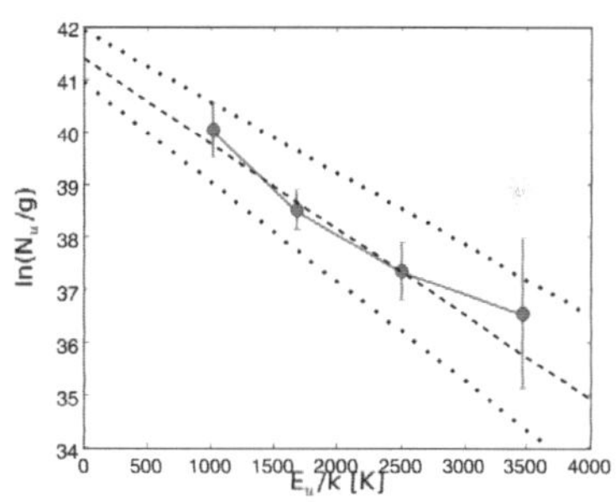

Figure 4. H_2 rotational diagram for a north-east region of the ring. $N(H_2)=10^{18}$ cm^{-2} and $T_{H_2} = 620$ K.

5. Bottom-up formation of PAHs: photochemistry of small hydrocarbons

In the bottom-up PAH formation scenario, CO is dissociated which produces abundant free atomic carbon which can then – under the right conditions – form small hydrocarbons [6, 3]. In the case of the NGC 6720, N(C) > 10 × N(CO) [18, 19]. C, CO and H_2 have a clumpy distribution with the presence of relatively dense knots ($n_H > 10^4$ cm^{-3}) exposed to a moderate radiation field of $G_0 \sim 200$.

Carbon chemistry reactions with moderate activation energies (such as $C_2 + H_2 \rightarrow C_2H +$ H, $C_2H + H_2 \rightarrow C_2H_2 +$ H, and $C^+ + H_2 \rightarrow CH^+ +$ H) proceed slowly in the cold ISM, but can become rapid enough to control C-bearing species at relatively high temperature and density prevailing in PDRs of (P)PN, such as we find for the knots in NGC 6720. How much these reactions compete with the oxygen chemistry needs to be evaluated.

6. Conclusions/Summary

A peculiar PAH emission spectrum is detected in the main ring of the planetary nebula NGC 6720. The PAH emission appears co-spatial with the clumpy H_2 distribution. This detection is unexpected from previous studies as well as the lack of PAH emission in the other oxygen-rich extended PNe investigated here. We propose that an efficient in-situ production of small hydrocarbons in the dense knots of NGC 6720 leads to the bottom-up production of PAHs. The specific conditions and mechanisms of forming PAHs from small hydrocarbons in dense knots, and why this leads to the peculiar observed PAH spectrum, needs to be investigated. In future work we will focus on mapping individual knots in millimetre molecular lines, tracing, for example, the photo-chemistry, in NGC 6720. With JWST we can map the distribution of H_2 and PAHs/eVSG in extended PNe (NIRCam) and study the PAH spectrum of individual knots and their micro-PDRs (MIRI).

Acknowledgments

The research leading to these results has received funding from the European Research Council under the European Union's Seventh Framework Programme (FP/2007-2013) ERC-2013-SyG, Grant Agreement n. 610256 NANOCOSMOS. PP acknowledges financial support from the Centre National Etudes Spatiales (CNES).

References

[1] Perea-Calderón J V, García-Hernández D A, García-Lario P, Szczerba R and Bobrowsky M 2009 *Astron. Astrophys.* **495** L5–L8
[2] Cerrigone L, Hora J L, Umana G and Trigilio C 2009 *Astrophys. J.* **703** 585–600
[3] Matsuura M, Zijlstra A A, Molster F J, Hony S, Waters L B F M, Kemper F, Bowey J E, Chihara H, Koike C and Keller L P 2004 *Astrophys. J.* **604** 791–799
[4] Guzman-Ramirez L, Zijlstra A A, Níchuimín R, Gesicki K, Lagadec E, Millar T J and Woods P M 2011 *Mon. Not. R. Astron. Soc.* **414** 1667–1678
[5] Guzman-Ramirez L, Lagadec E, Jones D, Zijlstra A A and Gesicki K 2014 *Mon. Not. R. Astron. Soc.* **441** 364–377
[6] Cernicharo J 2004 *Astrophys. J. Lett.* **608** L41–L44
[7] Houck J R *et al* 2004 *Astrophys. J. Suppl. Series* **154** 18–24
[8] Cox N L J, Pilleri P, Berné O, Cernicharo J and Joblin C 2016 *Mon. Not. R. Astron. Soc.* **456** L89–L93
[9] Cohen M and Barlow M J 2005 *Mon. Not. R. Astron. Soc.* **362** 1199–1207
[10] Pottasch S R and Bernard-Salas J 2010 *Astron. Astrophys.* **517** A95
[11] Delgado-Inglada G and Rodríguez M 2014 *Astrophys. J.* **784** 173
[12] Latter W B, Kelly D M, Hora J L and Deutsch L K 1995 *Astrophys. J. Suppl. Series* **100** 159
[13] Speck A K, Meixner M, Jacoby G H and Knezek P M 2003 *Publ. Astron. Soc. Pac.* **115** 170–177
[14] Wang M Y, Muthumariappan C and Kwok S 2006 *Proc. Int. Astron. Union* (S234) **2** 537–538
[15] Manchado A, Stanghellini L, Villaver E, García-Segura G, Shaw R A and García-Hernández D A 2015 *Astrophys. J.* **808** 115
[16] Compiègne M, Abergel A, Verstraete L, Reach W T, Habart E, Smith J D, Boulanger F and Joblin C 2007 *Astron. Astrophys.* **471** 205–212
[17] Matsuura M, Speck A K, McHunu B M, Tanaka I, Wright N J, Smith M D, Zijlstra A A, Viti S and Wesson R 2009 *Astrophys. J.* **700** 1067–1077
[18] Bachiller R, Huggins P J, Cox P and Forveille T 1994 *Astron. Astrophys.* **281** L93–L96
[19] Sahai R, Morris M R, Werner M W, Güsten R, Wiesemeyer H and Sandell G 2012 *Astron. Astrophys.* **542** L20

Journal of Physics: Conference Series **728** (2016) 032012 doi:10.1088/1742-6596/728/3/032012

The dusty side of planetary nebulae: a HerPlaNS view

Toshiya Ueta[1], **Djazia Ladjal**[2] **and the HerPlaNS team**

[1] Dept. of Physics & Astronomy, University of Denver, Denver, CO 80208, USA

[2] Gemini Observatory, La Serena, Chile

E-mail: [1] `tueta@du.edu`, [2] `dladjal@gemini.edu`

Abstract. HerPlaNS (Herschel Planetary Nebula Survey) is a far-IR imaging/spectroscopic survey of planetary nebulae (PNe) using the Herschel Space Observatory. In this presentation, we review our investigation into the physical properties of the cold dust component of the target PNe. We find that the far-IR surface brightness emission from PNe is generally dominated by thermal dust emission, which exhibits particular characteristics in terms of the dust emissivity and dust temperature compared with dust grains found elsewhere. The PN dust displays little variation in the emissivity while a large spread in the temperature, suggesting the presence of rather homogeneous dust chemistry and size distribution in the circumstellar environs.

1. The HerPlaNS Survey

The planetary nebula (PN) phase marks the last throes of stellar evolution for stars of about 0.8–8 $M_\odot$ initial mass [1]. These stars develop the circumstellar shell of gas and dust via mass loss in the asymptotic giant branch (AGB) and post-AGB phases. During the PN phase, the circumstellar shell undergoes a dramatic transformation (i.e., ionization, photo-dissociation, and dynamical shaping) caused by the fast wind and the intense radiation from the central star and by the less powerful, but often significant interstellar radiation field from the interstellar space. As a consequence, a variety of physical conditions are showcased within PNe, from fully ionized hot plasma to dusty cold atomic/molecular clouds,which exist in a (at least to first order) stratified manner around the central star. Therefore, PNe provide excellent astrophysical laboratories to test theories of stellar evolution and gas-dust dynamical processes in interacting stellar winds that can also interact with the surrounding interstellar medium (ISM).

We conducted a comprehensive far-IR imaging and spectroscopic mini-survey of PNe with the Herschel Space Observatory (*Herschel*; [2]), dubbed the Herschel Planetary Nebula Survey (HerPlaNS; [3]). Nearly 200 hours of observing time was used to take advantage of *Herschel*'s mapping capabilities – broadband and spectral imaging as well as spatio-spectroscopy – at high spatial resolutions made possible by its 3.5 m primary mirror. Our chief objective was to examine both the dust and gas components of the 11 target PNe[1] in the far-IR and investigate the energetics of these gas-dust systems in a spatially-resolved manner. In this contribution, we briefly summarize the HerPlaNS survey by focusing on the dusty side of the target PNe.

[1] The HerPlaNS sample consists of NGC 40, 2392, 3242, 6445, 6543, 6720, 6781, 6826, 7009, and 7026, plus Mz3.

2. Broadband Mapping

Far-IR broadband images of the target PNe (covering at most $7' \times 7'$) were taken at 70, 100, and 160 μm with PACS [4] and at 250, 350, and 500 μm with SPIRE [5]. Dust emission distribution in the far-IR in these PNe is generally consistent with that of ionized gas line emission in the optical. However, dust emission is generally more extended than ionized gas emission, indicating that the colder dusty molecular/atomic part of the circumstellar shell goes well beyond the "PN proper" region of hot ionized plasma.

The total far-IR fluxes (F_ν) were computed by aperture photometry with the 3-$\sigma_{\rm sky}$ contour in the 70 μm map, yielding uncertainties up to 5% for the PACS bands and 15% for the SPIRE bands after color-correction: the resulting spectral energy distributions (SEDs) are shown in Figure 1 (open triangles and asterisks), together with fluxes derived from *WISE* and *Spitzer*/IRAC observations (open stars). As seen in Figure 1, the *Herschel* bands cover the Rayleigh-Jeans shoulder of the thermal dust emission component and often the thermal dust emission peak. Therefore, the dust temperature, $T_{\rm dust}$, can be estimated by fitting the far-IR SED with a power-law dust emissivity, $\lambda^{-\beta} B_\nu(T_{\rm dust})$, where β defines the emissivity characteristics of dust grains at the temperature $T_{\rm dust}$. We note here that these measured broadband fluxes would include some contribution from far-IR emission lines. Thus, we assessed the degree of line contamination by using PACS spectra. We found that the level of line contamination was generally minor (median 4% and 2% at 70 and 160 μm, respectively and insignificant in other bands), while as high as 10–20% for PNe having a hot (> 100 kK) central star, and hence, line contamination can be safely ignored in the dust emissivity fitting.

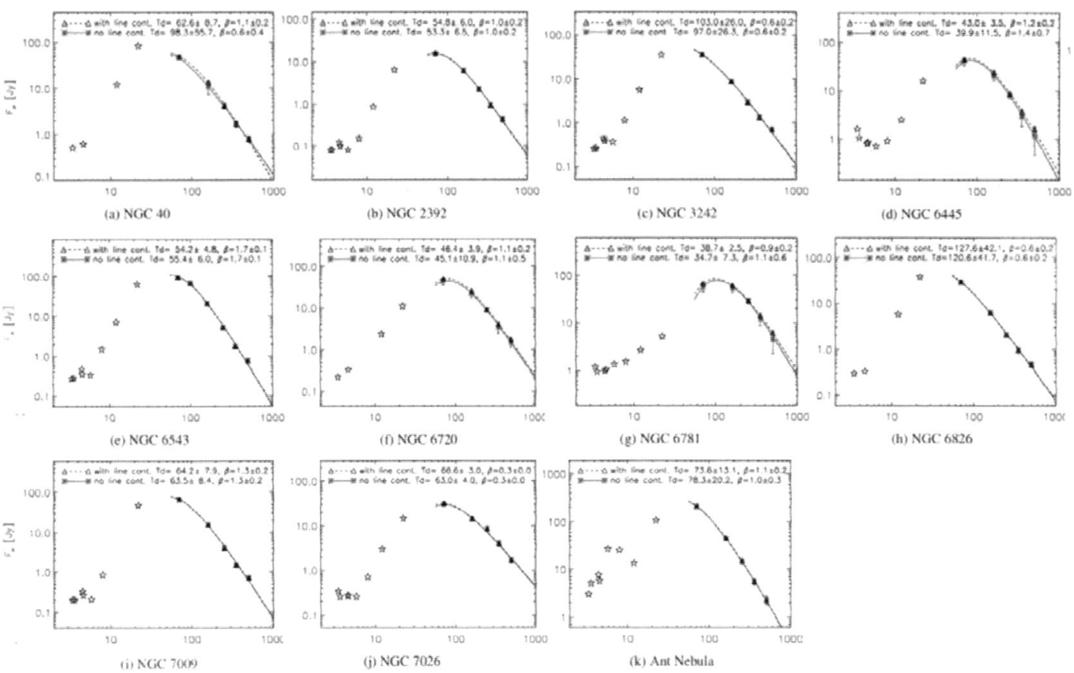

Figure 1. Thermal dust emission SEDs of the HerPlaNS target PNe

The derived $T_{\rm dust}$ varies from 38 K (NGC 6781) to more than 100 K (NGC 3242 and 6826). When the dust SED peak is well constrained within the *Herschel* bands, the uncertainty of $T_{\rm dust}$ is less than 10%. However, the uncertainty reaches 30% or more if $T_{\rm dust} > 65$ K (i.e., the dust SED peaks at $< 70\,\mu$m). Overall, there is no single $T_{\rm dust}$ that fits both the far-IR region (the Rayleigh-Jeans side) and mid-IR region (the Wien side) of the observed PN thermal dust emission. Hence, there exist distinct T_{dust} regimes over these PNe. This is consistent with the fact that the dust distribution in PNe is spread over well beyond the ionized gas region into the atomic/molecular gas regions. These SEDs also indicate that a significant amount of thermal dust emission arises in the far-IR and that a significant amount of cold dust exists in the outer regions of PNe beyond the central ionized region.

The emissivity power index β for all PNe we investigated turned out to be 1.1 or less except for NGC 6543 (1.7 ± 0.1). Typically, the value of β is roughly 2 for silicate dust grains and graphite grains and close to 1 for amorphous carbon grains (e.g., [6, 7, 8, 9] for theoretical/lab studies).

To investigate the variation of β in relation of T_{dust} within the PN volume for our sample, the derived values of T_{dust} and β at specific spatial locations within the target PNe are plotted (Figure 2). To compare the β–T_{dust} relationship in PNe and some other object types (ISM and star forming regions), similarly measured β and T_{dust} values from other objects are also shown in the same plot. The distribution of the β–T_{dust} points in the plot appears to show anti-correlation between the two parameters: there is a tendency that the points cluster around a very low T_{dust} value (~ 27 K) with high β values (between 0.1 and 3) and around a low β value (~ 0.85) with high T_{dust} values (between 25 and 88 K). In general, most PN points are found in the low-T_{dust} and high-β part of the distribution: NGC 7009 is the case in which a large spread of values is seen in both T_{dust} and β. From the spatial distribution of these points, we see that the high-T_{dust} and low-β points are found close to the PN central stars, while low-T_{dust} and high-β points are found away from the PN central stars. On the other hand, dust in the ISM (starless dark cloud cores) is found at low T_{dust} with a range of β (from 0.5 to 1.4).

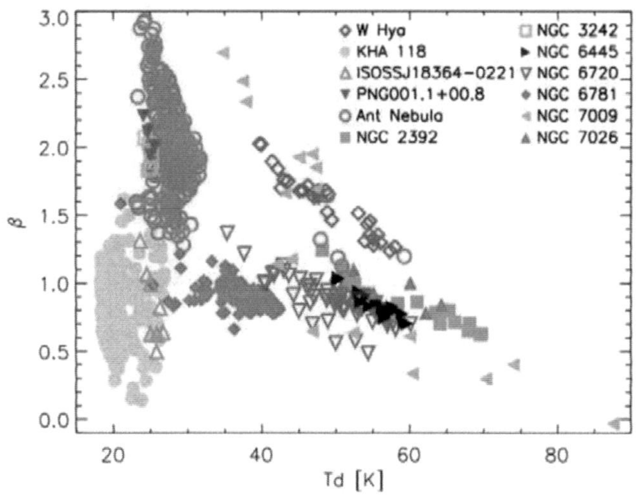

Figure 2. Spatially-resolved distributions of the dust temperature T_{dust} and emissivity index β in the target PNe and other astronomical objects (W Hya: O-rich AGB, KHA 118 and ISOSSJ18364−0221: starless dark cloud cores, PNG001.8+00.8: bulge PN).

From this result, we can conclude that (1) dust in the inner regions of the circumstellar shell tends to show low-β and high-T_{dust} characteristics and higher-β, (2) lower-T_{dust} dust is found toward the outer regions in the circumstellar shell, and (3) lowest-T_{dust} dust is eventually found in the ISM.

Dupac et al. [10] found a strong anti-correlation between β and T_{dust} using balloon-borne sub-mm observations of different ISM regions with $10\,\text{K} \geq T_{\text{dust}} \geq 80\,\text{K}$ and $0.8 \leq \beta \leq 2.4$. Paradis et al. [11] and Anderson et al. [12] both investigated the β–$T_{\rm dust}$ distribution of star forming regions using *Herschel* data and found an anti-correlation for a smaller range of $T_{\rm dust}$ and β. This trending can be explained by the presence of different different temperature components along the line of sight and/or a change in dust chemistry or crystallinity, as corroborated by laboratory analysis (e.g. [13, 14, 15]). On the other hand, some authors claim that the observed anti-correlation is caused by incorrect treatment of noise with the least-square minimization fitting (which assumes that $T_{\rm dust}$ and β are independent, but they are not really independent; e.g. [16, 17]). While we are presently investigating the $T_{\rm dust}$-β dependence and its effects in the analysis, we tentatively consider the observed trending is real and represents the change of the physical properties of the far-IR emitting dust (interstellar type vs. circumstellar type) because our finding is found from high signal-to-noise data (i.e., bright regions of PNe) and is still consistent with the previous ISM results.

3. Summary

We conducted a far-IR imaging/spectroscopic survey of PNe using the Herschel Space Observatory. Far-IR broadband imaging probed the dust component of the target PNe at the highest spatial resolution ever used to image these objects, while far-IR spectroscopy permitted detection of ionized, atomic, and molecular line emission from the gas component. Here, we summarized the results of a far-IR broadband imaging mini-survey of 11 PNe [18].

Cold thermal dust emission in the far-IR (50–500μm) detected from about a dozen PNe observed as part of the Herschel Planetary Nebula Survey revealed that

(i) in these PNe the largest fraction of thermal dust emission arises in the far-IR part of the spectrum, indicating that the coldest dust component in PNe representing dust grains ejected as part of ancient AGB winds cannot be ignored in accounting for the total mass and energetics, and

(ii) by fitting far-IR broadband fluxes with a power-law dust emissivity, $\lambda^{-\beta}B_\lambda(T_{\rm dust})$, the spatially-resolved β–$T_{\rm dust}$ relationship can separate the circumstellar (with high $T_{\rm dust}$ and low, specific β) and interstellar (with low, specific $T_{\rm dust}$ and high β) dust components.

References

[1] Kwok S 2000 The Origin and Evolution of Planetary Nebulae, (Cambridge:CUP)
[2] Pilbratt G L *et al* 2010 *Astron. Astrophys.* **518** L1
[3] Ueta T *et al* 2014 *Astron. Astrophys.* **565** A36
[4] Poglitsch A *et al* 2010 *Astron. Astrophys.* **518** L2
[5] Griffin M J *et al* 2010 *Astron. Astrophys.* **518** L3
[6] Bohren C F and Huffman D R 1983 Absorption and scattering of light by small particles (New York:Wiley)
[7] Draine B T and Lee H M 1984 *Astrophys. J.* **285** 89
[8] Rouleau F and Martin P G 1991 *Astrophys. J.* **377** 526
[9] Mennella V, Colangeli L and Bussoletti E 1995 *Astron. Astrophys.* **295** 165
[10] Dupac X *et al* 2003 *Mon. Not. R. Astron. Soc.* **344** 105
[11] Paradis D *et al* 2010 *Astron. Astrophys.* **520** L8
[12] Anderson L D *et al* 2010 *Astron. Astrophys.* **518** L99
[13] Agladze N I, Sievers A J, Jones S A, Burlitch J M and Beckwith S V W 1996 *Astrophys. J.* **462** 1026
[14] Mennella V, Brucato J R, Colangeli L, Palumbo P, Rotundi A and Bussoletti E 1998 *Astrophys. J.* **496** 1058
[15] Coupeaud A, Demyk K, Meny C, Nayral C, Delpech F, Leroux H, Depecker C, Creff G, Brubach J-B, Roy P 2011 *Astron. Astrophys.* **535** A124
[16] Kelly B C, Shetty R, Stutz A M, Kauffmann J, Goodman A A and Launhardt R 2012 *Astrophys. J.* **752** 55
[17] Juvela M and Ysard N 2012 *Astron. Astrophys.* **539** A71
[18] Ladjal D 2016 in preparation

Journal of Physics: Conference Series **728** (2016) 032013 doi:10.1088/1742-6596/728/3/032013

Pulsational variability in proto-planetary nebulae and other post-AGB objects

Bruce J. Hrivnak[1]

[1] Department of Physics & Astronomy, Valparaiso University, Valparaiso, IN, 46383, USA

E-mail: [1] bruce.hrivnak@valpo.edu

Abstract. Light and velocity curves of several classes of pulsating stars have been successfully modeled to determine physical properties of the stars. In this observational study, we review briefly the pulsational variability of the main classes of post-AGB stars. Our attention is focused in particular on proto-planetary nebulae (PPNe), those in the short-lived phase from AGB stars to the planetary nebulae. New light curves and period analyses have been used to determine the following general properties of the PPNe variability: (a) periods range from 35 to 160 days for those of F−G spectral types, with much shorter periods ($\leq$ 1 day) found for those of early-B spectral type; (b) there is a correlation between the pulsation period, maximum amplitude, and temperature of the star, with cooler stars pulsating with longer periods and larger amplitudes; (c) similar correlations are found for carbon-rich, oxygen-rich, and lower-metalicity PPNe; and (d) multiple periods are found for all of them, with $P_2/P_1 = 1.0\pm0.1$. New models are needed to exploit these results.

1. Introduction

After spending most of their life cycle as main sequence stars, intermediate- (2−8 $M_\odot$) and solar-mass stars subsequently evolve through the red giant, asymptotic giant branch (AGB), and planetary nebula (PN) phases on their way to becoming white dwarfs. Stars lose large amounts of mass during the red giant and especially during the end of the AGB phases. Between the AGB and PN phases there is an interval of time when the intensive mass loss has ended and the star is in the post-AGB phase, but when the temperature of the star is too low to photo-ionize the nebula, signaling the beginning of the PN phase. These transitional objects are called proto-planetary nebulae (PPNe) or pre-planetary nebulae [1].

During this PPNe phase the star is luminous (1−10 x 10^3 $L_\odot$) and increasing in temperature from 4000 K (end of AGB) to 30,000 K (start of PN). The study of these objects blossomed after the identification of many candidates in the *IRAS* database. They are found to be stars with spectral types ranging from G to B, surrounded by a dusty, molecular envelope visible in scattered light. Spectral energy distributions showed a characteristic double-peak, with a peak in the visible arising from the reddened photosphere and a peak in the mid-infrared arising from a cool, detached dust shell [2, 3].The timescale for this evolutionary phase is short, $\sim 10^3$ years [4, 5].I estimate the number of PPNe confirmed thus far to be about 60 in the Milky Way Galaxy, with a similar number of additional candidates. (The Torun catalog includes these among its post-AGB stars [6]). Detailed follow-up studies of individual objects have yielded information on the chemistry of the gas and the dust [7, 8, 9, 10, 11],the shapes of the nebulae (from *HST* images [12, 13]), the kinematics of the gas [14, 15], and the chemistry of the central stars [16].

In this paper, we will discuss the observed pulsational properties of the central stars of PPNe. Analysis of the pulsational properties (period, amplitude) can yield information on the physical properties (mass, luminosity) and possibly the rate of evolution of the central stars during this transitional phase. Pulsation is known to be a driver of mass loss in evolved stars, and these studies can also shed light on the level of on-going mass loss in the post-AGB phase.

2. Pulsational light curve studies of post-AGB stars

Pulsation is a common feature in post-AGB stars. We will very briefly review the results from post-AGB stars in general in order to provide a context and comparison for the results from the studies of PPNe.

2.1. Pulsation in other post-AGB stars

RV Tauri variable stars are post-AGB stars that have light curves characterized by alternating deep and shallow minima, with light curves ranging in amplitude up to 3 mags. They are of spectral types F–K, with periods of 30–150 days [17, 18]. A recent study shows that many RV Tauri variables possess an infrared excess matching that of a disk, with evidence strongly suggesting that these are all binaries [19].

Other post-AGB stars show light curves that are semi-regular (SRV) or those of Type II Cepheids. Many of these have been found to be in binaries systems [20], with pulsation periods of 50–120 days and amplitudes ranging from 0.1 to 1.5 mags [21, 22].Kiss et al., in their study of these post-AGB stars, found correlations between pulsation amplitude and T_{eff} and between luminosity and T_{eff}, with many of the objects lying within the classical instability strip. Similar objects have been observed in the nearby Large Magellanic (LMC) and Small Magellanic Cloud (SMC) galaxies [23, 24]. Recently Kamath et al. (2016) [25] discovered in the LMC and SMC a new class of post-red giant branch stars; these are dusty, of low mass, and evolved. They are found to significantly outnumber the post-AGB stars in these galaxies, suggesting that some of the SRV and RV Tauri variables in the Milky Way Galaxy (MWG) are also post-RGB rather than the post-AGB stars.

The rare R Coronae Borealis (RCB) stars are thought to also be post-AGB objects that are hydrogen-poor but carbon-rich. In addition to their large drops in brightness (up to 8 mags!), attributed to dust ejections, they also show pulsational variations on the order of 40–100 days with amplitudes of a few tenths mag [26].

2.2. Pulsation in proto-planetary nebulae

Pulsational studies of PPNe have been carried out primarily by two groups. Arkhipova and collaborators at the Sternberg Institute in Moscow have been observing the brighter of these objects since their earliest identification, beginning in some cases as early as 1991. In a series of papers they have presented and discussed light curves of ~20 PPNe and other post-AGB objects [27, 28, 29].

The second is our group is at Valparaiso University. Observations began in 1994 and have continued to the present time, carried out primarily by supervised undergraduate students. More recently, we have begun observing some fainter and some southern hemisphere PPNe candidates using the SARA (Southeast Association for Research in Astronomy) consortium telescopes in Arizona (0.9 m) and Chile (0.6 m). Presently we are observing ~50 PPNe and candidates with V = 8–15 mag.

These two studies are complementary, and in some cases we have combined the published data of Arkhipova et al. with ours to increase the data sets for analysis.

3. Results

3.1. Initial results

Initially we analyzed the light curves for 12 C-rich, F–G spectral type PPNe obtained from 1994–2007. All of them were found to vary, with maximum amplitudes ranging from 0.15 to 0.6 mags. The light curves of each changed in amplitude, in some cases resembling a beat phenomenon such as is produced by the combination of two rather similar periods. They each were redder (cooler) when fainter. Period analysis determined that each of them varied in a periodic manner, with periods (P) of 38 to 153 days. Several trends were found in the results. Stars with lower temperatures have longer periods and larger amplitudes, and these decreased monotonically with stars of higher temperatures to a minimum range of values by $T_{eff} \approx 6000$ K (these are shown graphically by Hrivnak et al. 2010 [30], Figs. 18 and 19; Arkhipova et al. 2011 [29] found similar results). This suggested that we were seeing evolutionary effects, and that the trends in this sample display the evolution of a typical C-rich PPNe. The linear relation found for $P-T_{eff}$ over this temperature range of 5000–8000 K, coupled with a post-AGB evolutionary model [31] suggested a decrease in period of ~2 day per decade. This should be measurable and presented the possibility of measuring evolution in real time and using the results to constrain the evolutionary rate, at least through this range in temperature.

3.2. More recent results

This initial study of C-rich PPNe has been expanded in several directions to include objects that are O-rich, fainter, hotter, or of a differently metallicity. We have also investigated the velocity variations during pulsation. These results are discuss below.

In the analysis of four O-rich, F spectral type PPNe, we were able to increase the temporal baseline by the inclusion of our data up to 2012 and inclusion of data from Arkhipova et al. (2010) [28] and others. This resulted in the determination of a dominant period in each ranging from 41 to 113 days, with additional multiple periods [32]. The maximum amplitudes ranged from 0.14 to 0.21 mag, and thus the periods and amplitudes were in the same range as those of the larger C-rich sample. In all cases, the secondary period was close to the dominant period (±10 %); a similar result was found for two of the previously studied C-rich PPN based on an analysis of an enlarged data set [33].

With the addition of the SARA telescopes and a new CCD at Valparaiso University Observatory, we were able to include some fainter (V = 13–15 mag) and more southerly PPNe candidates. The light curves of a sample of these are shown in Figure 2. The maximum amplitudes range from 0.12 to 0.58 mags. Periods have been determined for 6 of the 9, with dominant periods in the range of 25 to 135 days and multiple periods for each. This study is in progress.

We have also monitored some of the hotter, B spectral type PPNe, both O-rich and C-rich. Again, all of them vary. However, they are found to vary on a shorter time scale, less than a few days, and no periods have been found for the sample. Arkhipova et al. (2006) [34] found similar results. It is suspected that the cadence of our observing, every few days at most, and the low amplitudes combined with multiple periods, make it difficult to detect any periods present in the data. A more intensive study of 2 PPNe observed continuously over several hours on several non-consecutive days revealed some short-term cyclical variability (< 1 day) but no dominant period was detected. This study is in progress.

The use of the *Spitzer Space Telescope* has allowed the identification of many dusty evolved stars in the relatively nearby Large and Small Magellanic Cloud (LMC, SMC) galaxies [23, 24, 35]. These galaxies have lower metallicities, with that of the LMC about 0.4 of the solar metalicity. Mid-IR spectra of these show features that are attributed to O-rich or C-rich dust chemistries. Twenty-two of these were identified as C-rich [36] PPNe. In most cases, OGLE light

Journal of Physics: Conference Series **728** (2016) 032013 doi:10.1088/1742-6596/728/3/032013

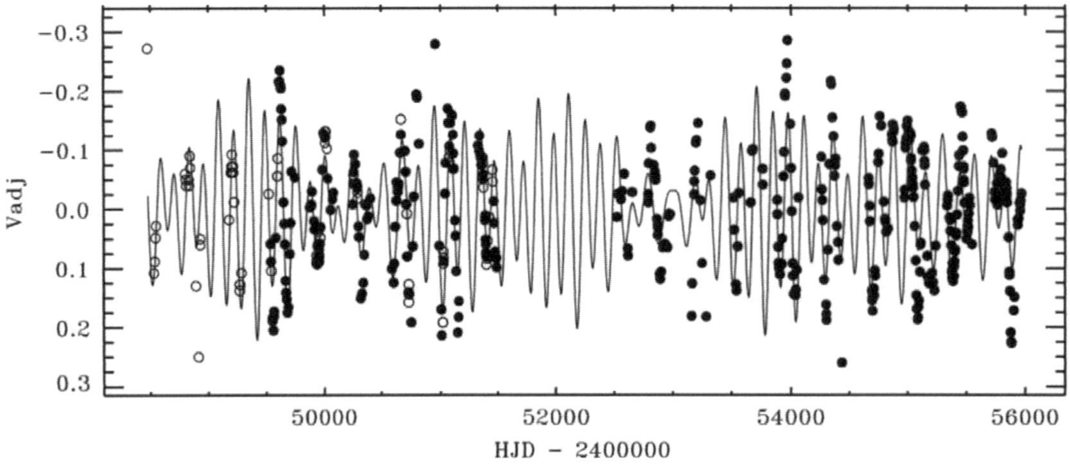

Figure 1. Light curve of IRAS 22272+54435, fitted by four periods, with P_1 = 132 day. The open and filled circles represent data from Arkhipova et al. and from the Valparaiso University Observatory, respectively.

curves were available ranging from 8 to 13 years. For those without OGLE data, MACHO data covering eight years were available to be used. Periods were determined for the light variations in eight of these, with periods ranging from 49 to 157 days, similar to those found in the MWG. Multiple periods were again found, with secondary periods close to the primary periods. These are primarily F stars. For those without period determinations, the variations were of shorter timescale, with some and perhaps all of them being of earlier spectral type. Similar to those in the MWG, there is a trend of decreasing period with increasing temperature and a trend of decreasing amplitude with decreasing period. While the sample of those with period and temperature determinations in the LMC/SMC is small, there is a suggestion that the trend of decreasing period with increasing temperature is offset to lower temperatures or shorter periods than in the MWG [37].

In Figure 3 are plotted these parameter correlations for the various objects discussed above. One can see the general trend of decreasing periods with increasing temperature, as was found for the initial sample of 12 C-rich PPNe. The possible shift for the LMC objects to lower temperatures (or lower amplitude pulsations) can also be seen. Instead of a plot of maximum amplitude with temperature, we have plotted instead the correlation with period since some of the targets do not have temperature determinations. They show that the maximum amplitude decreases steeply with increasing temperature and decreasing period, reaching the low values of 0.15–0.25 mag (V) when the temperature increases to ~6000 K or the period decreases to ~120 days (Fig. 3).

3.3. Radial velocity studies

We began a radial velocity monitoring program of the brighter PPNe in 1991 to investigate the possible presence of binary companions. None have been found thus far [38]. However, we did detect periodic variations with periodicities similar to those found in the light curves. This initial study extended to 1995 and then the study was re-initiated in 2007 and is continuing. Seven bright (V=7–11 mag) PPNe have been monitored for velocity variations. Three of these have longer periods (88–132 days) and show consistent light, color, and velocity correlations, with the objects being brightest when hottest and smallest [33]. An example is shown in Figure 4. The pattern is not so clear for the three with short periods, 35–40 days. This may be due to

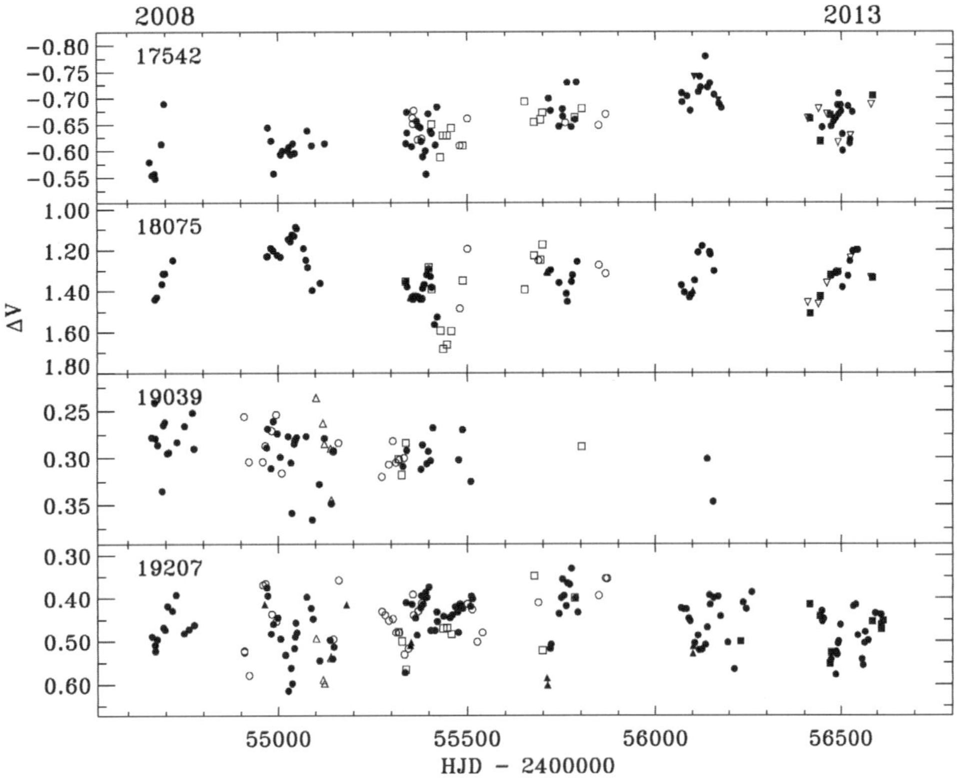

Figure 2. New light curves for a sample of fainter galactic PPN candidates. The different symbols represent data from different telescopes and detectors.

the presence of shocks in these pulsating objects, which lead to line spitting at certain phases and to Hα emission features [39, 40].

4. Discussion & Conclusions

Since the central stars are clearly visible in these PPNe, intensive mass loss has ended by the time that P$\sim$160 days (late-G spectral types). Their spectral energy distributions show that they are surrounded by detached dust shells. By the time that P$\sim$120 days (early-G spectral types), the pulsation amplitudes are small (0.15−0.25 mag) and the pulsation-driven post-AGB mass loss rate is presumably low. It is this post-AGB mass loss that dominates the evolution of the star rather than the fusion of hydrogen at the base of the photosphere.

Fokin et al. (2001) [41] have run non-linear radiative models to fit the light curve of one of the PPN with P$\sim$40 days. The results indicate the way that core mass, luminosity, and temperature effect the period, amplitude, and regularity of the resultant light curves. However, a good fit required an unusually high core mass (0.8 $M_\odot$) and these models would not produce the longer period pulsations. New models are needed to exploit this present wealth of light and velocity curve data.

From the study thus far of about 35 PPN in the MWG and 22 in the LMC/SMC, several patterns have emerged.

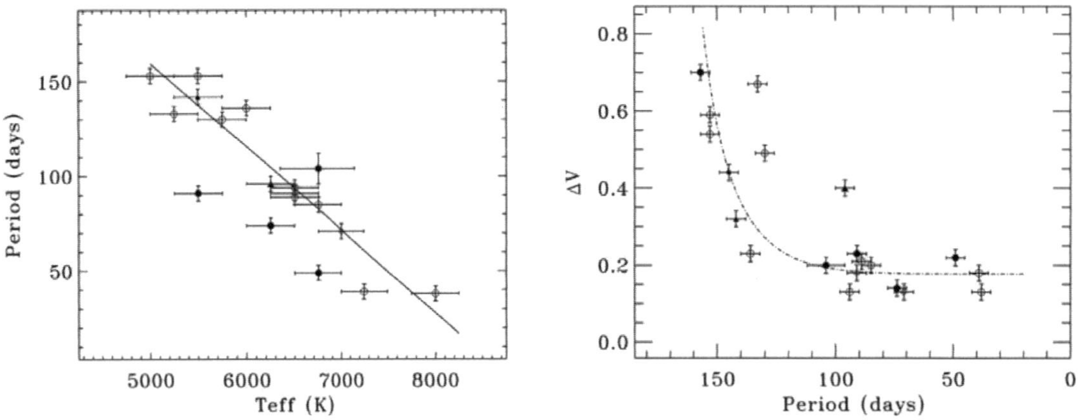

Figure 3. Plots of pulsation period vs. T_{eff} (left) and maximum amplitude vs. pulsation period (right). Plotted are the initial 12 C-rich MWG PPNe (open circles) and 6 C-rich LMC/SMC PPNe (filled circles). The lines represent general trends in the data based on the initial 12 C-rich PPNe. (From Hrivnak et al. 2015a [37].)

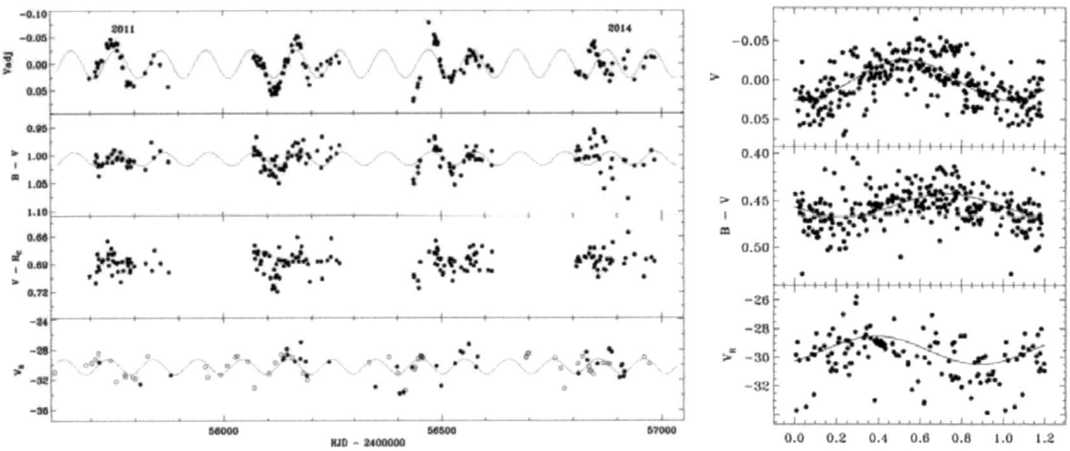

Figure 4. IRAS 18095+2704: (Left) Plots showing (from top to bottom) the normalized V light curve, the B−V and V−R_C color curves, and the radial velocity curve, all fitted with curves based on P = 113 day. (Right) The V, B−V, and radial velocity curves phased with the same period.

1. PPNe with F−G spectral types pulsate with periods ranging from 35 to 160 days. Those that are hotter pulsate on shorter timescales.

2. PPNe have pulsation amplitudes ranging from 0.6 mag (V) for late-G spectral types with P≈150 days to 0.15−0.25 mag for early-F spectral types at P≈40 days. Those at earlier B spectral types vary on timescales of a few days or less.

3. PPNe as an ensemble show evolution to shorter periods and smaller amplitudes, reaching minimum values in amplitude at T ∼ 6000 K and at P ∼ 120 days.

4. PPNe have multiple periods to their pulsations, with the secondary close to the primary period ($P_2/P_1 = 1.0 \pm 0.1$).

5. The evidence for decreasing period with time over the two decades of this study is

not found. This was investigated by dividing the data with longer time strings into subsets and determining their periods. However, the investigation of this effect is complicated by the multiple periods found in the data. It is expected that evidence for this will be found with longer temporal baselines, but it is nevertheless disappointing that it was not found in our two decades of observations.

6. Evidence based on the radial velocity data of three PPNe with longer periods, 88–132 days, shows clear correlations of the brightness, temperature, and size of the stars, with the stars being brightest and hottest when smallest.

7. There is a suggestion of slight differences in the pulsational properties of PPNe due to composition. However, an increased sample size is needed to make more certain claims.

Acknowledgments

I begin by acknowledging Sun Kwok, who introduced me to the study of PPNe and with whom I have had a long-term collaboration on many aspects of the observational properties of these interesting objects. I wish to thank the many people who have contributed to the success of this long-term monitoring project. These include W. Lu with the data reduction, G. Hanson and T. Hillwig with the SARA light curve monitoring, D. Bohlender, H. Van Winckel, G. Van de Steene, J. Sperauskus with the radial velocity monitoring, and the many Valparaiso University students who have assisted in the long-term photometric monitoring of these PPNe. The comments of the referee were appreciated. Financial support for this study has come from a series of NSF grants, most recently AST-1413660.

References

[1] Kwok S 1993 *Annu. Rev. Astron. Astrophys.* **31** 63
[2] Parthasarathy M and Pottasch S R 1986 *Astron. Astrophys.* **154** L16
[3] Hrivnak B J, Kwok S and Volk K M 1989 *Astrophys. J.* **346** 265
[4] Vassiliadis E and Wood P R 1994 *Astrophys. J. Suppl. Series* **92** 125
[5] Blöcker T 1995 *Astron. Astrophys.* **297** 755
[6] Szczerba R *et al* 2011 *Proc. Int. Astron. Union* (S283) **7** 506
[7] Likkel L 1989 *Astrophys. J.* **344** 350
[8] Omont A, Loup C, Forveille T, te Lintel Hekkert P, Habing H and Sivagnanam P 1993 *Astron. Astrophys.* **267** 515
[9] Hrivnak B J, Volk K and Kwok S 2000 *Astrophys. J.* **535** 255
[10] Molster F J, Waters L B F M, Tielens A G G M and Barlow M J 2002 *Astron. Astrophys.* **382** 222
[11] Manchado A 2016 these proceedings
[12] Sahai R, Morris M, Sánchez Contreras C and Claussen M 2007 *Astron. J.* **134** 2200
[13] Siódmiak N, Meixner M, Ueta T, Sugerman B E K, Van de Steene G C and Szczerba R 2008 *Astrophys. J.* **677** 382
[14] Bujarrabal V, Castro-Carrizo A, Alcolea J and Sánchez Contreras C 2001 *Astron. Astrophys.* **377** 868
[15] Balick B 2016 these proceedings
[16] Van Winckel H 2003 *Annu. Rev. Astron. Astrophys.* **41** 391
[17] Pollard K R, Cottrell P L, Kilmartin P M and Gilmore A C 1996 *Mon. Not. R. Astron. Soc.* **279** 949
[18] Pollard K R, Cottrell P L, Lawson W A, Albrow M D and Tobin W 1997 *Mon. Not. R. Astron. Soc.* **286** 1
[19] Gezer I, Van Winckel H, Bozkurt Z, De Smedt K, Kamath D, Hillen M and Manick R 2015 *Mon. Not. R. Astron. Soc.* **453** 133
[20] Van Winckel H 2007 *Baltic Astron.* **16** 112
[21] Kiss L L, Derekas A, Szabó M, Bedding T R and Szabados L 2007 *Mon. Not. R. Astron. Soc.* **375** 1338
[22] Van Winckel H *et al* 2009 *Astron. Astrophys.* **505** 1221
[23] van Aarle E, Van Winckel H, Lloyd Evans T, Ueta T, Wood P R and Ginsburg A G 2011 *Astron. Astrophys.* **530** A90
[24] Kamath D, Wood P R and Van Winckel H 2014 *Mon. Not. R. Astron. Soc.* **439** 2211
[25] Kamath D, Wood P R, Van Winckel H and Nie J D 2016 *Astron. Astrophys.* **586** L5
[26] Clayton G 1996 *Publ. Astron. Soc. Pac.* **108** 225
[27] Arkhipova V P, Ikonnikova N P, Noskova R I and Sokol G V 2000 *Astronomy Lett.* **26** 609
[28] Arkhipova V P, Ikonnikova N P and Komissarova G V 2010 *Astronomy Lett.* **36** 269

[29] Arkhipova V P, Ikonnikova N P and Komissarova G V 2011 *Astronomy Lett.* **37** 635
[30] Hrivnak B J, Lu W, Maupin R E and Spitzbart B D 2010 *Astrophys. J.* **709** 1042
[31] Steffan M, Szczerba R and Schönberner D 1998 *Astron. Astrophys.* **337** 149
[32] Hrivnak B J, Lu W and Nault K A 2015b *Astron. J.* **149** 184
[33] Hrivnak B J, Lu W, Sperauskas J, Van Winckel H, Bohlender D and Začs L 2013 *Astrophys. J.* **766** 116
[34] Arkhipova V P, Ikonnikova N P, Komissarova G V and Noskova R I 2006 *Proc. Int. Astron. Union* (S234) 357
[35] Kamath D, Wood P R and Van Winckel H 2015 *Mon. Not. R. Astron. Soc.* **454** 1468
[36] Volk K *et al* 2011 *Astrophys. J.* **735** 127
[37] Hrivnak B J, Lu W, Volk K, Szczerba R, Soszyński I and Hajduk M 2015 *Astrophys. J.* **805** 78
[38] Hrivnak B J, Lu W, Wefel K L, Bohlender D, Morris S C, Woodsworth A W and Scarfe C D 2011 *Astrophys. J.* **734** 25
[39] Lèbre A, Mauron N, Gillet D and Barthès D 1996 *Astron. Astrophys.* **310** 923
[40] Začs L, Musaev F, Kaminsky B, Pavlenko Y, Grankina A, Sperauskas J and Hrivnak B J 2016 *Astrophys. J.* **816** 3
[41] Fokin A B, Lèbre A, Le Coroller H and Gillet D 2001 *Astron. Astrophys.* **378** 546

Journal of Physics: Conference Series **728** (2016) 032014 doi:10.1088/1742-6596/728/3/032014

The discovery and characterisation of binary central stars in planetary nebulae

David Jones[1,2]

[1] Instituto de Astrofísica de Canarias, E-38205 La Laguna, Tenerife, Spain

[2] Departamento de Astrofísica, Universidad de La Laguna, E-38206 La Laguna, Tenerife, Spain

E-mail: `djones@iac.es`

Abstract. Close binary central stars of planetary nebulae are key in constraining the poorly-understood common-envelope phase of evolution, which in turn is critical in understanding the formation of a wide-range of astrophysical phenomena (including cataclysmic variables, low-mass X-ray binaries and supernovae type Ia). Here, I present the results of our on-going, targeted search for close-binaries in planetary nebulae which has led to the discovery of more than ten new central binaries in just the last few years (almost the same as the total discovered during the 1980s and 1990s together). This success has been rooted in the targeted selection of objects for study, based on morphological features deemed typical of binarity, as well as novel observing strategies (including the employment of narrow-band filters for photometry to minimise nebular contamination), both of which are discussed. These new discoveries coupled with the painstaking characterisation of both newly discovered systems and those from the literature mean that we are now in a position to begin to probe the poorly understood common-envelope phase.

1. Introduction

While central star binarity has long been considered a key factor in the formation of aspherical morphologies in planetary nebulae (PNe), it is only recently that it has been shown observationally that a close-binary pathway is responsible for a significant fraction of PNe [1] [1]. This lag was, in part, due to the painstakingly difficult nature of the search for binary central stars – with photometric monitoring generally limited to the brightest central stars and fainter nebulae for ease of photometry with modestly sized telescopes. This all changed with the advent of the OGLE survey, an i-band monitoring survey of the Galactic bulge primarily in search of gravitational lensing events, but which also resulted in the discovery of more than ten new close-binary central stars – approximately doubling the sample known at the time [1] (see figure 1 of [5]). With the resulting sample, it was possible to identify morphological features which were particularly prevalent amongst those nebulae found to host close-binary central stars [6]. This features included jets, rings and filamentary structures, and as such any nebula displaying these structures were considered strong candidates for photometric monitoring in search of variability consistent with central star binarity [6].

[1] Note that in these proceedings, I will only cover the close-binary pathway. However, long-term radial velocity surveys have also produced very promising results in the search for longer period binaries [2], as have searches for infrared excess associated with lower temperature companions at all separations [3, 4].

Journal of Physics: Conference Series **728** (2016) 032014 doi:10.1088/1742-6596/728/3/032014

2. A targeted search

Based on the morphological features identified by [6] as being dominant amongst PNe found to host binary central stars, a campaign of photometric monitoring focusing on nebulae displaying those structures was initiated by the author and collaborators. Many of the best candidate nebulae were bright and irregular presenting obvious problems for photometric monitoring of the central star, where contamination from the nebular background can dilute any observed variability or, worse, introduce spurious variability [7]. The nebular contamination can often be minimised by careful choice of filter in order to avoid those whose bandpass includes the brightest nebular emission lines (e.g. avoiding R-band which includes the H-α emission line), in many cases the I-band was found to present sufficiently low contamination for successful photometry. However in several cases there was still significant residual nebular emission, for these nebulae it was found that the novel employment of off-emission-line continuum filters reduced the nebula contamination to a minimum acceptable for photometric monitoring. This approach was successfully employed for several nebulae including, for example, Hen 2-155 [8], where the difference in nebular contamination between broadband filters and the narrowband H-β continuum filter is shown in figure 1.

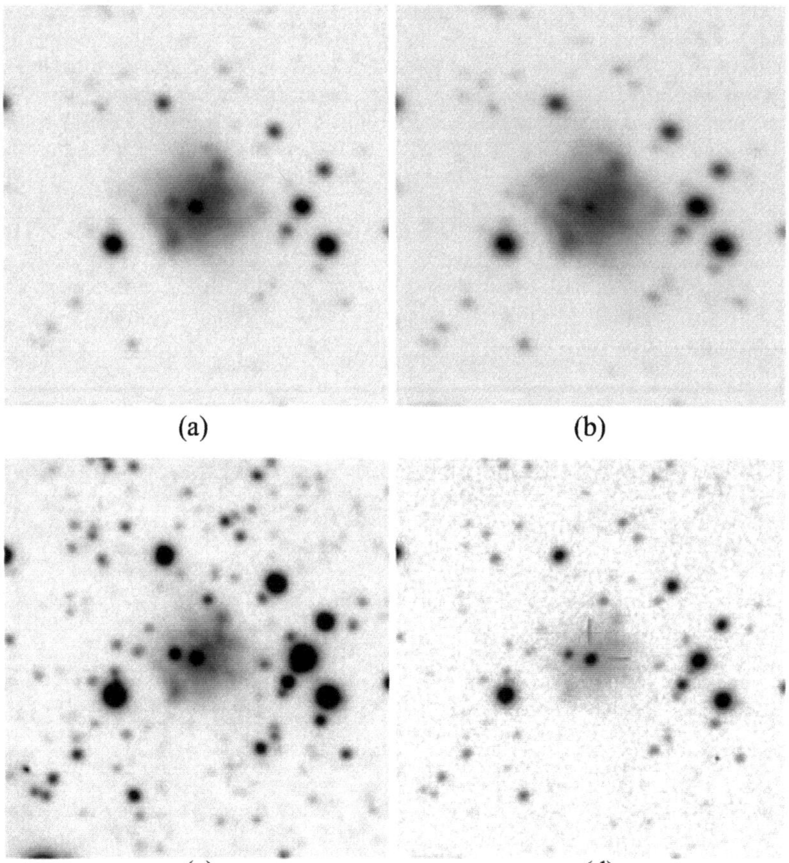

(a) (b) (c) (d)

Figure 1. Images of the PN Hen 2-155 in (a) B-band, (b) V-band, (c) I-band and (d) H-β continuum, demonstrating the value of the use of narrowband, off-emission-line filters in minimising nebular contamination [8]. The location of the central star is marked in the H-β continuum image (d).

Ultimately, the programme of targeted monitoring has been extremely successful, adding more than ten new binaries to the sample and reaching ever closer to the milestone of 50 known close-binary central stars. Taking the sample as a whole there are two striking results. Firstly, the the period distribution is heavily biased towards shorter periods which even when accounting for observational biases does not seem to fit well with population synthesis models [9]. Secondly, there seems to be an over-abundance of double-degenerate systems (classifying the secondary type based on the shape of the lightcurve). Double-degenerate central stars should be intrinsically more difficult to detect than those with main sequence companions, and so the fact that roughly 20% of the known sample of close-binary central stars are double-degenerate has striking consequences for their frequency, in general.

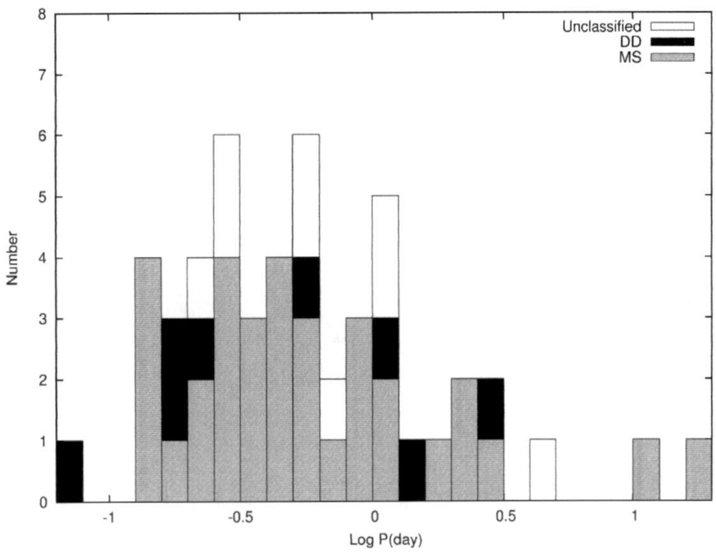

Figure 2. The period distribution for all known post-common-envelope binary central stars. Where known the nature of the secondary is highlighted (main sequence, MS, or degenerate, DD), highlighting the relatively high fraction of double degenerate central binaries. For full details and references, please see http://drdjones.net/bCSPN

Examining individually the newly discovered systems, there are several examples which also have further critical implications for our understanding of common-envelope evolution. Perhaps the most interesting is the central star system of the Necklace [10], where the secondary component of the binary was found to be contaminated with AGB material from the primary indicating that significant mass transfer has occurred in the system [11] – something which is not predicted by current common-envelope models [12].

3. Characterisation

While some properties of the central binaries can be discerned from light curves alone, full characterisation requires complementary radial velocity studies (with the level of degeneracy in the final parameters also dependent on the presence of eclipses and the single/double-lined nature of the observed radial velocity curve). These detailed studies are time-consuming and complicated by the presence of nebular emission lines which coincide with many of the central star absorption features which would otherwise be ideal for radial velocity measurements. As such, only eight central binary systems have been subjected to detailed modelling based on

Journal of Physics: Conference Series **728** (2016) 032014 doi:10.1088/1742-6596/728/3/032014

a combination of photometric and radial velocity observations, with six of those being found to host main sequence secondaries which the remaining two are double-degenerate systems [8]. In each of the systems with a main sequence secondary, the secondary is found to be highly inflated with respect to its ZAMS radius, consistent with a period of intense mass transfer prior to entering the common-envelope phase to which the secondary has yet to thermally adjust – again, something which is not predicted by current models. Of the two double-degenerate systems, one is found to be the best supernova type Ia candidate to-date. Hen 2-428 has been shown to play host to a pair of degenerate stars, whose total mass exceeds the Chandrasekhar mass, which due to their close orbit will merge in less than a Hubble time [13].

4. Conclusions

Recent works have shown that a common-envelope evolution is responsible for a significant fraction of PNe, with the number of known close-binary central stars now up to approximately 50. The continued work to search for and characterise close-binary central stars in PNe is critical in understanding not only the formation and evolution of PNe but poorly-understood common-envelope phase, and important phase in the formation of many other astrophysical phenomena. Even at this early stage, the investigation has revealed many problems with our understanding of the common-envelope phase, perhaps the most interesting being the apparent over-abundance of double-degenerate post-common-envelope systems (particularly interesting in the context of supernova type Ia formation [14]) and the clear evidence of a phase of pre-common-envelope mass transfer, neither of which fit with current theoretical models. The study of these central stars is also critical in the context of understanding the nature of post-common-envelope PNe and how they relate to the general PN population and other related phenomena [15].

References

[1] Miszalski B, Acker A, Moffat A F J, Parker Q A and Udalski A 2009 *Astron. Astrophys.* **496** 813
[2] Van Winckel H, Jorissen A, Exter K, Raskin G, Prins S, Perez Padilla J, Merges F and Pessemier W 2014 *Astron. Astrophys.* **563** 10
[3] De Marco O, Passy J-C, Frew D J, Moe M and Jacoby G H 2013, *Mon. Not. R. Astron. Soc.* **428** 2118
[4] Douchin D, De Marco O, Frew D J, Jacoby G H, Jasniewicz G, Fitzgerald M, Passy J-C, Harmer D, Hillwig T and Moe M 2015 *Mon. Not. R. Astron. Soc.* **448** 3132
[5] Jones D 2015 *EAS Publ. Series* **71-72** 113
[6] Miszalski B, Acker A, Parker Q A and Moffat A F J 2009 *Astron. Astrophys.* **505** 249
[7] Miszalski B, Jones D, Rodríguez-Gil P, Boffin H M J, Corradi R L M and Santander-García M 2011 *Astron. Astrophys.* **531** 158
[8] Jones D, Boffin H M J, Rodríguez-Gil P, Wesson R, Corradi R L M, Miszalski B and Mohamed S 2015 *Astron. Astrophys.* **580** 19
[9] De Marco O, Hillwig T C and Smith A J 2008 *Astron. J.* **136** 323
[10] Corradi R L M, Sabin L, Miszalski B, Rodríguez-Gil P, Santander-García M, Jones D, Drew J E, Mampaso A, Barlow M J, Rubio-Díez M M, Casares J, Viironen K, Frew D J, Giammanco C, Greimel R and Sale S E 2011 *Mon. Not. R. Astron. Soc.* **410** 1349
[11] Miszalski B, Boffin H M J and Corradi R L M 2013 *Mon. Not. R. Astron. Soc.* **428** 39
[12] Passy J-C, De Marco O, Fryer C L, Herwig F, Diehl S, Oishi J S, Mac Low M-M, Bryan G L and Rockefeller G 2012 *Astrophys. J.* **744** 52
[13] Santander-García M, Rodríguez-Gil P, Corradi R L M, Jones D, Miszalski B, Boffin H M J, Rubio-Díez M M and Kotze M M 2015 *Nature* **519** 63
[14] Tsebrenko D and Soker N 2015 *Mon. Not. R. Astron. Soc.* **447** 2568
[15] Corradi R L M 2016 *these proceedings*

The Hα surface brightness – radius plane as a diagnostic tool for photoionized nebulae

David J Frew[1,2], Ivan S Bojičić[1,2] and Quentin A Parker[1,2]

[1] Department of Physics, The University of Hong Kong, Hong Kong SAR, China
[2] Laboratory for Space Research, The University of Hong Kong, Hong Kong SAR, China

E-mail: djfrew@hku.hk

Abstract. The Hα surface brightness – radius ($S_{\mathrm{H}\alpha}$–r) relation is a robust distance indicator for planetary nebulae (PNe), further enhanced by different populations of PNe having distinct loci in $S_{\mathrm{H}\alpha}$–r space. Other types of photoionized nebulae also plot in quite distinct regions in the $S_{\mathrm{H}\alpha}$–r plane, allowing its use as a diagnostic tool. In particular, the nova shells and massive star ejecta (MSE) plot on relatively tight loci illustrating their evolutionary sequences. For the MSE, there is potential to develop a distance indicator for these objects, based on their trend in $S_{\mathrm{H}\alpha}$–r space. As high-resolution, narrowband surveys of the nearest galaxies become more commonplace, the $S_{\mathrm{H}\alpha}$–r plane is a potentially useful diagnostic tool to help identify the various ionized nebulae in these systems.

1. Introduction

This paper explores the potential of the $S_{\mathrm{H}\alpha}$–r plane as a diagnostic tool for the study of photoionized emission nebulae. As part of our work on determining distances for Galactic planetary nebulae (PNe), we developed and calibrated the Hα surface brightness – radius ($S_{\mathrm{H}\alpha}$–r) relation [1], which has been shown to be more accurate than previous statistical distance methods, particularly for the demographically common 'senile' PNe. Its application requires an angular size, an integrated Hα flux, and the reddening to the PN. From these quantities, an intrinsic radius is calculated, which when combined with the angular size, yields the distance directly. Rather than using the radio domain, we chose the Hα emission-line, firstly as it best represents the nebular ionized mass, and secondly because a number of narrowband Hα imaging surveys have recently become available [2][3][4][5], from which accurate integrated fluxes and/or diameters can be determined [6][7][8]. Ongoing Hα surveys [9] will continue to be useful, as part of our ongoing efforts to characterize the entire Galactic PN population [10].

2. The Hα relation for planetary nebulae

In [1] we described in detail the construction of a catalogue of Hα fluxes, angular diameters, and distances for both Galactic and extra-galactic PNe, to be used as primary calibrators for the $S_{\mathrm{H}\alpha}$–r relation. This relation was refined from earlier versions [11][12][13][14][15] published previously, based on continued improvement in the input data, with consequent reduction of the uncertainties. A range of criteria to further improve the precision of the $S_{\mathrm{H}\alpha}$–r relation were investigated by [15], with [1] eventually separating PNe into two broad groups based on spectroscopic criteria. Optically thick PNe (with strong [O II] and [N II] lines) are systematically

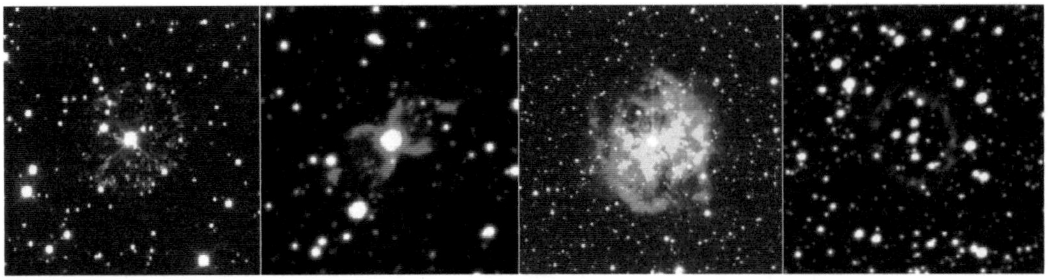

Figure 1. Color images of four 'PN mimics' taken from [10]: from left, the nova shell around GK Per, the bipolar symbiotic outflow Hen 2-104, and the WR ejecta M 1-67 and PCG 11. This mosaic illustrates the morphological confusion between PNe and other kinds of stellar ejecta.

more massive than the optically-thin PNe that fall along the lower part of the PN locus. Using sub-trends has allowed more precision in determining distances, as good as 18 per cent in the case of optically-thin PNe. Recently, [16] analysed the commonly-used SSV distance scale [17] in some depth. In particular, this scale has a substantial scale error at large PN radii, meaning that the distances for evolved (demographically-common) PNe are considerably underestimated, by a factor of two-or-so. Our revised scale [1] is largely free of this problem, having improved on the mean $S_{\mathrm{H}\alpha}$–r relation of [15], which in turn had been independently validated [16][18][19] as the most reliable statistical distance scale to date. We expect our distances to remain useful even after the expected data avalanche from the Gaia satellite [21], as many PN central stars are fainter than the Gaia magnitude limit, or have confusion problems in compact PNe of high surface brightness [22]. Of course the Gaia parallaxes will allow the refinement of our proposed sub-trends in the S–r plane, enhancing its ability both as a diagnostic tool, and as a robust measure for the many PNe without trigonometric distances.

3. Other Emission Nebulae

Besides the PNe discussed in [1], we are also interested in the ability of the $S_{\mathrm{H}\alpha}$–r diagram to discriminate between bona fide PNe, transitional objects, and the zoo of PN-like nebulae and outright mimics (see Fig. 1) that have been confused with them [23][24], both in the Milky Way and in the nearest galaxies. It is well known that some bipolar PNe have similar morphologies to the outflows around D-type symbiotic stars (SyS), and their similarities and differences have been discussed several times in the literature, e.g. [25][26][27][28]. To investigate these nebulae, we have adopted Hα fluxes from the literature, supplemented with our own data [6][7][10], while distances have been taken from the sources given in [1], supplemented with a few distances from [20]. Note that the locations of the SyS nebulae in the $S_{\mathrm{H}\alpha}$–r plot are only indicative, as their fluxes and dimensions are difficult to measure precisely. For the ejecta around luminous blue variable (LBV) and Wolf-Rayet stars [29][30], we adopt the data directly from [1], as we do for the low-mass H II regions in the ISM. We also plot a range of nova shells where we utilize the rather heterogeneous data compiled by [31][32][33]. Finally, we add the faint bipolar nebula surrounding CK Vul [34][35], the bowshock nebulae around the nova-like cataclysmic variables (CVs) BZ Cam [36][37] and V341 Ara [15][38], as well as the peculiar object Te 11 [39][40].

4. Discussion

In Fig. 2, we plot a range of different photoionized nebulae in $S_{\mathrm{H}\alpha}$–r space. Besides the primary PN locus, some transitional PNe and miscellaneous PN-like objects described in [1] are also plotted. Particularly interesting are the locations of the core and lobes of the outflow KjPn8 [41], showing the large range in ejecta mass between outflow episodes. The massive star ejecta (MSE) generally plot above the PNe, reflecting their larger ionized masses in the mean. For the

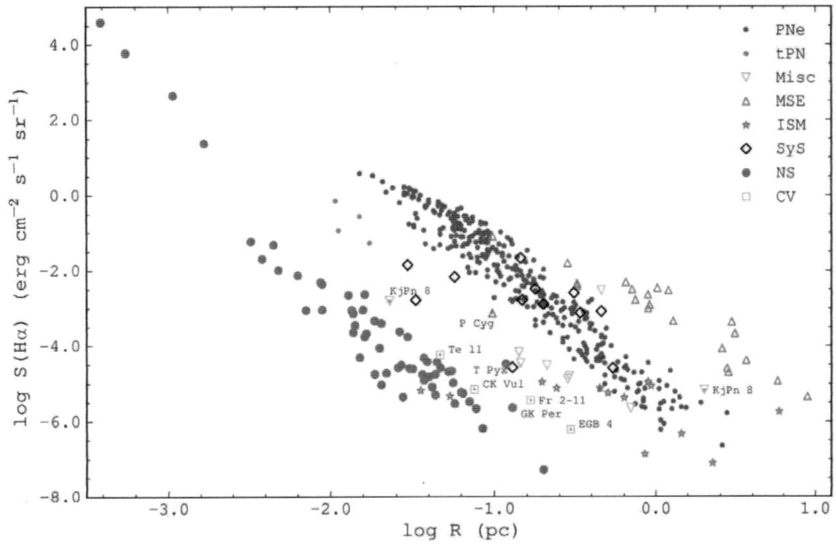

Figure 2. PNe and mimics plotted in the $S_{\rm H\alpha}$–r plane. Transitional PNe (tPN), miscellaneous objects (Misc), massive star ejecta (MSE), low-mass H II regions in the ISM, symbiotic (SyS) outflows, nova shells (NS), and CV nebulae have been plotted separately. Individual objects discussed in the text are labelled.

MSE, a reasonably tight evolutionary trend is seen, if we exclude the young, low-mass outflow around the historical LBV, P Cygni. The points can be fit by a relation with a slope of −2.3, markedly shallower than the PN locus, indicating the substantial amount of circumstellar and/or interstellar material being swept up in the evolved MSE. The data suggest an approximate distance scale can be developed for ejecta around LBVs and WN stars, at least for those shells that are not dominated by snow-plowed interstellar matter.

The nova shells, as expected, show a steep relation (power law slope $\simeq -5$) at young ages due to adiabatic expansion, before the trend flattens at larger radii as additional circumstellar material is swept up. The unusual (massive) shells around GK Per and T Pyx [42][43][44] are seen to be above the mean shell trend. Note there is at least one PN hosting a classical nova [45], as well a very faint bipolar nebula [43][46] surrounding the nova ejecta from GK Per, with properties similar to a low-mass PN [1]. The CV bowshock nebulae (EGB 4 and Fr 2-11) are clearly seen to be of lower ionized mass than PNe, though apparently distinct to classical nova shells. Te 11 has affinities with these nebulae, based on its nebular abundances and very low expansion velocity; while it plots close to the ordinary nova shells, its distinct morphology (i.e. a high volume filling factor) suggests a greater ionized mass and different origin, cf. [40]. The bipolar nebula around CK Vul is seen to have a mass comparable to a nova shell, but its origin remains uncertain [35]. As shown in [1] the diffuse H II regions in the ISM ionized by low-mass stars are generally of low to very-low surface brightness and plot on and around the PN locus at medium to large radii. What is somewhat unexpected is the substantial overlap in phase space between the symbiotic outflows and PNe, with some evolved SyS nebulae being as massive as typical PNe. Along with the emission-line diagnostic diagrams we have developed [23][47][48], the $S_{\rm H\alpha}$–r plane will also be a useful adjunct to help identify ionized nebulae in and beyond the Milky Way, particularly with the advent of deep hydrogen-line surveys using the next generation of telescopes. A more detailed discussion will be published elsewhere.

References

[1] Frew D J, Parker Q A and Bojičić I S 2016 *Mon. Not. R. Astron. Soc.* **455** 1459

[2] Gaustad J E, McCullough P R, Rosing W and Van Buren D 2001 *Publ. Astron. Soc. Pac.* **113** 1326
[3] Haffner L M, Reynolds R J, Tufte S L, Madsen G J, Jaehnig K P and Percival J W 2003 *Astrophys. J. Suppl. Series* **149** 405
[4] Drew J E *et al* 2005 *Mon. Not. R. Astron. Soc.* **362** 753
[5] Parker Q A *et al* 2005 *Mon. Not. R. Astron. Soc.* **362** 689
[6] Frew D J, Bojičić I S and Parker Q A 2013 *Mon. Not. R. Astron. Soc.* **431** 2
[7] Frew D J, Bojičić I S, Parker Q A, Pierce M J, Gunawardhana M L and Reid W A 2014a *Mon. Not. R. Astron. Soc.* **440** 1080
[8] Madsen G J, Frew D J, Parker Q A, Reynolds R J and Haffner L M 2006 *Proc. Int. Astron. Union* (S234) **2** 455
[9] Drew J E *et al* 2014 *Mon. Not. R. Astron. Soc.* **440** 2036
[10] Parker Q A *et al* 2015 *Amer. Astron. Soc. Meeting* **225** 108.06
[11] Pierce M J, Frew D J, Parker Q A and Köppen J 2004 *Publ. Astron. Soc. Australia* **21** 334
[12] Frew D J, Parker Q A and Russeil D 2006 *Mon. Not. R. Astron. Soc.* **372** 1081
[13] Frew D J and Parker Q A 2006 *Proc. Int. Astron. Union* (S234) **2** 49
[14] Frew D J and Parker Q A 2007 *Proc. APN IV Conf.* eds Corradi R L M *et al* id. 68 (online at `http://www.iac.es/proyect/apn4`)
[15] Frew D J 2008 *PhD thesis* (Macquarie University, Sydney)
[16] Smith H 2015 *Mon. Not. R. Astron. Soc.* **449** 2980
[17] Stanghellini L, Shaw R A and Villaver E 2008 *Astrophys. J.* **689** 194
[18] Jacob R, Schönberner D and Steffen M 2013 *Astron. Astrophys.* **558** A78
[19] Ali A, Ismail H A and Alsolami Z 2015 *Astrophys. Space Sci.* **357** 21
[20] Vickers S B, Frew D J, Parker Q A and Bojičić I S 2015 *Mon. Not. R. Astron. Soc.* **447** 1673
[21] Cacciari C, Pancino E and Bellazzini M 2016 *Astron. Nachr.* in press (arXiv:1512.03658)
[22] Manteiga M, Fustes D, Dafonte C and Arcay B 2014 *Proc. APN VI Conf.* eds Morisset C *et al* id. 57 (online at `http://www.astroscu.unam.mx/apn6/PROCEEDINGS/`)
[23] Frew D J and Parker Q A 2010 *Publ. Astron. Soc. Australia* **27** 129
[24] Frew D J, Madsen G J, O'Toole S J and Parker Q A 2010 *Publ. Astron. Soc. Australia* **27** 203
[25] Corradi R L M and Schwarz H E 1995 *Astron. Astrophys.* **293** 871
[26] Corradi R L M 2003 in *ASP Conf. Proc.* vol 303 *Symbiotic Stars Probing Stellar Evolution*, ed R L M Corradi, J Mikolajewska and T J Mahoney (San Francisco: Astronomical Society of the Pacific) p 393
[27] Kwok S 2003 in *ASP Conf. Proc.* vol 303 *Symbiotic Stars Probing Stellar Evolution*, ed R L M Corradi, J Mikolajewska and T J Mahoney (San Francisco: Astronomical Society of the Pacific) p 428
[28] Miszalski B., Boffin H M J, Frew D J, Acker A, Köppen J, Moffat A F J and Parker Q 2012 *Mon. Not. R. Astron. Soc.* **419** 39
[29] Chu Y-H 2003 *Proc. IAU* **S212** 585
[30] Weis K 2003 *Astron. Astrophys.* **408** 205
[31] Downes R A and Duerbeck H W 2000 *Astron. J.* **120** 2007
[32] Downes R A, Duerbeck H W and Delahodde C E 2001 *J. Astronomical Data* **7** 6
[33] Gill C D and O'Brien T J 2000 *Mon. Not. R. Astron. Soc.* **314** 175
[34] Hajduk M *et al* 2007 *Mon. Not. R. Astron. Soc.* **378** 1298
[35] Hajduk M, van Hoof P A M and Zijlstra A A 2014 *Mon. Not. R. Astron. Soc.* **432** 167
[36] Greiner J, Tovmassian G, orio M, Lehmann H, Chavushyan V, Rau A, Schwarz R, Casalegno R and Scholz R-D 2001 *Astron. Astrophys.* **376** 1031
[37] Hollis J M, Oliversen R J, Wagner R M and Feibelman W A 1992 *Astrophys. J.* **393** 217
[38] Frew D J, Madsen G J and Parker Q A 2006 *Proc. Int. Astron. Union* (S234) **2** 395
[39] Jacoby G H *et al* 2010 *Publ. Astron. Soc. Australia* **27** 156
[40] Miszalski B *et al* 2016 *Mon. Not. R. Astron. Soc.* **456** 633
[41] López J A, Vázquez R and Rodríguez L F 1995 *Astrophys. J.* **455** L63
[42] Liimets T, Corradi R L M, Santander-García M, Villaver E, Rodríguez-Gil P, Verro K and Kolka I 2012 *Astrophys. J.* **761** 34
[43] Shara M M, Zurek D, De Marco O, Mizusawa T, Williams R and Livio M 2012 *Astron. J.* **143** 143
[44] Shara M M *et al* 2015 *Astrophys. J.* **805** 148
[45] Wesson R *et al* 2008 *Astrophys. J.* **688** L21
[46] Bode M F, Roberts J A, Whittet D C B, Seaquist E R and Frail D A 1987 *Nature* **329** 519
[47] Sabin L, Parker Q A, Contreras M E, Olguín L, Frew D J, Stupar M, Vázquez R, Wright N J, Corradi R L M and Morris R A H 2013 *Mon. Not. R. Astron. Soc.* **431** 279
[48] Frew D J, Bojičić I S, Parker Q A, Stupar M, Wachter S, DePew K, Danehkar A, Fitzgerald M T and Douchin D 2014b *Mon. Not. R. Astron. Soc.* **440** 1345

Journal of Physics: Conference Series **728** (2016) 032016 doi:10.1088/1742-6596/728/3/032016

A chemically peculiar post-AGB star in the Small Magellanic Cloud

D. Kamath[1], H. Van Winckel[1], K. De Smedt[1] and P.R.Wood[2]

[1] Instituut voor Sterrenkunde, K.U.Leuven, Celestijnenlaan 200D bus 2401, B-3001 Leuven, Belgium

[2] Research School of Astronomy and Astrophysics, Mount Stromlo Observatory, Weston Creek, ACT 2611, Australia

E-mail: [1] devika.kamath@ster.kuleuven.be

Abstract. Post-Asymptotic Giant Branch (post-AGB) stars bear signatures of the entire chemical and morphological changes that occur prior to and during the AGB phase of evolution. These objects also provide vital clues on the ultimate fate of the star. Detailed chemical abundance studies of some of these objects have shown that they are chemically much more diverse than anticipated. As expected, some are the most *s*-process enriched objects known to date while others are not *s*-process enriched. Our recent study has revealed a star in the Small Magellanic Cloud, J005252.87-722842.9, which displays a peculiar chemical signature that does not correspond to the expected chemical diversity observed in these objects. This unique object reveals the possibility of a new stellar evolutionary channel where the star evolves without any third dredge-up episodes or during its evolution becomes devoid of its nucleosynthetic history.

1. Introduction

Post-AGB stars are low- to intermediate-mass stars that have evolved off the AGB because a strong dusty mass loss removed almost the whole stellar envelope. This mass loss occurs either by binary interaction or, for single stars, by a phase of very high mass-loss rate called the superwind [1]. Their atmospheres hence contain signatures of the chemical enrichment from internal nucleosynthesis that has occurred prior to and during their entire AGB lifetime. During the post-AGB phase, the warm stellar photosphere makes it possible to quantify the chemical abundances in post-AGB stars for a very wide range of elements from CNO up to the most heavy *s*-process elements, well beyond the Ba peak [2], that are brought to the stellar surface during the AGB phase [3].

The observationally well-studied Galactic sample of post-AGB stars resulted in some interesting discoveries. Firstly, the Galactic post-AGB stars have been found to be chemically much more diverse than anticipated: 21 micron sources, *s*-process enriched single stars, non *s*-process enriched objects, and binaries that display photospheric depletion. Secondly, Galactic post-AGB can be classified into two groups based on their spectral energy distributions (SEDs). The first group of objects (shell-sources), display a double peaked SED. This type of SED is most likely characteristic of single post-AGB stars. The second group of objects (disc-sources) display SEDs with an onset in the near-IR, indicating the presence of hot-dust which represents a stable compact circumbinary disc, characteristic of binary post-AGB stars. See [4] for a review on post-AGB stars.

Though the Galactic sample is observationally well studied the poorly known distances (and hence luminosities and masses) of the Galactic post-AGB sample, hamper the interpretation of their abundances in the broader theoretical context of stellar (chemical) evolution. Due to the known distances to the Magellanic Cloud (MC) post-AGB stars, these objects provide excellent constraints and offer unprecedented tests for AGB theoretical models of low- and intermediate-mass stars, as a function of initial mass and metallicity. Our recent search for MC post-AGB stars [5, 6], in the SMC and LMC, respectively. yielded a spectroscopically verified catalogue of optically visible post-AGB stars.

We are currently studying the MC post-AGB objects in detail. Our studies so far have revealed that the MC objects also display a chemical and morphological diversity similar to the Galactic post-AGB stars [7] and references therein. Our recent study reveled a peculiar object, J005252.87-722842.9, a SMC post-AGB star, that does not follow any of the observed chemical diversity trends mentioned above and does not conform to the standard single or binary evolution and nucleosynthesis scenarios.

2. Data and Observations

J005252.87-722842.9 is a SMC post-AGB star first discovered in our extensive low-resolution spectroscopic survey of optically bright post-AGB stars in the SMC [5]. The photometry covering the optical, near-IR and mid-IR bands are presented in [5]. Based on the visual inspection of the position of the peak of the dust excess in the SED of this object (see Figure 1), it is likely that it is a shell-type source with an expanding dusty circumstellar shell.

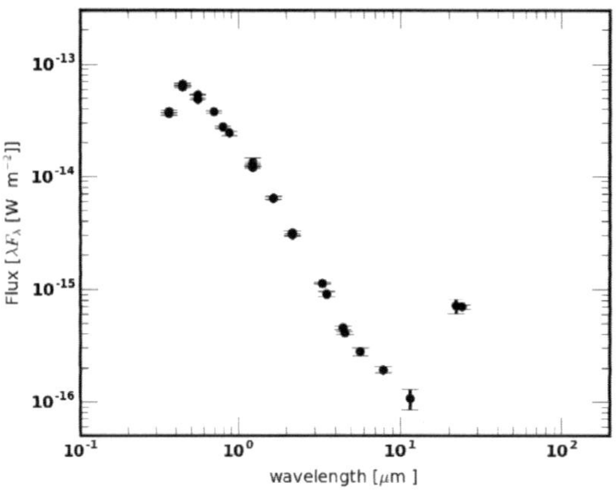

Figure 1. Spectral energy distribution of J005252. The black symbols indicate the broadband photometry corrected for foreground extinction.

We obtained high-resolution spectra using the UVES echelle spectrograph, mounted on the 8m UT2 Kueyen Telescope of the VLT array at the Paranal Observatory of ESO in Chili. Full details of the spectra will be presented Kamath et al., 2016, in prep. We note that the data reduction procedure is presented in [7].

3. Spectral analysis

To study the photospheric composition of J005252, we performed a systematic spectral analysis for this object. Using the UVES spectra, we first determined accurate atmospheric parameters for this object and then we carried out a detailed abundance analysis study. The atmospheric parameter determination and the detailed abundance analysis of J005252 will be presented in Kamath et al., 2016, in prep. The determined stellar parameters of J050221 are listed in Table 1.

Table 1. Spectroscopically determined atmospheric parameters of J005252. N_{FeI} and N_{FeII} show the number of lines used for Fe I and Fe II respectively. Note: Assumed solar [Fe/H] = 7.50 dex [8]. $L_{ph}/L_{\odot}$ represents the photospheric luminosity of J005252 and $M/M_{\odot}$ is the derived mass of J00525.

Stellar Parameters	$J05252_{UVES}$
T_{eff} (K)	8500 ± 125
$\log g$ (dex)	1.50 ± 0.25
ξ_t (kms^{-1})	2.0 ± 0.25
[FeI/H]	-1.11 ± 0.01
[FeII/H]	-1.02 ± 0.07
[Fe/H]	-1.20 ± 0.15
N_{FeI}	34
N_{FeII}	39
E(B-V)	0.05
$L_{ph}/L_{\odot}$	9000
$M/M_{\odot}$	0.65

Our detailed abundance analyses showed that J005252 has rather intriguing abundances. It is relatively metal poor (with [Fe/H] =-1.2 ± dex) with [O/Fe] = 0.6 ± dex. It shows no signs of Carbon (including the strong C I line at 6587 Å, see Figure 2). It also shows no signs of s-process enrichments. All post-AGB stars that are s-process enriched shows the presence of Barium (see [7] and references therein). However no traces of Ba was found in the spectra of J005252 (see Figure 2). Furthermore, we also find that the photosphere of J005252 does not show any signs of photospheric depletion (see [9] and references therein).

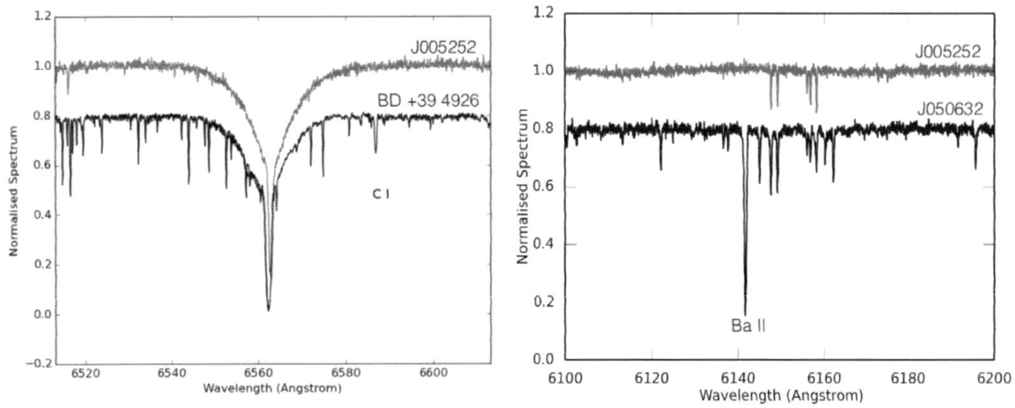

Figure 2. Right Panel: Normalised spectra of J005252 (in green) and of BD +39 4926, a carbon-enhanced Galactic object (in black), covering the region that contains a strong C I line at 6587 Å. Left Panel: Normalised spectra of J005252 (in green) and of J050632, a s-process rich LMC object (in black), covering the region that contains a strong BaII line at 6142 Å.

4. Understanding the photospheric chemistry of J005252

Assuming, that J005252 is a single star (due to its non-depleted photospheric chemistry and also based on the shell-type SED of the object), stellar evolution and nucleosynthesis models [10, 11] that are appropriate for J005252 (based on its luminosity, initial mass and metallicity) predict this object to have undergone TDU events and therefore be C-rich and likely to be *s*-process enhanced. Synthetic models of TP-AGB stars that are calculated to fit the SMC CSLF also predict a star with the luminosity and metallicity of J005252 to be C-rich. We consider the possibilities of nucleosynthetic process, such as hot bottom burning (HBB) [12, 13] that can prevent the formation of high-luminosity carbon stars. However, HBB is only active in stars with $M \gtrsim 4-5\,M_{\odot}$. Since J005252 is star with $M \approx 1.5-2\,M_{\odot}$ and $Z=0.001$, it is likely that the temperature at the base of the convective envelope during the interpulse period is not high enough to activate HBB.

Therefore, the only way to explain the intriguing chemistry of J005252 is via some mechanism that restricts the third dredge-up, possibly a mass-loss mechanism that results in an AGB life without dredge-up episodes. Alternatively, this object could be a product of a merger scenario wherein it is not easy to trace the merger process and therefore understand the nucleosynthetic history of the end-product.

5. Conclusions

In this paper, we present the discovery of J005252, a chemically peculiar star in the SMC. This object is luminous (9000 $L_{\odot}$), metal-poor ([Fe/H] = -1.2 ± 0.08 dex) and has a shell-type SED. This object shows no traces of carbon, indicating that it has not undergone any third-dredge up episode during its entire AGB phase. This is contradictory to what is expected from a object with low-metallicity, low initial mass (1.5 to 2.0 $M_{\odot}$) and a shell-type SED (which indicates that it is likely to be a single star). Furthermore, J005252 shows neither signs of an *s*-process enrichment, which is characteristic of single post-AGB stars, nor a depleted photospheric chemistry which is characteristic of binary post-AGB stars. Therefore, this object provides the first observational evidence for a star that fails the third dredge-up.

Acknowledgments

DK and HVW acknowledge the support of the KU Leuven contract GOA/13/012. PRW has received support from the Australian Research Council Discovery Project DP120103337.

References

[1] Nie J D, Wood P R and Nicholls C P 2012 *Mon. Not. R. Astrophys. Soc.* **423** 2764
[2] Reyniers M and Van Winckel H 2003 *Astron. Astrophys* **408** 33
[3] Karakas A I and Lattanzio J C 2007 *Publ. Astron. Soc. Australia* **24** 103
[4] Van Winckel H 2003 *Astron. Astrophys. Rev.* **41** 391
[5] Kamath D, Wood P R and Van Winckel H 2014 *Mon. Not. R. Astrophys. Soc.* **439** 2211
[6] Kamath D, Wood P R and Van Winckel H 2015 *Mon. Not. R. Astrophys. Soc.* **454** 1468
[7] De Smedt K, Van Winckel H, Kamath D and Wood P R 2015 *Astron. Astrophys* **583** 56
[8] Asplund M, Grevesse N, Sauval A J and Scott P 2009 *Astron. Astrophys. Rev.* **47** 481
[9] Maas T, Van Winckel H and Lloyd Evans T 2005 *Astron. Astrophys* **429** 297
[10] Karakas A I 2010 *Mon. Not. R. Astrophys. Soc.* **403** 1413
[11] Fishlock C K, Karakas A I, Lugaro M and Yong D 2014 *Astrophys. J.* **797** 44
[12] Boothroyd A I, Sackmann I J and Ahern S C 1993 *Astrophys. J.* **416** 762
[13] Lattanzio J, Frost C, Cannon R and Wood P R 1996 *Mem. Soc. Astron. Italiana* **67** 729

Journal of Physics: Conference Series **728** (2016) 042001 doi:10.1088/1742-6596/728/4/042001

Transient classification and novae ejecta

Robert Williams

Space Telescope Science Institute, Baltimore, MD 21218, USA

E-mail: wms@stsci.edu

Abstract. A third parameter, in addition to luminosity and rate of brightness decline, derived from the spectra of transients is suggested as a means of more accurately classifying objects in outburst. Principal component analysis of the spectra of transients is suggested as the best way to determine the third parameter. A model is suggested for novae ejecta that is based on the ballistic ejection of an ensemble of clouds having a wide range of sizes. Short term brightness fluctuations of novae, the formation of dust, and the production of X-ray emission follow naturally from such a picture of the evolving clouds.

1. Introduction

Although the basics of the nova outburst have been understood for some time there is still uncertainty about some aspects of the phenomenon. Outbursts are the result of steady accretion but there may be other triggers. What causes some novae to vary 1-2 magnitudes in brightness over a few weeks time? What are the key parameters that determine how and when some novae develop copious dust? In recent years new luminous transients have been discovered that do not fit into either the normal nova or supernova paradigm, including some that are observed only in the infrared. It is important to determine what observations may be most useful in understanding the physical mechanism driving these outbursts.

2. Classifying Transients

The standard plot used to describe transients is one of luminosity vs. brightness decline time. The significance of this plot derives from Arp's classic study of the novae in M31 [1], which he undertook immediately after completing his doctoral thesis at Caltech with the support of W. Baade, A. Sandage, and E. Hubble shortly before the latter's death. Observing 290 clear nights on Mt. Wilson over a 1.5 yr interval in a Herculean effort Arp found that rapidly declining novae were more luminous than slow decliners. The resulting Maximum Magnitude vs. Rate of Decline (MMRD) relation, calibrated independently from the Cepheid, RR Lyrae, H II region distances for M31, was useful to Sandage as a more refined primary extragalactic distance indicator compared to the less accurate average absolute magnitude of the few novae determined by Curtis [2] in his famous debate with Shapley over the size of the Milky Way and distance to M31.

The MMRD relation has an intrinsic dispersion that has steadily increased over time with the larger sample of novae that have been observed within nearby galaxies since Arp's original survey. Until the past two decades a plot of visible luminosity vs. time of decline for transients was dominated primarily by novae and SNe that occupied a relatively small region of space in the plot. However, as seen in figure 1 whose data come primarily from the Palomar Transient

Factory survey [3] [4], the MMRD relation is at best a 'tendency', with many variant outlier novae. And transients of all types occupy virtually every part of the luminosity-decay time diagram, as can be seen in figure 2.

Historically, the position of a transient on a plot of maximum magnitude vs. light decay time has been a standard way of expressing the nature or type of the transient. However, it is clear from figure 2 that there is so much overlap of different transient types in the diagram that little information on the nature of a transient results from its position in this diagram. Since many transients have a broad distribution of decay times following outburst the decline time is not much of a defining characteristic of a transient. Rather, some other parameter would be much more useful in distinguishing between different transient phenomena.

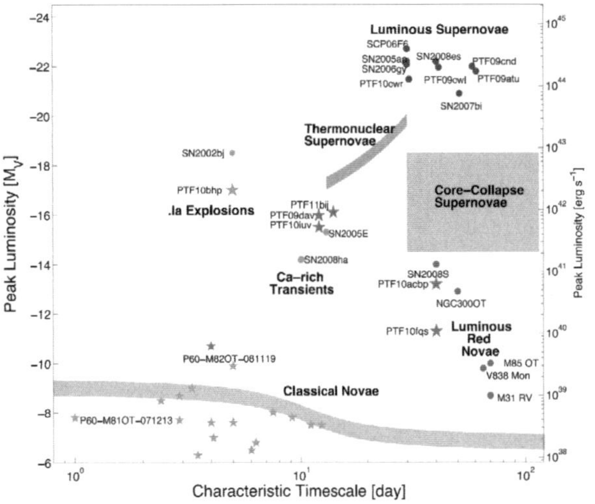

Figure 1. MMRD relationship with transients, from [4]

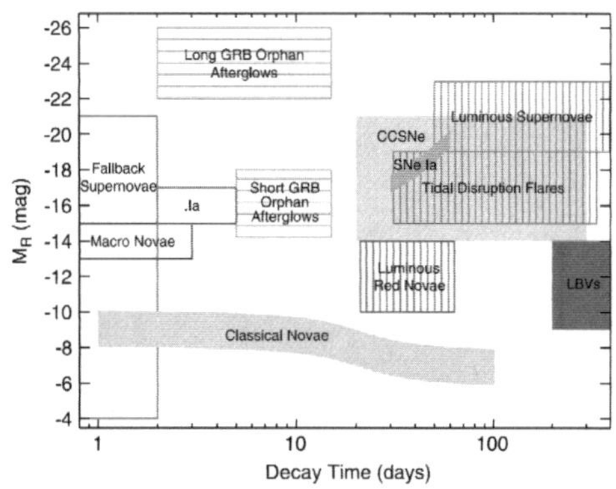

Figure 2. Luminosity vs. decay time plot for transients, from [3]

A similar situation confronted stellar population studies after studies of galaxies showed that three parameters were important for defining the stellar population of a galaxy: age, metal abundance, and star formation history. This led to the concept of the 3-dimensional population box [5]. The same situation appears to apply to transients, where the identification of a third parameter, in addition to the luminosity and decline time, would be useful in distinguishing between different classes of transients.

Spectroscopy of transients offers easily obtained information that allows kindred objects to be identified and differentiated among the different types of objects. Thus, the spectroscopic characteristics of a transient are arguably the best way to determine what the object is and how it may be similar to or different from other transients. This same problem has been addressed by the quasar community, who have applied principal component analysis (PCA) to quasar spectra [6] [7] to define those properties most useful in discriminating between different quasar properties and types.

Our understanding of transients would be greatly improved by subjecting the spectra of all such objects, i.e., novae, supernovae, stellar mergers, LBVs, [Ca II] transients, etc., to PCA. The breakdown of the main spectral features into principal components, or eigenvectors, could then be used to identify the largest PC (pc1, or eigenvector1 to use the nomenclature of the quasar community) that could be used as the third parameter to create a 3-dimensional 'transient cube', as shown in figure 3. Pc1 might or might not correspond to some clear spectroscopic property such as the continuum slope or intensity of H-alpha or strength of narrow absorption lines, etc., however it almost certainly would better enable transients that are similar to be defined and differentiated from other types of transients. Classifying transients by plotting them on a diagram of luminosity vs. a principal component, such as pc1, would be much more instructive than the current practice of showing their position in a plot of luminosity vs. rate of brightness decline.

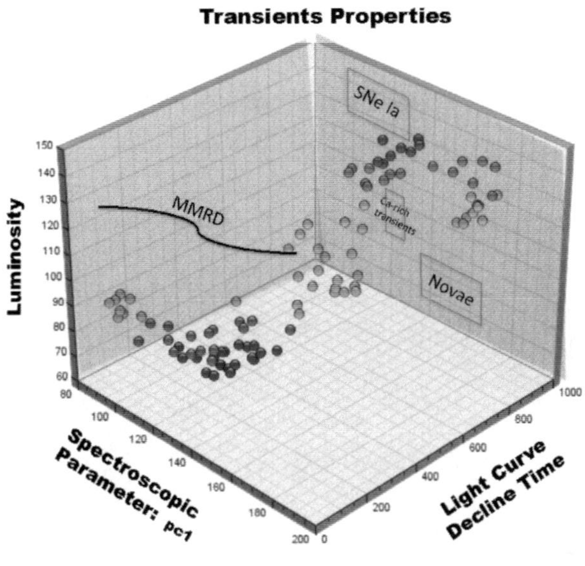

Figure 3. Consideration of a third parameter that characterizes the spectrum of a transient would provide important information that better allows identification of the class of the transient.

3. Novae Ejecta Geometry

Many novae, especially those having slower declines in brightness from maximum, show variation in visible brightness of several magnitudes over timescales of weeks. Multiple secondary maxima sometimes occur in addition to short-lived up-and-down fluctuations in brightness, i.e., jitter, that last a few days. These are no doubt due to changing conditions in the ejecta, but no systematic explanation for this behaviour has emerged.

Many novae form dust which produces an infrared continuum that is typically weaker than the optical continuum, although in the case of substantial dust formation can almost completely absorb the visible continuum, transferring most of the visible luminosity into the IR. High density regions in the ejecta that are protected to some extent from high UV radiation so that molecular species are plentiful are believed to define the locale of the dust formation.

Repeated observations of postoutburst novae by the *Swift* satellite have detected X-ray emission in most novae [8]. Initially, the X-rays have hard 2 − 10 keV energies which progressively weaken until eventually soft X-rays emerge as the expanding ejecta reveal the hot post-TNR WD surface.

All three features relating to the brightness variations, dust formation, and the evolution of the X-ray emission may be explained by constraining various ejecta parameters which do evolve over time due to their expansion. However, there has not been a coherent explanation of all of these phenomena that relates them to each other and explains how they differ so much between different novae.

We propose here that fluctuations in brightness, the formation of dust, and observed X-ray emission from novae in postoutburst decline all follow naturally from ejecta that are formed from an ensemble of dense clouds of different sizes that are ballistically ejected by the TNR. The small clouds collide with each other, emitting hard X-rays as they rapidly dissipate to form a low density ambient medium. Less common large clouds have dense cores surrounded by a

Journal of Physics: Conference Series **728** (2016) 042001 doi:10.1088/1742-6596/728/4/042001

photosphere that provides protection for the outer dense regions where dust can form. And when a large cloud emerges from the lower density ambient region of ejecta its radiation will produce a variation in brightness of the ensemble. The large variation in characteristics of different novae can then be understood in terms of differing size distribution of the ejected clouds from object to object. An example of the geometrical model for novae ejecta proposed here is shown in figure 4.

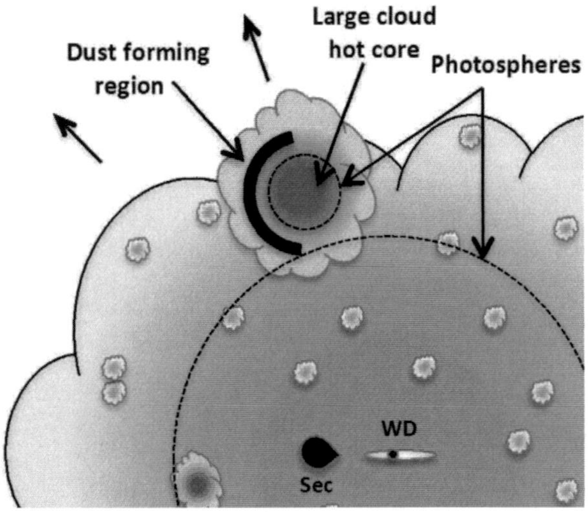

Figure 4. An expanding large cloud emerges from the ejecta and causes a brightness variation as it dissipates. Its dense, cooler layers form dust. X-rays are produced by collisions among the numerous small clouds.

Acknowledgments
Finally, I wish to pay tribute to our colleague Dr. Sun Kwok, who having done such a fine job of organizing this conference is now stepping back from his administrative responsibilities as HKU Dean for the past decade to return to full time research. Sun is a fine example of scientist who has succeeded in combining a career of significant research results with effective international leadership. At the same time he has authored a number of highly regarded books on a variety of subjects that have served to educate the professional community. A model worth emulating, we wish Sun continued success and longevity within our ranks.

References
[1] Arp H C 1956 *Astron. J.* **61** 15
[2] Curtis H D 1917 *Publ. Astron. Soc. Pac.* **29** 206
[3] Rau A *et al* 2009 *Publ. Astron. Soc. Pac.* **121** 1334
[4] Kasliwal M 2011 *Bull. Astron. Soc. India* **39** 375
[5] Hodge P 1989 *Annu. Rev. Astron. Astrophys.* **27** 139
[6] Boroson T and Green R 1992 *Astrophys. J. Suppl. Series* **80** 109
[7] Francis P J and Wills B J 1999 *Quasars and Cosmology* (Astronomical Society of the Pacific Conference Series vol 162) ed Ferland G and Baldwin J p 363
[8] Schwarz G J *et al* 2011 *Astrophys. J. Suppl. Series* **197** 31

eMERLIN imaging of γ-ray nova V959 Mon's surprising evolution

F. Healy[1], T, J. O'Brien[2] and R. Beswick[2]

[1,2] Jodrell Bank Centre for Astrophysics, Alan Turing Building, University of Manchester, Oxford Road, Manchester M13 9PL

E-mail: [1] fiona.healy@postgrad.manchester.ac.uk

Abstract. High resolution radio images of γ-ray nova V959 Mon using e-MERLIN are presented here, at epochs in September 2012, November 2012, February 2013, October 2013 and February 2014. A light curve at C-Band frequency has been produced which is in good agreement with contemporaneous VLA observations. While early e-MERLIN observations of V959 Mon appear to show an east-west elongation in the ejecta morphology, subsequent observations show that the ejecta are now elongated in the north-south direction. Our high-resolution observations of the structure of V959 Mon can assist us in further understanding the mechanisms of mass ejection from novae.

1. Introduction

1.1. *Classical Novae*

A nova is a cataclysmic variable star in which the white dwarf primary undergoes a thermonuclear runaway on its surface, as a result of build-up of accreted material from the secondary. This leads to a large expulsion of matter from the WD surface, as well as a dramatic increase in the optical magnitude of the system. Novae are also radio sources, primarily due to thermal bremsstrahlung, although synchrotron emission can be detected if there are shocks present.

Models based on the radio light curves of novae assume a uniformly dense shell of expanding ejecta. The Hubble flow model assumes that all the material is ejected at once and expands as a thick, spherically symmetric shell, with the slowest parts of the gas (velocity V_0) at the inner radius, and the fastest parts (velocity V_1) at the outer radius. This model predicts a t^2 rise in radio intensity while the ejecta are optically thick, and a t^{-3} decay once the source has become optically thin. However, some observations of novae indicate poor agreement with such simple models (see for example [9], [6], [7]).

1.2. *V959 Mon*

V959 Mon (Nova Mon 2012) was first detected in June 2012 by the Fermi Large Area Telescope, as a GeV γ-ray transient source [1]. The presence of γ rays is unexpected in a nova as the Hubble flow model does not predict the high energies required to generate them. VLA spectra shortly after this time had a spectral index of $\alpha \approx -0.1$, indicating synchrotron emission rather than the expected optically thick emission. Ribeiro et al. 2013 report that V959 Mon has a bipolar ejecta structure, with an angle of inclination of $82° \pm 6°$ [13]. Its distance from Earth is 1.5 kpc [10]. Resolved 5 GHz EVN observations of V959 Mon (from September 2012) at 2-7 milliarcsecond (mas) resolution by [11] detected 2 compact components extending north-west and south-east, and moving apart with a proper motion of 0.45 mas/day.

Journal of Physics: Conference Series **728** (2016) 042002 doi:10.1088/1742-6596/728/4/042002

Table 1. *Total integrated fluxes of V959 Mon for each image presented in this section. The days since outburst are given with respect to the date of the FERMI detection.*

Epoch	Days Since Outburst	Total Flux (mJy)
18th Sept 2012	91	8 ± 2
12th Nov 2012	146	31 ± 1
22nd Nov 2012	156	27 ± 1
26th Feb 2013	252	14 ± 1
11th Oct 2013	479	11 ± 3
21st Feb 2014	612	5 ± 1

Due to V959 Mon's unusual emission and morphology, the Hubble flow model is unlikely to provide a complete description of its behaviour. Chomiuk et al. 2014 [4] suggest an alternative model in which the ejecta consists of two perpendicular bi-polar components: a slower-moving component which is accelerated by transfer of energy from the orbiting binary system, and a faster-moving component accelerated by winds from the WD surface, and emanating primarily from the polar regions. The faster-moving wind driven component would initially dominate radio images, before becoming optically thin, revealing the slower orbital driven component. At early times, shocks would occur at the interface between the components. This model would explain V959 Mon's unusual morphology and emission of γ rays.

2. Observations

The observations presented here were made with the e-MERLIN array over 6 epochs: September 18th 2012, November 12th 2012, November 22nd 2012, February 21st 2013, October 11th 2013 and February 21st 2014. The observations were made in the C Band (central frequency $\sim$ 5 GHz), providing a resolution of 40 mas, and with a bandwidth of 512 MHz. The bandwidth was divided into 4 intermediate frequency bands (IFs), each containing 512 channels. The data were averaged such that each IF effectively contained 128 channels.

The reduction and imaging of each epoch was carried out with AIPS. J0645+0541 was observed as a phase calibrator, 3C286 as a flux calibrator and OQ208 as a band-pass calibrator. The images were constructed using the different CLEAN restoring beams fitted to each one. They were then re-constructed using the same 70 mas $\times$ 70 mas circular beam for each epoch.

3. Results

Table 1 shows the total integrated flux for each epoch. Fluxes were calculated using two different methods; using the AIPS tasks JMFIT and IMSTAT. The average of the two results was taken, and the uncertainty was taken to be the difference between them. The data in Table 1 were used to construct light curves for V959 Mon, shown in Figure 1. While all the observations were made in the C-band, two different central frequencies were used; 5.7 GHz in 2012 and 5.0 GHz post-2012.

Intensity contour plots of V959 Mon at each epoch are shown in Figure 3. In the first four epochs, the ejecta appear to be elongated in the east-west direction. The elongation appears to become more pronounced with each successive epoch, with the east-west diameter expanding from an estimated 141 milliarcseconds in September 2012 to 280 milliarcseconds in February 2013. In the subsequent epochs (October 2013 and February 2014), the emission is elongated north-south rather than east-west. Again, an estimated increase in north-south diameter was observed between the epochs, with a diameter of 388 milliarcseconds in October 2013 increasing to a diameter of 458 milliarcseconds in February 2014. It is possible that this north-south

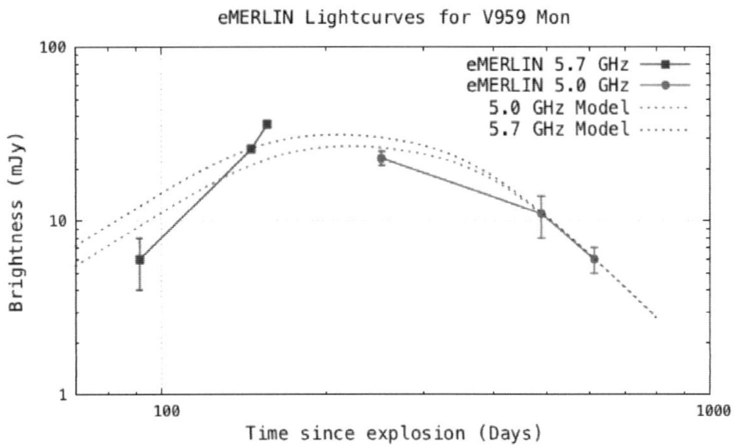

Figure 1. *e-MERLIN lightcurve for V959 Mon plotted alongside a model light curve calculated using the Hubble flow model (best fit parameters given in Discussion.*

structure can be seen to be emerging in the February 2013 epoch, which, though predominately elongated from east to west, also appears to have a north-south component.

4. Discussion

4.1. V959 Mon Light Curve

In Figure 1, a comparison is shown between the light curve of V959 Mon as observed by e-MERLIN, and a simulated light curve fit to the data using the Hubble flow model. When fitting the model to the observations, the distance to the nova (1.5 kpc, [10]) and the electron temperature of the ejecta (2×10^4 K, [4]) were fixed. The model fitting program returned best fit parameters of $1.1 \times 10^{-4} M_\odot$ for the ejecta mass, 2300 km s^{-1} for the outer velocity of the shell and 470 km s^{-1} for the inner velocity of the shell. Our observations appeared consistent with the light curves predicted by the Hubble flow model. However, only one frequency has been considered here; when multiple frequencies are considered, seeming agreements between model light curves and light curves constructed from observations can break down.

4.2. Morphology

Our resolved images of V959 Mon (Figure 3) display an extension from east to west (perpendicular to V959 Mon's orbital plane) in the first four epochs. In the final two epochs shown in Figure 2, an emerging north-south structure can be seen. This evolution in the morphology is consistent with the predictions of Chomiuk et al. (2014) [4]. For a further discussion of V959 Mon's complex morphology, see Healy et al. (2016, in prep).

5. Conclusion

High resolution eMERLIN observations have been presented here of V959 Mon at six epochs ranging from September 2012 to February 2014. The observations were made at C-band (~5 GHz). Resolved images show a complex, aspherical, evolving morphology, which changes its orientation from being elongated east-west to north-south. The variable morphology of V959 Mon is consistent with the hypothesis of Chomiuk et al. 2014 [4] that nova ejecta consists of a slow component driven by orbital energy transferred from the rotating binary, and a faster-moving polar component driven by winds on the white dwarf surface.

To further our understanding of V959 Mon's unusual behaviour, we will test the model proposed by [4] using radio emission models. By constructing density distributions representative

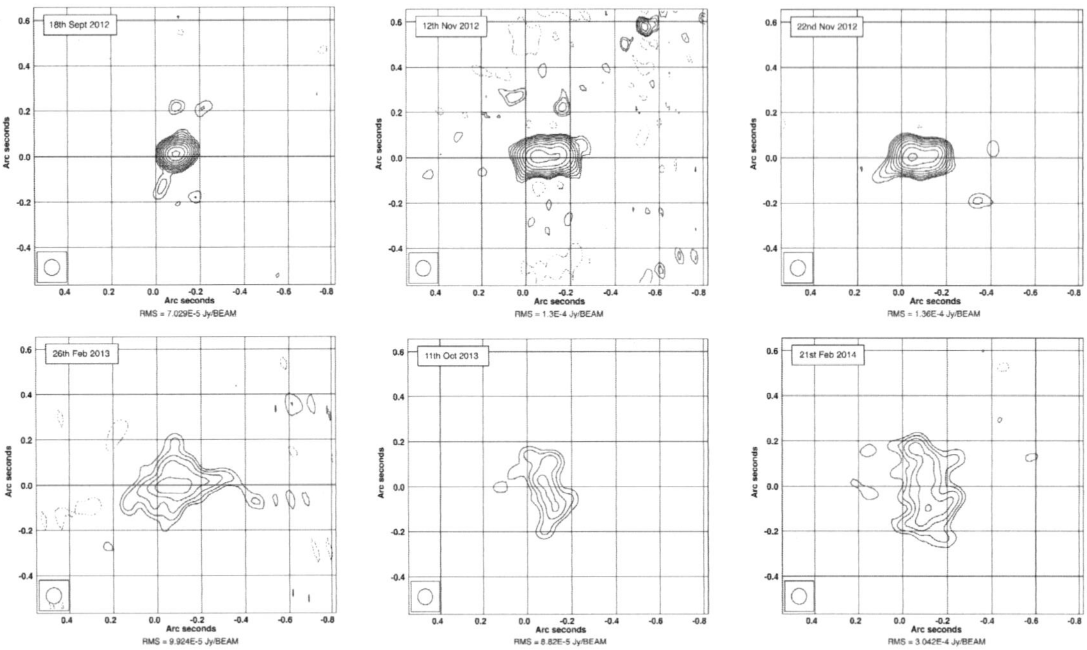

Figure 2. *Intensity contour plots of V959 Mon; from top-left: September 2012 (day 91), 12th November 2012 (day 146), 22nd November 2012 (day 156), February 2013 (day 252), October 2014 (day 479), February 2014 (day 612). The morphology can be seen to change from an east-west orientation, in the first four epochs, to what seems to be a north-south orientation, in the last two epochs. Each image was constructed using a 70 × 70 mas beam. The lowest contour level in each case is equal to three times the rms noise level (according to AIPS task IMSTAT) in the image.*

of the ejecta behaviour proposed by [4], we can investigate whether such behaviour results in the radio emission we have observed.

References

[1] Cheung, C.C. et al., 2012, The Astronomer's Telegram, 4224
[2] Cheung, C.C. et al., 2012, The Astronomer's Telegram, 4310
[3] Chomiuk, Laura, et al., 2012, The Astronomer's Telegram, 4352
[4] Chomiuk, L., et al., 2014, Nature, 514(7522), 339-342
[5] Fermi-LAT Collaboration, 2014, Science, 345(6196), 554-558.
[6] Eyres, S. P. S., Bode, M. F., O'Brien, T. J., Watson, S. K., & Davis, R. J., 2000, Monthly Notices of the Royal Astronomical Society, 318(4), 1086-1092
[7] Heywood, I., O'Brien, T. J., Eyres, S. P. S., Bode, M. F., & Davis, R. J., 2005, V723 Cas, Monthly Notices of the Royal Astronomical Society, 362(2), 469-474.
[8] Hjellming, R. M., Wade, C. M., Vandenberg, N. R., & Newell, R. T., 1979, The Astronomical Journal, 84, 1619-1631.
[9] Krauss, M. I., et al. 2011, The Astrophysical Journal Letters 739.1 (2011): L6.
[10] Munari, U., Dallaporta, S., Castellani, F., Valisa, P., Frigo, A., Chomiuk, L., & Ribeiro, V. A. R. M., 2013, Monthly Notices of the Royal Astronomical Society, stt1340.
[11] O'Brien, T. J., et al., 2012, The Astronomer's Telegram, 4408, 1.
[12] Osborne, J. P., Beardmore, A., & Page, K., 2013, The Astronomer's Telegram, 4727, 1.
[13] Ribeiro, V. A. R. M., Munari, U., & Valisa, P. 2013, The Astrophysical Journal, 768(1), 49.
[14] Wagner, R. M., Woodward, C. E., & Starrfield, S. 2013, The Astronomer's Telegram, 4737, 1.

Journal of Physics: Conference Series **728** (2016) 042003 doi:10.1088/1742-6596/728/4/042003

Variable X-Ray and UV emission from AGB stars: Accretion activity associated with binarity

Raghvendra Sahai[1], Jorge Sanz-Forcada[2] and Carmen Sánchez Contreras[2]

[1] Jet Propulsion Laboratory, MS 183-900, California Institute of Technology, Pasadena, CA 91109, US

[2] Astrobiology Center (CSIC-INTA), ESAC campus, E-28691 Villanueva de la Canada, Madrid, Spain

E-mail: [1] sahai@jpl.nasa.gov

Abstract. Almost all of our current understanding of the late evolutionary stages of $(1-8)\,M_{\odot}$ stars is based on single-star models. However, binarity can drastically affect late stellar evolution, producing dramatic changes in the history and geometry of mass loss that occurs in stars as they evolve off the AGB to become planetary nebulae (PNe). A variety of binary models have been proposed, which can lead to the generation of accretion disks and magnetic fields, which in turn produce the highly collimated jets that have been proposed as the primary agents for the formation of bipolar and multipolar PNe. However, observational evidence of binarity in AGB stars is sorely lacking simply these stars are very luminous and variable, invalidating standard techniques for binary detection. Using an innovative technique of searching for UV emission from AGB stars with GALEX, we have identified a class of AGB stars with far-ultraviolet excesses (fuvAGB stars), that are likely candidates for active accretion associated with a binary companion. We have carried out a pilot survey for X-ray emission from fuvAGB stars. The X-ray fluxes are found to vary in a stochastic or quasi-periodic manner on roughly hour-long times-scales, and simultaneous UV observations show similar variations in the UV fluxes. We discuss several models for the X-ray emission and its variability and find that the most likely scenario for the origin of the X-ray (and FUV) emission involves accretion activity around a main-sequence companion star, with confinement by strong magnetic fields associated with the companion and/or an accretion disk around it.

1. Introduction

Intermediate mass stars lose half or more of their mass during the AGB evolutionary phase in via slowly expanding, spherical molecular outflows. In contrast, most planetary nebulae (PNe), the end products of the AGB phase, are not round. Morphologically unbiased surveys of young PNe with the Hubble Space Telescope (HST) reveal that half of these objects show bipolar or multipolar morphologies, and a third are elongated. The number of round objects is less than 4%. The search for the fundamental causes of this transition has been a subject of intense inquiry for a long time and a series of international conferences (*Asymmetrical Planetary Nebulae 1-VI*).

It has long been argued that binarity is responsible, directly or indirectly, for the dramatic and poorly-understood changes in the history and geometry of mass loss that occurs during late stellar evolution (e.g., [1, 2, 3, 4]), especially for AGB stars evolving into PNe. A variety of binary models have been proposed, which can lead to the generation of accretion disks and

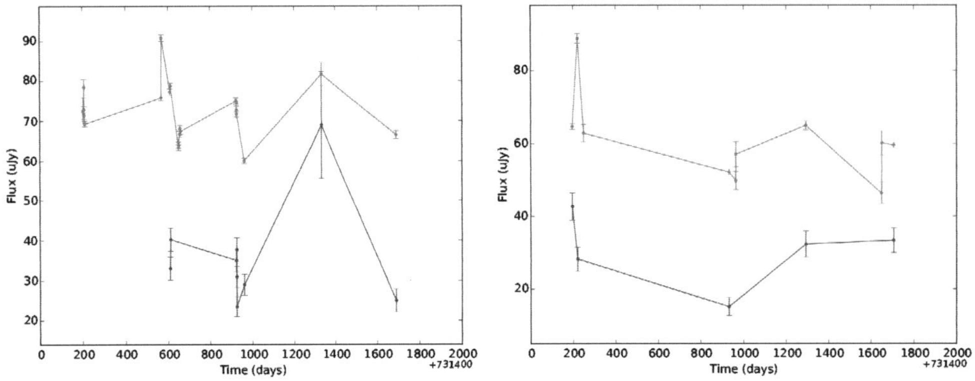

Figure 1. Two fuvAGB stars observed with GALEX that show strong variability in their FUV and NUV emission.

magnetic fields (e.g., [5]). The latter are likely the underlying physical cause for the highly collimated jets that have been proposed as the primary agents for the formation of bipolar and multipolar PNe [6, 7].

However, observational evidence of binarity in AGB stars has been lacking because AGB stars are very luminous and variable, invalidating standard techniques for binary detection such as radial-velocity and photometric variations due to a companion star. Sahai et al. [8] therefore used a new technique of searching for UV emission from AGB stars with GALEX that exploits the favorable secondary-to-primary photospheric flux contrast ratios reached in the UV for companions of spectral type hotter than about G0 (T_{eff}=6000 K) and luminosity, $L \gtrsim 1 L_{\odot}$. [8] detected emission from 9/21 objects in the GALEX FUV (1344 – 1786 Å) and NUV (1771 – 2831 Å) bands; since these objects (hereafter fuvAGB stars) also showed significant UV variability, [8] concluded that the UV source was unlikely to be solely a companion's photosphere, and was dominated by emission from variable accretion activity.

2. UV and X-Ray Emission from fuvAGB Stars

2.1. UV Observations

We searched the MAST[1]/GALEX archive for fuvAGB stars. We developed an efficient python-based pipeline that used a comprehensive input catalog of ~4000 AGB stars (includes O-rich, C-rich and S-stars) collected from various published sources and searches for counterparts in the GALEX archive within a user-specified search radius. The pipeline created a photometric database that provides (i) relevant parameters such as FUV and NUV fluxes, errors, and exposure times, for detected objects for each "visit", and (ii) the background "sky-noise" for stars that were observed but not detected. A summary of variability properties for each of the detected objects was also created.

We detected about 100 fuvAGB stars in the FUV band at a $\gtrsim 5\sigma$ level. Even for the hottest sources in our catalog (sp. type M4), the detected FUV fluxes ($\gtrsim 20\,\mu$Jy) correspond to a significant excess above photospheric emission. We find that many fuvAGB stars show extreme UV variability as well (e.g., Fig. 1, and Y Gem: [9]). We find that such sources show time periods where the FUV and NUV variations are correlated, suggesting changes in the emission measure (EM) and/or the obscuring column (N_H), and where they are anti-correlated, suggesting changes in the temperature of the emitting material.

[1] Mikulski Archive for Space Telescopes

In order to understand the nature of the UV emission (e.g., contuinuum emission from a hot accretion disk or line emission from an accretion inflow), we carried out low-resolution (grism) spectroscopy with GALEX for a sample of 10 fuvAGB stars. Unfortunately, the UV detector broke before soon after our Cycle 5 program was scheduled, so we were able to obtain both the FUV and NUV spectrum in only one object, EY Hya. The spectrum of this object shows both emission lines and an apparent weak continuum; the latter is stronger in the FUV than in the NUV. The spectrum can be fit with a composite model consisting of blacbody emission from an accretion-disk hot spot ($T \sim 27000$ K, $L \sim 3.4 \times 10^{-3} L_{\odot}$) and a nebular component likely associated with an accretion flow ($T \sim 1.9 \times 10^{4}$ K). The size of the latter is comparable to that of a low-mass main-sequence star ($\sim 10^{11}$ cm). However, the spectral resolution is low (10 Å in the FUV, 25 Å in the NUV), so one cannot rule out the possibility that the continuum has a significant contribution from a blend of weak emission lines; higher resolution spectra with HST could help resolve this issue.

Other objects, such as V Hya (with only NUV spectra) show a weak continuum with strong Mg h & k (λ2798) lines and a strong CII (λ2332) line. Such emission may arise in a nebular component that may be associated with an accretion flow (as for EY Hya) or a chromosphere around the primary AGB star.

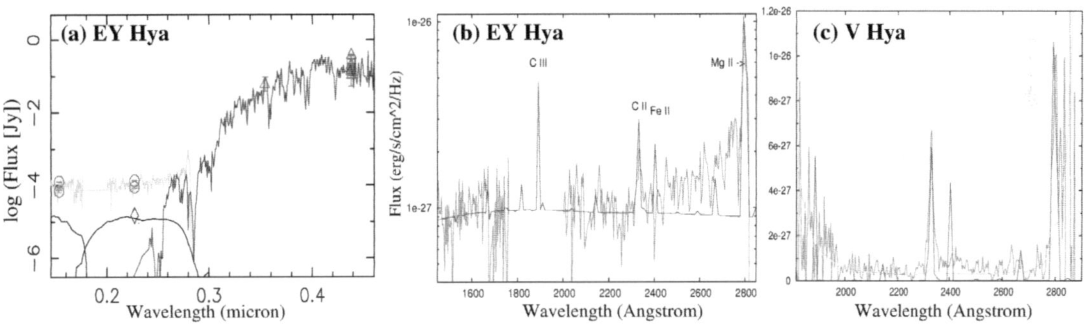

Figure 2. (a) GALEX grism spectrum (*green*), and broad-band fluxes – GALEX NUV & FUV and ground-based optical (*red symbols*) of fuvAGB star EY Hya – and a M7III model fit to the optical SED (*blue*). The expected UV fluxes (*blue diamonds*) from the AGB star, obtained by convolving the model with the GALEX bandpasses (*black*) lie far below the observed values (FUV lies below x-axis). (b) GALEX FUV+NUV grism spectrum of EY Hya (*red*), and a composite model (*blue*) consisting of an accretion-disk hot spot and an accretion flow. (c) GALEX NUV grism spectrum of V Hya (*red*, & noise *green*), and a nebular component (*blue*) with $T \sim 2 \times 10^{4}$ K.

2.2. X-Ray Observations

We carried out a pilot search for X-ray emission from a small subset of fuvAGB stars. A sensitive search for such emission in two AGB stars by Kastner & Soker [10] only yielded upper limits. An archival study using XMM and ROSAT revealed 2 AGB stars with X-ray emission, however the low signal-to-noise (S/N) ratio prevented an accurate estimate of the emission properties [11].

We observed a total of 5 (13) targets with XMM in observing cycles AO-12 (AO-13), and 6 targets with CXO cycle 15 (3 of which were common with the XMM observations). Typical integration times were 10 − 15 ksec. Our analysis of the XMM AO-12 and CXO cycle 15 data yielded high-quality X-ray spectra for 3 objects (Fig. 3a,b), and sensitive upper limits for the remaining. The Optical Monitor (OM) was used to obtain simultaneous UV observations, providing us UV light-curves for these 3 sources. We also obtained X-ray and UV light curves. We summarize the main results from this study [12] below.

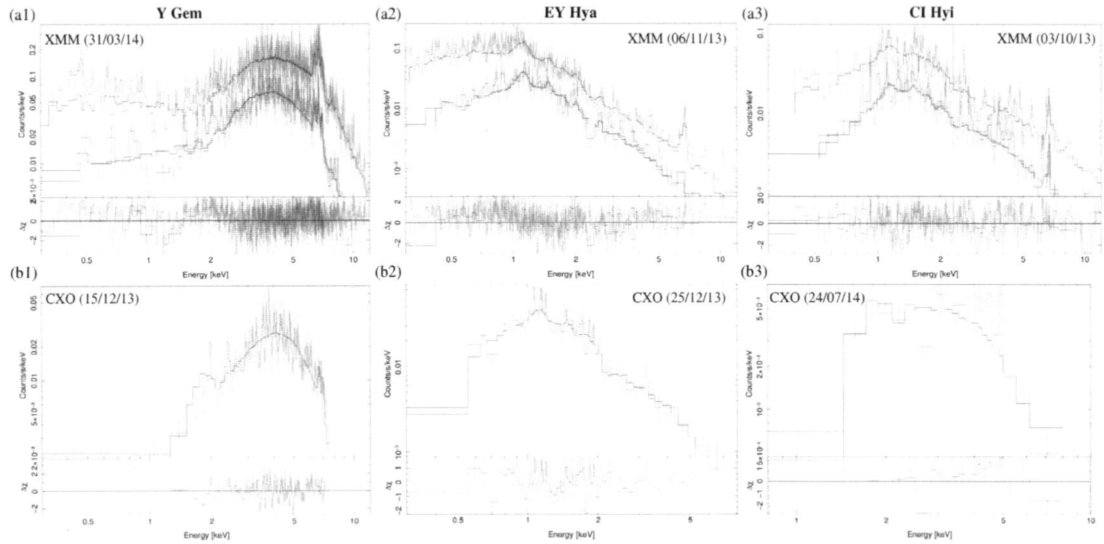

Figure 3. X-ray spectra (colored curves) and model fits (black curves) of the fuvAGB stars Y Gem, EY Hya and CI Hyi. Panels rows show XMM/EPIC (pn: red, MOS1: green, MOS2: blue), and CXO (ACIS-S). Spectra shown include the background; the ISIS model fitting procedure takes the latter into account (see text for details).

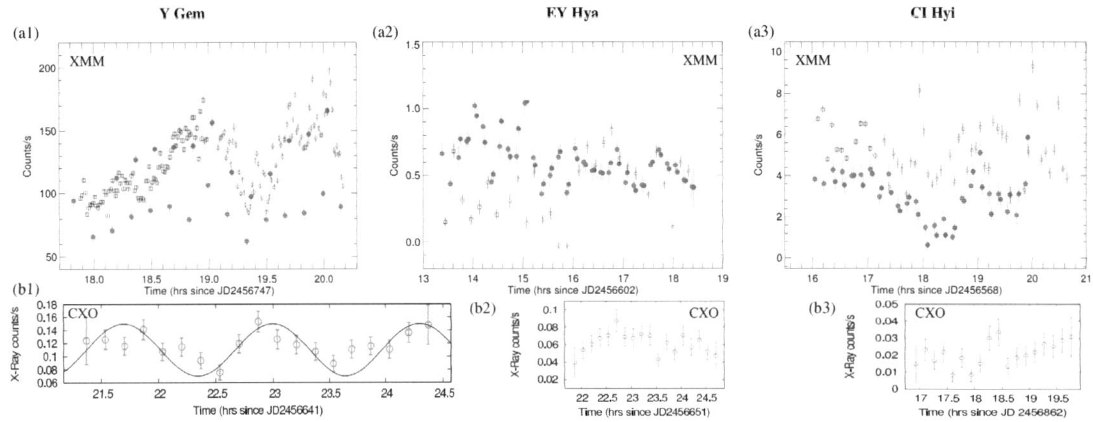

Figure 4. X-ray and UV light-curves of the fuvAGB stars Y Gem, EY Hya and CI Hyi. Panels show data from XMM (EPIC=pn+MOS1+MOS2: red, MOS=MOS1+MOS2: green, UVM2: black squares, UVW2: black circles), and CXO (ACIS-S). The EPIC, MOS and UVW2 data have been respectively re-scaled as follows: 80, 130, 2.5 (Y Gem), 2, 5, 0.8 (EY Hya), 20, 30, 2.5 (CI Hyi). A sinusoidal fit (by eye) with period, P=1.35 hr is shown for Y Gem. All data are background-subtracted; error bars are $\pm 1\sigma$. In order to facilitate the comparison between the XMM and CXO light curves, the widths of the two panels for each source have been adjusted so that similar time-intervals have similar lengths (on the horzontal axes).

Table 1. Models of X-Ray Emission for fuvAGB Stars

Name	Sp. Typ.	n(H) 10^{22} cm^{-2}	log(Tx) K	Lx $10^{-3} L_\odot$	d kpc
CI Hyi	M6	0.33 – 2.8	7.87	7.8 – 7.9	0.58
EY Hya	M7	0.05 – 0.095	7.57 – 7.74	2.0 – 2.4	0.35
Y Gem	M8	8.1 – 15.6	8.1 – 8.2	226 – 115	0.58
T Dra	C6,2	0.21	7.8	8.7	0.42

Short-term flux X-ray variations are seen during the course of each observation, spanning a factor $\sim (2-3.5)$ on hour-long timescales (Fig. 4a,b). For CI Hyi and EY Hya, the variations appear to be stochastic or quasi-periodic. In Y Gem, the light-curves give stronger evidence of a discrete period of about 1.2–1.4 hr, best seen in the CXO/ACIS-S light curve (Fig. 4b1) that spans a longer period than the EPIC one. For the longer time-scales between the XMM and CXO observational epochs for each source, we find changes in the observed fluxes up to a factor 1.8 (in Y Gem and EY Hya); however a direct comparison cannot be made between the spectra as these were taken with different instruments.

The UV fluxes also show short-term variations like the X-ray fluxes (Fig. 4a). Since the UVM2 and UVW2 data were taken sequentially (the OM can take data only through one filter at a time), we scaled the UVW2 count rate to match that of UVM2 at the time of transition between the two. We find that the UV and X-ray variations appear reasonably well correlated in the case of Y Gem and CI Hyi. For EY Hya, the correlation appears to be weaker.

The ISIS package [13] and the Astrophysics Plasma Emission Code (APEC, [14]) were used to fit the spectra with thermal models. For XMM, we simultaneously fit the data from the EPIC pn, MOS 1 and MOS 2 detectors.

All 3 sources also show a weak line feature at $\sim 6.3-6.8$ keV in their XMM spectra, that is likely due to emission from (a) the Fe XXV, XXVI lines complexes at ~ 6.7 keV from the very hot plasma that we have found in these objects, and/or (b) the 6.4 keV Fe I Kα line. The origin of Fe I Kα line may be the same as in Young Stellar Objects (YSOs), where it has been inferred to be flourescent emission from cold, neutral (<1 MK) material in a disk irradiated with energetic ($E \gtrsim 7.1$ keV) X-rays [15].

The most robust parameters from our models are the X-ray temperature (Tx) and the emission measure (EM), followed by the hydrogen absorption column density (NH) – these provide us with the intrinsic X-ray luminosity (Lx) of each of our objects. We find that Lx $\sim (0.002-0.2)\,L_\odot$ and Tx $\sim (35-160)$ MK for our detected objects (Table 1). For the minor Fe I line component, our derived values of the line flux are sensitive to the emission measure (EM) of the hot plasma that contributes to the nearby Fe XXV, XXVI complex, and are less robust than indicated by the formal uncertainties derived from the least-squares fitting.

3. Models for X-ray Emission from fuvAGB Stars

Based on the time-scale ($\sim$1.3 hr) of the quasi-periodic variations – similar to the period of material orbiting close to the inner radius of an accretion disk around a sub-solar mass companion, i.e., with $M_c \lesssim 0.35\,M_\odot$ (implying a semi-major axis $a \lesssim 3\times10^{10}$ cm). [12] argue that the most likely model for the X-ray emission from fuvAGB stars is that it arises at or near the magnetospheric radius in a truncated disk, or the boundary layer between the disk and star.

Direct accretion onto a stellar or brown dwarf companion is ruled out as such accretion produces a maximum temperature of $T_s = 3.44\times10^6\,K(M_c/R_c)$, too low to explain the observed high Tx values; WD (or more compact) companions are needed. However, none of our X-ray emitting AGB stars is known to be a symbiotic star with a WD companion. For EY Hya, an optical spectrum taken with the Palomar 5m telescope shows no emission lines in the 3889–5436 Å region. Thus, if accretion onto WD companions is responsible

Table 2. UV Variability & X-Ray Emission for fuvAGB Stars

Name	FUV σ/median	FUV (max-min)/median	X-Ray detection?
Y Gem	1.9	5.13	Y
EY Hya	0.39	0.93	Y
CI Hyi	0.67	1.33	Y
R Uma	0.15	0.36	Y
T Dra	0.069	0.20	Y
V Hya	0.27	0.54	N
del01 Aps	0.052	0.024	N
NU Pav	0.13	0.26	N

for the X-ray emission, then these must be rather cool (i.e. with effective temperatures $T_{eff} < 20,000$ K.) However, if an adequate mass of accreting gas can be pre-accelerated to high speeds (v_s) by magnetic reconnection, then the shock temperature is, $T_s = (3/16\,k_B)\,\mu\,m_H\,v_s^2 \sim 70\,MK\,(v_s/1500\,\text{km s}^{-1})^2$, assuming a mean molecular weight, $\mu = 1.33$, and is not constrained by the mass-to-radius ratio of the companion.

The relatively high values of Tx that we have found in our fuvAGB sources argues against the X-ray emission coming from stellar coronae, which typically show values of Tx in the range $\sim (2-10)$ MK and rarely as high as the lowest Tx value in our sample, ~ 40 MK (e.g., [16]). In addition, coronal emission from a main-sequence companion is strongly ruled out because the large observed values of Lx imply unrealistically large companion masses, using the Lx/Lc relationship determined by Pizzolato et al. [17].

If coronal gas is present on the AGB star, it has to be magnetically confined and heated, requiring a relatively strong field, $B_{av} = (0.3-15)\,\text{G}/f_{agb}$, where f_{agb} is the fraction of the AGB's star's surface covered by the B field ($f_{agb} \sim 1$ for a large-scale field and $f_{agb} << 1$ for localized fields). Since we also need to produce substantial amounts of plasma implied by our derived values of the X-ray emission measure, around the fuvAGB stars, and since these objects are cool late-type stars, with three of them known to drive cool, dusty molecular winds (EY Hya, T Dra, R Uma), the surface filling factor of the coronal gas (and thus f_{agb}) is likely to be relatively small. Although recent studies show that such values of B_{av} are plausible [18], a sensitive XMM search for X-ray emission in two AGB stars with known or suspected strong B-fields yielded null detections [10], making coronal emission from the AGB star a less likely candidate for producing the observed X-ray emission.

4. Emerging Trends from X-ray and UV-observations

(i) X-ray emission is found only in fuvAGB stars. For example, all 8 AGB stars, including 2 ROSAT detections reported by Ramstedt et al. [11], that have been confidently detected in X-rays are fuvAGB stars. SS Vir, also a fuvAGB star, shows a tentative XMM detection (4.8 ± 1.3 counts ks^{-1}) that may be real or may be due to optical loading [11].
(ii) The GALEX FUV/NUV flux ratio $R_{fuv/nuv}$ tentatively appears to be a better indicator of X-ray emission than the FUV (or NUV) flux. We find $R_{fuv/nuv} > 0.17$ in the fuvAGB stars with X-ray emission. Amongst the stars not detected in X-rays, we find $R_{fuv/nuv} < 0.12$ (Fig. 5), except for V Hya and V Eri which were not detected.
(iii) A tabulation of FUV variability indicators – standard deviation/median and (maximum-minimum)/median – versus the presence/absence of X-ray emission shows that the former is a good indicator of X-ray emission (Table 2), and supports our model in which the X-ray and FUV emission is related to a variable accretion process.
(iii) The (non-coeval) UV flux does not appear to be a good predictor of the X-ray flux – e.g., both T Dra and R Uma, detected in X-rays, are fairly weak FUV sources with fluxes of 8.4 and

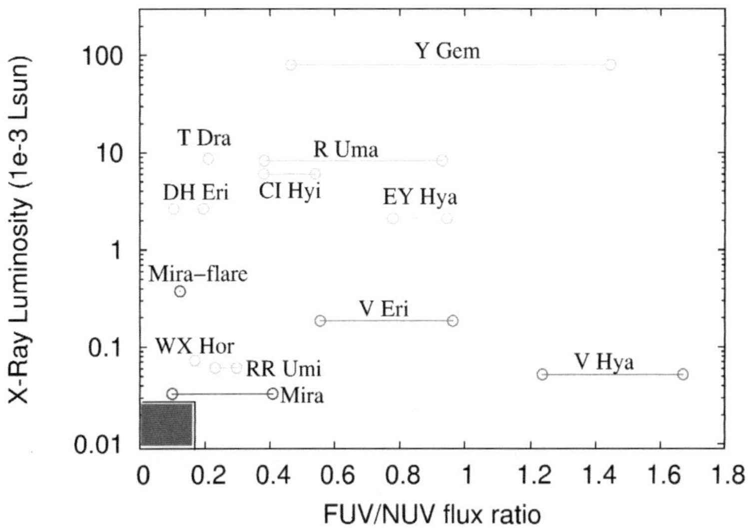

Figure 5. X-ray luminosity, Lx, versus the FUV-to-NUV flux ratio, $R_{fuv/nuv}$. Symbols show fuvAGB stars with $R_{fuv/nuv} > 0.17$ that have been observed in X-rays (*red* symbols show 3σ upper limits for Lx). The red box covers a region with 7 additional fuvAGB stars that are not detected in X-rays – for these, $R_{fuv/nuv} < 0.17$. For stars with multiple X-ray (UV) observations, the mean (full range of $R_{fuv/nuv}$) is shown. The plot also shows data for the symbiotic star, Mira – its quiescent Lx versus the $R_{fuv/nuv}$ range, as well as Lx during a flare versus $R_{fuv/nuv}$ from a near-contemporaneous epoch.

11 μJy, whereas our 2 non-detections from XMM AO-12, del01 Aps and NU Pav, have much stronger FUV emission (326 and 385 μJy). This may be partly because the UV and X-ray observations are not contemporaneus, and both are variable – we see UV variability on the time-scale of ~1-year in many stars with multi-epoch GALEX data. The XMM data from our survey show that X-ray and UV fluxes vary in a stochastic or quasi-periodic manner on roughly hour-long times-scales. Comparison with the CXO spectra obtained for CI Hyi, EY Hya, & Y Gem at a 2nd epoch show variations on several month-long time-scales as well.

5. Implications for Binarity, Accretion & Magnetic Fields

Our current working hypothesis, first proposed to explain our serendipitous discovery of extremely strong and variable UV emission in the late-M star, Y Gem [9], and that is consistent with these trends, is (a) fuvAGB stars have binary companions and associated variable (possible episodic) accretion which produces variable UV and X-ray emission, and (b) AGB stars without UV-excesses are either single stars, or if they have companions, the associated accretion activity when they were observed was too weak to produce detectable UV emission.

The observed short-term X-ray and UV light-curves, especially that of EY Hya, show that, in addition to the quasi-periodic variations presumably associated with orbital motion at the inner radius of an accretion disk, there is a second mechanism that produces stochastic variability. A plausible mechanism for the latter is flare activity, that in the case of YSOs, is believed to arise from reconnection events associated with long magnetic loops connecting the star and a circumstellar disk that get sheared due to orbital motion in the disk. However, the variations we find in fuvAGB stars do not look like the flares seen in YSOs that have short rise times and long decay time (e.g. [19]). One possibility is that fuvAGB stars have multiple weak, partly overlapping flares that wash out the individual rise and fall signatures. However, with only a couple of stars where such light curves are available currently, it is not possible to draw general conclusions.

6. Broad Impact on Binary Stellar Astrophysics

FuvAGB stars have a strong synergy and evolutionary connection with two other well-studied classes of stars. The first of these are the central stars of planetary nebulae (CSPNe). The "ChanPlaNS" survey of X-ray emission from a sample of nearby PNe has resulted in the surprising and exciting discovery of relatively hard ($\gtrsim$0.5 keV) X-ray emission from CSPNe in 20/59 objects, that is not photospheric but results either from self-shocking stellar winds from the CSPNe or coronal emission from magnetically active companions [20]. The origin of the activity in such companions is unknown, but could result from them being "rejuvenated" during the AGB phase, e.g., by wind accretion from the primary [21]. If so, then fuvAGB stars are the progenitors of the PNe with magnetically-active companions.

The second class consist of symbiotic stars (red giants with WD companions) that are strong UV and X-ray emitters (e.g., [22]), and a classification scheme for the spectra that is diagnostic of the emission processes has been developed. FuvAGB stars with "cold" WD companions may represent the evolutionary progeny of symbiotic stars.

No quantitative theoretical models exist as yet to explain the X-ray and FUV emission that we have discovered in fuvAGB stars. Our current data is too sparse to robustly test our favored hypothesis or others, and it is eminently possible that more than one mechanism is responsible for the X-ray and UV emission in these objects.

An extended survey of a statistically significant sample of fuvAGB stars, with co-eval X-ray and UV observations, can robustly confirm the tentative trends described earlier, and provide the crucial "missing" data on the earliest stages of binary interaction in AGB stars with companions and the necessary observational impetus for theoretical work in this field.

References

[1] Lee C-F 2016 paper presented at this conference
[2] Corradi R 2016 paper presented at this conference
[3] Williams R 2016 these proceedings
[4] Olofsson H 2016 these proceedings
[5] Balick B and Frank A 2000 *Annu. Rev. Astron. Astrophys.***40** 439
[6] Sahai R and Trauger J T 1998 *Astron. J.* **116** 1357
[7] Sahai R, Morris M R and Villar G G 2011 *Astron. J.***141** 134
[8] Sahai R, Findeisen K, Gil de Paz A and Sánchez Contreras C 2008 *Astrophys. J.***689** 1274
[9] Sahai R, Neill J D, Gil de Paz A and Sánchez Contreras C 2011 *Astrophys. J. Lett.***740** L39
[10] Kastner J H and Soker N 2004 *Astrophys. J.***608** 978
[11] Ramstedt S, Montez R, Kastner J and Vlemmings W H T 2012 *Astron. Astrophys.***543** A147
[12] Sahai R, Sanz-Forcada J, Sánchez Contreras C and Stute M 2015 *Astrophys. J.***810** 77
[13] Houck J C and Denicola L A 2000 *Astronomical Data Analysis Software and Systems IX* **216** 591
[14] Smith R K, Brickhouse N S, Liedahl D A and Raymond J C 2001 *Astrophys. J. Lett.***556** L91
[15] Favata F 2005 *Mem. S.A.It.***76** 337
[16] Schmitt J H M M, Collura A, Sciortino S, Vaiana G S, Harnden F R Jr and Rosner R 1990 *Astrophys. J.***365** 704
[17] Pizzolato N, Maggio A, Micela G, Sciortino S and Ventura P 2003 *Astron. Astrophys.***397** 147
[18] Leal-Ferreira M L, Vlemmings W H T, Kemball A and Amiri N 2013 *Astron. Astrophys.***554** AA134
[19] Favata F, Flaccomio E, Reale F, Micela G, Sciortino S, Shang H, Stassun K G and Feigelson E D 2005 *Astrophys. J. Suppl. Series***160** 469
[20] Montez R, Jr *et al* 2015 *Astrophys. J.***800** 8
[21] Montez R, Jr, De Marco O, Kastner J H and Chu Y-H 2010 *Astrophys. J.***721** 1820-1828
[22] Luna G J M, Sokoloski J L, Mukai K and Nelson T 2013 *Astron. Astrophys.***559** A6

Stellar evolution in the outer Galaxy

Ryszard Szczerba[1], **Natasza Siódmiak**[1], **Aleksandra Leśniewska**[2], **Agata Karska**[2] **and Marta Sewiło**[3,4]

[1] N. Copernicus Astronomical Center, PAS, Rabiańska 8, 87-100 Toruń, Poland
[2] Astronomical Observatory, Adam Mickiewicz University, Słoneczna 36, 60-268 Poznań, Poland
[3] NASA Goddard Space Flight Center, 8800 Greenbelt Rd., Greenbelt, MD 20771, USA
[4] Astronomical Observatory of the Jagiellonian University, Orla 171, 30-244 Kraków, Poland

E-mail: [1] `szczerba@ncac.torun.pl`

Abstract. We investigate the distribution of different classes of spectroscopically identified sources and theoretical models in the color-color diagrams (CCDs) combining the near-infrared (NIR) and mid-infrared (MIR) data to develop a method to classify Outer Galaxy sources detected with the *Spitzer* Space Telescope (hereafter *Spitzer*) SMOG survey in the IRAC 3.6–8.0 μm and MIPS 24 μm bands. We supplement the *Spitzer* data with the data from other satellite and ground-based surveys. The main goal of our study is to discover and characterize the population of intermediate- and low-mass young stellar objects (YSOs) in the Outer Galaxy and use it to study star formation in a significantly different environment than the Galaxy inside the solar circle. Since the YSOs can be confused with evolved stars in the MIR, these classes of objects need to be carefully separated. Here we present the initial results of our analysis using the K_s-[8.0] vs. K_s-[24] CCD as an example. The evolved stars separated from YSOs in the YSO selection process will be investigated in detail in the follow-up study.

1. Introduction

Star formation studies in the Galaxy has been concentrating on the nearby molecular clouds (e.g., [1]; [2]; [3]) or distant massive star formation regions (e.g., [4]), leaving vast clouds of low- and intermediate-mass star formation at larger distances undiscovered. We have initiated a systematic study of star formation in the Outer Galaxy to uncover the population of YSOs in these previously unstudied clouds and investigate the impact of the environment on the star formation process. The *Spitzer* IRAC and MIPS cameras are ideally suited to investigate dusty envelopes of YSOs. We use the data from the "*Spitzer* Mapping of the Outer Galaxy" survey (SMOG; PI Sean Carey) that covered ~24 deg^2 region in the Outer Galaxy: l = (102°, 109°), b = (-0.2°, 3.2°) in the IRAC 3.6–8.0 μm and MIPS 24 μm bands (see Figure 1). This region that we refer to as "L105", samples a range of environments and star formation activities. In Figure 1 we marked the boundary of the Cep OB2 association with a blue/black (color electronic version/black and white printout). The H II regions from the [5] catalog are shown as blue/black circles and identified by their numbers.

In the Outer Galaxy we expect a contamination from evolved stars such as Asymptotic Giant Branch (AGB) stars, post-AGB objects and planetary nebulae (PNe), or more massive ones such as Red SuperGiants (RSGs), or even Yellow HyperGiants (YHGs). Here, as an example, we concentrate on identifying the location of the low- and intermediate-mass evolved stars (AGBs,

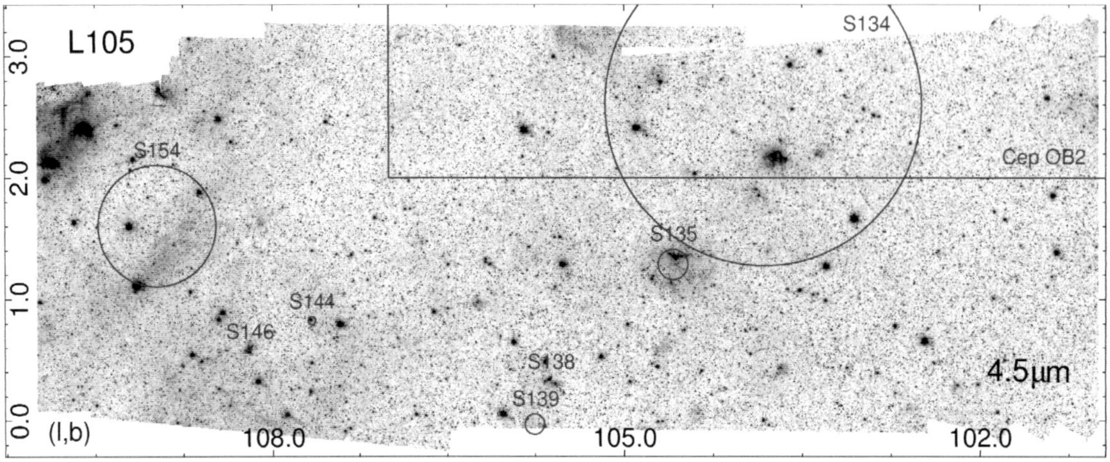

Figure 1. The IRAC image (~2" resolution) of L105 region at 4.5 μm. See text for details.

post-AGBs and PNe) in the K_s-[8.0] vs. K_s-[24] color-color space. We are developing a method to disentagle these classes of sources from the sample of YSO candidates. A detailed description of the YSO selection process will be presented in detail in a future paper.

2. The K_s-[8.0] vs. K_s-[24] color–color diagram

The quite robust method to separate different classes of objects, especially when distances to them are unknown, are CCDs and a comparison of new objects with already classified sources or with theoretical models. One of such diagrams, the K_s-[8.0] vs. K_s-[24] color–color diagram, is known to separate the C-rich from O-rich AGB stars (e.g. [6]). In addition, theoretical models from the Grid of RSG and AGB ModelS (GRAMS, [7], [8]) are available. The GRAMS grid consists of 66 000 O-rich and 12 000 C-rich models that span the range of stellar and dust shell parameters expected for evolved stars in the Large Magellanic Cloud (LMC). In the Outer Galay we expect metallicities more typical for the LMC or even for the Small Magellanic Cloud (SMC) than for the Galaxy Bulge, so the GRAMS models can be safely applied. On the other hand, there are also available self-consistent time-dependent hydrodynamic (HD) radiative transfer calculations for gaseous dusty circumstellar shells around C-rich and O-rich stars in the final stages of their AGB/post-AGB evolution ([9]). These calculations are based on one particular stellar evolutionary track and single dust size grain (a=0.05 μm) for its both chemical compositions: amorphous carbon (AC: [10]) and *astronomical* silicates (ASil: [11]).

Figure 2 shows the K_s–[8.0] vs. K_s–[24] CCD and compares the GRAMS models for C-rich stars (black dots) with the HD evolutionary track for AC dust (dashed red/dark-gray line), and those for O-rich ones (blue/gray dots) with the HD track for ASil dust (dashed green/light-gray line). The end of the AGB/beginning of the post-AGB phase of evolution (time "0": see [9] for a definition) is marked by the corresponding big star symbol. The large temporal luminosity and mass loss variations, which are associated with thermal pulses during the AGB phase can be recognized as bluer "loops" (see [9] for details). Despite the fact that HD models were not adjusted for any specific metallicity, the agreement between the GRAMS models and the HD evolutionary tracks is quite impressive. The advantage of the HD models is an additional coverage of the post-AGB phase of stellar evolution (to the left from the big star symbols).

Figure 3 shows the same CCD as Figure 2 and compares HD models (same colors as in Figure 2, but know the last 200 years of AGB and about 1000 years of post-AGB evolution

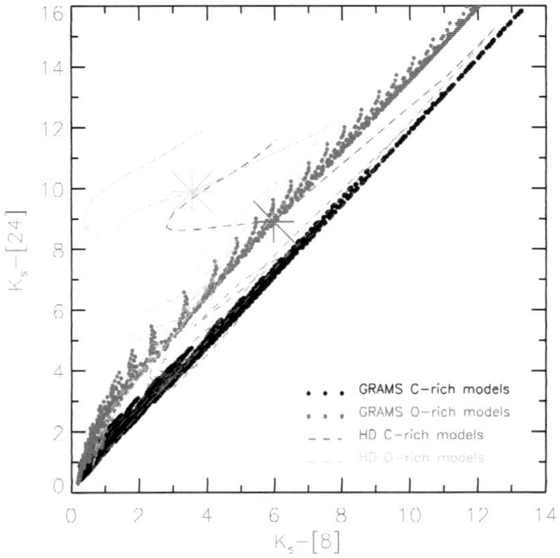

Figure 2. A comparison between the GRAMS models for RSGs and AGBs from [7] and [8] with the HD models of AGB and post-AGB phases from [9] on the K_s-[8.0] vs. K_s-[24] CCD. See the legend and text for details.

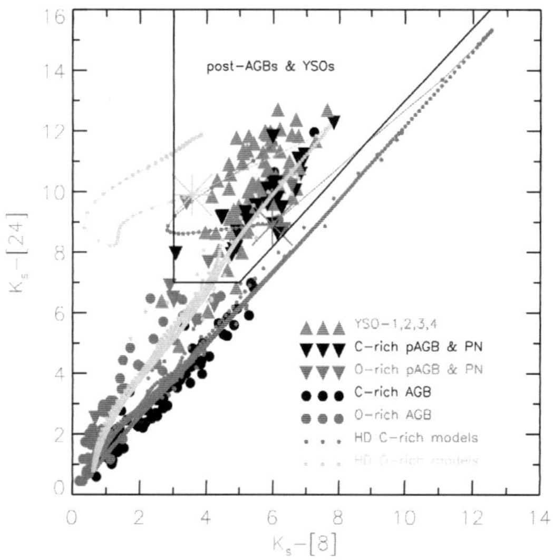

Figure 3. A comparison between the HD models of AGB and post-AGB evolution from [9] with sources from the LMC and SMC classified by [12] and [13]. See the legend and text for details.

are connected by the solid lines) with sources from the LMC and SMC classified by [12] and [13]. The C-rich evolved stars are shown by black symbols, while O-rich ones are represented by blue/gray symbols. All types of YSOs defined by [12] and [13] based on their spectra are shown by magenta/gray triangles. The HD models during the AGB phase are plotted in intervals of 100 years (HD models of the post-AGB phase are not equidistant in time), so their number density is a direct measure of the probability of finding objects in different parts of the diagram. Figure 3 shows that there is a clear separation between the C-rich and O-rich AGB objects, which can be explained well by the HD models, and that post-AGB objects overlap completely with YSOs. The O-rich post-AGB phase of our original HD models seems to be too blue. The reason is too fast drop of flux in [8.0] band in case of *astronomical* silicates due to too large optical depth during the transition from AGB to post-AGB phase. This is caused by too small velocity of O-rich dust shell (see Figure 21 in [9]). By black solid lines we have delimited a region of K_s–[8.0] vs. K_s–[24] CCD, where majority of post-AGB sources and YSOs are located (compare to Figure 10 of [6]).

In Figure 4, the K_s-[8.0] vs. K_s-[24] CCD for all SMOG sources with good quality data at all bands is shown. Due to the small number of good quality measurements by MIPS at 24 μm the total number of plotted objects is only 15 311. Note, that the total number of sources detected at shorter wavelengths by IRAC is 2 836 618. The SMOG sources, which are located in the "post-AGBs & YSOs" region on this CCD (249) are shown by larger symbols then the rest of the SMOG objects. One characteristic result can be seen for SMOG objects located in the L105 Outer Galaxy region: there is relatively small number of C-rich AGB sources located around positions suggested by the HD track for C-rich dust (red/dark gray solid line), and especially those with large mass loss rates (the reddest) are missing. In addition, the reddening vector shown in Fig.4 constructed for $A_V = 10$ demonstrates that our HD tracks computed for the lowest mass loss of the order of 10^{-7} $M_\odot\,yr^{-1}$, will move out of the clump of the SMOG points around K_S-[8]~ 0.8.

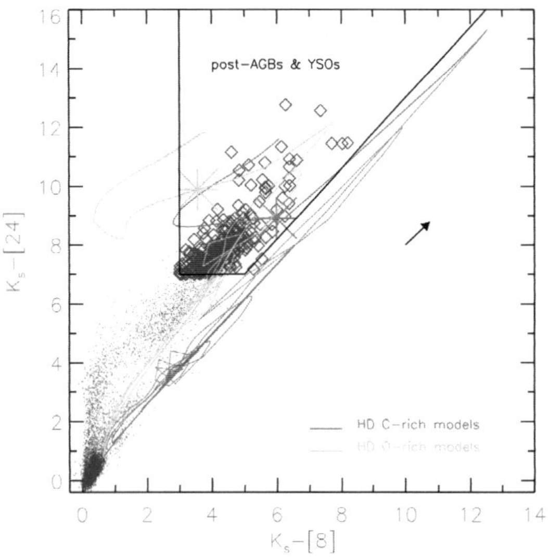

Figure 4. The K_s-[8.0] vs. K_s-[24] CCD for all SMOG sources (15 311). SMOG sources are shown by dots, which are replaced by open large diamonds in the "post-AGBs & YSOs" region. HD models for the AGB and post-AGB phases are shown by red/dark gray solid line for C-rich dust and by green/light gray line for O-rich dust. The end of the AGB/beginning of post-AGB phase of evolution is marked by the corresponding big star symbol. The mass loss at the end of the AGB evolution started to decrease about 200 years before moment "0". The "post-AGBs & YSOs" region is marked with the black solid lines (contains 249 objects). The black arrow shows the reddening vector for $A_V = 10$.

3. Summary

To investigate star formation in the Outer Galaxy we have started analyzing the SMOG data obtained by IRAC and MIPS instruments onboard *Spitzer*. We demonstrated that the K_s-[8.0] vs. K_s-[24] CCD separated the C-rich and O-rich AGBs well (see also [6]), but this CCD is not able to disentangle YSOs from post-AGB objects. Our color–color cuts in the K_s-[8.0] vs. K_s-[24] CCD select ~250 post-AGB objects and YSO candidates. Color-color cuts based on a set of NIR/MIR CCDs will allow us to select an initial sample of YSO and evolved star candidates. To classify these sources and remove contaminants from each sample, we will perform the spectral energy distribution fitting and visual inspection the sources' environments using the multiwavelength images.

Acknowledgments

Authors would like to acknowledge financial support from the Polish National Science Center grant 2014/15/B/ST9/02111. R.Sz. and N.S. acknowledge support from the EU FP7-PEOPLE-2010-IRSES program in the framework of project POSTAGBinGALAXIES (Grant Agreement No. 269193) and NCN grant 2011/01/B/ST9/02031. A.K. acknowledges support from the Foundation for Polish Science (FNP) and the Polish National Science Center grant 2013/11/N/ST9/00400. We thank Sean Carey for providing us with the MIPS 24 μm catalog.

References

[1] Gutermuth R A, Myers P C, Megeath S T, Allen L E, Pipher J L, Muzerolle J, Porras A, Winston E and Fazio G 2008 *Astrophys. J.* **674** 336
[2] Evans N. J II *et al* 2009 *Astrophys. J. Suppl. Series* **181** 321
[3] Megeath S T *et al* 2012 *Astron. J.* **144** 192
[4] Povich M S *et al* 2009 *Astrophys. J.* **696** 1278
[5] Sharpless S 1959 *Astrophys. J. Suppl. Series* **4** 257
[6] Matsuura M *et al* 2014 *Mon. Not. R. Astron. Soc.* **439** 1472
[7] Srinivasan S, Sargent B A and Meixner M 2011 *Astron. Astrophys.* **532** A54
[8] Sargent B A, Srinivasan S and Meixner M 2011 *Astrophys. J.* **728** 93
[9] Steffen M, Szczerba R and Schoenberner D 1998 *Astron. Astrophys.* **337** 149
[10] Rouleau F and Martin P G 1991 *Astrophys. J.* **337** 526
[11] Draine B and Lee H M 1984 *Astrophys. J.* **285** 89
[12] Woods P M *et al* 2011 *Mon. Not. R. Astron. Soc.* **411** 1597
[13] Ruffle P M E *et al* 2015 *Mon. Not. R. Astron. Soc.* **451** 3504

Journal of Physics: Conference Series **728** (2016) 042005 doi:10.1088/1742-6596/728/4/042005

An ALMA view of the post-AGB object HD 101584

H. Olofsson[1], W. Vlemmings[1], M. Maercker[1], E. Humphreys [2], M. Lindqvist [1], L. Nyman[3] and S. Ramstedt[4]

[1] Dept. of Earth and Space Sciences, Chalmers Univ. of Technology, Gothenburg, Sweden
[2] European Southern Observatory, Garching, Germany
[3] Joint ALMA Observatory, Santiago, Chile
[4] Dept. of Physics and Astronomy, Uppsala University, Uppsala, Sweden

E-mail: [1] hans.olofsson@chalmers.se

Abstract. ALMA cycles 1 and 3 observations of CO isotopologues and 1.3 mm continuum are used in a study of the circumstellar environment of the binary HD 101584, a post-AGB star and a low-mass companion that is most likely a post-common-envelope-evolution system. These data are supplemented with new information from OH maser emission. It is inferred that the large-scale circumstellar medium has a bipolar hour-glass structure, seen almost pole-on, formed by an energetic, $\geq 150\,\mathrm{km\,s^{-1}}$, jet. Significant amount of material still resides in the central region. It is proposed that the circumstellar morphology is related to an event which took place ≤ 500 yr ago, possibly a capture event where the companion spiralled in towards the AGB star. Several observed features remain to be explained, and may hint to a more complicated scenario.

1. Introduction

HD 101584 (V885 Cen, IRAS 11385–5517) is of spectral type A6Ia and bright at optical wavelengths ($V \approx 7$ mag.) [1, 2]. The optical and infrared characteristics suggest that it evolved from the asymptotic giant branch (AGB) at most a few hundred years ago [3]. The presence of a binary companion is inferred from photometric and radial velocity variations with a periodicity of ≈ 200 days [4, 5]. The absence of spectroscopic emission from the companion indicates that it is a low-mass main sequence star. The HST images show diffuse circumstellar dust, and a ring-like structure of radius ≈ 1.5", which is roughly centered on the star [6].

Using ^{12}CO and ^{13}CO $J = 1$–0 and 2–1 single-dish map data, [7] inferred that the highest-velocity gas (covering $\approx 300\,\mathrm{km\,s^{-1}}$) show an east-west bipolar outflow, the most blue- and red-shifted emission ≈ 5" to the W and the E, respectively. The velocity gradient is Hubble-like. The rest of the circumstellar gas was found to reside within a few arc seconds of the center. On the contrary, a double-peaked OH 1667 MHz maser line (covering $\approx 80\,\mathrm{km\,s^{-1}}$) shows a symmetry axis along a position angle PA $\approx -60°$ with the most blue- and red-shifted emission ≈ 2" to the SE and the NW, respectively [8].

We present here interpretations of the cycle 1 ^{12}CO, ^{13}CO, and C^{18}O $J = 2$–1 line and 1.3 mm continuum ALMA data on HD 101584, as well as present some indications from the new cycle 3 ALMA data (ADS/JAO.ALMA#2012.1.00248.S and #2015.1.00078.S). In addition to these lines, we have detections of line emission from SiO, CS, H_2S, SO, SO_2, OCS, H_2CO, and some of their isotopologues. We have also re-observed the OH 1667 MHz line, originally observed by [8] using ATCA.

2. Observations

The cycle 1 data were obtained with 35 antennas of the ALMA 12 m array in two frequency settings of Band 6, and the cycle 3 data are obtained with ≈ 40 antennas in one frequency setting. The data contains four spectral windows of 1.875 GHz width and with 3840 channels per frequency setting. The FWHM of the synthesized beam is $\approx 0.6"$ and $\approx 0.05"$ (corresponds to a resolution of 35 AU at the distance of the object) for cycles 1 and 3, respectively. The OH observations with ATCA cover the four $^2\Pi_{3/2}$ Λ-doublet lines at a resolution of $\approx 3"$ and $0.5\,\mathrm{km\,s^{-1}}$ velocity resolution.

3. Results

3.1. Global characteristics

The cycle 1 1.3 mm continuum map, as well as an initial discussion of the ALMA cycle 1 data, are presented in [9]. The total flux, 112 mJy, is almost a factor of four lower than expected from the extrapolated SED model of [3], suggesting a significant loss of extended flux. There are faint structures surrounding an intense inner region, where $\approx 70\%$ of the total flux comes from inside a central 2" diameter circle. The cycle 3 data shows that a significant fraction of the flux from the central region is still unresolved at an angular scale of 0.05".

The line data points to a systemic velocity, $\upsilon_{\rm sys}$, of $42 \pm 1\,\mathrm{km\,s^{-1}}$ (LSR). The cycle 1 ALMA total fluxes are 690, 150, and $18\,\mathrm{Jy\,km\,s^{-1}}$ for the ^{12}CO, ^{13}CO, and C^{18}O $J=2$–1 lines, respectively. Based on a detailed comparison of the ALMA global and the single-dish line profiles, [9] conclude that no flux is missed in the central $20\,\mathrm{km\,s^{-1}}$ and at the extreme velocity peaks ($|\upsilon - \upsilon_{\rm sys}| \geq 90\,\mathrm{km\,s^{-1}}$), while $\approx 25\%$ of the flux is missed in the intermediate velocity range. At 0.05" resolution significant amount of flux is lost, but it remains to combine these data with the 0.6" and 0.3" (to be obtained during cycle 3) data.

3.2. The morphology and kinematics

The cycle 1 ^{12}CO($J=2$–1) channel maps are shown in Fig. 1. The morphology of the emission is very symmetric with respect to the systemic velocity. The high-velocity outflow is narrow and directed at PA $\approx 90°$. It covers almost $300\,\mathrm{km\,s^{-1}}$ and has a Hubble-like velocity gradient. The CO line emission, and in particular that of SiO($J=5$–4), appear to come from hot spots, presumably regions where a high-velocity jet interacts with the circumstellar medium. The ring-like emission is interpreted in the form of an hour-glass structure, probably formed as a consequence of a high-velocity jet. The inclination angle is uncertain, but it must be small, $\approx 10°$. Adopting a tilt angle of 10°, a projected expansion length of 4" for the high-velocity flow, and an expansion velocity of $150\,\mathrm{km\,s^{-1}}$, the age of the high-velocity outflow is estimated to be ≈ 540 yr. Since this structure is presumably formed by a jet, this gives an upper limit to the time since the jet was launched.

3.3. Mass, energy, and momentum estimates

An estimate of the dust mass using the flux density at 1.3 mm results in $M_{\rm d} \approx 0.007\,M_\odot$ (with some considerable uncertainty). This suggests about a solar mass of gas in the central region, assuming a dust-to-gas ratio of 200. A simple estimate of the gas mass using the C^{18}O(2–1) line intensity results in $M_{\rm g} \approx 0.3\,M_\odot$ (also with some considerable uncertainty). More than half of it lies in the central $20\,\mathrm{km\,s^{-1}}$ range, and hence in the central region.

Using the C^{18}O(2–1) line, the kinetic energy and momentum of the gas accelerated to velocities $\geq 10\,\mathrm{km\,s^{-1}}$ is estimated to be $\approx 4 \times 10^{45}$ erg and $\approx 10^{39}\,\mathrm{g\,cm\,s^{-1}}$, respectively. The estimated momentum rate over 500 yr is $\approx 7 \times 10^{28}\,\mathrm{erg\,cm^{-1}}$, substantially higher than that available from radiation (L/c) using $L = 4000\,L_\odot$, $\approx 5 \times 10^{26}\,\mathrm{erg\,cm^{-1}}$, a common situation for post-AGB objects [10].

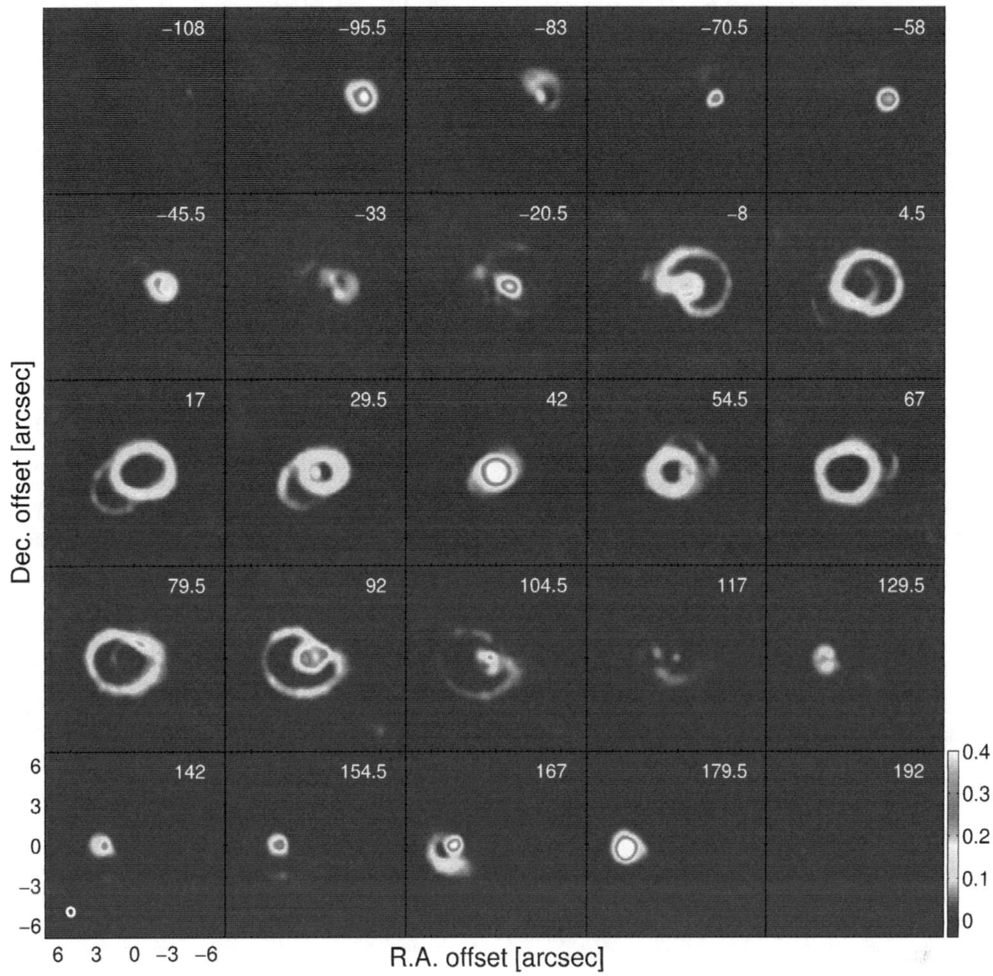

Figure 1. ALMA cycle 1 velocity-channel maps ($\Delta v = 12.5\,\mathrm{km\,s^{-1}}$) of the ^{12}CO($J = 2$–1) emission towards HD 101584 (synthesized beam in the lower left corner). The intensity scale unit is Jy beam^{-1} (noise rms 2 mJy beam^{-1}).

3.4. A scenario

We propose that what we observe is the effect of a capture event in which a fair fraction of the mass of the AGB star was released, and the AGB evolution was terminated. Most of the material remains in the vicinity of the binary, presumably in the binary plane, but some of it has been accelerated up to high velocities through the action of a jet. Using information from [3] and [4] and our estimated inclination angle, we infer that the mass of the companion is $\approx 0.6\,M_\odot$ and it lies at a distance of ≈ 0.7 AU from the primary, suggesting a common-envelope evolution. The amount of orbital energy released is estimated to be $\approx 2 \times 10^{45}$ erg. There are considerable uncertainties in both the orbital and energy estimates, and the energy transfer efficiency of the common-envelope evolution is uncertain. Taken at face values the released orbital energy is not enough, and another mechanism must augment, or even dominate, the ejection event.

4. Remaining issues

There exists a number of features in the data that remain to be explained. We briefly identify some of them here. At $\approx \pm 30\,\mathrm{km\,s^{-1}}$ from $v_{\rm sys}$, the ring-like emission breaks up into a complex structure, yet symmetric w.r.t. the center, resembling a spiral linking the ring to the centre. It is possible that this structure has its origin in the capture and jet-launching events.

In the central velocity range $|v\text{-}v_{\rm sys}| \leq 10\,\mathrm{km\,s^{-1}}$, the $H_2S(2_{20}\text{–}2_{11})$ line emission, in particular, outlines an essentially circular structure of radius ≈ 1" that has a sharp outer edge, possibly a truncated disc or a torus. The cycle 3 data resolve out most of this emission, but a significant fraction of it remains in an unresolved region at 0.05" resolution.

At lower brightness level, in the velocity range $|v\text{-}v_{\rm sys}| \leq 30\,\mathrm{km\,s^{-1}}$, there is evidence for another bipolar structure. Its velocity gradient is opposite to that of the CO outflow and the position angle is different (PA $\approx -45°$).

We have re-observed the OH $^2\Pi_{3/2}$ Λ-doublet lines (Vlemmings et al., in prep). The results are peculiar. We confirm the spatial structure found by [8] with a symmetry axis along PA $\approx$ −60° and a velocity gradient opposite to that of the CO outflow. The most reasonable explanation for this is that the maser emission comes from the walls of the hour-glass structure where the amplification is highest along the line of sight. Curiously, only the 1667 MHz line is detected, and its flux has not varied on a time scale of 25 yr (the S/N ratio of the data is high at both epochs). The line shows no polarisation, and the emission avoids the inner circular structure discussed above.

References

[1] Sivarani T, Parthasarathy M, García-Lario P, Manchado A and Pottasch S R 1999 *Astron. Astrophys. Suppl. Series* **137** 505–519
[2] Kipper T 2005 *Baltic Astronomy* **14** 223–233
[3] Bakker E J, Lamers H J G L M, Waters L B F M, Waelkens C, Trams N R and Van Winckel H 1996 *Astron. Astrophys.* **307** 869–890
[4] Bakker E J, Lamers H J G L M, Waters L B F M and Waelkens C 1996 *Astron. Astrophys.* **310** 861–871
[5] Díaz F, Hearnshaw J, Rosenzweig P, Guzman E, Sivarani T and Parthasarathy M 2007 *IAU Symposium* vol 240 ed Hartkopf W I, Harmanec P and Guinan E F p 127
[6] Sahai R, Morris M, Sánchez Contreras C and Claussen M 2007 *Astron. J.* **134** 2200–2225
[7] Olofsson H and Nyman L Å 1999 *Astron. Astrophys.* **347** 194–202
[8] te Lintel Hekkert P T L, Chapman J M and Zijlstra A A 1992 *Astrophys. J. Lett.* **390** L23–L26
[9] Olofsson H, Vlemmings W H T, Maercker M, Humphreys E M L, Lindqvist M, Nyman L and Ramstedt S 2015 *Astron. Astrophys.* **576** L15
[10] Bujarrabal V, Castro-Carrizo A, Alcolea J and Sánchez Contreras C 2001 *Astron. Astrophys.* **377** 868–897

Journal of Physics: Conference Series **728** (2016) 042006 doi:10.1088/1742-6596/728/4/042006

Interactions between planets and evolved stars

Qian Shengbang[1,2,3], Han Zhongtao[1,2,3], Fernández Lajús, E[4,5,*], Zhu liying[1,2,3], Liao Wenping[1,2], Zejda Miloslav[6], Li Linjia[1,2], Irina Voloshina[7], Liu Liang[1,2,3] and He Jiajia.[1,2]

[1] Yunnan Observatories, Chinese Academy of Sciences, P.O. Box 110, 650216 Kunming, P.R. China (e-mail: qsb@ynao.ac.cn)
[2] Key Laboratory for the Structure and Evolution of Celestial Objects, Chinese Academy of Sciences, P.O. Box 110, 650216 Kunming, P. R. China
[3] University of the Chinese Academy of Sciences, Yuquan Road 19#, Shijingshan Block, 100049 Beijing, P. R. China
[4] Facultad de Ciencias Astronómicas y Geofísicas, Universidad Nacional de La Plata, 1900 La Plata, Buenos Aires, Argentina
[5] Instituto de Astrofisica de La Plata (CCT La plata - CONICET/UNLP), Argentina
[6] Department of Theoretical Physics and Astrophysics, Masaryk University, Kotlářská 2, CZ-611 37 Brno, Czech Republic
[7] Sternberg Astronomical Institute, Moscow State University, Universitetskij prospect 13, Moscow 119992, Russia
[*] Visiting Astronomer, Complejo Astronómico El Leoncito operated under agreement between the Consejo Nacional de Investigaciones Científicas y Técnicas de la República Argentina and the National Universities of La Plata, Córdoba and San Juan.

E-mail: [1] `qsb@ynao.ac.cn`

Abstract. Searching for planetary companions to evolved stars (e.g., white dwarfs (WD) and Cataclysmic Variables (CV)) can provide insight into the interaction between planets and evolved stars as well as on the ultimate fate of planets. We have monitored decades of CVs and their progenitors including some detached WD binaries since 2006 to search for planets orbiting these systems. In the present paper, we will show some observational results of circumbinary planets in orbits around CVs and their progenitors. Some of our findings include planets with the shortest distance to the central evolved binaries and a few multiple planetary systems orbiting binary stars. Finally, by comparing the observational properties of planetary companions to single WDs and WD binaries, the interaction between planets and evolved stars and the ultimate fate of planets are discussed.

1. Introduction

White dwarfs (WD) are the final evolutionary state of low- and medium-mass stars (including our Sun). It is thought that over 97% of the stars in the Milky Way will end their life as white dwarfs [1]. Exoplanets surveys demonstrate that they are prevalent in the Solar neighborhood as well as in the Milky Way [2, 3]. However, little is known about the evolution of planetary systems in orbiting around white dwarfs. Searching for exoplanets around WDs could give some constraints on late stages of stellar evolution (e.g., red giant stars, asymptotic giant branch stars, and planetary nebula). The investigations could provide us more knowledge on the interacting between planets and evolved stars especially on the ultimate fate of planets.

Both the radial velocity and transit methods are extensively used to search for planets around solar-type main-sequence stars. As a WD has a size similar to that of a planet, the transit method could be used to search for earth-like planets or even small objects because the transits would produce strong eclipses [4]. However, the possibility of the transit between a WD and a planet is extremely small. To data, only a minor planet in a debris cloud is detected to transit the metal-rich white dwarf WD 1145+017 that was generated by the disintegration of the planetesimal [5]. As for the other method, the radial velocities of white dwarfs can not be determined in a high precision because of the high surface gravities. Only a brown dwarf circling WD 0137-349 was found by measuring the radial velocities of a WD [6]. No planets were detected to orbit white dwarfs by using this method. When a WD is a member of an eclipsing binary star, eclipses provide a good chance to search for planets. Thanks to the compact structures of the WD component, the eclipse times can be measured in a high precision and therefore its small wobbles caused by the presence of exoplanets can be discovered. By monitoring the changes of eclipse times due to binary wobble, circumbinary planets can be found [7, 8, 9]. We have monitored some WD binaries since 2006 by using several 2.0-m class telescopes and some small telescopes. The hosting binary stars that we interested include detached WD+dM binaries, polars (magnetic cataclysmic variables, MCVs) and total eclipsing normal cataclysmic variables (CVs).

2. Detached white dwarf binary systems

Detached white dwarf binaries (DWDB) are post-common envelope systems with orbital periods from a few hours to a week or so [10]. They usually contain a WD primary and a red dwarf secondary and both components are well within their critical Roche lobes [11]. Because of the detached configurations, they are a good source to search for and to investigate circumbinary substellar objects by analyzing the light travel-time effect. To search for substellar objects orbiting WD binaries, we monitored more than 20 selected eclipsing DWDB photometrically [12, 13, 8]. Some of those targets were later investigated by other authors and the parameters of circumbinary substellar objects were revised [14, 15].

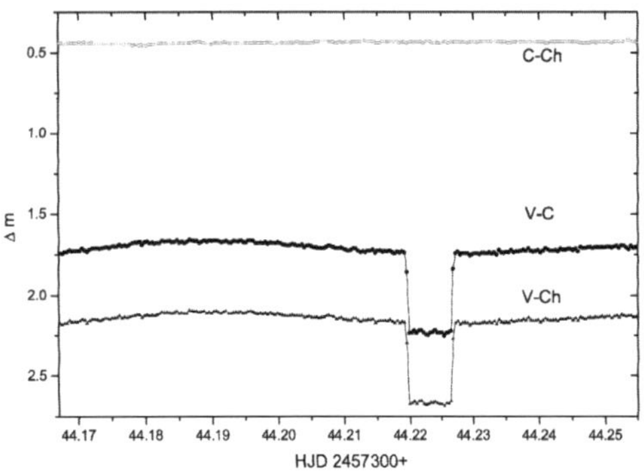

Figure 1. White-light CCD photometric light curve of J 030308 observed on November 17, 2015 by using the 2.4-m TNT telescope in Thailand.

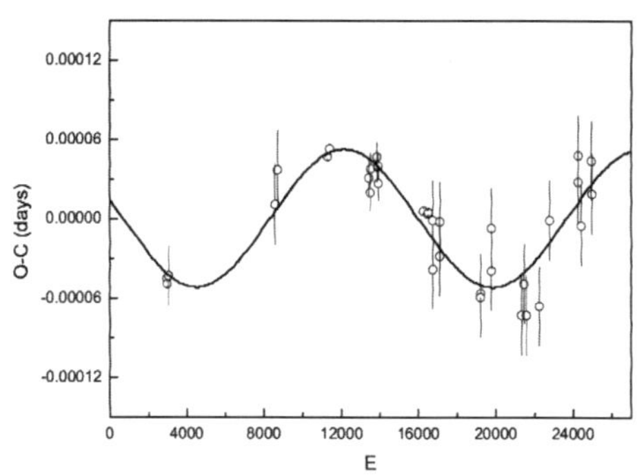

Figure 2. The cyclic variation of the O-C curve for J 030308 after an upward parabolic change is subtracted from the O-C diagram.

SDSS J030308.35+005444.1 (henceforth J 030308) is one of our monitored DWDB that contains a cool ($\sim$ 8000 K) WD and an M4.5 dwarf star with a 3.2 h orbit [16]. The magnetic field strength of the WD is 8 MG and it is a pre-CV that will evolve into an intermediate polar [17]. J 030308 has been monitored for several years by using several 2-m class telescopes. The eclipse profiles shown in Fig. 1 was obtained by using the 2.4-m Thai National Telescope (TNT) of National Astronomical Research Institute of Thailand (NARIT) on November 17, 2015. This telescope is located on one of the highest ridges of Doi Inthanon (about 2457-m high from the sea level). The TNT is a Ritchey-Chrètien with two Nasmyth focuses and A 4K CCD photometer with BVRI filter system was applied. The camera is a cryogenically cooled (liquid nitrogen, -110 C) dewar holding a E2V232-84 thinned, astronomy broadband AR coated, grade one CCD. The O-C diagram was constructed with our new determined eclipse times together with those compiled from literature [17]. A cyclic oscillation with a period of 5.64 years is found to be superimposed on an upward parabolic variation. The cyclic change is shown in Fig. 2 that is explained by the light-travel time effect via the presence of a circumbinary planet with a mass of $M_3 \sin i' = 3.1\,M_{Jup}$. The orbital separation of the planet is only 3.2 AU indicating that it is the shortest distance to the central WD binary among our targets.

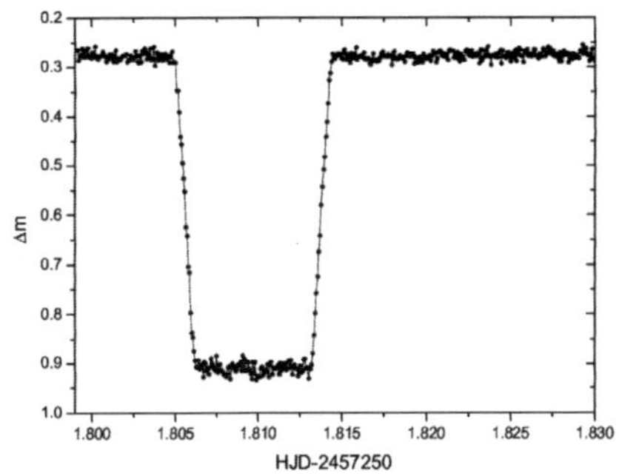

Figure 3. The eclipse profiles of RR Cae obtained with with the 2.15-m Jorge Sahade telescope in Argentina.

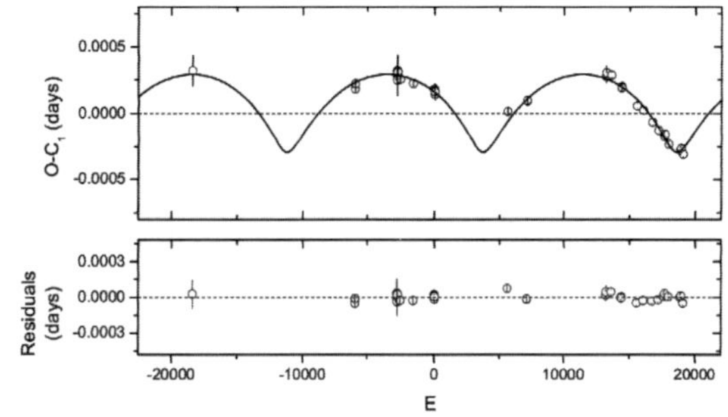

Figure 4. The theoretical light-travel time effect orbit of the circumbinary planet in RR Cae. After both the parabolic variation and the cyclic oscillation are subtracted, the residuals are displayed in the lower panel.

The other monitored example is RR Cae that is a DWDB with a period of 7.3 h. It is a double-lined eclipsing binary containing a cool WD primary with a mass of $0.44\,M_{\odot}$ and an M_4-type secondary with a mass of $0.182\,M_{\odot}$[18]. An upward parabolic variation in the O-C curve was found to be superimposed on a cyclic change with a period of 11.9 years and an amplitude of 14.3 s. The cyclic change was explained by the light-travel time effect via the presence of giant planet with a mass of $M_3 \sin i' = 4.2\,M_{Jup}$[8]. This binary was monitored continuously with the 2.15-m Jorge Sahade telescope at Complejo Astronomico El Leoncito (CASLEO), San Juan, Argentina. The eclipse profile observed on August 16, 2015 is shown in Fig. 3. During the observation, no filters were used and a Versarray 1300B camera with a thinned EEV CCD36-40 de 1340×1300 pix CCD chip was applied. After the upward parabolic change was removed the cyclic oscillation is plotted in Fig. 4. By considering a general case with eccentricity, the mass of the planet is revised as $M_3 \sin i' = 7.0\,M_{Jup}$. The planet is orbiting the central WD binary in eccentric orbit with an eccentricity of 0.61 at an orbital separation of 5.4 AU.

3. Eclipsing polars

In eclipsing MCVs (polars), the magnetic field of the WD primary is strong enough to prevent materials from the main-sequence companion to form an accretion disc [19]. Therefore, the accretion stream and the hot spot on the WD can be isolated during the eclipses, times of ingress and egress of eclipsing polars can be determined with a high precision. Therefore, they are a good source to search for substellar objects orbiting CVs by analyzing the eclipse times. Some substellar objects were recently discovered to be orbiting the polars [7]. We have selected about ten eclipsing polars (e.g., DP Leo, V2301 Oph, EK UMa, HU Aqr, UZ For, and MN Hya) and monitored them photometrically since 2009 by using the 1.0-m and the 2.4-m telescopes at Yunnan observatories in China, the Danish 1.54-m telescope at La Silla, the 2.15-m "Jorge Sahade" telescope in Argentina, and the 2.4-m telescope in Thailand [20].

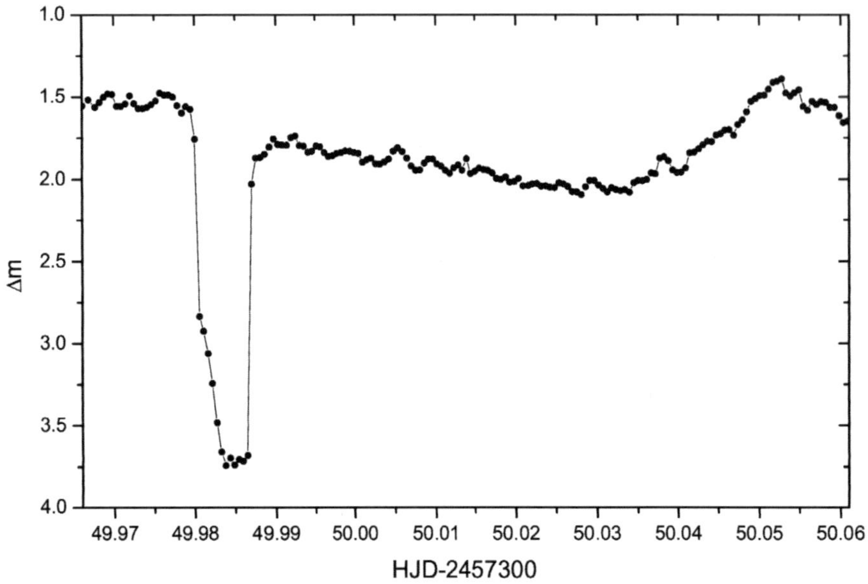

Figure 5. White-light CCD photometric light curve of HU Aqr observed on November 23, 2015 with the 2.4-m telescope in Yunnan observatories.

We pay more attention on the most interesting target HU Aqr. It is a total eclipsing polar with an orbital inclination of $i = 87°$ and has only one accretion pole [21]. The times of mid-egress were used for period investigation because the profiles of the egress are stable. Qian et al. (2011) proposed two planetary objects orbiting the eclipsing polar [22]. Then this polar was monitored continuously. One white-light eclipse profile of HU Aqr is shown in Fig. 5. This eclipse profile was obtained observed on November 23, 2015 by using the 2.4-m telescope in Lijiang observational station of Yunnan observatories. As shown in the figure, the eclipse starts with the limb of the secondary star eclipsing the accretion hot spot and the WD. The accretion stream is then the dominant source of the brightness with a small contribution of the secondary. Finally only the red-dwarf component star is visible and provides a constant contribution. The sequence of the egress is approximately reversed. The O-C diagram is constructed by including new data (see Fig. 6). Details analysis will confirm and revise the parameters of those planets.

4. Total eclipsing normal cataclysmic variables

4.1. Deeply eclipsing dwarf novae

In general, CVs are binary systems with a WD accreting matter from a stellar companion and form an accretion disk around the WD. Dwarf novae are one of the subtypes of CVs that have a very low rate of mass transfer [23]. They are characterized by their recurrent outbursts with Δm ~3-5 magnitudes and the typical duration is about 5-10 days. Those outbursts are explained by the structure change of their discs [24]. Some deeply eclipsing dwarf novae (e.g., GY Cnc, EX Dra, V2051 Oph, U Gem, OY Car, Z Cha, HT Cas and V893 Sco) were selected and monitored [25, 26, 27, 28]. One of the good examples is V2051 Oph that is one of a few ultrashort-period dwarf novae [29]. We started to monitor it since 2008 by using the 1.0-m and the 2.4-m telescopes in Yunnan Observatories, the 0.6-m Helen Sawyer Hogg (HSH) telescope and the 2.15-m Jorge Sahade telescope in Argentina. Two eclipse profiles of V2051 Oph obtained in June 2015 with the 2.15-m Jorge Sahade telescope are shown in Fig. 7. As shown in the figure, the eclipse profile is changing with time indicating that both the hot spot and the accretion disc are variable.

To determine the eclipse times of the WD, the derivative technique proposed by Wood et

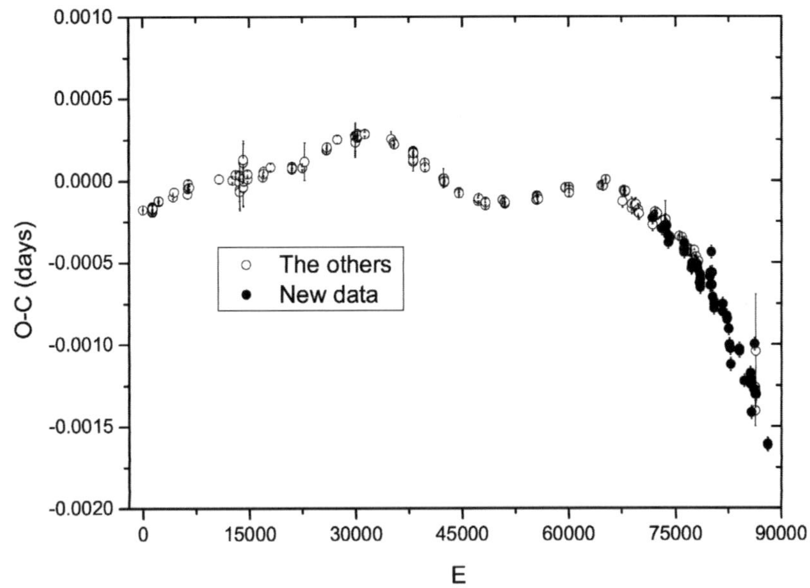

Figure 6. O-C diagram of HU Aqr. Solid dots refer to new observations, while open circles to those data collected from the literature.

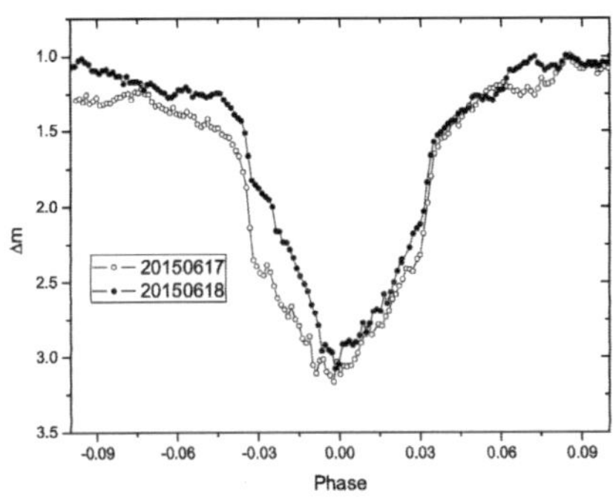

Figure 7. Two eclipse profiles of V2051 Oph obtained with the 2.15-m Jorge Sahade telescope in Argentina.

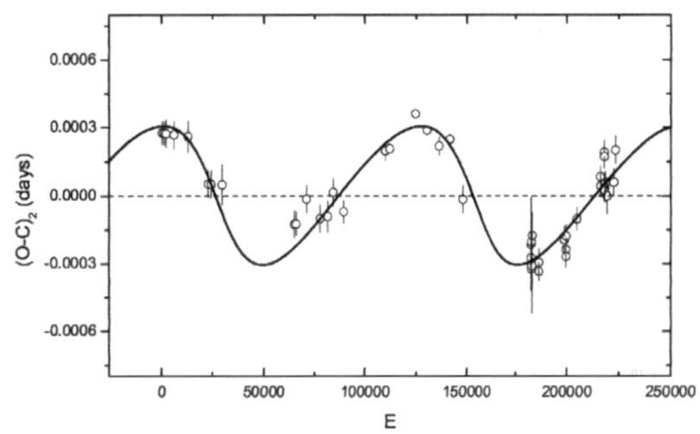

Figure 8. The theoretical light-travel time effect orbit of the circumbinary planet orbiting the dwarf nova V2051 Oph.

al. (1985) was used [30]. By analyzing the O-C diagram, a continuous period decrease was discovered to be superimposed on a cyclic variation with a small amplitude of $0.^{d}000329$ and a period of 21.64 years [20]. The cyclic oscillation can not be explained by magnetic activity cycles of the secondary (i.e., the Applegate mechanism) because the required energy is much larger than that radiated from the secondary in 10 years [31]. After the continuous period decrease was subtracted from the O-C diagram, the cyclic oscillation is displayed in Fig. 8 that is caused by the presence of a giant planet with a mass of $M_3 \sin i' = 7.3(\pm 0.7)$ Jupiter masses and an eccentricity of $e' = 0.37$. The giant circumbinary planet is orbiting around the dwarf nova at an orbital separation of about 9.0 AU.

4.2. Nova-like cataclysmic variables

Nova-like CVs usually have higher mass-transfer rates and longer orbital periods. We have monitored some deeply eclipsing nova-like CVs (e.g., SW Sex, V363 Aur, QZ Aur, V347 Pup, V348 Pup, LX Ser, PX And, BH Lyn and RW Tri) since 2006 [32, 33]. Here we show the results of one of nova-like CVs, V363 Aur. It is an eclipsing nova-like CV with an orbital period of 7.7 h that The was found by Lanning (1973) as an ultraviolet-bright source during a survey in the galactic plane by using Palomar 48 inch Schmidt [34]. This nova-like CV consists of a WD primary with a mass of $0.90\,M_{\odot}$ and a G7-type secondary with a mass of $1.06\,M_{\odot}$ [35]. We started to monitor V363 Aur since January 15, 2007 with the PI 1024 TKB CCD photometric system attached to the 1.0-m reflecting telescope at Yunnan Observatories. The eclipse profiles obtained with the 85-cm telescope is shown in Fig. 9. By using all available eclipse times, the O-C diagram was constructed. After the long-term period decrease was removed from the O-C curve, the cyclic variation is shown in Fig. 10. The preliminary results indicate that the mass of the third body is $M_3 \sin i' = 61.6(\pm 6.6) M_{Jup}$. It is may be a circumbinary brown dwarf orbiting the nova-like CV at a distance of about 9.9 AU.

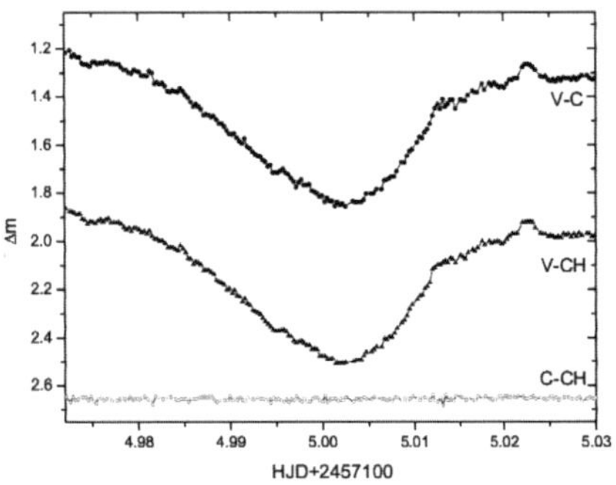

Figure 9. The eclipse profile of V363 Aur obtained with the 85-cm telescope in Xinglong observational station of national astronomical observatories (NAOs). "V" refers to V363 Aur, "C" to the comparison star, and "Ch" to the check star. Magnitude differences between V363 Aur, the comparison and the check stars are also shown in the figure.

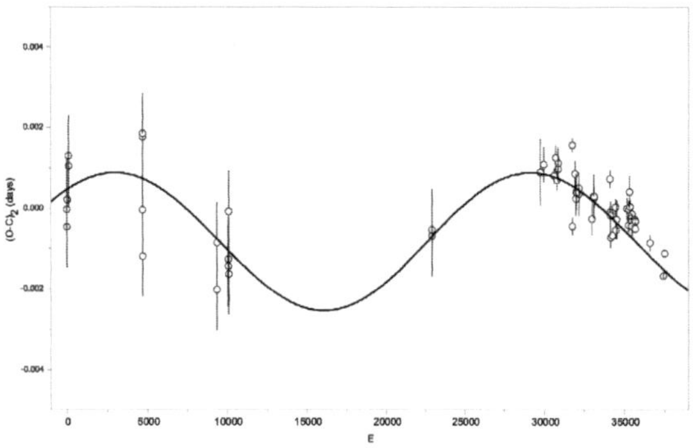

Figure 10. The cyclic variation of V363 Aur with an amplitude of 0.00172 days and a period of 23.0 years. The solid line refers to the light travel-time effect orbit caused by the presence of a brown dwarf with a mass of $M_3 \sin i' = 61.6\,M_{Jup}$.

5. Discussions and conclusions

Cyclic variations in the O-C diagrams were found for different types of WD binaries. Most of them can not be explained by the Applegate mechanism because of the energy problem. The plausible mechanism to cause those cyclic oscillations is the light-travel time effect via the presence of circumbinary planets. For some systems, e.g., J 030308 and RR Cae, the upward parabolic changes may be only a part of a long-period cyclic variation indicating that there is another planet in the systems. As shown by other investigators [36, 37, 38], our studies suggest that planets can exist in a completely different kind of host stars, i.e., WD binaries. Several exoplanets are found to be orbiting several kinds of WD binaries. However, substellar objects orbiting single white dwarfs are very rare. To date, only a few brown dwarfs companion to WD were found [39, 40, 41, 42, 6]. The project of direct imaging searching for exoplanets around white dwarfs also obtained null results [43]. The reasons for lack of detections may be (i) those methods are not very suitable (or not sensitive) or have a low efficiency to detect plants orbiting white dwarfs and (ii) the original planets may have been destroyed or escaped during the post main-sequence evolution [44, 45].

A very interesting question is why are the exoplanets orbiting WD binaries so common? Those WD binaries have undergone the common envelope (CE) evolution. The low-mass stars in the original binaries spiraled in the CE after the more massive stars evolve into a red giant or a AGB star. The ejection of the CE removed a large amount of angular momentum, and then the short-period WD binaries were formed. It is possible that the spiraling process of the low-mass stars in CE protected those circumbinary exoplanets [46]. The other possibility is that they are second generation exoplanets formed during the late evolution of binary stars [47]. Apart from searching for exoplanets, those long-term photometric data could provide valuable information on CV evolution and outburst [32, 48, 20, 28, 49]. One of the good examples is that the long-term period decrease in the eclipsing dwarf nova V2015 Oph. It indicates that the orbital evolution of short-period CVs is about 5 times faster than that predicted by the standard CV theory. This result suggests that additional angular momentum loss is required for CVs below the period gap [20].

Acknowledgments

This work is supported by the Chinese Natural Science Foundation (Nos. 11133007, 11325315, 11403095), Yunnan Natural Science Foundation (2014FB187), and the Strategic Priority Research Program "The Emergence of Cosmological Structures" of the Chinese Academy of Sciences, Grant No. XDB09010202. New CCD photometric observations were obtained with the 60-cm, the 1.0-m, and the 2.4-m telescopes of Yunnan Observatories, the 0.6-m Helen Sawyer Hogg telescope and the 2.15-m Jorge Sahade telescope in Argentina, the 85-cm and the 2.16-m telescope in Xinglong observational station of NAOs, the 2.4-m TNT of NARIT.

References

[1] Fontaine G, Brassard P and Bergeron P 2001 *Publ. Astron. Soc. Pac.* **113** 409
[2] Howard A W, Marcy G W, Johnson J A, Fischer D A, Wright J T Isaacson H, Valenti J A, Anderson J, Lin D N C and Ida S 2010 *Science* **330** 653
[3] Cumming A, Butler R P, Marcy G W, Vogt S S, Wright J T and Fischer D A 2008 *Publ. Astron. Soc. Pac.* **120** 531
[4] Agol E 2011 *Astrophys. J. Lett.* **635** L31
[5] Vanderburg A, Johnson J A, Rappaport S, Bieryla A, Irwin J, Lewis J A, Kipping D, Brown W R and Dufour P 2015 *Nature* **526** 546
[6] Maxted P F L, Napiwotzki R, Dobbie P D and Burleigh M R 2006 *Nature* **442** 543
[7] Qian S-B, Liao W-P, Zhu L-Y and Dai Z-B 2010a *Astrophys. J.* **708** L66
[8] Qian S-B, Liu L, Zhu L-Y, Dai Z-B, Fernández Lajús E and Baume G L 2012 *Mon. Not. R. Astron. Soc.* **422** L24
[9] Qian S-B, Zhu L-Y, Liao W-P, Zejda M, Mikulásek Z, Fernández Lajús E, Zola S, Zhou X and Han Z-T 2015a *Living Together: Planets, Host Stars and Binaries* eds Rucinski S M *et al* (ASP Conference Series vol 496) p 388
[10] Nebot Gómez-Morán A *et al* 2011 *Astron. Astrophys.* **536** A43
[11] Parsons S G, Gänsicke B T, Marsh T R, Drake A J, Dhillon V S, Littlefair S P, Pyrzas S, Rebassa-Mansergas A and Schreiber M R 2013 *Mon. Not. R. Astron. Soc.* **429** 256
[12] Qian S-B, Dai Z-B, Liao W-P, Zhu L-Y, Liu L and Zhao E G 2009a *Astrophys. J.* **706** L96
[13] Qian S-B, Liao W-P, Zhu L-Y, Dai Z-B, Liu L, He J-J, Zhao E-G and Li L-J 2010b *Mon. Not. R. Astron. Soc.* **401** L34
[14] Almeida L A and Jablonski F 2011 *Proc. Int. Astron. Union* (S276) **6** 495
[15] Marsh T R, Parsons S G, Bours M C P, Littlefair S P, Copperwheat C M, Dhillon V S, Breedt E, Caceres C and Schreiber M R 2014 *Mon. Not. R. Astron. Soc.* **437** 475
[16] Pyrzas S, Gänsicke B T, Marsh T R, Aungwerojwit A, Rebassa-Mansergas A, Rodríguez-Gil P, Southworth J, Schreiber M R, Gomez-Moran A N and Koester D 2009 *Mon. Not. R. Astron. Soc.* **394** 978
[17] Parsons S G, Marsh T R, Gänsicke B T, Schreiber M R, Bours M C P, Dhillon V S and Littlefair S P 2013 *Mon. Not. R. Astron. Soc.* **436** 241
[18] Maxted P F L, O'Donoghue D, Morales-Rueda L, Napiwotzki R and Smalley B 2007 *Mon. Not. R. Astron. Soc.* **376** 919
[19] Giovannelli F and Sabau-Graziati L 2012 *Memorie della Societa Astronomica Italiana* **83** 446
[20] Qian S-B, Zhu L-Y, Zhao E-G, Fernández Lajús E, Zhang J, Shi G and Han Z-T 2015b *Acta Polytechnica* **2** 152
[21] Schwope A D, Schwarz R, Sirk M and Howell S B 2001 *Astron. Astrophys.* **375** 419
[22] Qian S-B, Liu L, Liao W-P, Li L-J, Zhu L-Y, Dai Z-B, He J-J, Zhao E-G, Zhang J and Li K 2011 *Mon. Not. R. Astron. Soc.* **414** L16

[23] Warner B 1995 *Cataclysmic Variable Stars* Cambridge Astrophysics Series 28 (Cambridge: Cambridge University Press)
[24] Saito R K and Baptista R 2006 *Astron. J.* **131** 2185
[25] Dai Z-B, Qian S-B and Fernández Lajús E 2009a *Astrophys. J.* **703** 109
[26] Dai Z-B and Qian S-B 2009b *Astrophys. Space Sci* **321** 91
[27] Han Z-T, Qian S-B, Fernández Lajús E, Liao W-P and Zhang J 2015 *New Astron.* **34** 1
[28] Qian S-B, Han Z-T, Fernández Lajús E, Zhu L-Y, Li L-J, Liao W-P and Zhao E-G 2015c *Astrophys. J. Suppl. Series* **221** 17
[29] Baptista R, Catalan M S, Horne K and Zilli D 1998 *Mon. Not. R. Astron. Soc.* **300** 233
[30] Wood J H, Irwin M J and Pringle J E 1985 *Mon. Not. R. Astron. Soc.* **214** 475
[31] Applegate J H 1992 *Astrophys. J.* **385** 621
[32] Qian S-B, Soonthornthum B, Dai Z B, Zhu L Y, He J-J, Liao W P and Li L J 2009b *The Eighth Pacific Rim Conference on Stellar Astrophysics: A Tribute to Kam-Ching Leung* (ASP Conference Series vol 404) p 248
[33] Shi G and Qian S-B 2014 *Publ. Astron. Soc. Japan* **66** 41
[34] Lanning H 1973 *Publ. Astron. Soc. Pac.* **85** 70
[35] Thoroughgood T D, Dhillon V S, Watson C A, Buckley D A H, Steeghs D and Stevenson M J 2004 *Mon. Not. R. Astron. Soc.* **353** 1135
[36] Beuermann K *et al* 2011, *Astron. Astrophys.* **526** A53
[37] Bruch A 2014 *Astron. Astrophys.* **566** A101
[38] Goździewski K, Slowikowska A, Dimitrov D, Krzeszowski K, Żejmo M, *et al* 2015 *Mon. Not. R. Astron. Soc.* **448** 1118
[39] Becklin E F and Zuckerman B 1988 *Nature* **336** 656
[40] Farihi J and Christopher M 2004 *Astron. J.* **128** 1868
[41] Dobbie P D, Burleigh M R, Levan A J, Barstow M A, Napiwotzki, Holberg J B, Hubeny I and Howell S B 2005 *Mon. Not. R. Astron. Soc.* **357** 1049
[42] Burleigh M R, Hogan E, Dobbie P D, Napiwotzki R and Maxted P F L 2006 *Mon. Not. R. Astron. Soc.* **373** L55
[43] Burleigh M R *et al* 2008 *Mon. Not. R. Astron. Soc.* **386** L5
[44] Nordhaus J, Spiegel D S, Ibgui L, Goodman J and Burrows A 2010 *Mon. Not. R. Astron. Soc.* **408** 631
[45] Veras D and Raymond S N 2012 *Mon. Not. R. Astron. Soc.* **421** L117
[46] Qian S-B, Han Z-T, Soonthornthum B, Zhu L-Y, He J-J, Rattanasoon S, Aukkaravittayapun S, Liao W-P, Zhao E-G, Zhang J and Fernández Lajús E 2016 *Astrophys. J.* **817** 151
[47] Völschow M, Banerjee R and Hessman F V 2014 *Astron. Astrophys.* **562** 19
[48] Qian S-B, Zhu L-Y, Fernández Lajús E, He J-J, Liao W-P, Zhao E-G, Liu L and Yang Y-G 2014 *The Tenth Pacific Rim Conference on Stellar Astrophysics* (ASP Conference Series vol 482) 171
[49] Qian S-B, Han Z-T, Zhu L-Y, Liao W-P, Fernández Lajús E, Zejda M, Liu L, Soonthornthum B and Zhou X 2015d *Publ. Korean Astron. Soc.* **30** 175

Journal of Physics: Conference Series **728** (2016) 042007 doi:10.1088/1742-6596/728/4/042007

Profile disparity of Raman-scattered O VI in symbiotic stars

Hee-Won Lee

Department of Physics and Astronomy, Sejong University, Seoul, Korea

E-mail: hwlee@sejong.ac.kr

Abstract. Symbiotic stars are wide binary systems consisting of a hot compact star (usually a white dwarf) and a mass losing giant. Symbiotic activities are believed to occur through gravitational capture of a fraction of the slow stellar wind from the giant. Raman scattered features of O VI resonance doublet 1032 and 1038 appearing at around 6825 Å and 7082 Å are a unique spectroscopic diagnostic tool to probe the mass transfer process in symbiotic stars. The Raman O VI features often exhibit multiple peak structures and in many cases the blue peak of 7082 features is relatively more suppressed than that of 6825 features. We propose that the disparity of the two profiles is attributed to the local variation of optical depths of O VI, implying that the accretion flow is convergent in the red emission region and divergent in the blue emission region. It is argued in this presentation that Raman scattering by atomic hydrogen is a natural mirror to provide an edge-on view of the accretion disk and a lateral view of the bipolar outflow in symbiotic stars. We discuss the spectropolarimetric implications of this interpretation.

1. Introduction

Symbiotic stars are binary systems consisting of a hot white dwarf and a mass losing giant. They exhibit prominent emission lines with a broad range of ionization and excitation. The symbiotic activities are attributed to the mass transfer processes involving a gravitational capture of some fraction of the slow stellar wind from the giant component. One unique aspect of the spectra of symbiotic stars is the presence of the spectral features formed through Raman scattering of far UV radiation more energetic than Lyα by atomic hydrogen.

The Raman scattering process is the combination of the annihilation of the incident radiation with the excitation of the hydrogen atom and the creation of the Raman scattered radiation accompanied by the de-excitation of the hydrogen atom into the $2s$ state. For example, an O VIλ1032 photon is Raman scattered to appear at 6825 Å, and in a similar way an O VIλ1038 is Raman scattered to be emergent at 7082 Å. These two Raman O VI features at 6825 Å and 7082 Å are observed in about half of symbiotic stars and were first identified by Schmid [1]. Raman scattered line features observed thus far include He II$\lambda\lambda$ 1025, 972 and 949, Ne VIIλ973 and C II$\lambda\lambda$1036 and 1037.

The scattering cross section for resonance doublet O VI$\lambda\lambda$ 1032 and 1038 is of order 10^{-22} cm^2. The cross section for O VIλ1032 is larger than for O VIλ1038, which results in about 5-10 times stronger Raman O VI feature at 6825 Åthan 7082 Å. The small cross section for these Raman features implies that the operation of Raman scattering requires a very special condition of the existence of a highly thick H I region in the vicinity of a strong far UV source.

Journal of Physics: Conference Series **728** (2016) 042007 doi:10.1088/1742-6596/728/4/042007

This condition is ideally met in symbiotic stars where the giant component provides a highly neutral region illuminated by the far UV nebular region surrounding the white dwarf. A number of young planetary nebulae are known to exhibit Raman scattered He II features (e.g. Pequignot et al. [2], Groves et al. [3], Kang et al. [4]).

The energy conservation relates the wavelength λ_o to the that of the incident radiation by

$$\lambda_o^{-1} = \lambda_i^{-1} + \lambda_\alpha^{-1}, \tag{1}$$

where λ_α is the wavelength of Lyα. Variation of this relation yields

$$\frac{\Delta\lambda_o}{\lambda_o} = \left(\frac{\lambda_o}{\lambda_i}\right)\frac{\Delta\lambda_i}{\lambda_i}, \tag{2}$$

which leads to a broadened profile exhibited in the Raman scattered features. Therefore, the Raman O VI features exhibit broader profiles by a factor of ~ 7 than their parent emission lines in the far UV region. This relation is very important because the profile of a Raman scattered feature reflects only the kinematics between the far UV emission source and the H I region and is almost independent of the observer's lie of sight. The profile broadening attributed to the inelasticity of scattering leads to Raman scattering to play a role of the natural mirror allowing the accretion disk from the vantage point of the giant.

2. Accretion of the Slow Stellar Wind in Symbiotic Stars

According to Mastrodemos & Morris [5] some fraction of the slow stellar wind from the giant can be captured gravitationally by the white dwarf to form an accretion disk. Accretion disks are often accompanied by the bipolar outflows. For example, Angeloni et al. [6] found that Sanduleak's star exhibits a huge outflow extending up to 17 pc. However, the overall mass transfer processes in symbiotic stars are only poorly known.

Raman O VI features at 6825 Å and 7082 Å are known to exhibit complicated profiles including double-peak and triple-peak profiles. When we analyze the Raman scattered O VI at 7082 Å along with its stronger twin feature at 6825 Å, we find an interesting fact that the profiles of the 7082 and 6825 features differ in a systematic way that the blue part of the 7082 feature is relatively more suppressed than its counterpart of the 6825 feature. Because O VI 1032 and 1038 are formed in the same region, we may find the physical mechanism of the profile difference in the O VI emission region. Arising from transition $2S_{1/2} - 2P_{1/2,3/2}$, the statistical factor for O VIλ1032 is twice larger than that for O VIλ1038. In an optically thin emission region the flux ratio of O VIλ1032 and O VIλ1038 is two. As the optical depth increases, the flux ratio will approach the thermal limit of unity. In this way, the flux ratio may differ depending on the local variation of physical conditions in the O VI emission region.

In Fig. 1, we illustrate schematically the O VI emission region identified with the accretion disk around the white dwarf, where the accretion disk is in turn decomposed into two subcomponents of the blue emission region and the red emission region. We denote these two components by BER (blue emission region) and RER (red emission region). It is expected that the accretion flow is convergent in the red emission region leading to a flux ratio near 1, whereas the accretion flow is divergent in the blue emission region resulting in a flux ratio approaching two. In this way when we normalize the two Raman features so that the red peaks coincide, the blue peak of the Raman 6825 feature will be larger than that of the Raman 7082 feature.

3. Profile Disparity of the Raman Scattered O VI Features

Fig. 2 shows the line profiles of the Raman scattered O VI at 6825 Å and 7082 Å of the symbiotic nova V1016 Cygni presented by Heo & Lee [7]. In this spectrum the red peaks of the two features are stronger than the blue peaks, which implies that the accretion flow is convergent on the side

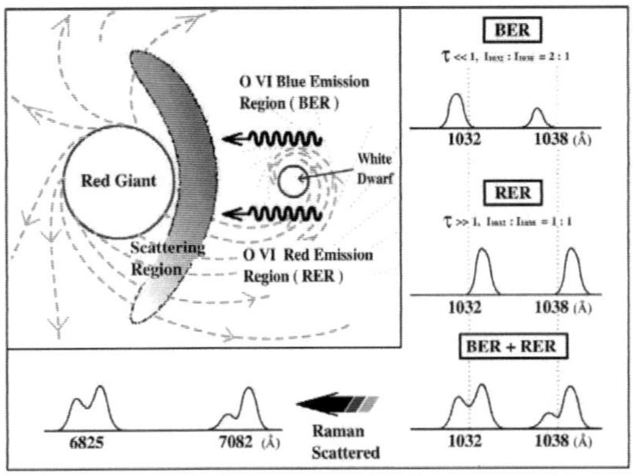

Figure 1. A schematic illustration of the formation of Raman scattered O VI at 6825 Å and 7082 Å in a symbiotic star. The O VI emission region is identified with the accretion disk formed via gravitational capture of a fraction of the slow stellar wind from the giant. Lying near the giant, the scattering neutral region commands an edge-on view of the O VI emission region. In this model, when the profiles are normalized so that the red peaks coincide, the blue peak of Raman 7082 is weaker than that of Raman 6825.

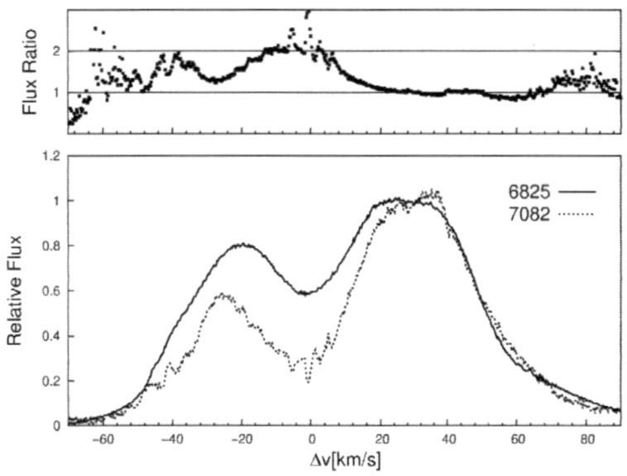

Figure 2. The line profiles of the Raman scattered O VI at 6825 Å and 7082 Å of the symbiotic star V1016 Cygni obtained with the BOES (Bohyunsan Optical Echelle Spectrograph) installed on the 1.8 m telescope at the Bohyunsan Optical Observatory on the night of 2005 November 7. The horizonatl axis is the Doppler factor in units of km s^{-1}. The upper panel shows the flux ratio and the lower panel shows the line profiles.

of the red emission region leading to enhanced emissivity. We assume that the H I region is stationary and that the effect of the observer's line of sight is neglected. We transform from the observed wavelength space to the Doppler factor space, where the Doppler factor is measured with respect to the stationary H I region.

The two profiles are normalized in such a way that the red peaks of the same strength noting that the lower bound of the O VI flux ratio is 1:1. It happens that the flux ratio at the Doppler factor near zero is almost 2:1, which is the theoretical upper bound of the OVI flux ratio. The part of the O VI emission region with negligible Doppler factors may be identified with the region extended significantly far from the white dwarf. Lying in the shallow gravitational potential with low density, this region is plausibly characterized by the flux ratio of 2:1. We also note that the blue emission region exhibits the flux ratio of ~ 1.5. It turns out that the normalization of the equal strength of the red peaks allows the flux ratio to vary in the full range between one and two, which means that this normalization is the only theoretically admissible one.

We propose that the O VI emission region of V1016 Cyg is characterized by three regions, one is the blue emission region with the intermediate flux ratio of 1.5:1, another is the extended region with zero Doppler factor showing the flux ratio of 2:1 and the other is the red emission region with the flux ratio of 1:1. The flux ratio indicates that the red emission region is of the highest density and the centrally extended region of the lowest density. It is an interesting possibility that the red emission region with the highest density can be identified with a hot spot that is commonly observed in cataclysmic variables.

4. Discussion

Schmid [8] proposed that multiple peak profiles exhibited in the Raman O VI are attributed to the expansion of the H I region, where the blue peaks are formed in the part of the H I that approaches the O VI emission region. In this picture, in order to obtain stronger red peak, the H I region that recedes from the O VI should extend much further than the approaching part. In this model where the O VI emission region is assumed to be point-like, it is very difficult to reconcile the fact that the profiles of the two Raman O VI features are different. One explanation may be that the scattering optical depth is very different for each velocity interval of H I, which would require a very contrived distribution of neutral material around the giant. Furthermore, the distribution of H I with respect to the point-like O VI emission region is axially symmetric, which makes it difficult to explain the phenomena of polarization flip.

The accretion disk component naturally leads to Raman O VI features with a double-peaked profile. With an additional contribution from the bipolar outflow, we expect that the profile will be augmented by another peak located redward of the accretion disk profile. In case the velocity scale of the bipolar outflow exceeds that of the accretion disk, the resultant profile will be triply peaked. One consequence of this view is the flip of the position angle of the polarization of the Raman O VI features in the reddest part.

One can find the phenomena of polarization flip from the Spectropolarimetric data compiled by Harries & Howarth [9], which appears to be in support of this view. Furthermore, Lee & Kang [10] proposed that the double peak profiles of the Raman O VI at 6825 in the symbiotic stars V1016 Cyg and HM Sge are consistent with the Keplerian motion of the O VI emission region around the white dwarf. Our future work includes a combined study of the hydrodynamics and radiative transfer of the slow stellar wind accretion in symbiotic stars.

Acknowledgments

This research was supported by the Basic Science Research Program of the National Research Foundation (NRF-2014R1A1A2054887) and the Korea Astronomy and Space Science Institute under the R&D program (Project No. 2015-1-320-18).

References

[1] Schmid H M 1989 *Astron. Astrophys.* **211** L31
[2] Pequignot D, Baluteau J-P, Morisset C and Boisson C 1997 *Astron. Astrophys.* **323** 217
[3] Groves B, Dopita M A, Williams R E and Hua C-T *Publ. Astron. Soc. Australia* **19** 425
[4] Kang E-H, Lee H-W and Lee B-C 2009 *Astrophys. J.* **695** 542
[5] Mastrodemos N and Morris M 1998 *Astrophys. J.* **497** 303
[6] Angeloni R, Di Mille F, Bland-Hawthorn J and Osip D J 2011 *Astrophys. J. Lett.* **743** L8
[7] Heo J-E and Lee H-W 2015 *J. Korea Astron. Soc.* **48** 105
[8] Schmid H M 1989 *Mon. Not. R. Astron. Soc.* **282** 511
[9] Harries T J and Howarth I D 1996 *Astron. Astrophys. Suppl. Series* **119** 61
[10] Lee H-W and Kang S 2007 *Astrophys. J.* **669** 1156

Journal of Physics: Conference Series **728** (2016) 052001 doi:10.1088/1742-6596/728/5/052001

Chemistry in AGB stars: successes and challenges

T J Millar

Astrophysics Research Centre, School of Mathematics and Physics, Queen's University Belfast, University Road, Belfast BT7 1NN, UK

E-mail: tom.millar@qub.ac.uk

Abstract. Emission and absorption line observations of molecules in late-type stars are a vital component in our understanding of stellar evolution, dust formation and mass loss in these objects. The molecular composition of the gas in the circumstellar envelopes of AGB stars reflects chemical processes in gas whose properties are strong functions of radius with density and temperature varying by more than ten and two orders of magnitude, respectively. In addition, the interstellar UV field plays a critical role in determining not only molecular abundances but also their radial distributions. In this article, I shall briefly review some recent successful approaches to describing chemistry in both the inner and outer envelopes and outline areas of challenge for the future.

1. Introduction

The circumstellar envelopes (CSEs) of AGB stars have long been known to present a rich molecular chemistry dominated by the interaction of external, interstellar FUV photons with parent species formed by thermal equilibrium processes near the photosphere [1, 2, 3, 4, 5, 6]. These, and more recent models [7, 8, 9, 10], have included more accurate descriptions of the physical conditions through the inclusion of clumps and density-enhanced rings in the CSE around the carbon-rich AGB star IRC+10216 (CW Leo) and through the addition of an extensive chemistry to describe the anions recently detected therein. The result is a consensus that the chemistry of the external envelope of IRC+10216, and by extension all AGB CSEs, is a photon-dominated process, a process whose final molecular products give information on mass-loss history, wind acceleration, dust formation, dredge-up and nucleosynthesis.

As well as studies of the chemical processes in the outer CSE, there have also been investigations of the the interaction between physics and chemistry in the inner CSE. For example, pioneering work on the chemistry induced by shock waves driven by stellar pulsations, [11, 12], has been extended [13, 14] to include the formation of new species and dust grain formation. These papers show that if shocks are strong then any molecules formed in thermodynamic equilibrium (TE) are rapidly destroyed in the immediate post-shock gas and that 'parent' species available for chemistry in the outer CSE are the end products of shock chemistry coupled with dust nucleation and growth. Challenges in understanding astrophysics and astrochemistry are, as ever in astronomy, driven by advances in observational techniques, instruments and facilities, most recently from the *Herschel Space Observatory* and *ALMA*.

Journal of Physics: Conference Series **728** (2016) 052001 doi:10.1088/1742-6596/728/5/052001

2. Circumstellar chemistry

In this section I will briefly review some progress made in the chemistries of both O-rich and C-rich AGB stars and give some indication of current challenges.

2.1. O-rich late-type stars

Herschel and *ALMA* have both given us remarkable new information on the chemistry of O-rich CSEs, especially in the internal layers close to the photosphere. A large number of detailed observational studies [15, 16, 17, 18, 19] have both increased the range of species detected near the photosphere and provided much improved abundance estimates. These include studies of the refractory species SiO, TiO, TiO_2, AlO and AlOH, thought to be involved in the creation of silicate grains in these stars. Gobrecht et al. [14] have presented a very detailed model of dust formation in the O-rich star IK Tau. They considered chemical reactions in the shocked gas created by periodic stellar pulsations including the formation of small magnesium silicate and alumina clusters. Their best-fit model to observations starts with a periodic shock wave at a velocity of 32 km s^{-1} at a radius of 1 $R_\star$ which propagates outward with a velocity proportional to r^{-2}, where r is the radial distance from the photosphere. For radii less than 2 $R_\star$ the immediate post-shock densities are very high, greater than 10^{13} cm^{-3}, and solidly in the regime where three-body reactions must be considered. The immediate post-shock temperature is also very high, more than 4000 K. At these densities and temperatures any molecules formed at thermal equilibrium (TE) in the photosphere are destroyed. Over a pulsational phase the density and temperature both fall and new molecules, whose compositions and abundances depend on chemical kinetics in a cooling, expanding flow, form. In this post-shock gas the chemistry is dominated by high temperature neutral-neutral reactions.

It is, of course, not surprising that in O-rich AGB stars, oxides, dioxides and hydroxides form readily through gas-phase chemistry. What is surprising, and certainly not predicted by the TE models is the presence of carbon-bearing molecules since all available carbon is expected to be locked up in CO. HCN, CS and CO_2, hoever, have relatively large abundances in the inner CSE. Gobrecht et al.[14] show that the chemistry, while complex, occurs on very fast time-scales. Thus CS forms in hot gas via:

$$S + H_2 \longrightarrow SH + H \quad (1)$$
$$C + SH \longrightarrow CS + H \quad (2)$$
$$CO + SH \longrightarrow OCS + H \quad (3)$$
$$OCS + H \longrightarrow CS + OH \quad (4)$$

while CN and HCN form via:

$$N + CO \longrightarrow CN + O \quad (5)$$
$$N + CS \longrightarrow CN + S \quad (6)$$

followed by

$$CN + H_2 \longrightarrow HCN + H \quad (7)$$

Many of these reactions produce atomic hydrogen and the reverse reactions can be significant particularly when the abundance of H atoms is high. This occurs in the zone where dust precursors form since their formation converts some H_2 to H, as discussed below. Gobrecht et al. show that the gas-phase abundances calculated in their pulsational shock model at 6–8 $R_\star$ match to within an order of magnitude those observed in the inner CSE, including HCN and CS.

Some molecules, not surprisingly, do not fit as well, for example, SO and SO_2. Danilovich et al. [20] have recently observed many transitions of these two molecules in R Dor and supplemented these with HIFI and other observations toward another four O-rich AGB stars. Using a detailed radiative transfer model they have constrained the distributions of these two species, both in abundance and in radial extent. In all five stars they find that SO_2 peaks on the stellar position whereas SO has a shell-like structure in IK Tau and R Cas, with a peak fractional abundance of $\sim 10^{-6}$ at 1.3 $\times 10^{16}$ cm.

When the peak position of the SO abundance is plotted against the wind density, $\propto \dot{M}/v_{exp}$, the results for R Cas, TX Cam and IK Tau follow a power-law dependence consistent with a circumstellar chemistry that is dominated by photodissociation in the outer envelope. SO_2, on the other hand, appears to constrained to the inner envelope where it should be formed in high-temperature chemistry. Although the latter fact is broadly consistent with the models of Gobrecht et al., the observed SO_2 abundances in R Dor and W Hya are an order of magnitude larger than predicted. Indeed all models fail to predict the very large abundances of SO and SO_2 which, in total, approach the elemental sulphur abundance. Model calculations generally predict sulphur to be in atomic form. A failure to agree in every respect with the observations should, however, be qualified by noting that an accurate description of the chemical kinetics occurring at high density and temperature is very difficult, especially in the case of O-rich AGB stars since inorganic chemistry is not so well studied in the laboratory. In addition, the balance between forward and reverse reactions, such as several of those above, is controlled by the H:H_2 abundance ratio which is not well determined either observationally or theoretically in the inner envelope.

In addition to the synthesis of molecules such as CO, H_2O, PN and HCl, which do not participate in dust formation, many other species are produced, several of which are likely to be intimately connected with the process of cluster formation and grain growth. Goumans and Bromley [21] discussed the detailed energetics of the formation of the dimers of enstatite $(MgSiO_3)_2$ and fosterite $(Mg_2SiO_4)_2$ from an initial gas of SiO, Mg, H_2 and H_2O at 1000 K. Although the initial dimerisation of SiO is an endoergic process, its equilibrium constant is 5.5×10^{-4} at 1000 K, giving rise to a low abundance of Si_2O_2. Subsequent reactions with H_2O, which result in O-atom addition, followed by addition reactions with Mg are exothermic and can rapidly build dimers of both enstatite and forsterite. Since reactions of H_2O and Mg are exoergic for reactions with the dimers and larger clusters, silicate dust grains will form. Gobrecht et al [14] find dimer fractional abundances of 10^{-11} for enstatite and 5×10^{-8} for forsterite at 3.5 $R_\star$. Dimer formation becomes very efficient outside 3 $R_\star$ with growth of silicate grains occurring between 3 and 6 $R_\star$. The authors follow the diffusion and coagulation of these particles to determine the grain size distribution as particles propagate outwards from 3.5 to 10 $R_\star$. Their results show that, in most cases, gas-phase abundances agree well with those determined for the inner wind of IK Tau, that forsterite grains are much more abundant than enstatite and metal oxides such as MgO and SiO, and that the overall dust-to-gas mass ratio is $\sim$ (1–6) $\times 10^{-3}$, in reasonable agreement with observations.

Dust grains grow and their size distribution evolves over a number of pulsations as the gas is lifted slowly away from the photosphere. Due to their high binding energy, clusters of alumina form readily in the hottest gas near the photosphere in O-rich stars. For a radial drift velocity of 0.5 km s^{-1}, it takes 12 pulsations for the gas in IK Tau to move between 1 and 2 $R_\star$. Gobrecht et al. find that the size distribution of alumina favours larger particles and that growth of alumina grains stops beyond 2 $R_\star$ as all available aluminium is tied up in dust at that point. Such grains make only a minor contribution to the overall dust-to-gas mass ratio.

Silicates, on the other hand, form at larger radii, out to about 10 $R_\star$, since formation of the underlying dimer population cannot occur at high temperatures close to the star. The drift velocity in the silicate dust zone is larger, perhaps 1.5 km s^{-1}, than that in the alumina dust

Journal of Physics: Conference Series **728** (2016) 052001 doi:10.1088/1742-6596/728/5/052001

zone since radiation pressure on the alumina grains begins to drive the mass loss. In this case, it takes about 10 pulsations to move material from 3.5 to 6 $R_\star$ where the majority of silicate grains form. As these grains form further from the star where densities are lower, silicates tend to have a smaller size distribution that alumina grains. They do, however, because of the abundance of silicon, magnesium and oxygen, contribute more to the dust mass. At 10 $R_\star$, the dust-to-gas mass ratio is $\sim 2\times10^{-3}$, similar to those observed in O-rich AGB stars, with $\sim$ 22% of elemental silicon contained in the dust [14].

The outer CSE chemistry of O-rich AGB stars is dominated by effects produced by irradiation of the outer envelope by interstellar FUV photons, that is, outer CSEs are examples of photon-dominated regions (PDRs). Li et al. [22] have presented a detailed model of the outer CSE chemistry of O-rich CSEs including, for the first time, shielding of N_2 in addition to the usual self- and mutual-shielding of H_2 and CO. The authors assume an extensive list of parent species, some 18, with initial conditions derived from either observation or from the shock-induced abundances calculated at 6 $R_\star$ by Gobrecht et al. [14]. They determine the chemistry of some 467 species using the latest release of the The UMIST Database for Astrochemistry [23]. Li et al. calculate radial abundances and column densities for mass-loss rates between 10^{-8} and 10^{-4} $M_\odot$ yr^{-1} and expansion velocities of 10-40 km s^{-1} and make specific comparison with the observed abundances in IK Tau ($\dot{M} = 4.5 \times 10^{-6}$ $M_\odot$ yr^{-1}, $v_{exp} = 24$ km s^{-1}).

Li et al. were able to include a detailed consideration of N_2 photodissociation due to over 25 years of laboratory and theoretical studies [24, 25, 26]. The rate coefficient is determined primarily by the overlap of the N_2 absorption bands with those of H_2 (mutual shielding) in the 912–1000 Å wavelength range together with self-shielding of N_2. The overall shielding is thus a complex function of gas temperature and column density [27]. For parameters appropriate to IK Tau, for example, the radius at which the fractional abundance of atomic nitrogen, produced by the photodissociation of N_2, reaches 10^{-5}, increases from 1 to 6 $\times$ 10^{16} cm; at a fractional abundance of 10^{-4} the increase is from 2.8 to 18 $\times$ 10^{16} cm. Thus the atomic N abundance increases over an appreciable volume of the outer envelope. It has, however, only a limited impact on molecular abundances of species other than N_2, in part because N is a fairly unreactive element at low temperatures and because the gas number density is low, $\sim 2 \times 10^4(r/10^{16})^{-2}$ cm^{-3}, and hence collision times are long. One molecule that shows a large difference when the N_2 shielding is modelled correctly is NO, produced by the N + OH reaction, with its peak fractional abundance decreasing by over an order of magnitude from $\sim 10^{-6}$ to 6 $\times10^{-8}$ at a radius of 2.5 $\times10^{16}$ cm. For parameters appropriate to IK Tau, the increased abundance of N_2 in the outer CSE leads to an increased N_2H^+ abundance due to the proton transfer reaction, $H_3^+ + N_2 \longrightarrow N_2H^+ + H_2$, at $r < 10^{17}$ cm, and, at larger radii, the reaction $He^+ + N_2 \longrightarrow N_2^+$ + He followed by $N_2^+ + H_2 \longrightarrow N_2H^+ + H$. The abundance of N_2H^+ is directly correlated with the initial (unknown) abundance adopted for N_2 but its use as a tracer of N_2 is limited since its predicted column density is low, only 3.4 $\times$ 10^{10} cm^{-2}.

2.2. *C-rich late-type stars*

Carbon-rich AGB star envelopes experience the same physical processes as those around O-rich stars but their molecular content is very different, in both composition and complexity, primarily due to the reactive nature and unique bonding properties of the carbon atom.

Cherchneff [13] has produced the most detailed model for the non-equilibrium chemistry of the inner dust formation region of IRC+10216, by far the most well observed carbon-rich AGB star. This star has a mass-loss rate of 1.5 $\times$ 10^{-5} $M_\odot$ yr^{-1}, a terminal expansion velocity of 14.5 km s^{-1} and is known to contain at least 80 molecules in its CSE. The vast majority of these molecules are hydrocarbons, well understood because of the high abundance of carbon relative to oxygen in this star. A surprise discovery, however, was the presence of cold water [28], OH [29] and H_2CO [30]. A number of explanations were put forward including the evaporation of icy bodies within the CSE [31] and the formation of water on metallic grains [32] but all mechanisms had their problems. Subsequent to these observations, the *Herschel* satellite was used to survey water in a number of C-rich AGB stars and it was found that warm water was present in many [33, 34], indicating that abundant water was present close to the dust-forming regions in these stars. In this scenario, alternative formation mechanisms become possible, most importantly, shock chemistry following stellar pulsations [13], similar to that in O-rich stars, and photon-driven chemistry following deep penetration of interstellar photons through a clumpy envelope [10].

More recently, Lombaert et al. [35] present *Herschel* observations of H_2O toward 18 C-rich AGB stars to look for correlations between abundances, dynamics and physical conditions. They find warm H_2O emission from all stars and conclude that water is located close to or inside the wind acceleration zone, i.e., the dust formation zone since the wind is driven by momentum transfer from the dust to the gas [36]. Detailed excitation and radiative transfer calculations indicate that the fractional abundance of water lies in the range 10^{-6}–10^{-4}, at maximum some two to three orders of magnitude larger than predicted by the UV photodissociation [10] or shock chemistry [13] models. In the latter model, a fraction of parent CO is collisionally destroyed in the immediate post-shock gas and the O atoms released take part in fast neutral reactions that either reform CO or form oxides, most importantly H_2O and SiO. One should note that the shock model does predict a high fractional abundance, 10^{-4}, of H_2O inside the dust formation zone at less than 2.5 $R_\star$ for the specific case of IRC+10216. In addition, the large abundance variations, some six orders of magnitude, predicted within a pulsational phase at these small radii should lead to variable emission in the high-energy water transitions.

Furthermore, these non-equilibrium shock models reproduce the abundances of several other species, including NaCl, AlCl and KCl, to within an order of magnitude, remarkably well given the uncertainties in many of the rate coefficients. Cherchneff [13] has also calculated the abundance of simple hydrocarbons up to benzene, C_6H_6, which is known to be necessary for the production of polycyclic aromatic hydrocarbons (PAHs) and, perhaps more generally, for the formation of carbonaceous dust grains in C-rich AGB stars. The models find that a large fractional abundance, $\sim 10^{-6}$, of benzene forms late in the pulsation, at phases greater than 0.8, when the gas is cool and the abundances of H_2O and OH are low, less than 10^{-6}, since both species oxidise benzene and prevent the growth of larger PAH-like molecules. Her calculations, under the assumption that all C_6H_6 is converted to coronene, $C_{24}H_{12}$, through reactions involving acetylene, C_2H_2, and that the total mass of coronene ends up in dust, gives reasonable agreement with the dust-to-gas mass ratio observed in IRC+10216.

The UV photodissociation model [10] produces H_2O with a fractional abundance of (2–10) $\times 10^{-7}$ at 2–10 $R_\star$, much lower than observed for high mass-loss rate stars and also depends critically on a significant degree of clumping and/or scattering of UV photons to allow a few percent of the interstellar UV flux to reach radii less than 10^{15} cm. At this radius a spherically symmetric uniform outflow, with a mass-loss rate equivalent to that of IRC+10216, would have a radial UV extinction of more than 50 magnitudes. The challenge for both the shock and the UV models is that in order to produce very high abundances of water, O atoms must be liberated efficiently from CO and processed by the chemistry away from CO to H_2O. For the UV model, the main isotopologue of CO, $^{12}C^{16}O$, has a very small photodissociation rate since it self-shields efficiently and is mutually shielded by H_2, the same is likely true also for $^{13}C^{16}O$. As a result water should be enhanced in ^{17}O and ^{18}O but this does not seem to be the case [37]. We note that isotope effects are not expected in the shock model since CO is destroyed collisionally and not radiatively. Photodissociation of SiO may provide O atoms but, because of the cosmic abundance of silicon, cannot account for water fractional abundances much greater than 10^{-5}.

Observations at high spatial resolution with *ALMA* are now providing a remarkable view of the inner envelope of IRC+10216, particularly those molecules that appear to be related to dust formation. *ALMA* has been used to observe SiS, SiO and SiC_2 [38] with different radial distributions pointing to different formation mechanisms. SiS emission comes from a small region, radius $\sim$ 1" centred on the star, with SiO also peaking there but with a more extended distribution, $\sim$ 3–3.5" in radius. SiC_2, on the other hand, shows both a central peak but also a ring of emission with radius around 10", or 2×10^{16} cm, consistent with a photochemical origin in the outer envelope. Of these molecules, SiS is the most abundant, with the total abundance of the three molecules accounting for a significant fraction of elemental silicon. Recently, some 112 rotational detections of SiCSi were detected, the first disilicon molecule discovered in space [39] and a molecule predicted to be abundant in TE calculations.

One molecule detected in the inner CSE but not expected from TE or shock chemistry, is CH_3CN which has a hollow shell distribution with inner and outer radii of 1" and 2", respectively [40]. In interstellar clouds CH_3CN is formed by the fast radiative association of CH_3^+ with HCN followed by dissociative recombination with electrons, with likely a minor contribution from ice chemistry in regions where that is important. The very large abundance of HCN in the inner envelope clearly helps produce CH_3CN but if the ultimate source of the ionisation in the inner envelope is cosmic ray protons, as it is in the dense interstellar clouds in which CH_3CN is observed, then the abundance of CH_3^+ is likely to be vanishingly low for two reasons. One is that the ionisation rate cannot be larger than 10^{-17} s^{-1}, a constraint imposed by the very low abundance of HCO^+ detected in IRC+10216 [41]. The second is that the ionisation fraction generally decreases as $1/n$ in dense gas so that the formation of ions is less efficient in the inner envelope than further out [40]. An alternative explanation has been considered [40], namely that a few percent of interstellar FUV photons incident on the external envelope penetrate down to or close to the photosphere. In this case the CH_3CN abundance increases by about two orders of magnitude inside 8", although the distribution is centrally peaked on the star rather than distributed in a hollow shell [40].

The spectacular and complex hydrocarbon chemistry of the outer CSE in IRC+10216 has been explored by a number of authors ([3, 4, 6, 42, 7, 9, 43]). Here, the most important species are parents such as C_2H_2 and HCN whose photodissociation and photoionisation provides a rich reactive soup of radicals, atoms and ions that rapidly build long-chain hydrocarbons. Photodissociation of parent molecules gives rise to the ring distributions seen in daughter species such as C_2H and CN. These radicals react with other radicals as well as parents to build complexity, e.g., the reactions

$$C_2H + C_2H_2 \longrightarrow C_4H_2 + H \tag{8}$$

$$C_2H_2^+ + C_2H_2 \longrightarrow C_4H_3^+ + H \tag{9}$$

$$C_2H_2^+ + C_2H_2 \longrightarrow C_4H_2^+ + H_2 \tag{10}$$

$$CN + C_2H_2 \longrightarrow HC_3N + H \tag{11}$$

rapidly form abundant C_4-bearing hydrocarbon neutrals such as C_4H and H_2CCCC, and cyanoacetylene, HC_3N.

In a model calculation containing molecules with up to 23 carbon atoms, Millar et al. [7] show that simple synthetic pathways give rise to efficient growth in molecular size and to ring distributions as observed. For specific classes of molecules, such as the cyanopolyynes or alkenes, they find that peak fractional abundances and column densities typically fall by a factor of 2–3 as the number of carbon atoms increase. For a constant mass-loss rate and expansion velocity, the increased time to make larger molecules from smaller species results in radial distributions in which the position of the peak abundance generally increases as molecular size increases. Thus, for example, the peak fractional abundance of C_2H is reached at 4.0×10^{16} cm while that for C_7H occurs at 7.1×10^{16} cm. This type of behaviour is not always seen in the observations

[44] indicating that either the chemistry is more complex, occurring in parallel rather than sequentially, or that molecules are being produced by processes involving grains. The molecular shells of HC_3N and HC_5N are found to be clumpy, co-spatial and with a distribution that closely matches that of dust shells and arcs in the outer CSE of IRC+10216 [45]. These shells are also seen in CO emission out to a radius of around 180" where CO is photodissociated [46]. Surprisingly, these shells are not centred on the star itself suggesting that these periods of enhanced mass loss are induced at periastron by a companion star. A more physically realistic model of this envelope, taking into account the presence of enhanced density shells in both gas and dust was produced by Cordiner and Millar [9]. They based their idealised model on observations and added eight shells, each 2" thick, with an overdensity of 5 compared to the normal $1/r^2$ distribution, and an intershell spacing of 12". Assuming a distance of 130 pc to IRC+10216 and the observed expansion velocity of 14.5 km s^{-1}, this corresponds to an enhanced mass-loss rate occurring for 90 years every 530 years.

The inclusion of shells, not surprisingly, causes the radial distributions of molecules to be better aligned to one another and to the dust. Chemistry is enhanced within the shells since the reaction time goes as n^{-2}. In addition, the shells provide additional extinction to the penetration of external UV photons and move the inner edge of the molecular ring distributions outward. For example, the peak fractional abundances of HC_3N and HC_5N move from 8" to 15" when shells are included. Cordiner and Millar [9] find that the shell at 15" dominates the emission characteristics of a number of molecules, that is, the model predicts rings of co-spatial emission from C_2H, C_4H and C_6H and from HC_3N and HC_5N, as observed.

One of the major successes of the photochemical modelling of IRC+10216 has been the prediction and subsequent detection of several large anions in the outer CSE. In the past ten years or so, laboratory measurements of the microwave spectra of anions has led to the identification of C_4H^-, C_6H^-, C_8H^-, CN^-, C_3N^- and C_5N^- in IRC+10216. Such anions were predicted with abundances that could be a significant fraction of their neutral analogues, for example the C_8H^-/C_8H column density ratio was predicted to be 0.25 [7] and observed to be 0.26 [47]. These anions are formed predominantly through the radiative attachment of electrons to neutral hydrocarbons which possess large electron affinities. For molecules with five or more carbon atoms, the attachment occurs on almost every collision. The abundance of anions in IRC+10216 is so large that there are regions in the envelope in which the anion abundance exceeds that of free electrons. They are also very reactive and play an important role in the synthesis of even larger hydrocarbon species [9].

The most recent release of the UMIST Database for Astrochemistry (www.udfa.net) now contains over 20 anions involved in some 1300 gas-phase reactions. The full UDfA database, some 6173 reactions among 467 species, was used to study chemistry in a model of IRC+10216 assuming a constant mass-loss rate [23]. Some 31 out of 47 of the 'daughter' species were found to have column densities that agreed to within an order of magnitude of those observed, indicating that we understand in broad terms the nature of the chemistry in carbon-rich circumstellar envelopes in AGB stars. This sort of agreement with observation implies that we have a fairly complete knowledge of the gas-phase chemical kinetics that occurs in the outer envelope of IRC+10216. Remaining uncertainties are linked either to unknown rate coefficents, primarily photodissociation rates, and reactions involving large hydrocarbon ions and neutrals, or to uncertain or unknown abundances of parent molecules or to physical structures within the CSE. The situation in in the inner CSE, roughly defined here as interior to 10^{16} cm, densities greater than 10^{5-6} cm^{-3} and temperatures greater than 100K is still open to significant improvement in understanding. We have already mentioned some areas in relation to both shock chemistry and FUV-dominated chemistry near the photosphere. The role of dust grains, once formed and driven outward by radiation pressure, is unexplored and there is observational evidence that they can act both as sinks and sources of gas-phase molecules.

Journal of Physics: Conference Series **728** (2016) 052001 doi:10.1088/1742-6596/728/5/052001

3. Discussion

As outlined above, the discovery of H_2O in C-rich CSEs, and more recently CS and HCN in O-rich CSEs, indicates that there are processes that perturb the TE chemistry that is expected to dominate at and close to the stellar photosphere. Shock chemistry induced by stellar pulsations is clearly important as may be the detailed chemistry associated with dust formation and growth. Despite the advances that have been made recently, this still remains a poorly understood area with a lack of critical experimental data appropriate to the densities and temperatures found in the dust formation zone. The penetration of FUV photons deep into the CSE is another possible mechanism. The models discussed above require that a significant fraction, some 2.5% [10, 40], of interstellar photons need to avoid around 30 magnitudes of FUV extinction that lie between 1" (2×10^{15} cm) and the edge of the CSE. If this occurs then it has profound effects on the composition of the gas in the inner regions of both C-rich and O-rich CSEs [10, 40]. Figure 1 shows the radial distributions of the fractional abundances of some hydrocarbon anions calculated under standard conditions, i.e. a spherically symmetric outflow at constant mass-loss rate, with interstellar UV photons incident on the outer CSE. I have, in figure 2, adopted this model to allow the same percentage of FUV photons to reach 10^{15} cm unaffected by dust extinction [7, 9], a useful exercise since I calculate the FUV radiative transfer in a different way to that of Agúndez and collaborators. It can be seen that the radial distributions of the anions, a representative class of the hydrocarbons, show significant differences particularly inside (3–4) $\times$ 10^{16} cm, when FUV photons are allowed to penetrate. The fractional abundances typically increase by about two orders of magnitude inside 10^{16} cm although the change in column density is less pronounced, typically 2–3 for these species and generally less than a factor of five for most others.

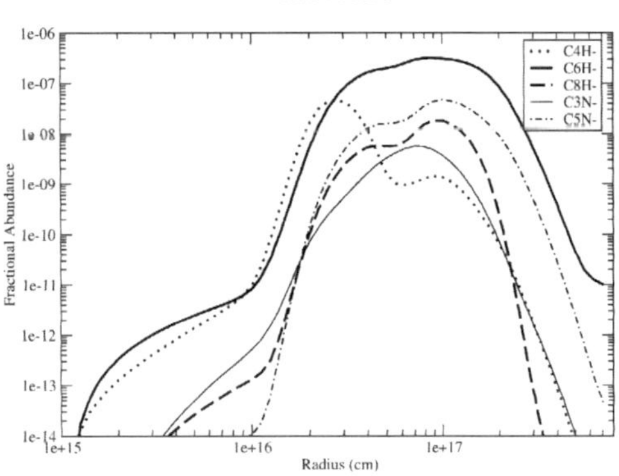

Figure 1. Fractional abundance of anions versus radius for the standard spherically symmetric outflow.

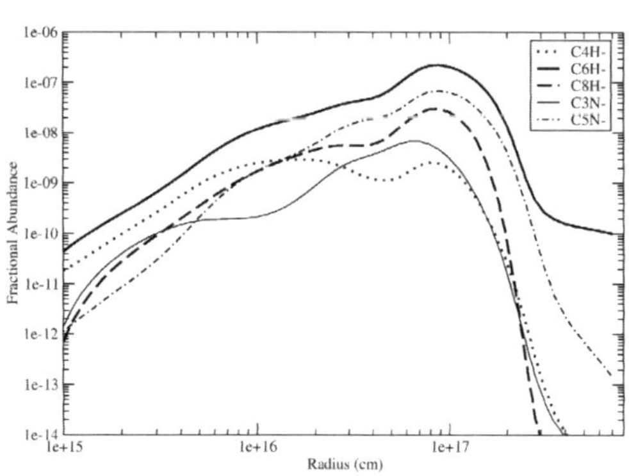

Figure 2. As figure 1 but with 2.5% of interstellar photons able to penetrate deep into the inner CSE.

Although FUV photons do tend to increase the fractional abundances down to a few 10^{15} cm, the distributions do not show the sharp inner boundaries evident in some emission maps [45, 40]. Could these sharp inner edges be an indicator that stellar photons are responsible? To date, the role of such photons has been ignored on two grounds. The first is that IRC+10216 is too cool (T_{eff} = 2330 K) to produce UV photons, the second that the dust-forming zone will provide a large amount of extinction, several hundred magnitudes at UV wavelengths, to photons generated by the star. If, however, the mass loss and dust formation processes themselves produce the clumpy structures that are inferred beyond 1" then it remains a possibility that some stellar photons do leak out to 50 $R_\star$.

The fact that the star is cool will indeed imply that the flux of photons at wavelengths less than 2000 Å is small. There are, however, a number of molecules which have relatively small bond energies and which can be destroyed by photons at longer wavelengths. Examples include the hydrocarbon anions, $C_{2n}H^-$, n = 1–4, which have electron affinities (EA) that range from 3.02 eV (C_2H^-) to 3.96 eV (C_8H^-). The photodetachment cross-sections of these anions may be calculated using the empirical formula [48]:

$$\sigma(\epsilon) = \sigma_\infty \left(1 - \frac{EA}{\epsilon}\right)^{1/2} \tag{12}$$

where ϵ is the photon energy and σ_∞ is the asymptotic cross-section at large energies. Data for EA and σ_∞ have been provided experimentally [49, 50]. At a distance of 50 $R_\star$ and assuming no extinction due to dust, the electron photodetachment rates vary from 10^{-5} s^{-1} (CN^-) to 1.98 × 10^{-6} s^{-1} (C_6H^-), Other species that have large photodissociation rates include CH, l-C_3H, C_5H and NaCl. Thus, if even a small fraction of these stellar photons penetrate the dust-formation region and beyond, they could have a significant effect on the radial distribution of some species. Detailed calculations investigating this are underway.

Acknowledgments

I should like to that the organisers for inviting me to participate in this conference. Astrophysics at QUB is supported by a grant from the STFC.

References

[1] Huggins P J and Glassgold A E 1982 *Astrophys. J.* **252** 201
[2] Nejad L A M, Millar T J and Freeman A *Astron. Astrophys.*
[3] Huggins P J, Glassgold A E and Morris M 1984 *Astrophys. J.* **279** 284
[4] Nejad L A M and Millar T J 1987 *Astron. Astrophys.* **183** 279
[5] Cherchneff I, Glassgold A E and Mamon G A 1993 *Astrophys. J.* **410** 188
[6] Cherchneff I and Glassgold A E 1993 *Astrophys. J. Letts.* **419** L41
[7] Millar T J, Herbst E and Bettens R P A 2000 *Mon. Not. R. Astron. Soc.* **316** 195
[8] Agúndez M, Fonfría J P, Cernicharo J, Pardo J R and Guélin M 2008 *Astron. Astrophys.* **479** 493
[9] Cordiner M A and Millar T J 2009 *Astrophys. J.* **697** 68
[10] Agúndez M, Cernicharo J and Guélin M 2010 *Astrophys. J. Letts.* **724** L133
[11] Cherchneff I, Barker J R and Tielens A G G M 1992 *Astrophys. J.* **401** 269
[12] Duari D, Cherchneff I and Willacy K 1999 *Astron. Astrophys.* **341** L47
[13] Cherchneff I 2012 *Astron. Astrophys.* **545** A12
[14] Gobrecht D, Cherchneff I, Sarangi A, Plane J M C and Bromley S T 2016 *Astron. Astrophys.* **585** A6
[15] Decin L, Beck E D, Brünken S, Müller H S P, Menten K M, Kim H, Willacy K, de Koter A and Wyrowski F 2010 *Astron. Astrophys.* **516** A69
[16] Kim H, Wyrowski F, Menten K M and Decin L 2010 *Astron. Astrophys.* **516** A68
[17] Tenenbaum E D and Ziurys L M 2010 *Astrophys. J. Letts.* **712** L93
[18] Tenenbaum E D, Dodd J L, Milam S N, Woolf N J and Ziurys L M 2010 *Astrophys. J. Suppl. Ser.* **190** 348
[19] Beck E D, Kamiński T, Patel N A, Young K H, Gottlieb C A, Menten K M and Decin L 2013 *Astron. Astrophys.* **558** A132
[20] Danilovich T, Beck E D, Black J H, Olofsson H and Justtanont K 2016 *ArXiv e-prints* (*Preprint* 1602.00517)
[21] Goumans T P M and Bromley S T 2012 *Mon. Not. R. Astron. Soc.* **420** 3344
[22] Li X, Millar T J, Heays A N, Walsh C, van Dishoeck E F and Cherchneff I 2016 *Astron. Astrophys.* **588** A4
[23] McElroy D, Walsh C, Markwick A J, Cordiner M A, Smith K and Millar T J 2013 *Astron. Astrophys.* **550** A36
[24] Ajello J M, James G K, Franklin B O and Shemansky D E 1989 *Phys. Rev.* A **40** 3524
[25] Ndome H, Hochlaf M, Lewis B R, Heays A N, Gibson S T and Lefebvre-Brion H 2008 *J. Chem. Phys.* **129** 164307
[26] Heays A N, Dickenson G D, Salumbides E J, de Oliveira N, Joyeux D, Nahon L, Lewis B R and Ubachs W 2011 *J. Chem. Phys.* **135** 244301
[27] Li X, Heays A N, Visser R, Ubachs W, Lewis B R, Gibson S T and van Dishoeck E F 2013 *Astron. Astrophys.* **555** A14

Journal of Physics: Conference Series **728** (2016) 052001 doi:10.1088/1742-6596/728/5/052001

[28] Melnick G J, Neufeld D A, Ford K E S, Hollenbach D J and Ashby M L N 2001 *Nature* **412** 160
[29] Ford K E S, Neufeld D A, Goldsmith P F and Melnick G J 2003 *Astrophys. J.* **589** 430
[30] Ford K E S, Neufeld D A, Schilke P and Melnick G J 2004 *Astrophys. J.* **614** 990
[31] Ford K E S and Neufeld D A 2001 *Astrophys. J. Letts.* **557** L113
[32] Willacy K 2004 *Astrophys. J. Letts.* **600** L87
[33] Decin L *et al* 2010 *Nature* **467** 64
[34] Neufeld D A *et al* 2010 *Astron. Astrophys.* **521** L5
[35] Lombaert R *et al* 2016 *ArXiv e-prints* (*Preprint* 1601.07017)
[36] Kwok S 1975 *Astrophys. J.* **198** 583
[37] Neufeld D, Tolls V, Agúndez M, González-Alfonso E, Decin L, Daniel F, Cernicharo J, Melnick G J, Schmidt M and Szczerba R 2013 *Astrophys. J. Letts.* **767** L3
[38] Prieto L V *et al* 2015 *Astrophys. J. Letts.* **805** L13
[39] Cernicharo J *et al* 2015 *Astrophys. J. Letts.* **806** L3
[40] Agúndez M, Cernicharo J, Quintana-Lacaci G, Velilla Prieto L, Castro-Carrizo A, Marcelino N and Guélin M 2015 *Astrophys. J.* **814** 143
[41] Pulliam R L, Edwards J L and Ziurys L M 2011 *Astrophys. J.* **743** 36
[42] Doty S D and Leung C M 1998 *Astrophys. J.* **502** 898
[43] Li X, Millar T J, Walsh C, Heays A N and van Dishoeck E F 2014 *Astron. Astrophys.* **568** A111
[44] Guélin M, Neininger N, Lucas R and Cernicharo J 1999 *The Physics and Chemistry of the Interstellar Medium* ed Ossenkopf V, Stutzki J and Winnewisser G p 326
[45] Dinh-V-Trung and Lim J 2008 *Astrophys. J.* **678** 303
[46] Cernicharo J, Marcelino N, Agúndez M and Guélin M 2015 *Astron. Astrophys.* **575** A91
[47] Remijan A J, Hollis J M, Lovas F J, Cordiner M A, Millar T J, Markwick-Kemper A J and Jewell P R 2007 *Astrophys. J. Letts.* **664** L47
[48] Millar T J, Walsh C, Cordiner M A, Chuimín R N and Herbst E 2007 *Astrophys. J. Lett.* **662** L87
[49] Best T *et al* 2011 *Astrophys. J.* **742** 63
[50] Kumar S S, Hauser D, Jindra R, Best T, Roučka Š, Geppert W D, Millar T J and Wester R 2013 *Astrophys. J.* **776** 25

Journal of Physics: Conference Series **728** (2016) 052002 doi:10.1088/1742-6596/728/5/052002

Search for complex organic molecules in space

Masatoshi Ohishi[1]

[1] National Astronomical Observatory of Japan, 2-21-1, Osawa, Mitaka, Tokyo, 181-8588 Japan

E-mail: [1] masatoshi.ohishi@nao.ac.jp

Abstract. It was 1969 when the first organic molecule in space, H_2CO, was discovered. Since then many organic molecules were discovered by using the NRAO 11 m (upgraded later to 12 m), Nobeyama 45 m, IRAM 30 m, and other highly sensitive radio telescopes as a result of close collaboration between radio astronomers and microwave spectroscopists. It is noteworthy that many famous organic molecules such as CH_3OH, C_2H_5OH, $(CH_3)_2O$ and CH_3NH_2 were detected by 1975. Organic molecules were found in so-called hot cores where molecules were thought to form on cold dust surfaces and then to evaporate by the UV photons emitted from the central star. These days organic molecules are known to exist not only in hot cores but in hot corinos (a warm, compact molecular clump found in the inner envelope of a class 0 protostar) and even protoplanetary disks.

As was described above, major organic molecules were known since 1970s. It was very natural that astronomers considered a relationship between organic molecules in space and the origin of life. Several astronomers challenged to detect glycine and other prebiotic molecules without success. ALMA is expected to detect such important materials to further consider the gexogenous deliveryh hypothesis.

In this paper I summarize the history in searching for complex organic molecules together with difficulties in observing very weak signals from larger species. The awfully long list of references at the end of this article may be the most useful part for readers who want to feel the exciting discovery stories.

1. Introduction

Molecules exist in a variety of physical conditions of the Universe. It was in 1940 when unidentified UV absorption lines were attributed to CH and CN in diffuse interstellar clouds [1]. This was the first report of molecules in space. In early 1930s radio astronomy opened new eyes to the Universe. In 1963 the OH molecule was discovered by a radio telescope [2]. As of December 2015 more than 190 molecules were detected or reported, primarily by means of radio astronomical observations, in interstellar clouds, circumstellar envelopes, and even external galaxies. The smallest and the lightest molecule is H_2. The longest and the heaviest molecule is $HC_{11}N$ with its molecular weight of 147 [3]. A list of molecules in space may be found at, e.g., the Cologne Database for Molecular Spectroscopy (CDMS) [4].

Molecules in space may be classified into several categories: simple molecules, molecular ions, radicals, ring molecules and stable molecules. Simple molecules can be seen in a variety of sources. Among them, the most abundant molecule is H_2. Indeed around 99.99 % of molecules in space are H_2. The next abundant molecules are H_2O and CO whose relative abundances to H_2 are about 10^{-4}. H_2O, CO and CO_2 are ubiquitous molecules in the solid phase; these molecules in dust mantles are often observed in absorption towards bright IR sources. Molecular ions and radicals are characteristic for molecules in space, since these are so-called "short-lifetime" molecules. Under the terrestrial physical conditions where the mean free time is the order of milli-seconds such "short-lifetime" molecules react and are converted into other species immediately they are formed. However, under the interstellar and circumstellar conditions where the mean free time is typically of the order of a year, such "short-lifetime" molecules can survive for a long time. It is now known that molecular ions play an important role in the gas-phase where "ion-molecule" reactions can successfully explain the formation of many major molecules in space.

Large organic molecules, the primary topics of this paper, belong to "Stable Molecules". Many of large organic molecules are the closed-shell molecules, and many of them, such as CH_3OH and C_2H_5OH, exist stably under the standard terrestrial conditions. Past studies have shown that many of large organic species are formed on cold, dust grains through the hydrogenation of small gas-phase molecules. For example, gas-phase CO is adsorbed onto dust grains. Since the hydrogen atoms on the dust surface can move quickly through the "tunneling effect", CO is converted to H_2CO, and ultimately to CH_3OH. When a star is formed in the center of a molecular cloud core, the solid-phase CH_3OH molecules may be heated by the UV photons from the central star and evaporated back to the gas-phase.

From now I will summarize the detection history of such large organic molecules in space. In this paper organic molecules refer primarily to carbon-bearing species except for a few species (e.g., CO) which have been regarded "inorganic" traditionally.

2. Detection History: Between 1969 and 1979

Table 1. Organic Molecules Detected between 1969 and 1979.

Year Reported	Species				
1969	H_2CO [5]				
1970	CH_3OH [6]				
1971	CH_3CN [7]	HC_3N [8]	HCOOH [9]	NH_2CHO [10]	HCN [11]
1972	HNCO [12]				
1973	CH_3CHO [13]	H_2CS [14]	CH_2NH [15]		
1974	$(CH_3)_2O$ [16]	CH_3NH_2 [17]	C_2H [18]		
1975	CH_2CHCN [19]	C_2H_5OH [20]	$HCOOCH_3$ [21]	NH_2CN [22]	
1976	HC_5N [23]	C_2H_2 [24]			
1977	C_2H_5CN [25]	H_2CCO [26]	C_3N [27]		
1978	HC_7N [28]	HC_9N [29]			
1979	CH_3SH [30]	HNCS [31]			

Table 1 shows a list of molecules reported detected between 1969 and 1979. 1969 is the year when the first organic polyatomic molecule, H_2CO, was detected [5] towards many galactic and extragalactic sources. Prior to the detection of H_2CO, the first "large" millimeter-wave telescope, the 11 m radio telescope of the National Radio Astronomy Observatory (NRAO), started its operation in 1968. The NRAO 11 m radio telescope was one of pioneering telescopes; it discovered CO [32], CH_3CN [7], HCN [11], HNCO [12], $(CH_3)_2O$ [16], C_2H [18], C_2H_5OH [20], NH_2CN [22], C_2H_5CN [25], H_2CCO [26], C_3N [27], and HNCS [31] in the 1970s. It was in the 1970s when other countries joined the "molecular hunting". Australia used the Parkes 64 m radio telescope in detecting H_2CS [14], CH_2NH [15], CH_2CHCN [19] and $HCOOCH_3$ [21]. Canada succeeded in detecting "long carbon-chain molecules" such as HC_5N [23], HC_7N [28], and HC_9N [29]. Japan constructed the 6 m millimeter-wave telescope, and succeeded to discover CH_3NH_2 [17].

It should be stressed that many organic molecules were already detected in the 1970s. It is well known that CH_3OH is widespread in a variety of sources; CH_3OH masers are commonly observed towards young, star forming regions. Large organic molecules, such as CH_3CN, CH_3CHO, $(CH_3)_2O$, NH_2CHO, are usually observed towards dense and hot regions, *i.e.*, hot cores and/or hot corinos.

There was an important reason why such successful results were obtained in the early days of radio astronomy – interdisciplinary collaboration between radio astronomy and microwave spectroscopy. In the reference list at the bottom of this paper, we can find many famous microwave spectroscopists: Gottlieb, Thaddeus, Brown, Godfrey, Lovas, Winnewisser, Kroto and Oka. Oka made a great contribution in predicting transition frequencies for HC_9N and $HC_{11}N$ [33]. Immediately after the discovery of HC_7N, Oka extrapolated the molecular constants of HCN, HC_3N, HC_5N and HC_7N toward predicting molecular constants of longer cyanopolyynes HC_9N and $HC_{11}N$. HC_9N was successfully discovered based on the prediction![29] Such a prediction was not possible without detailed knowledge on the microwave spectroscopy, and it was an epoch-making proof that close interdisciplinary collaboration was so fruitful not only for radio astronomy but microwave spectroscopy. For microwave spectroscopists, the extreme condition of the interstellar space is ideal in studying very unstable "short-lifetime" molecules, such as molecular ions and radicals. It was possible for microwave spectroscopists to measure, for example, very small hyperfine splittings, to study very detailed intra-molecular interactions that were not possible in the laboratories.

3. Detection History: 1980s

Table 2. Organic Molecules Detected in the 1980s.

Year Reported	Species			
1980	none			
1981	C_2H_4 [34]	C_4H [35]	CH_3C_2H [35]	$HOCO^+$ [36]
1982	none			
1983	none			
1984	CH_3C_3N [37]	CH_3C_4H [38, 39]		
1985	C_3O [40]	*c*-C_3H_2 [41]	*l*-C_3H [42]	
1986	C_5H [43]	C_6H [44]	$HCNH^+$ [45]	
1987	$(CH_3)_2CO$ [46]	CCS [47]	C_3S [48]	*c*-C_3H [49]
1988	C_3 [50]	CH_3NC [51]	HCCCHO [52]	CH_2CN [53]
1989	C_5 [54]	C_4Si [55]		

Table 2 shows a list of molecules reported detected between 1980 and 1989. As is seen in this table, fewer molecules were detected for a few years in the early 80s. Comparing with the exciting ten years in the 70s, some people rumored that no more new molecules would be detected. Such the situation was changed after the three very large milimmeter-wave telescope started to operate: the Onsala 20 m telescope in Sweden, the Nobeyama 45 m telescope in Japan and the IRAM 30 m telescope in Spain. The Onsala group made the first extensive "spectral line surveys" towards Orion KL and IRC+2016 between 72.2 and 91.1 GHz [56], which reported approximately 170 lines from 24 known molecules towards Orion KL and 45 lines from 12 molecules towards IRC+10216.

Followed by the successful spectral line survey project at the Onsala Space Observatory, the Nobeyama and the IRAM groups started their wide-band line survey projects. The Nobeyama 45 m telescope covered very wide frequency ranges between 1.4 GHz and 116 GHz. However the Nobeyama group concentrated their spectral line surveys between 8.8 and 116 GHz. Especially the 7 mm range was covered only by the 45 m telescope. Further the telescope was equipped with the 16 000 channel radiospectrometers with frequency resolutions of 37 kHz (high-resolution) and 250 kHz (wide-band). Thus it was possible for the 45 m telescope to observe cold, dark clouds, such as TMC-1 where the typical line width was around 0.5 kms^{-1}. The spectral line survey results towards TMC-1, that covered between 8.8 and 50 GHz, contained 414 lines from 38 species including 11 new interstellar molecules [57]. The Nobeyama group was successful in detecting many new interstellar molecules: C_3O [40], C_6H [44], CCS [47], C_3S [48], c-C_3H [49], HCCCHO [52], CH_2CN [53] and C_4Si [55] only in the 80s.

The IRAM 30 m telescope group made the extensive spectral line survey between 1 and 3 mm ranges towards IRC+10216. In the 80s they discovered C_5H [43], $(CH_3)_2CO$ [46], CH_3NC [51] as well as several metal halides (NaCl, AlCl, KCl and AlF) [58]. Many unidentified lines observed by the IRAM 30 m telescope were later identified to be c-SiC_2 [59]. c-SiC_2 was the first molecular ring. It was curious to see that some microwave spectroscopists resisted to accept the identification since three-membered ring molecules were thought to have much higher energy than their linear forms. It was demonstrated that the ring form of the SiC_2 molecule has much lower energy than its linear form; it is the lowest energy isomer.

As is easily seen many molecules discovered by the Nobeyama 45 m and the IRAM 30 m telescopes are linear carbon-chain species. It was established in the 80s that linear carbon-chain molecules are important constituents in the interstellar and circumstellar environment.

4. Detection History: 1990s

Table 3. Organic Molecules Detected in the 1990s.

Year Reported	Species				
1990	none				
1991	C_2O [60]	l-C_3H_2 [61]	l-C_4H_2 [62]	HCCN [63]	
1992	HCCNC [64]	HNCCC [65]			
1993	none				
1994	C_{60}^+ [66]	H_2CN [67]	HC_3NH^+ [68]		
1995	CH_2 [69]				
1996	C_8H [70]	H_2COH^+ [71]			
1997	l-C_6H_2 [72]	C_7H [73]	c-C_2H_4O [74]	CH_3COOH [75]	$HC_{11}N$ [3]
1998	C_5N [76]				
1999	none				

Table 3 shows a list of molecules reported detected between 1990 and 1999. Many new molecules were discovered primarily by using the Nobeyama 45 m and the IRAM 30 m telescopes. Researchers were able to detect less abundant species since these telescopes equipped with very sensitive Superconductor-Insulator-Superconductor (SIS) receivers since late 1980s. In the late 90s several complex organic molecules were detected as well as linear carbon-chain molecules. In some cases such complex organic molecules are higher energy isomers of already known species. For example, c-C_2H_4O [74] is a higher energy isomer to C_2H_5OH and CH_3COOH [75] is an higher

Journal of Physics: Conference Series **728** (2016) 052002 doi:10.1088/1742-6596/728/5/052002

energy isomer to $HCOOCH_3$. Further the detection of HCCNC [64] and HNCCC [65], which are isomers to HC_3N, prompted detailed studies regarding the formation of these carbon-chain species.

5. Detection History: Between 2000 and 2015

Table 4. Organic Molecules Detected between 2000 and 2015.

Year Reported	Species			
2000	CH_2OHCHO [77]	CH_3 [78]		
2001	*c*-C_6H_6 [79]	CH_2CHOH [80]	*l*-HC_4H [79]	*l*-HC_6H [79]
2002	$(CH_2OH)_2$ [81]			
2003	none			
2004	CH_2CHCHO ? [82]	CH_3CH_2CHO [82]	*l*-HC_4N [83]	
2005	$C_2H_5OCH_3$? [84]			
2006	*c*-H_2C_3O [85]	CH_2CCHCN [86]	CH_2CNH ? [87]	CH_3C_5N [88]
	CH_3CONH_2 [89]	C_6H^- [90]		
2007	C_4H^- [91]	C_8H^- [92]	CH_2CHCH_3 [93]	
2008	HCOCN [94]	NH_2CH_2CN [95]		
2009	C_2H_5OCHO [96]	HCNO [97]	HOCN [98]	HSCN [99]
	n-C_3H_7CN [96]			
2010	C_{60} [100]	C_{70} [100]		
2011	none			
2012	CH_3O [101]	HNCNH [102]		
2013	CH_3CHNH [103]	CH_3COOCH_3 [104]	H_2CNO^+ ? [105]	HNCHCN [106]
2014	C_2N [107]	CH_3CH_2SH ? [108]	*i*-C_3H_7CN [109]	
2015	CH_3NCO [110]	HCCO [111]	$NCCNH^+$ [112]	

Table 4 shows a list of molecules reported detected between 2000 and 2015. Note that a question mark (?) in this Table denotes that the detection was claimed but not yet confirmed. It is readily seen that many very complex organic molecules are detected since the Green Bank Telescope (GBT) started its operation in 2000. Further IRAM 30 m telescope has been very active in the "molecular hunting"; it has detected many new molecules even after its 30th anniversary

An epoch-making event was the discovery of glycolaldehyde (CH_2OHCHO) in 2000 [77]. According to the detection report [77], the generic form of sugar can be expressed as $(H_2CO)_n$ ($n \geq 2$). Because glycolaldehyde has the form of $(H_2CO)_2$, it was regarded as the simplest sugar in space. Thus this discovery had led astronomers to seriously search for other "prebiotic molecules" in space. It was unfortunate that the correct generic form of sugar is $(H_2CO)_n$ ($n \geq 3$), not 2 !, *i.e.*, glycolaldehyde is not a sugar.

A large complex organic molecule generally shows very weak line intensities due to its very large rotational partition function. Thus the sensitivity of a telescope is crucial in searching for such weak lines. Further it is required to detect as many lines in a wide frequency range as possible to enable reliable excitation analysis prior to claiming secure detection. In this regard the GBT group has conducted the PRIMOS project [113], which has resulted in detecting many large complex organic molecules.

6. Search for Prebiotic Molecules

Nontherless many astronomers have been searching for prebiotic molecules based on a hypothesis that simple bio-molecules are formed in the interstellar space, which would then be delivered to early Earth (planets) by comets and/or meteorites ("exogenous delivery"). Indeed the StarDust mission successfully detected the simplest amino acid, glycine (NH_2CH_2COOH), in the ejecta from Comet Wild 2 [114].

Unfortunately astronomical searches for interstellar glycine and other prebiotic molecules have not yet been successful as of today. Since we already know that proposed precursors to glycine (*e.g.*, NH_2CH_2CN and CH_3NH_2) exist in the interstellar space, it is expected to detect amino acids, precursors to nucleobases, and other prebiotic molecules by means of the collecting power by the ALMA telescope.

Acknowledgments

I am grateful all the members of Astronomy Data Center, the National Astronomical Observatory of Japan (NAOJ), for their support throughout this work. I also thank Prof. Bill Irvine for his critical reading of this paper. This research was supported by the Department of Astrobiology, Center for Novel Science Initiatives, National Institutes of Natural Sciences (NINS) and by the JSPS Kakenhi Grant Numbers 15H03646 and 14J03618. We utilized the Japanese Virtual Observatory (JVO; http://jvo.nao.ac.jp/) in finding relevant reference papers. This research has made use of NASA's Astrophysics Data System.

References

[1] McKellar A 1940 *Pub. Astron. Soc. Pacific* **52** 187-192
[2] Weinreb S, Barrett A H, Meeks M L and Henry J C 1963 *Nature* **200** 829-831
[3] Bell M B, Feldman P A, Travers M J, McCarthy M C, Gottlieb C A and Thaddeus P 1997 *Astrophys. J.* **483** L61-L64
[4] The Cologne Database for Molecular Spectroscopy, http://www.astro.uni-koeln.de/cdms/molecules
[5] Snyder L E, Buhl D, Zuckerman B and Palmer P 1969 *Phys. Rev. Lett.* **22** 679-681
[6] Ball J A, Gottlieb C A, Lilley A E and Radford H E 1970 *Astrophys. J.* **162** L203-L210
[7] Solomon P M, Jefferts K B, Penzias A A and Wilson R W 1971 *Astrophys. J.* **168** L107-L110
[8] Turner B E 1971 *Astrophys. J.* **163** L35-L39
[9] Zuckerman B, Ball J A and Gottlieb C A 1971 *Astrophys. J.* **163** L41-L45
[10] Rubin R H, Swenson G W Jr., Solomon R C and Flygare H L 1971 *Astrophys. J.* **169** L39-L43
[11] Buhl D and Snyder L E 1971 *Astrophys. J.* **163** L47-L52
[12] Snyder L E and Buhl D 1972 *Astrophys. J.* **177** 619-623
[13] Gottlieb C A, 1973 "Detection of Acetaldehyde in Sagittarius", *Molecules in the Galactic Environment* eds. M. A. Gordon and L. E. Snyder, Wiley-Interscience, New York, 181-186
[14] Sinclair M W, Fourikis N, Ribes J C, Robinson R B, Brown R D and Godfrey P D 1973 *Aust. J. Phys.* **26** 85-91
[15] Godfrey P D, Brown R D, Robinson B J and Sinclair M W 1973 *Astrophys. Lett.* **13** 119-121
[16] Snyder L E, Buhl D, Schwartz P R, Clark F O, Johnson D R, Lovas F J and Giguere P T 1974 *Astrophys. J.* **191** L79-L82
[17] Kaifu N, Morimoto M, Nagane K, Akabane K, Iguchi T and Takagi K 1974 *Astrophys. J.* **191** L135-L137
[18] Tucker K D, Kutner M L and Thaddeus P 1974 *Astrophys. J.* **193** L115-L119
[19] Gardner F F and Winnewisser G 1975 *Astrophys. J.* **195** L127-L130
[20] Zuckerman B, Turner B E, Johnson D R, Lovas F J, Fourikis N, Palmer P, Morris M, Lilley A E, Ball J A and Clark F O 1975 *Astrophys. J.* **196** L99-L102
[21] Brown R D, Crofts J G, Godfrey P D, Gardner F F, Robinson B J and Whiteoak J B 1975 *Astrophys. J.* **197** L29-L31
[22] Turner B E, Liszt H S, Kaifu N and Kisliakov A G 1975 *Astrophys. J.* **201** L149-L152
[23] Avery L W, Broten N W, MacLeod J M and Oka T 1976 *Astrophys. J.* **205** L173-L175
[24] Ridgway S T, Hall D N B, Wojslaw R S, Kleinmann S G and Weinberger D A 1976 *Nature* **264** 345-346
[25] Johnson D R, Lovas F J, Gottlieb C A, Gottlieb E W, Litvak M M, Thaddeus P and Guélin M 1977 *Astrophys. J.* **218** L370-L376
[26] Turner B E 1977 *Astrophys. J.* **213** L75-L79

[27] Guélin M and Thaddeus P 1977 *Astrophys. J.* **212** L81-L85
[28] Kroto H W, Kirby C, Walton D R M, Avery L W, Broten N W, MacLeod J M and Oka T 1978 *Astrophys. J.* **219** L133-L137
[29] Broten N W, Oka T, Avery L W, MacLeod J M and Kroto H W 1978 *Astrophys. J.* **223** L105-L107
[30] Linke R A, Frerking M A and Thaddeus P 1979 *Astrophys. J.* **232** L139-L142
[31] Frerking M A, Linke R A and Thaddeus P 1979 *Astrophys. J.* **234** L143-L145
[32] Wilson R W, Jefferts K B and Penzias A A 1970 *Astrophys. J.* **161** L43-L44
[33] Oka T 1978 *J. Mol. Spectrosc.* **72** 172-174
[34] Betz A L 1981 *Astrophys. J.* **244** L103-L105
[35] Irvine W M, Höglund B, Friberg P, Askne J and Elldér J 1981 *Astrophys. J.* **248** L113-L117
[36] Thaddeus P, Guélin M and Linke R A 1981 *Astrophys. J.* **246** L41-L45
[37] Broten N W, MacLeod J M, Avery L W, Irvine W M, Höglund P, Friberg P and Hjalmarson Å1984 *Astrophys. J. Lett.* **276** L25-L29
[38] MacLeod J M, Avery L W and Broten N W 1984 *Astrophys. J.* **282** L89-L92
[39] Walmsley C M, Jewell P R, Snyder L E and Winnewisser G 1984 *Astron. Astrophys.* **134** L11-L14
[40] Brown R D, Godfrey P D, Cragg D M, Rice E H N, Irvine W M, Friberg P, Suzuki H, Ohishi M, Kaifu N and Morimoto M 1985 *Astrophys. J.* **297** 302-308
[41] Thaddeus P, Vrtilek V M and Gottlieb C A 1985 *Astrophys. J.* **299** L63-L66
[42] Thaddeus P, Gottlieb C A, Hjalmarson Å, Johansson L E B, Irvine W M, Friberg P and Linke R A 1985 *Astrophys. J. Lett.* **294** L49-L53
[43] Cernicharo J, Kahane C, Gómez-Gnzález J and Guélin M 1986 *Astron. Astrophys.* **164** L1-L4
[44] Suzuki H, Ohishi M, Kaifu N, Ishikawa S-I and Kasuga T 1986 *Publ. Astron. Soc. Japan* **38** 911-917
[45] Ziurys L M and Turner B E 1986 *Astrophys. J.* **302** L31-L36
[46] Combes F, Gerin M, Wooten A, Wlodarczak G, Clausset F and Encrenaz P J 1987 *Astron. Astrophys.* **180** L13-L16
[47] Saito S, Kawaguchi K, Yamamoto S, Ohishi M, Suzuki H and Kaifu N 1987 *Astrophys. J.* **317** L115-L118
[48] Yamamoto S, Saito S, Kawaguchi K, Kaifu N, H. Suzuki and Ohishi M 1987 *Astrophys. J.* **317** L119-L121
[49] Yamamoto S, Saito S, Ohishi M, Suzuki H, Ishikawa S-I, Kaifu N and Murakami A 1987 *Astrophys. J.* **322** L55-L58
[50] Hinkle K H, Keady J J and Bernath P F 1988 *Science* **241** 1319-1322
[51] Cernicharo J, Kahane C, Guélin M and Gómez-González, J 1988 *Astron. Astrophys.* **189** L1-L2
[52] Irvine W M, Brown R D, Cragg D M, Friberg P, Godfrey P D, Kaifu N, Matthews H E, Ohishi M, Suzuki H and Takeo H 1988 *Astrophys. J.* **335** L89-L93
[53] Irvine W M *et al* 1988 *Astrophys. J.* **334** L107-L111
[54] Bernath P F, Hinkle K H and Keady J J 1989 *Science* **244** 562-564
[55] Ohishi M, Kaifu N, Kawaguchi K, Murakami A, Saito S, Yamamoto S, Ishikawa S-I, Fujita Y, Shiratori Y and Irvine W M 1989 *Astrophys. J.* **345** L83-L86
[56] Johansson L E B, Andersson C, Elldér J, Friberg P, Hjalmarson Å, Höglund B, Irvine W M, Olofsson H and Rydbeck G 1984 *Astron. Astrophys.* **130** 227-256 (1984A&A...130..227J)
[57] Kaifu N *et al* 2004 *Publ. Astron. Soc. Japan* **56** 69-173
[58] Cernicharo J and Guélin M 1987 *Astron. Astrophys.* **183** L10-L12
[59] Thaddeus P, Cummins S E and Linke R A 1984 *Astrophys. J.* **283** L45-L48
[60] Ohishi M, Ishikawa S-I, Yamada C, Kanamori H, Irvine W M, Brown R D, Godfrey P D, Kaifu N and Suzuki H 1991 *Astrophys. J.* **380** L39-L42
[61] Cernicharo J, Gottlieb C A, Guélin M, Killian T C, Paubert G, Thaddeus P and Vrtilek J M 1991 *Astrophys. J.* **368** L39-L41
[62] Cernicharo J, Gottlieb C A, Guélin M, Killian T C, Paubert G, Thaddeus P and Vrtilek J M 1991 *Astrophys. J.* **368** L43-L45
[63] Guélin M and Cernicharo J 1991 *Astron. Astrophys.* **244** L21-L24
[64] Kawaguchi K, Ohishi M, Ishikawa S-I and Kaifu N 1992 *Astrophys. J.* **386** L51-L53
[65] Kawaguchi K *et al* 1992 *Astrophys. J.* **396** L49-L51
[66] Foing B H and Ehrenfreund P 1994 *Nature* **369** 296-298
[67] Ohishi M, McGonagle D, Irvine W M, Yamamoto S and Saito S 1994 *Astrophys. J.* **427** L51-L54
[68] Kawaguchi K, Kasai Y, Ishikawa S-I, Ohishi M, Kaifu N and Amano T 1994 *Astrophys. J.* **420** L95-L97
[69] Hollis J M, Jewell P R and Lovas F J 1995 *Astrophys. J.* **438** 259-264
[70] Cernicharo J and Guélin M 1996 *Astron. Astrophys.* **309** L27-L30
[71] Ohishi M, Ishikawa S-I, Amano T, Oka H, Irvine W M, Dickens J E, Ziurys L M and Apponi A J 1996 *Astrophys. J.* **471** L61-L64
[72] Langer W D, Velusamy T, Kuiper T B H, Peng R, McCarthy M C, Travers M J, Kovacs A, Gottlieb C A

Journal of Physics: Conference Series **728** (2016) 052002 doi:10.1088/1742-6596/728/5/052002

and Thaddeus P 1997 *Astrophys. J.* **480** L63-L66
[73] Guélin M, Cernicharo J, Travers M J, McCarthy M C, Gottlieb C A, Thaddeus P, Ohishi M, Saito S and Yamamoto S 1997 *Astron. Astrophys.* **317** L1-L4
[74] Dickens J E, Irvine W M, Ohishi M, Ikeda M, Ishikawa S-I, Nummelin A and Hjalmarson Å1997 *Astrophys. J.* **489** 753-757
[75] Mehringer D M, Snyder L E, Miao Y and Lovas F J 1997 *Astrophys. J.* **480** L71-L74
[76] Guélin M, Neininger N and Cernicharo J 1998 *Astron. Astrophys.* **335** L1-L4
[77] Hollis J M, Lovas F J and Jewell P R 2000 *Astrophys. J.* **540** L107-L110
[78] Feuchtgruber H, Helmich F P, van Dishoeck E F and Wright C M 2000 *Astrophys. J.* **535** L111-L114
[79] Cernicharo J, Heras A M, Tielens A G G M, Pardo J R, Herpin F, Guélin M and Waters L B F M 2001 *Astrophys. J.* **546** L123-L126
[80] Turner B E and Apponi A J 2001 *Astrophys. J.* **561** L207-L210
[81] Hollis J M, Lovas F J, Jewell P R and Coudert L H 2002 *Astrophys. J.* **571** L59-L62
[82] Hollis J M, Jewell P R, Lovas F J, Remijan A and Møllendal H 2004 *Astrophys. J.* **610** L21-L24
[83] Cernicharo J, Guélin M and Pardo J R 2004 *Astrophys. J.* **615** L145-L148
[84] Fuchs G W, Fuchs U, Giesen T F and Wyrowski F 2005 *Astron. Astrophys.* **444** 521-530
[85] Hollis J M, Remijan A J, Jewell P R and Lovas F J 2006 *Astrophys. J.* **642** 933-939
[86] Lovas F J, Remijan A J, Hollis J M, Jewell P R and Snyder L E 2006 *Astrophys. J.* **637** L37-L40
[87] Lovas F J, Hollis J M, Remijan A R and Jewell P R 2006 *Astrophys. J.* **645** L137-L140
[88] Snyder L E, Hollis J M, Jewell P R, Lovas F J and Remijan A 2006 *Astrophys. J.* **647** 412-417
[89] Hollis J M, Lovas F J, Remijan A R, Jewell P R, Ilyushin V V and Kleiner I 2006 *Astrophys. J.* **643** L25-L28
[90] McCarthy M C, Gottlieb C A, Gupta H and Thaddeus P 2006 *Astrophys. J.* **652** L141-L144
[91] Cernicharo J, Guélin M, Agúndez M, Kawaguchi K, McCarthy M C and Thaddeus P 2007 *Astron. Astrophys.* **467** L37-L40
[92] Brünken S, Gupta H, Gottlieb C A, McCarthy M C and Thaddeus P 2007 *Astrophys. J.* **664** L43-L46
[93] Marcelino N, Cernicharo J, Agúndez M, Roueff E, Gerin M, Martin-Pintado J, Mauersberger R and Thum C 2007 *Astrophys. J.* **665** L127-L130
[94] Remijan A J, Hollis J M, Lovas F J, Stork W D, Jewell P R and Meier D S 2008 *Astrophys. J.* **675** L85-L88
[95] Belloche A, Menten K M, Comito C, Müller H S P, Schilke P, Ott J, Thorwirth S and Hieret C 2008 *Astron. Astrophys.* **482** 179-196
[96] Belloche A, Garrod R T, Müller H S P, Menten K M, Comito C and Schilke P 2009 *Astron. Astrophys.* **499** 215-232
[97] Marcelino N, Cernicharo J, Tercero B and Roueff E 2009 *Astrophys. J.* **690** L27-L30
[98] Brünken S, Gottlieb C A, McCarthy M C and Thaddeus P 2009 *Astrophys. J.* **697** 880-885
[99] Halfen D T, Ziurys L M, Brünken S, Gottlieb C A, McCarthy M C and Thaddeus P 2009 , *Astrophys. J.* **702** L124-L127
[100] Cami J, Bernard-Salas J, Peeters E and Malek S E 2010 *Science* **659** 1180-1182
[101] Cernicharo J, Marcelino N, Roueff E, Gerin M, Jiménez-Escobar A and Muñoz Caro G M 2012 *Astrophys. J. Lett.* **759** L43
[102] McGuire B A, Loomis R A, Charness C M, Corby J F, Blake G A, Hollis J M, Lovas F J, Jewell P R and Remijan A J 2012 *Astrophys. J.* **758** L33
[103] Loomis R A *et al* 2013 *Astrophys. J.* **765** L9
[104] Tercero B, Kleiner I, Cernicharo J, Nguyen H V L, López A and Muñoz Caro G M 2013 *Astrophys. J. Lett.* **770** L13
[105] Gupta H, Gottlieb C A, Lattanzi V, Pearson J C and McCarthy M C 2013 *Astrophys. J.* **778** L1
[106] Zaleski D P *et al* 2013 *Astrophys. J.* **765** L10
[107] Anderson J K and Ziurys L M 2014 *Astrophys. J.* **795** L1
[108] Kolesniková L, Tercero B, Cernicharo J, Alonso J L, Daly A M, Gordon B P and Shipman S T 2014 *Astrophys. J.* **784** L7
[109] Belloche A, Garrod R T, Müller H S P and Menten K M 2014 *Science* **345** 1584-1587
[110] Halfen D T, Ilyushin V V and Ziurys L M 2015 *Astrophys. J.* **812** L5
[111] Agúndez M, Cernicharo J and Guélin M 2015 *Astron. Astrophys.* **577** L5
[112] Agúndez M *et al* 2015 *Astron. Astrophys.* **579** L10
[113] PRIMOS project `http://www.cv.nrao.edu/~aremijan/PRIMOS/`
[114] Elsila J E, Glavin D P and Dworkin J P 2009 *Meteoritics and Planetary Science* **44** (9) 1323-1330

Journal of Physics: Conference Series **728** (2016) 052003 doi:10.1088/1742-6596/728/5/052003

Transitory O-rich chemistry in heavily obscured C-rich post-AGB stars

D. A. García-Hernández[1,2], P. García-Lario[3], J. Cernicharo[4], D. Engels[5] and J. V. Perea-Calderón[6]

[1] Instituto de Astrofísica de Canarias, C/ Via Láctea s/n, E-38205 La Laguna, Spain
[2] Departamento de Astrofísica, Universidad de La Laguna (ULL), E-38206 La Laguna, Spain
[3] Herschel Science Centre. European Space Astronomy Centre, Research and Scientific Support Department of ESA. Villafranca del Castillo, P.O. Box 50727. E-28080 Madrid. Spain
[4] Group of Molecular Astrophysics, ICMM, CSIC, C/Sor Juana Inés de La Cruz N3, E-28049 Madrid, Spain
[5] Hamburger Sternwarte, Gojenbergsweg 112, D−21029 Hamburg, Germany
[6] European Space Astronomy Centre, INSA S.A. P.O. Box 78, E−28080 Madrid, Spain

E-mail: agarcia@iac.es

Abstract. *Spitzer/IRS* spectra of eleven heavily obscured C-rich sources rapidly evolving from asymptotic giant branch (AGB) stars to Planetary Nebulae are presented. IRAM 30m observations for three of these post-AGBs are also reported. A few (3) of these sources are known to exhibit strongly variable maser emission of O-bearing molecules such as OH and H_2O, suggesting a transitory O-rich chemistry because of the quickly changing physical and chemical conditions in this short evolutionary phase. Interestingly, the *Spitzer/IRS* spectra show a rich circumstellar carbon chemistry, as revealed by the detection of small hydrocarbon molecules such as C_2H_2, C_4H_2, C_6H_2, C_6H_6, and HCN. Benzene is detected towards two sources, bringing up to three the total number of Galactic post-AGBs where this molecule has been detected. In addition, we report evidence for the possible detection of other hydrocarbon molecules like HC_3N, CH_3C_2H, and CH_3 in several of these sources. The available IRAM 30m data confirm that the central stars are C-rich - in despite of the presence of O-rich masers - and the presence of high velocity molecular outflows together with extreme AGB mass-loss rates ($\sim10^{-4}$ $M_\odot$/yr). Our observations confirm the polymerization model of Cernicharo [1] that predicts a rich photochemistry in the neutral regions of these objects on timescales shorter than the dynamical evolution of the central H II region, leading to the formation of small C-rich molecules and a transitory O-rich chemistry.

1. Introduction

Proto-planetary nebulae (PPNe) or post-AGB stars - especially those with the most massive progenitors - are usually heavily obscured by their thick circumstellar envelopes (CSE) as a consequence of the strong mass loss that they experience at the end of the asymptotic giant branch (AGB) phase (see e.g., [2]). Indeed, the departure from the AGB usually occurs when the star is completely invisible in the optical domain, being only detectable at IR wavelengths. Later, in the post-AGB phase, the central star reappears again in the optical and the UV photons ionize the CSE, forming a planetary nebula (PN). The UV radiation field is predicted to produce a rich photochemistry on very short timescales (from $\sim$0.2 to a few yrs) leading to the formation of hydrocarbon molecules such as C_nH_2, $HC_{2n+1}N$, and C_n [1]. Present observations show that carbon-chain polyynic radicals (C_nH) are the most abundant species in IRC +10216 [3]; a C-rich AGB star with no polycyclic aromatic hydrocarbon (PAH) emission bands in its mid-IR spectrum. However, polyynes (C_nH_2), cyanopolyynes, benzene (C_6H_6), and methyl-polyynes - the building blocks of PAHs - rather than carbon chain radicals, dominate the IR spectrum of more evolved objects such as CRL 618 [4,5].

CRL 618 - the prototype C-rich PPN CSE - is a unique object in the sense that it is a genuine factory of small aromatic molecules. Cernicharo et al. [5] reported the first detection, outside the Solar System, of the basic aromatic molecule benzene (C_6H_6) towards this source. The presence of a central compact H II region with a strong UV radiation field as well as the high-velocity (~200 km s^{-1}) outflows of molecular gas [6] indicate that the central star is rapidly evolving towards the PN stage; indeed the B0 central star has already reappeared in the optical range. O-bearing molecules such as OH and H_2O are also present in this C-rich PPN [7]. The presence of these O-bearing species in C-rich PPNe is naturally explained by the chemical models of Cernicharo [1] without invoking a change in the chemical composition (C/O) of the CSE. In addition, if the mass-loss rate is at least 10 times larger than that of CRL 618, then these models predict that C-rich PPNe could have masers from O-bearing species such as OH and H_2O. Here we report the detection of small hydrocarbon molecules such as polyynes (C_nH_2), benzene, etc. in a small sample (11) of Galactic C-rich post-AGB stars. The variable OH and H_2O maser emission seen in some of these strong mass-losing C-rich sources indicates a transitory O-rich chemistry, which confirms the chemical model predictions by Cernicharo [1].

2. The sample and the *Spitzer* and IRAM 30m observations

The sources in our sample were originally included in our *Spitzer/IRS* survey of hidden Galactic post-AGB candidates (GO program #30258; P.I.: P. García-Lario). This survey is intended to characterize the infrared properties of a large (88) and flux-limited sample of heavily obscured (no optical counterpart) post-AGB candidates selected from the 'GLMP catalog of IRAS sources with infrared colors similar to those of known PNe' [8]. Approximately half of this sample are non-variable OH/IR stars and the other half is composed by non-variable sources of unknown chemistry - sometimes with negative OH maser detections. The *Spitzer/IRS* spectra show that the CSE of most of these hidden post-AGB stars are O-rich - the evolutionary outcome of massive hot bottom burning (HBB) AGB stars [2,9] - although a few heavily obscured C-rich post-AGB stars are also found (see e.g., [10]). We concentrate here on the eleven hidden post-AGB stars, which turned out to be C-rich stars, as evidenced by our *Spitzer/IRS* spectra (see below). Eight sources in this sample do not show any optical counterpart, being even obscured at wavelengths shorter than 3 μm (e.g., not detected by 2MASS). Interestingly, three sources (IRAS 17552, IRAS 18529, and IRAS 19566) in our sample of heavily obscured C-rich stars are known to exhibit variable maser emission of O-bearing molecules such as OH and/or H_2O [11].

Figure 1 shows the reduced *Spitzer/IRS* spectra for all C-rich sources in our sample. The CSE of these post-AGBs are clearly C-rich as indicated by the detection of strong acetelyne (C_2H_2) absorption at 13.7 μm. Several sources show also strong absorption features attributed to polyynes (e.g., C_2H_2, C_4H_2, and C_6H_2; see [11]) and HCN. The dust spectral energy distributions indicate high mass-loss rates and show similar features than in CRL 618 [4]. Benzene (C_6H_6) is detected for the first time in IRAS 17552 and IRAS 19566 [11], bringing up to three the total number of benzene detections in Galactic post-AGB stars. Acetylene (C_2H_2) is usually detected in less evolved AGB stars such as IRAS 09024−5019 (the only AGB variable in our sample as indicated by the IRAS variability index) and its detection in all sources in our sample indicate that they are quickly evolving to the PN phase. In addition, the unidentified IR bands (e.g., those at 6.2, 7.7, 8.6 and 11.3 μm, generally attributed to PAHs) are only present in IRAS 15038 (Figure 1); likely the most evolved source in our sample. Remarkably, some post-AGB stars like IRAS 19566 display strong absorption features at ~15.75 and 16.43 μm that could be attributed

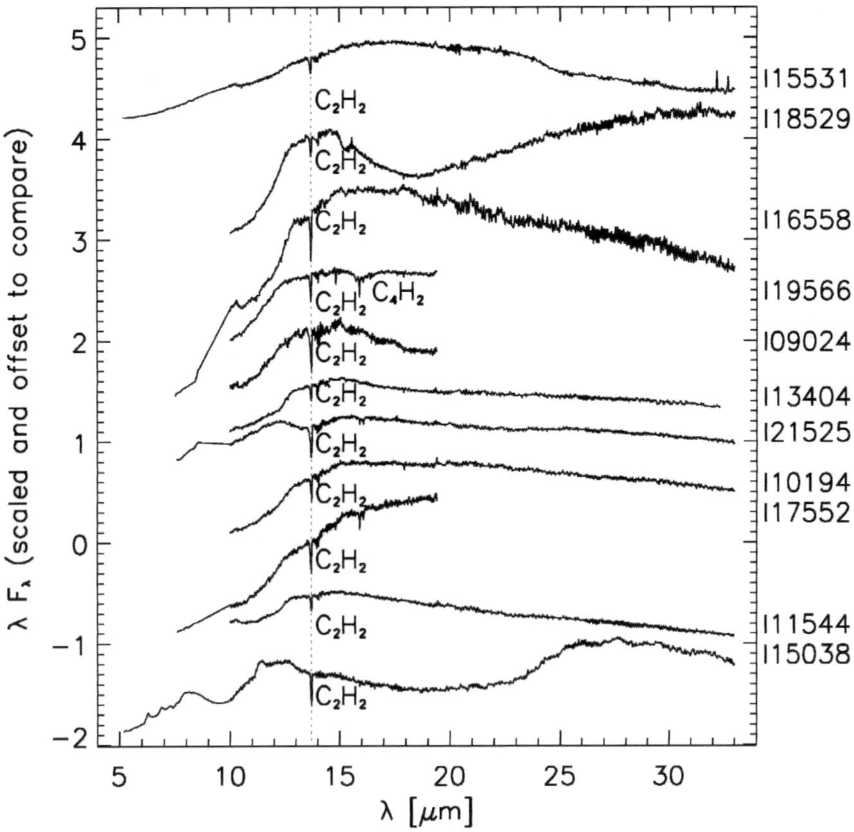

Figure 1. *Spitzer/IRS* spectra (~5–38 μm) of all heavily obscured C-rich post-AGB stars in our sample. The wavelength position of the acetylene (C_2H_2) absorption feature at 13.7 μm is marked with a dotted vertical line.

to HC_3N and CH_3, respectively. A few sources display also an unidentified absorption molecular feature at 17.92 μm, while one source (IRAS 17552) shows a strong absorption feature at ~7.5 μm that may be identified with CH_4 (see e.g., [12]).

We carried out complementary IRAM 30m observations of the few sources with variable maser emission. These observations confirm the C-rich nature of the central stars; as indicated by the detection of C-rich molecules such as CO and HCN together with the non detection of SiO. The IRAM 30m data reveal also the presence of high velocity (≥100 km/s) molecular outflows with extreme mass loss rates (~10^{-4} $M_\odot$/yr) in the previous AGB phase (see [11] for more details).

3. Formation of O-bearing molecules in C-rich PPNe

Our observations show that O-rich molecules may form in C-rich environments. As in the case of CRL 618, the prototypical C-rich PPN where O-bearing molecules are also detected, this can be explained by the chemical evolution models for the neutral layers of proto-PNe by Cernicharo [1], without the need to invoke a change in the chemistry (from O-rich to C-rich) of the central star. Our observations thus confirm the polymerization model of Cernicharo [1] that predicts a rich photochemistry in the neutral regions of these objects on timescales shorter than the dynamical evolution of the central H II region, leading to the formation of small C-rich molecules (small hydrocarbons and carbon clusters) and a transitory O-rich chemistry (formation of OH, H_2CO, and H_2O). This happens in zone II (see Figure 1 in [1]), where the CO is being photodissociated.

4. Conclusions

The analysis of heavily obscured transition sources with *Spitzer/IRS* confirms that most of these sources are non-variable OH/IR stars (O-rich), the result of the evolution of massive HBB AGB stars, but also a significant number of C-rich sources are found. We have detected small hydrocarbons, the building blocks of PAHs, in a small sample (11) of Galactic C-rich post-AGB stars [11]. In addition, benzene (C_6H_6) is clearly detected in two stars, bringing up to three the total number of Galactic post-AGB stars where this molecule has been detected [11]. The simultaneous detection of variable OH and H_2O maser emission seen in strong mass-losing C-rich sources confirms the chemical model predictions by Cernicharo [1]. As observed [11], these models predict the formation of small C-rich molecules and clusters, and a transitory O-rich chemistry as a result of time dependent chemistry and CO photodissociation.

Acknowledgments

D.A.G.H. was funded by the Ramón y Cajal fellowship number RYC−2013−14182 and he acknowledges support provided by the Spanish Ministry of Economy and Competitiveness (MINECO) under grant AYA−2014−58082-P. This work is based on observations made with the *Spitzer Space Telescope*, which was operated by the Jet Propulsion Laboratory (JPL), California Institute of Technology under a contract with NASA. D.A.G.H. also acknowledges support for this work provided by NASA through an award (#30258) issued by JPL/Caltech. J.C. thanks funding support from the European Research Council (ERC) under ERC−2013−SyG, G. A. 610256 NANOCOSMOS.

References

[1] Cernicharo J 2004 *Astrophys. J. Lett.* **608** L41-L44
[2] García-Hernández D A, Perea-Calderón J V, Bobrowsky M and García-Lario P 2007 *Astrophys. J. Lett.* **666** L33-L36
[3] Cernicharo J, Guélin M and Kahane C 2000 *Astron. Astrophys. Suppl. Series* **142** 181-215
[4] Cernicharo J, Heras A M, Tielens A G G M, Pardo J R, Herpin F, Guélin M and Waters L B F M 2001 *Astrophys. J. Lett.* **546** L123-L126
[5] Cernicharo J, Heras A M, Pardo J R, Tielens A G G M, Guélin M, Dartois E, Neri R and Waters L B F M 2001 *Astrophys. J. Lett.* **546** L127-L130
[6] Cernicharo J, Guélin M, Penalver J, Martin-Pintado J and Mauersberger R 1989 *Astron. Astrophys.* **222** L1-L4
[7] Herpin F and Cernicharo J 2000 *Astrophys. J. Lett.* **530** L129-L132
[8] García-Lario P, Manchado A, Pych W and Pottasch S R 1997 *Astron. Astrophys. Suppl. Series* **126** 479-502
[9] García-Hernández D A, García-Lario P, Plez B, Manchado A, D'Antona F, Lub J and Habing H 2007 *Astron. Astrophys.* **462** 711-730
[10] García-Hernández D A, Perea-Calderón J V, Engels D and García-Lario P 2007 *Asymmetric Planetary Nebulae IV* eds Corradi R L M *et al* p. 325
[11] García-Hernández D A, García-Lario P, Cernicharo J et al 2016 *Astrophys. J.* to be submitted
[12] Malek S E, Cami J and Bernard-Salas J 2012 *Astrophys. J.* **744** 16

Fullerenes and fulleranes in circumstellar envelopes

Yong Zhang[1,2], Sun Kwok[1] and SeyedAbdolreza Sadjadi[1,2]

[1] Laboratory for Space Research, Faculty of Science, The University of Hong Kong, Pokfulam Road, Hong Kong, China

[2] Department of Physics, The University of Hong Kong, Hong Kong, China

E-mail: [1,2] zhangy96@hku.hk

Abstract. Three decades of search have recently led to convincing discoveries of cosmic fullerenes. The presence of C_{60} and C_{60}^+ in both circumstellar and interstellar environments suggests that these molecules and their derivatives can be efficiently formed in circumstellar envelopes and survive in harsh conditions. Detailed analysis of the infrared bands from fullerenes and their connections with the local properties can provide valuable information on the physical conditions and chemical processes that occurred in the late stages of stellar evolution. The identification of C_{60}^+ as the carrier of four diffuse interstellar bands (DIBs) suggests that fullerene-related compounds are abundant in interstellar space and are essential for resolving the DIB mystery. Experiments have revealed a high hydrogenation rate when C_{60} is exposed to atomic hydrogen, motivating the attempt to search for cosmic fulleranes. In this paper, we present a short review of current knowledge of cosmic fullerenes and fulleranes and briefly discuss the implications on circumstellar chemistry.

1. Introduction

Fullerenes are carbon cage molecules consisting of pentagons and hexagons. C_{60}, the best known fullerene, was first discovered in the experiments of understanding the carbon-chain chemistry in circumstellar environments [1]. Given their remarkable stability, fullerenes have been long suspected to exist in the Universe. The detections of fullerenes in terrestrial rocks and meteorites [2, 3, 4] suggest that they can be formed in nature, although the formation mechanisms in various geological environments are still in considerable debate. However, albeit with extensive efforts, search for fullerenes in deep space has been frustrating until recently.

The electronic spectrum of C_{60} is dominated by a few UV features, and its visible bands are relatively weak. Highly reddened O-type stars can been targeted to search for electronic transitions from interstellar C_{60}. However, no positive detection has yet been made. Unsuccessful attempts [5, 6] only set an upper limit to the C_{60} abundance in the interstellar medium (ISM). C_{60} has a low ionization potential (7.6 eV), and thus its singly-ionized species, C_{60}^+, might be more abundant in the ISM. It has been long conjectured that two diffuse interstellar bands (DIBs) at 9632 Å and 9577 Å can be attributed to C_{60}^+ [7]. This was recently verified by experimental spectra of gas-phase C_{60}^+ [8]. The gas-phase laboratory spectra also exhibit two weaker bands at 9428.5 Å and 9365.9 Å, which were soon discovered in the spectra of DIBs [9]. The identification of C_{60}^+ as the carrier of four DIBs re-raises the possibility that fullerene derivatives might be responsible for the DIB phenomenon.

In infrared spectra, many vibrational modes of C_{60} are degenerate, resulting in four major features at 7.0, 8.5, 17.4, and 18.9 μm. Early attempts did not yield in certain detections [10, 11].

The first detection of C_{60} infrared bands was made in a young planetary nebula (PN) by the *Spitzer* Space Telescope [12]. Soon thereafter, C_{60} was detected in a variety of circumstellar environments [13, 14, 15, 16, 17, 18, 19, 20, 21], suggesting that it is a common molecule in the Universe and can survive in high-UV environments. About 5 percent of Galactic C-rich PNe have found to contain C_{60}, while Magellanic Cloud PNe show higher detection rate [22]. Infrared bands from the cation form C_{60}^{+} were also detected in some of the C_{60} sources [23, 24, 25]. There is no firm detection of the other charge states, although some efforts have been devoted to search for circumstellar C_{60}^{2+} and C_{60}^{-} [24, 25].

2. The Formation of Fullerenes

The discovery of C_{60} in a proto-PN [18] suggests that fullerenes can be formed on a timescale of 10^3 years. In low-density environments, fullerenes are unlikely to be formed through assembly of small carbon species on such a short timescale (i.e. the bottom-up process). Although fullerenes can be experimentally synthesized via several different routes, observations point to a top-down scenario in astronomical environments [14, 26]. In this scheme, photon- or shock-induced processes cause dehydrogenation and ejection of C_2 of large hydrocarbon molecules, providing a pathway for the formation of closed-caged carbon molecules. Spatially resolved observations of reflection nebulae and H II regions indicate that the formation of C_{60} is related to the radiation fields [13, 26, 27]. However, the detection of C_{60} in various evolved stars suggests that strong UV irradiation is not necessary for fullerene formation [17, 24].

What is the precursor of fullerenes? Hydrogenated amorphous carbon (HAC) particles and polycyclic aromatic hydrocarbons (PAHs) have been presented as the starting materials [14, 15, 26, 28]. All the C_{60} sources exhibit prominent plateau emission [22, 24], which can be attributed to stretching and bending modes of mixed aromatic-aliphatic organic nanoparticles (MAONs, [29]). Therefore, we propose that fullerenes might be formed through the sp^3 to sp^2 transformation in MAONs. MAONs have a three-dimensional structure, and thus can be processed into multi-layered fullerenes (buckyonions), which have been suggested as the carrier of the 217 nm extinction bump [30] and some DIBs [31].

3. Search for Fulleranes

When mixed with atomic hydrogen, fullerenes can be readily hydrogenated into fulleranes ($C_{60}H_m$). In experiments, $C_{60}H_{36}$ and $C_{60}H_{18}$ are the most significant products of the hydrogenation of C_{60}. Induced by UV photons or thermal energy, $C_{60}H_m$ can be converted back to C_{60} by releasing molecular hydrogen. Fullerenes and fulleranes, therefore, may play a crucial role in the formation of H_2 in astronomical environments [32]. However, a search for fulleranes in two PNe showing strong C_{60} emission has not met with success [33].

Early studies indicate that the experimental spectrum of $C_{60}H_{36}$ is able to match several infrared emission bands in proto-PNe [34]. IRAS 01005+7910, a C_{60}-containing proto-PN, shows a few features around $3.4\,\mu$m which have been tentatively ascribed to C-H stretching modes of $C_{60}H_m$ ($m \approx 36$) [24]. In the present work, we calculate the theoretical infrared (5–20 μm) spectra of several $C_{60}H_m$ using the quantum chemistry method [35], and compare the results with the observed spectrum of IRAS 01005+7910, as shown in Figure 1. Our theoretical calculations are reasonably close to the experimental spectra [36]. For slightly hydrogenated C_{60}, the four carbon skeletal vibration modes are still visible, but the relative intensities might vary compared to those of C_{60}. Therefore, it is possible that the previously detected C_{60} features actually stem in part from $C_{60}H_m$. This might help account for the discrepancy between the observed intensities and the predictions from excitation models of C_{60} [24]. For highly hydrogenated C_{60}, the carbon cage is significantly distorted by the repulsion between the hydrogen atoms, and can be disrupted if $m > 36$. The four features from the skeletal vibrations are relatively inconspicuous or substantially invisible in highly hydrogenated C_{60}, partly because the energy

Journal of Physics: Conference Series **728** (2016) 052004 doi:10.1088/1742-6596/728/5/052004

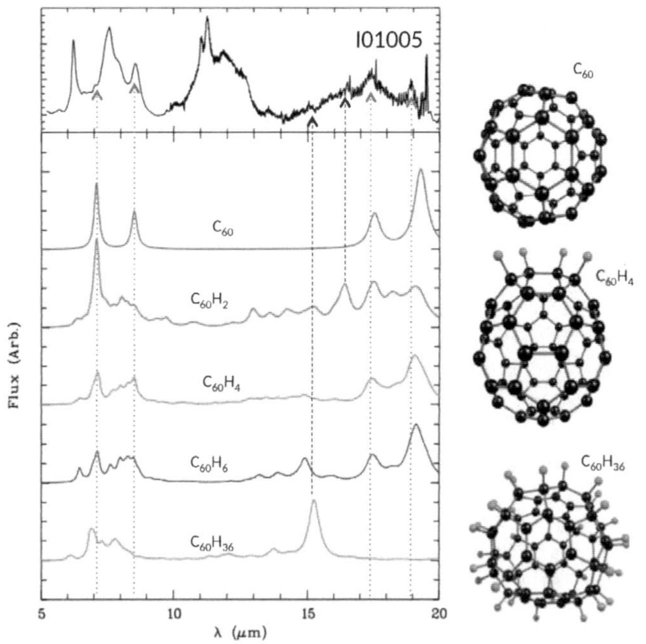

Figure 1. The graph on the left shows the observed spectrum of IRAS 01005+7910 (upper panel) and theoretical spectra of C_{60} and $C_{60}H_m$ (lower panel). The dotted arrows mark the four features arose from the skeleton vibrations of C_{60}. Hydrogenated C_{60} might be responsible for some weak features, as indicated by the dashed arrows. The energetically optimized structures of C_{60}, $C_{60}H_4$, and $C_{60}H_{36}$ are shown on the right.

is released largely through C-H vibrational modes. We therefore conjecture that fullerenes and fulleranes are rather common species in circumstellar envelopes, but hydrogenation will limit their detectability in infrared spectra. The main challenge in searching for circumstellar $C_{60}H_m$ is that the exact m value is unspecified and there are many isomers for a given m value. Nevertheless, we find that the observed spectrum might exhibit bands originating from $C_{60}H_m$ (Figure 1). In addition, it should be noted that many $C_{60}H_m$ have a ionization potential of $< 10\,\text{eV}$, and might exist as cation form in the ISM.

4. Summary and Perspectives

The detections of C_{60} and C_{60}^+ in circumstellar and interstellar environments indicate that fullerenes and their derivatives are abundant and widespread in the Universe. Given their high chemical activity and physical stability, fullerene-related compounds (FCs) are promising candidates for the carrier of DIBs. There are some potential matches between the properties of FCs and DIBs [37]. Strong C_{60}^+ absorption bands at 9632 Å and 9577 Å have been detected towards IRAS 01005+7910 [38]. Two DIBs are found to be enhanced in C_{60}-containing PNe [39]. These observations support the view that some of the DIB carriers might be FCs and have a circumstellar origin. Except fulleranes, some other FCs such as metallofullerenes [40] and fullerene/PAH adducts [41] can readily form together with fullerenes in experiments. Furthermore, if molecules like CO and H_2 are trapped inside the fullerene cage, their vibrational and rotational spectra may be strongly affected by the interaction with the wall. Future experimental and theoretical works on these FCs are highly valuable.

Acknowledgments
Financial support for this work was provided by the Research Grants Council of the Hong Kong under grants HKU7073/11P and HKU7062/13P.

References
[1] Kroto H W, Heath J R, Obrien S C, Curl R F and Smalley R E 1985 *Nature* **318** 162
[2] Buseck P R, Tsipursky S J and Hettich R 1992 *Science* **257** 215-217
[3] Becker L and Bada J L 1994 *Nature* **372** 507
[4] Heymann D, Chibante L P F, Brooks R R, Wolbach W S, Smit J, Korochantsev A, Nazarov M A and Smalley R E 1996 *The Cretaceous-Tertiary Event and Other Catastrophes in Earth History* (GSA Special Paper 307) pp 453-464
[5] Snow T P and Seab C G 1989 *Astron. Astrophys.* **213** 291-294
[6] Herbig G H 2000 *Astrophys. J.* **542** 334-343
[7] Foing B H and Ehrenfreund P 1994 *Nature* **369** 296-298
[8] Campbell E K, Holz M, Gerlich D and Maier J P 2015 *Nature* **523** 322-323
[9] Walker G A H, Bohlender D A, Maier J P and Campbell E K 2015 *Astrophys. J. Lett.* **812** L8
[10] Moutou C, Sellgren K, Verstraete L and Leger A 1999 *Astron. Astrophys.* **347** 949-956
[11] Kwok S, Volk K and Hrivnak B J 1999 *Astron. Astrophys.* **350** L35-L38
[12] Cami J, Bernard-Salas J, Peeters E and Malek S E 2010 *Science* **329** 1180
[13] Sellgren K, Werner M W, Ingalls J G, Smith J D T, Carleton T M and Joblin C 2010 *Astrophys. J. Lett.* **722** L54-L57
[14] García-Hernández D A, Manchado A, García-Lario P, Stanghellini L, Villaver E, Shaw R A, Szczerba R and Perea-Calderón J V 2010 *Astrophys. J. Lett.* **724** L39-L43
[15] Garcia-Hernandez D A, Kameswara Rao N and Lambert D L 2011 *Astrophys. J.* **729** 126
[16] García-Hernández D A, Iglesias-Groth S, Acosta-Pulido J A, Manchado A, García-Lario P, Stanghellini L, Villaver E, Shaw R A and Cataldo F 2011 *Astrophys. J. Lett.* **737** L30
[17] Gielen C, Cami J, Bouwman J, Peeters E and Min M 2011 *Astron. Astrophys.* **536** 54
[18] Zhang Y and Kwok S 2011 *Astrophys. J.* **730** 126
[19] Roberts K R G, Smith K T and Sarre P J 2012 *Mon. Not. R. Astron. Soc.* **421** 3277-3285
[20] Evans A *et al* 2012 *Mon. Not. R. Astron. Soc.* **421** L92-L96
[21] Otsuka M, Kemper F, Hyung S, Sargent B A, Meixner M, Tajitsu A and Yanagisawa K 2013 *Astrophys. J.* **764** 77
[22] García-Hernández D A, Villaver E, García-Lario P, Acosta-Pulido J A, Manchado A, Stanghellini L, Shaw R A and Cataldo F 2012 *Astrophys. J.* **760** 107
[23] Berné O, Mulas G and Joblin C 2013 *Astron. Astrophys.* **550** L4
[24] Zhang Y and Kwok S 2013 *Earth Planets Space* **65** 1069-1081
[25] Strelnikov D, Kern B and Kappes M M 2015 *Astron. Astrophys.* **584** 55
[26] Berné O and Tielens A G G M 2012 *Proc. National Acad. Sci. USA* **109** 401-406
[27] Castellanos P, Berné O, Sheffer Y, Wolfire M G and Tielens A G G M 2014 *Astrophys. J.* **794** 83
[28] Micelotta E R, Jones A P, Cami J, Peeters E, Bernard-Salas J and Fanchini G 2012 *Astrophys. J.* **761** 35
[29] Kwok S and Zhang Y 2011 *Nature* **479** 80-83
[30] Iglesias-Groth S 2004 *Astrophys. J. Lett.* **608** L37-L40
[31] García-Hernández D A and Diaz-Luis J J 2013 *Astron. Astrophys.* **550** L6
[32] Cataldo F and Iglesias-Groth S 2009 *Mon. Not. R. Astron. Soc.* **400** 291-298
[33] Díaz-Luis J J, García-Hernández D A, Manchado A and Cataldo F 2016 *Astron. Astrophys.* **589**, 5
[34] Cataldo F 2003 *Fullerenes, Nanotubes and Carbon Nanostructures* **11** 295-316
[35] Sadjadi S, Zhang Y and Kwok S 2015 *Astrophys. J.* **801** 34
[36] Iglesias-Groth S, García-Hernández D A, Cataldo F and Manchado A 2012 *Mon. Not. R. Astron. Soc.* **423** 2868-2878
[37] Alain O 2016 *Astron. Astrophys.* **590**, 52
[38] Iglesias-Groth S and Esposito M 2013 *Astrophys. J. Lett.* **776** L2
[39] Díaz-Luis J J, García-Hernández D A, Kameswara Rao N, Manchado A and Cataldo F 2015 *Astron. Astrophys.* **573** 97
[40] Dunk P W, Adjizian J-J, Kaiser N K, Quinn J P, Blakney G T, Ewels C P, Marshall A G and Kroto H W 2013 *Proc. National Acad. Sci. USA* **110** 18081-86
[41] García-Hernández D A, Cataldo F and Manchado A 2013 *Mon. Not. R. Astron. Soc.* **434** 415-422

Journal of Physics: Conference Series **728** (2016) 052005 doi:10.1088/1742-6596/728/5/052005

A search for hydrogenated fullerenes in fullerene-containing planetary nebulae

J J Díaz-Luis[1,2], D A García-Hernández[1,2], A Manchado[1,2,3] and F Cataldo[4,5]

[1] Instituto de Astrofísica de Canarias, C/ Via Láctea s/n, E−38205 La Laguna, Spain
[2] Departamento de Astrofísica, Universidad de La Laguna (ULL), E−38206 La Laguna, Spain
[3] Consejo Superior de Investigaciones Científicas, Madrid, Spain
[4] INAF- Osservatorio Astrofisico di Catania, Via S. Sofia 78, Catania 95123, Italy
[5] Actinium Chemical Research srl, Via Casilina 1626/A, 00133 Rome, Italy

E-mail: `jdiaz@iac.es`, `agarcia@iac.es`

Abstract. Detections of C_{60} and C_{70} fullerenes in planetary nebulae (PNe) of the Magellanic Clouds and of our own Galaxy have raised the idea that other forms of carbon such as hydrogenated fullerenes (fulleranes like $C_{60}H_{36}$ and $C_{60}H_{18}$), buckyonions, and carbon nanotubes, may be widespread in the Universe. Here we present VLT/ISAAC spectra (R ~600) in the 2.9-4.1 μm spectral region for the Galactic PNe Tc 1 and M 1-20, which have been used to search for fullerene-based molecules in their fullerene-rich circumstellar environments. We report the non-detection of the most intense infrared bands of several fulleranes around ~3.4-3.6 μm in both PNe. We conclude that if fulleranes are present in the fullerene-containing circumstellar environments of these PNe, then they seem to be by far less abundant than C_{60} and C_{70}. Our non-detections together with the (tentative) fulleranes detection in the proto-PN IRAS 01005+7910 suggest that fulleranes may be formed in the short transition phase between AGB stars and PNe but they are quickly destroyed by the UV radiation field from the central star.

1. Introduction

Fullerenes and fullerene-based molecules such as hydrogenated fullerenes and buckyonions may explain some astrophysical phenomena like the intense UV absorption band at 217 nm (e.g., [1]) or the so-called diffuse interstellar bands (DIBs; see e.g., [2, 3]). Their presence in astrophysical environments have been recently confirmed by the *Spitzer Space Telescope* via the detection of the mid-infrared C_{60} and C_{70} fullerene features in the spectrum of the young Planetary Nebula (PN) Tc 1 [4].

It has been demonstrated that fullerenes are efficiently formed in H-rich circumstellar environments only [5, 6], challenging our previous understanding of the fullerenes formation in space. In particular, the simultaneous detection of C_{60} fullerenes and PAHs in several PNe with normal hydrogen abundances has been reported [5, 7, 8, 9]. García-Hernández et al. [5] propose that both fullerenes and PAHs in H-rich circumstellar ejecta may form from the photochemical processing of hydrogenated amorphous carbon (HAC) in agreement with the experimental results of Scott et al. [10].

Duley & Williams [11] predict the production of fullerenes via thermal heating of HAC and this process may explain the detection of fullerenes in H-rich circumstellar environments (e.g.,

[5]). Interestingly, they predict also the formation of hydrogenated fullerenes (fulleranes). On the other hand, Iglesias-Groth et al. [12] found that the 3.44 and 3.55 μm bands of several fulleranes display molar extinction coefficients which are similar to those of the 17.4 and 18.8 μm bands of the isolated C_{60} molecule [13]. Thus, it could be expected to find fulleranes with line intensities similar to the ones already measured with *Spitzer* at 17.4 and 18.8 μm in fullerene-containing PNe.

2. Mid-infrared VLT/ISAAC spectroscopy of PNe with fullerenes

We acquired 3-4 μm infrared (IR) spectra of the fullerene PNe Tc 1 (S/N ~26 at the continuum), and M 1-20 (S/N ~11 at the continuum).

The M 1-20 spectrum includes the atomic hydrogen lines Pf_ϵ at 3.07 μm, a blend of the UIE feature at 3.31 with Pf_δ at 3.32 μm, Pf_γ at 3.75 μm, and Br_α at 4.05 μm. It also includes the tentatively identified H I (6-17), H I (6-15), and He I (4-5) lines at 3.76, 3.91, and 4.04 μm, respectively. In Figure 1, we display the M 1-20 spectrum (left panel) in the 2.9-4.1 μm range.

The Tc 1 spectrum includes the atomic hydrogen lines Pf_δ, Pf_γ, the tentatively identified H I (6-17) and H I (6-15) lines, and Br_α. It also includes the tentative He I (4-5) line at 4.04 μm. In Figure 1, we display the Tc 1 2.9-4.1 μm spectrum (right panel).

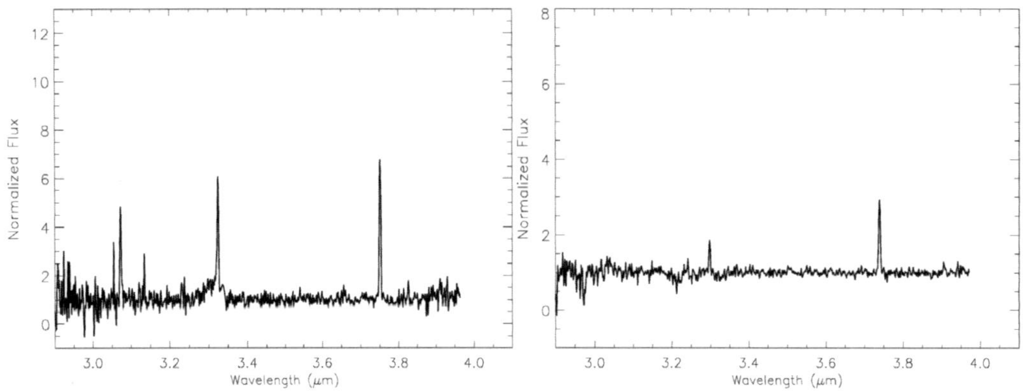

Figure 1. VLT/ISAAC spectra of M 1-20 (left panel) and Tc 1 (right panel).

3. The 3.3 μm UIE feature

We clearly detect the 3.3 μm UIE feature in M 1-20, while this feature is completely lacking in Tc 1. This is consistent with the *Spitzer* spectra; M 1-20 displays UIE features (PAH-like) at 6.2, 7.7, 8.6, and 11.3 μm, but these are absent in Tc 1 (e.g., [5]).

A wide variety of molecules have been proposed as possible carriers of the 3.3 and 11.3 μm UIE bands. Nowadays, the most accepted idea is that they originate from the C-H vibration modes of aromatic compounds or PAHs. However, still it is not clear if these bands are only due to aromatic compounds or to mixed aromatic/aliphatic structures characteristic of amorphous organic solids.

4. Non detection of the fullerane features

The IR laboratory spectra (R~500) of several fulleranes such as $C_{60}H_{18}$, $C_{60}H_{36}$ or $C_{70}H_{38}$ [12] show that the strongest features in the mid-IR (2-20 μm) are those at ~3.44, 3.51 and 3.54 μm, being the best IR bands for searching these molecules in the circumstellar environments of

Table 1. Summary table showing examples of the expected fluxes for the fullerane features. Quoted values are the measured central wavelengths of the fullerane features, the 2σ upper detection limits to the expected emission line fluxes in Tc 1, and the predicted fluxes for the fullerane bands in Tc 1 and M 1-20.

Fullerane	λ_c (μm)	FLUX (Tc 1)	Predicted fluxes (Tc 1) (10^{-14}ergcm^{-2}s^{-1}/"^{2})	Predicted fluxes (M 1-20) (10^{-13}ergcm^{-2}s^{-1}/"^{2})
$C_{60}H_{18}$ (isomer 1)	3.42	$\leq$3.00	9.43	2.39
	3.46	$\leq$5.13	10.20	2.58
	3.51	$\leq$7.28	13.25	3.36
$C_{60}H_{18}$ (isomer 2)	3.42	$\leq$3.13	12.74	1.42
	3.45	$\leq$4.85	13.92	1.56
	3.51	$\leq$7.85	18.21	2.06
$C_{60}H_{36}$ (+48°C)	3.44	$\leq$3.33	22.13	5.58
	3.54	$\leq$11.02	16.23	4.11
$C_{60}H_{36}$ (+250°C)	3.44	$\leq$3.10	30.11	3.39
	3.53	$\leq$14.38	22.08	2.47
$C_{70}H_{38}$ (+50°C)	3.44	$\leq$3.62	18.13	4.61
	3.54	$\leq$10.05	16.75	4.22

fullerene-rich PNe. However, none of these three emission features is detected in our VLT/ISAAC spectra of Tc 1 and M 1-20.

We could estimate approximate upper limits to the fluxes of the hydrogenated fullerene features (see Table 1). In order to obtain 2σ upper limits to the expected emission line fluxes of the fullerane features at $\sim$3.5 μm, we measure the rms in our flux calibrated VLT/ISAAC spectra; rms values of $\sim 4.28 \times 10^{-19}$ ergcm^{-2}s^{-1}Å^{-1} and $\sim 6.65 \times 10^{-18}$ ergcm^{-2}s^{-1}Å^{-1} for Tc 1 and M 1-20 are obtained, respectively. Then we multiply these values by the widths (FWHMs in the range 0.02-0.10 μm) of the fullerane bands measured in the infrared laboratory spectra. Finally, we divided these fluxes by the area ($\sim$0.6" arcsec^2 for both PNe) of the emission in the 3−4 μm range covered by our ISAAC observations.

On the other hand, by using the molar absorptivity values for the fullerenes and fulleranes reported by Iglesias-Groth et al. [13, 12], we may estimate the predicted fluxes for the fullerane features in Tc 1 and M 1-20 (see Table 1). By taking the molar absorptivity ratio (e.g., $\epsilon_{C60}/\epsilon_{fulleranes}$) of the fullerene and fullerane bands and the observed fluxes of the C_{60} and C_{70} infrared bands less contaminated by other species [8] we estimate the expected flux of the fullerane features at $\sim$3.5 μm.

Table 1 shows some examples of the estimated 2σ upper limits (in units of 10^{-16}ergcm^{-2}s^{-1}/"^{2}) and the predicted fluxes for the fullerane bands. The expected fluxes for all fullerane bands of $C_{60}H_{18}$, $C_{60}H_{36}$, and $C_{70}H_{38}$ (by using the observed *Spitzer* fluxes of the C_{60} 8.5 and 17.4 μm and C_{70} 14.9 μm bands) are in the range of $\sim$1.7-30 $\times 10^{-14}$ and $\sim$1.4-5.6 $\times 10^{-13}$ ergcm^{-2}s^{-1} /arcsec^2 in Tc 1 and M 1-20, respectively.

By comparing the values of the predicted fluxes with our 2σ upper limits (Table 1), we find that the expected fluxes are a factor of $\sim$20−1000 and $\sim$10−100 higher than the 2σ upper limits for Tc 1 and M 1-20, respectively. From these estimations we thus conclude that if fulleranes are present in Tc 1 and M 1-20, then they seem to be by far less abundant than C_{60} and C_{70}.

As we mentioned above, thermal heating via chemical reactions internal to HAC dust may explain the detection of fullerenes in the H-rich circumstellar environments of PNe and would potentially form fulleranes [11]. In addition, Duley & Hu [14] suggest an evolutionary sequence for the conversion of HAC to fullerenes, in which initial HAC de-hydrogenation is followed by

PFs formation and subsequent conversion of PFs to closed cage structures such as C_{60}. Under the Duley & Hu [14] scenario, Tc 1 would represent the last stage in the HAC de-hydrogenation process (i.e., only fullerenes are present), while the proto-PN IRAS 01005+7910 would represent an intermediate stage where the PFs are being converted to fullerenes. Remarkably, fulleranes may be also by-products of this conversion from HAC to fullerenes [14].

Interestingly, Zhang & Kwok [15] have (tentatively) detected fulleranes in the proto-PN IRAS 01005+7910; three strong C-H stretching bands at 3.48, 3.51, and 3.58 μm are apparently present in its ISO spectrum with fluxes comparable to the one of the 3.3 μm feature.

Our non-detection of fulleranes in more evolved PNe (such as Tc 1 and M 1-20 with $T_{eff}>$ 30,000 K) together with their detection in less evolved sources (like IRAS 01005+7910 with $T_{eff}\sim$21,500 K) thus suggest that fulleranes may be formed in the short transition phase between AGB stars and PNe but they are quickly destroyed; e.g., photochemically processed by the rapidly changing UV radiation from the central star.

Acknowledgments

We acknowledge Kameswara Rao for his help during the data analysis. J.J.D.L., D.A.G.H., and A.M. acknowledge support provided by the Spanish Ministry of Economy and Competitiveness (MINECO) under grants AYA−2014−58082−P and RyC−2013−14182 (DAGH). This work is based on observations obtained with ESO/VLT under the programme 290.D-5093(A).

References

[1] Cataldo F and Iglesias-Groth S 2009 *Mon. Not. R. Astron. Soc.* **400** 291
[2] García-Hernández D A and Díaz-Luis J J 2013 *Astron. Astrophys.* **550** L6
[3] Díaz-Luis J J, García-Hernández D A, Rao N K, Manchado A and Cataldo F 2015 *Astron. Astrophys.* **573** A97
[4] Cami J, Bernard-Salas J, Peeters E and Malek S E 2010 *Sci.* **329** 1180
[5] García-Hernández D A, Manchado A, García-Lario P, Stanghellini L, Villaver E, Shaw R A, Szczerba R and Perea-Calderón J V 2010 *Astrophys. J. Lett.* **724** L39
[6] García-Hernández D A, Rao N K and Lambert D L 2011 *Astrophys. J.* **729** 126
[7] García-Hernández D A, Iglesias-Groth S, Acosta-Pulido J A, Manchado A, García-Lario P, Stanghellini L, Villaver E, Shaw R A and Cataldo F 2011a *Astrophys. J. Lett.* **737** L30
[8] García-Hernández D A, Villaver E, García-Lario P, Acosta-Pulido J A, Manchado A, Stanghellini L, Shaw R A and Cataldo F 2012 *Astrophys. J.* **760** 107
[9] Otsuka M, Kemper F, Cami J, Peeters E and Bernard-Salas J 2014 *Mon. Not. R. Astron. Soc.* **437** 2577
[10] Scott A, Duley W W and Pinho G P 1997 *Astrophys. J. Lett.* **489** L193
[11] Duley W W and Williams D A 2011 *Astrophys. J. Lett.* **737** L44
[12] Iglesias-Groth S, García-Hernández D A, Cataldo F and Manchado A 2012 *Mon. Not. R. Astron. Soc.* **423** 2868
[13] Iglesias-Groth S, Cataldo F and Manchado A 2011 *Mon. Not. R. Astron. Soc.* **413** 213
[14] Duley W W and Hu A 2012 *Astrophys. J. Lett.* **745** L11
[15] Zhang Y and Kwok S 2013 *Earth, Planets and Space* **65** 1069

Properties of the fullerene C_{60}-containing PN Lin49 in the SMC; Explanations of strong near-IR excess

Masaaki Otsuka[1], Francisca Kemper[1], Marcelo L. Leal-Ferreira[2,†], Isabel Aleman[2], Jeronimo Bernard-Salas[3], Jan Cami[4,5], Bram Ochsendorf[2,6], and Els Peeters[4,5]

[1]ASIAA, 11F of Astronomy-Mathematics Building, AS/NTU. No.1, Sec. 4, Roosevelt Rd, Taipei 10617, Taiwan, R.O.C.
[2]Leiden Observatory, University of Leiden, PO Box 9513, 2300 RA, Leiden, The Netherlands
[3]Department of Physical Sciences, The Open University, Milton Keynes, MK7 6AA, UK
[4]Department of Physics and Astronomy, The University of Western Ontario, London, ON N6A 3K7, Canada
[5]SETI Institute, 189 Bernardo Ave, Suite 100, Mountain View, CA 94043, USA
[6]Space Telescope Science Institute, 3700 San Martin Drive, Baltimore, MD 21218, USA,
[†]CNPq/Brazil Fellow

E-mail: otsuka@asiaa.sinica.edu.tw

Abstract. We performed a detailed spectroscopic analysis of the fullerene C_{60}-containing planetary nebula (PN) Lin49 in the Small Magellanic Cloud (SMC). Lin49 is a C-rich and metal-deficient PN (Z∼0.0006) and its nebular abundances are in agreement with the AGB model for the initially 1.25 $M_{\odot}$ stars with the metallicity Z = 0.001 of Fishlock et al. (2014, [1]). By stellar absorption fitting with TLUSTY, we derived stellar abundances, effective temperature, and surface gravity. We constructed the photo-ionization model with CLOUDY in order to investigate physical conditions of Lin49. The model with the 0.005-0.1 μm radius graphite and a constant hydrogen density shell could not fit the ∼1-5 μm SED owing to the strong near-IR excess. We propose that the near-IR excess indicates (1) the presence of extremely small carbon molecules or (2) the presence of high-density structure surrounding the central star.

1. C_{60}-containing PN Lin49 in the SMC

Recently, *Spitzer*/IRS mid-IR spectrum revealed that Lin49 is a C-rich dust PN, showing strong C_{60} resonances at 17.4 and 18.9 μm and similar dust features such as the broad 11 μm and 30 μm features seen in the other C_{60}-containing Magellanic Cloud PNe [2]. However, physical properties of the central star and the dusty nebula had been unknown. Therefore, we characterized Lin49 through abundance analysis, stellar absorption fitting, and photo-ionization modelling based on our own ESO/VLT X-SHOOTER UV to near-IR spectrum, the archived *Spitzer*/IRS spectrum, and the UV to mid-IR photometry data.

2. Nebular abundances and Stellar absorption line analysis

The resultant elemental abundances are listed in Table 1. In comparison to heavy α-elements S and Ar, Fe is highly depleted. It seems that S is not locked by dust grains such as MgS because the [S/H] abundance is comparable to the noble gas [Ar/H] abundance. According to

Journal of Physics: Conference Series **728** (2016) 052006 doi:10.1088/1742-6596/728/5/052006

Table 1. Elemental abundances ($\log_{10}\epsilon(\mathrm{H}) = 12$). The values in the last two columns are the predictions in the AGB models by [1] for initially 1.0 $M_\odot$ and 1.25 $M_\odot$ stars with $Z = 0.001$.

X	$\log_{10}\epsilon(\mathrm{X})$	$\log_{10}\epsilon(\mathrm{X}_\odot)$	[X/H]	$\log_{10}\epsilon(\mathrm{X}_{1.0\,M_\odot \mathrm{model}})$	$\log_{10}\epsilon(\mathrm{X}_{1.25\,M_\odot \mathrm{model}})$
He	10.80 - 11.01	10.93 ± 0.01	−0.13 - +0.08	10.99	11.01
C	8.46 ± 0.24	8.39 ± 0.04	+0.07 ± 0.25	8.06	8.56
N	6.93 ± 0.02	7.86 ± 0.12	−0.93 ± 0.12	7.15	7.26
O	8.11 ± 0.01	8.73 ± 0.07	−0.62 ± 0.07	7.58	7.68
Ne	7.18 ± 0.05	8.05 ± 0.10	−0.89 ± 0.11	6.89	7.37
S	6.02 ± 0.01	7.16 ± 0.02	−1.15 ± 0.02	5.99	6.00
Cl	4.03 ± 0.05	5.25 ± 0.06	−1.22 ± 0.08	4.07	4.08
Ar	5.48 ± 0.11	6.50 ± 0.10	−1.02 ± 0.15	5.27	5.28
Fe	4.55 ± 0.04	7.46 ± 0.08	−2.91 ± 0.09	6.37	6.38

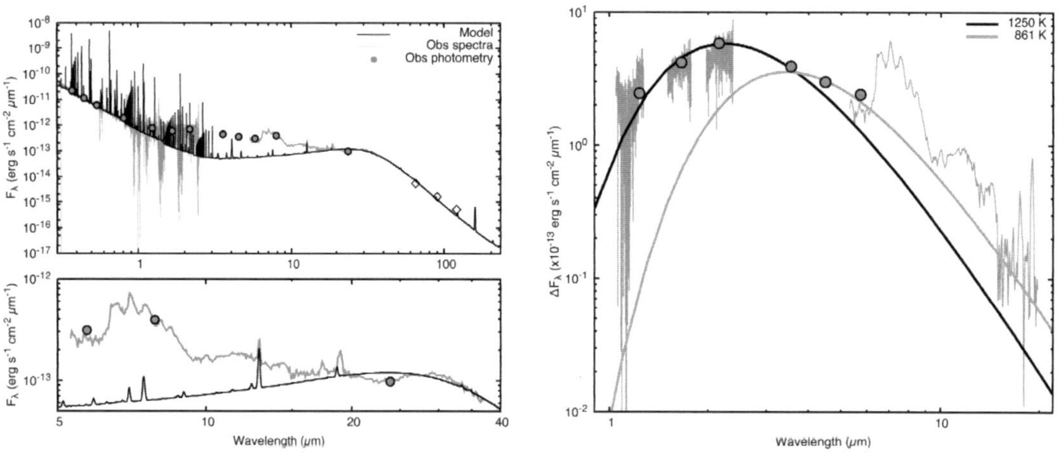

Figure 1. (*left panels*) (upper) Comparison between the CLOUDY model and observational data. The squares at 65, 90, and 120 μm are the *predicted* flux densities calculated by fitting the broad 30 μm feature. (lower) Closed-up plots for mid-IR wavelength. (*right panel*) Near-IR excess. The grey lines and the circles are the residual flux densities (ΔF_λ) between the observed X-SHOOTER and *Spitzer*/IRS spectra and photometry bands and the corresponding values obtained from the CLOUDY model.

the chemical evolution model of the Milky Way halo by [3], the [S/Fe] and [Ar/Fe] are ∼+0.4 and ∼+0.3 in the [Fe/H] <-1, respectively. By applying this prediction to Lin49, we obtained the [Fe/H] = −1.55 and −1.32 from the observed [S/H] and [Ar/H], respectively. From the average [Fe/H] of −1.4 between them, we estimated Z to be ∼0.0006. Over 96 % of the iron atoms in the nebula would exist as the dust grains. In the last two columns of Table 1, we list the predicted abundances in the AGB nucleosynthesis models by [1]. The observed abundances are in excellent agreement with the prediction of the 1.25 $M_\odot$ model. From the view of chemical abundances, we infer that the initial mass of the progenitor would be around 1.25 $M_\odot$.

We fit stellar absorption lines with the O-type star grid model OStar2002 by [4] using the non-LTE model stellar atmospheres modeling code TLUSTY [5]. We determined the effective temperature (T_{eff}) of 30 500 ± 500 K, the surface gravity ($\log g$) of 3.29 ± 0.06 cm s^{-2}, and the ϵ(He,C,N,O,Si) of 10.9 ± 0.3, 9.0 ± 0.3, 7.6 ± 0.3, 8.6 ± 0.1, and 6.8 ± 0.3, respectively.

3. Photo-ionization modeling; finding of the near-IR excess

Using the photo-ionization code CLOUDY [6], we investigated physical conditions of Lin49. As the SED of the ionization/heating source, we used the central star's spectrum synthesized by the TLUSTY and varied its intrinsic luminosity (L_*) to match the observed nebular elemental abundances, gas emission line fluxes, and broad band fluxes/flux densities. We supposed that the infrared continuum is due to graphite grains; we adopted the radius a = 0.005-0.1 μm and the $a^{-3.5}$ size distribution. We derived the gas and dust masses of 0.11 $M_\odot$ and 4.3×10^{-5} $M_\odot$, respectively. The dust temperature is in the range from 79 K to 825 K. We determined the core-mass of 0.53-0.57 $M_\odot$. The location of the derived L_* and $T_{\rm eff}$ on evolutionary tracks of post AGB stars with Z = 0.001 of [7] indicates that the initial mass is 1.0-1.5 $M_\odot$, consistent with our inference from the nebular abundances.

In Fig. 1 left, we present the observed and the modelled SEDs. The predicted SED could not fit the near-IR SED owing to the strong near-IR excess. To focus on the near-IR excess, we examined the differential spectra and photometry points between the observations and the CLOUDY model (Fig. 1 right); the near-IR excess peaks at ~2-3 μm and its SED can be fitted by a single blackbody temperature of 1250 K. Limited to the SMC C_{60} PNe, SMP24 shows the near-IR excess; [8] found that a hotter dust component (~1000 K) is necessary to fit the observed SED down to 1 μm, apart from the hot dust component (~270 K).

4. Interpretations for the near infrared excess

At the moment, we deduced two possibilities that the near-IR excess SED indicates (1) the presence of the *extremely* small molecules and/or (2) the high-density structure nearby the central star.

4.1. Stochastic heating by extremely small particles

Under the usual dust grain size (e.g., $\gtrsim$ 0.01 μm), dust temperatures are determined by solving an energy balance equation between the radiative heating owing to the central star and the cooling of grains. In such size grains, individual quantum events are not important. While, in extremely small grains which compose of 100 atoms or less, single photons would cause them to heat up significantly for very short time scale. This mechanism is known as stochastic heating, and has been proposed for the reflection nebulae NGC7023 and NGC2023 [9] and PNe e.g., IC418 [10]. Interestingly, these reflection nebulae and IC418 show mid-IR C_{60} band emission. We included the stochastic heating mechanism in the CLOUDY model.

According to [11], the thermal properties of solids are described by Debye's theory and the heat capacity C_V for T over Debye temperature Θ (the Θ in graphite is ~500 K) is $3Nk$ where N is the number of the atoms in a molecule and $3N$ is its number of degrees of freedom. For extremely small grains, a photon energy $E_{\rm ph}$ produces the difference between the maximum and minimum temperatures $\Delta T = E_{\rm ph}/3Nk$ K. The $T_{\rm eff}$ of the central star is 30 500 K, so the photon energy at the radiation peak is 13.05 eV. Thus, we obtained $\Delta T = 5.05\times10^4/N$ K.

By adopting the respective max and min temperatures of 1250 K and 20 K, we obtained N of be ~39, although our estimation is optimistic and also it depends on the min temperature. If such a molecule composing of ~39 C-atoms formed up a honeycomb structure sheet, the molecule's dimension is $\sim8\times10^{-4}$ μm$\times8\times10^{-4}$ μm square. The SED with ΔT of 841 K derived by keeping N = 60 can give more better fit to the differential SED plots in $\gtrsim$3.6 μm.

4.2. High density structure nearby the central star

In Section 3, we assumed that Lin49 does not have any sub-structures surrounding the central star but a normal density nebula only. However, the blackbody minimum emitting radius of the 1250 K component (~2 AU) suggests that near-IR component would emit from sub-structures nearby the central star. A similar idea was proposed for the near-IR excess at the central

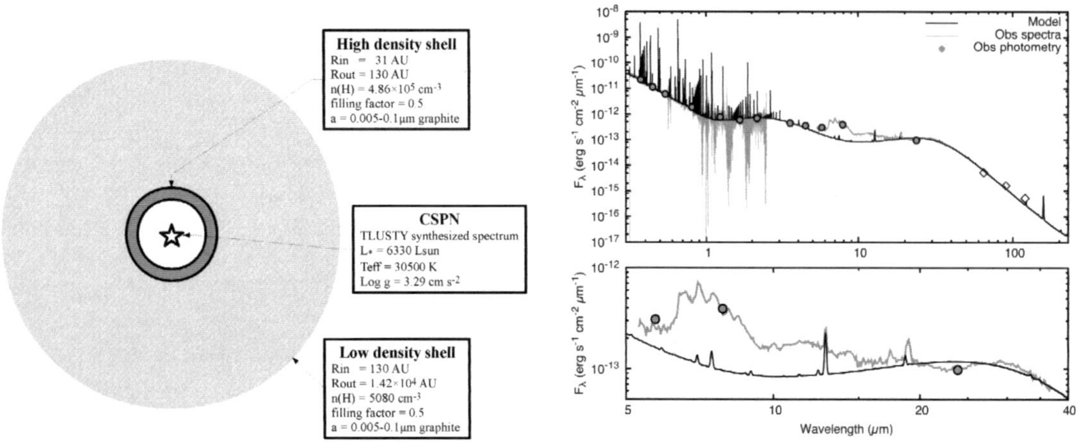

Figure 2. (*left panel*) The explanation of the two density shell model. The central star is located in the geometrical center of two spherical nebulae, so called *high density shell* and *low density shell*. As the ionization and heating source, we used the spectrum of the central star (CSPN) synthesized by the TLUSTY fitting. (*right two panels*) (upper) Comparison between the observed SED plots and the predicted SED by the two density shell model. (lower) Closed-up plot for mid-IR wavelength. The dust temperature was in the range from 79 K to 1453 K. The lines and symbols in both of panels are as defined in Fig. 1 left.

position of IC418; [12] took the near-IR *JHK* images of IC418 and found excess at the central position after subtracting out the contribution from the central star. They argued that the excess indicates a possible compact shell interior to the main shell. We tested this interpretation for the near-IR excess in Lin49. When we assumed two density shells as explained in Fig. 2 left, we can well fit the near-IR excess (Fig. 2 right). The (binary) disk rather than the shell could stably harbor the near-IR emitters nearby the central star for a long time, although we did not detect any narrower absorption lines from the cooler stellar components. Assuming the high density *disk*, the Keplerian rotational velocity was 4 km s^{-1} at 31 AU from the central star.

Several SMC and Galactic C_{60} PNe show the near-IR excess. It would be worth investigating these PNe in order to characterize properties of dust, small molecules, and the central star of C_{60} PNe and also understand the C_{60} formation in the circumstellar environment.

References

[1] Fishlock C K, Karakas A I, Lugaro M and Yong D 2014 *Astrophys. J.* **797** 44–68
[2] Sloan G C *et al* 2014 *Astrophys. J.* **791** 28–57
[3] Kobayashi C, Karakas A I and Umeda H 2011 *Mon. Not. R. Astron. Soc.* **414** 3231–3250
[4] Lanz T and Hubeny I 2003 *Astrophys. J. Suppl. Series* **146** 417–441
[5] Hubeny I 1988 *Computer Physics Communications* **52** 103–132
[6] Ferland G J, Porter R L, van Hoof P A M, Williams R J R, Abel N P, Lykins M L, Shaw G, Henney W J and Stancil P C 2013 *Revista Mexicana de Astronomia y Astrofisica* **49** 137–163
[7] Vassiliadis E and Wood P R 1994, *Astrophys. J. Suppl. Series*, **92**, 125–144
[8] Bojicic I S, Filipovic M D and Crawford E J 2010 *Serbian Astronomical Journal* **181** 63–68
[9] Sellgren K 1984 *Astrophys. J.* **277** 623–633
[10] Phillips J P, Sanchez Magro C and Martinez Roger C 1984 *Astron. Astrophys.* **133** 395–402
[11] Whittet D C B (ed) 2003 *Dust in the galactic environment*
[12] Hora J L, Deutsch L K, Hoffmann W F, Fazio G G and Shivanandan K 1993 *Astrophys. J.* **413** 304–316

Sulfur-bearing molecules observed in the massive star-forming regions, DR21(OH) and G33.92+0.11

Y. C. Minh

Korea Astronomy and Space Science Institute, 776 Daedeok-daero, Daejeon 34055, Korea

E-mail: minh@kasi.re.kr

Abstract. Recent high sensitive and high angular resolution observations are providing unprecedented amount of chemical data, especially, on the massive star-forming regions. It will greatly extend our understandings on the complicated star formation process, if we can digest those huge amount of information. We discuss here on the properties of the sulfur-bearing species observed with high angular resolutions toward two massive star-forming regions, DR21(OH) and G33.92+0.11. H_2S may not exist as a solid form in the grain mantles, but OCS is believed to be one of major solid sulfur species, as suggested before. In addition, the bipolar-like outflow of the H_2CS emission observed in DR21(OH) may suggest that H_2CS is also one of solid sulfur species on the grain mantles. Depending on the chemical environment, the competition between hydrogenation and oxidization on the grain surface may lead to formation of specific solid forms to dominate, which could be either H_2CS or OCS. SO and SO_2 are often observed to be associated with ionized gas, such as the UC HII regions. These species seem to be formed in the high temperature turbulent gas in a later stage of star formation after the hot core phase. Fractional abundances of these sulfur-bearing species appear to be consistent to a certain extent in several star-forming regions. The physical and chemical evolution of massive star formation seems to pass through very similar stages in most star-forming regions. Consequently, it may indicate that there exists a consistent and coherent pattern of processes experienced by the massive star formation, in spite of the large variations in small scale locational differences.

1. Introduction

Massive stars are forming in the dense molecular cores of GMCs. They used to form in clusters with a relatively short evolutionary time scale and include much more energetic activities compared to the low mass star formations [28, 39]. Massive star-forming regions often show the tremendous complexity associated with dust continuum emission, hot cores, outflows, various masers, complicated neutral and ionized gas components, etc. These regions show extremely rich chemistry associated with hot cores, shocks, and high temperature gas in addition to general dense gas phase chemistry. Since these sources are embedded deeply in the dense molecular cloud cores, the chemical signatures associated with the sources are one of the most useful and important probes to investigate the nature of the star-forming cores.

Complicated molecular emission features have been observed toward the dense cores in an early evolutionary stage of massive star formation [19, 21]. Among the observed species, sulfur-bearing species have been known to be chemical clocks in the early phase of the star formation evolution. It is mainly because that the interstellar sulfur has been observed to be largely depleted (by about a factor of 1000) in the quiet dense gas and to appear again in the gas phase associated with star-forming activities [6, 12]. Some of the sulfur-bearing species observed in the gas phase may have evaporated directly from the grain mantles, but some of them must have been formed later in the gas phase from the evaporated species. The particular form of solid state sulfur in the grain mantles is still a controversial issue, and the gas phase evolution of sulfur-bearing species is also not clear [5, 12, 27, 34, 35]

Recent high sensitivity and high angular resolution observations open a new era of interstellar chemistry. Here we discuss on the chemical properties of several sulfur-bearing species observed in the massive star-forming regions DR21(OH) and G33.92+0.11 with SMA (Submillimeter Array) and ALMA (Atacama Large Millimeter and submillimeter Array), respectively.

2. Emission fragmentation

2.1. DR21(OH)

DR21(OH) is a well known massive star-forming region located at a distance of 1.4 kpc in the Cyguse X North region [23, 29, 30]. This source consists of several dense clumps ($\sim 10^3$ $M_\odot$ and $n(H_2) \sim 10^6$ cm^{-3}) [18], which harbor young stellar objects of very early stages. Figure 1 shows two bright compact continuum cores, MM1 and MM2, separated by $\sim 8''$ from each other [37]. This source is associated with various masers, such as OH [2, 24], H_2O [10], and CH_3OH [1, 4, 9, 13, 14, 26], indicating ongoing star formation activities. It also shows complicated molecular emission, which are strongly concentrated toward MM1, especially to the subcores MM1a and b, separated by $\sim 1.6''$. Depending on the peak position, the molecular emission distributions can be grouped into two: one group of emission lines peaks toward MM1a, and a second group peaks toward MM1b [21].

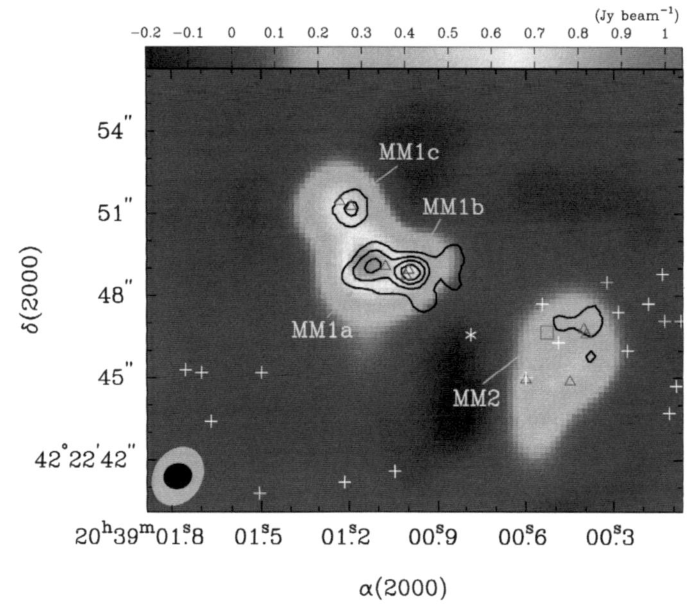

Figure 1. Continuum emission distribution of DR21(OH) at 230 GHz (contours) and 340 GHz (false color) observed with SMA [21]. Contours of the 230 GHz emission show the 20% levels of the peak intensity (0.13 Jy $beam^{-1}$). The beams are shown in the lower left as ellipses (grey for 340 GHz and black for 230 GHz). The asterisk is the H_2O maser position [10] and the crosses are methanol masers [1]. The MM2 position [37] is shown as a box (magenta). Triangles (purple) are the SMA sources [38].

Table 1. Physical parameters of subcores MM1a and b [21]

Subcore	Size(HPW)(P.A.) (″×″)	$M_{\rm core}$ ($M_\odot$)	N_{H_2} (cm^{-2})	n_{H_2} (cm^{-3})
MM1a	1.6×1.1 (−50°)	0.23	1.3×10^{23}	4×10^6
MM1b	1.9×1.4 (−80°)	0.43	1.1×10^{23}	2×10^6

The first group includes highly saturated molecules, such as CH_3OH, $HCOOCH_3$ and CH_3OCH_3, which suggests that MM1a is the site for hot cores. These saturated species are thought either to be evaporated from the grain mantles or to be formed in the hot core gas phase from the evaporated species. In contrast, the second group contains sulfur-bearing species such as SO and SO_2, which are thought to trace high temperature gas associated with UC HII regions [19, 21]. No close correlations between the emission peaks from these sulfur-bearing species and identified hot cores have been found [19, 21, 22]. These species are considered not to be evaporating directly from the grain mantles in hot cores, but to be formed in the dense gas phase from the evaporated sulfur or sulfur-bearing species.

One thing to note further is the discovery of the H_2CS bipolar-like outflow feature associated with MM1a of DR21(OH) (Figure 3) [20]. H_2CS can be formed in the gas phase [6], and its abundance has been reported to be increased in outflows associated with star formation [3, 7]. But it is also suggested as a grain mantle species with an abundance comparable to that of OCS [25]. The outflow feature of H_2CS emission in DR21(OH) indicates that H_2CS is either formed in the turbulent gas phase after the release of the sulfur species from the grains or evaporated directly from the grain mantles by the interaction with the outflow. Considering the formation

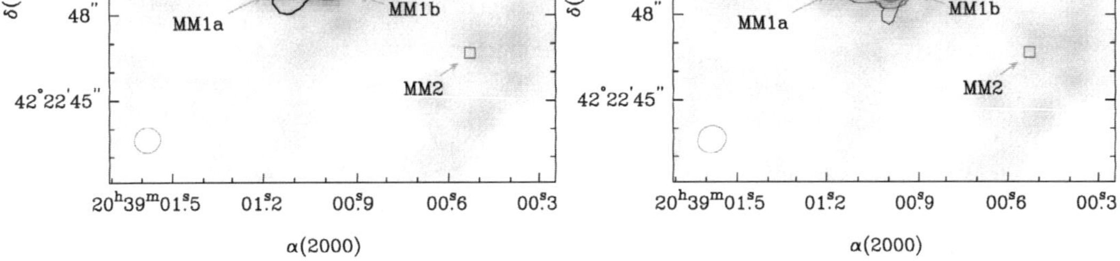

Figure 2. Integrated intensity maps of the $HCOOCH_3$ ($20_{2,19} - 19_{2,18}$ A) (*left*) and SO_2 ($22_{2,20} - 22_{1,21}$) (*right*) transitions [21]. The background is the 230 GHz continuum of Figure 1. The inset in the upper right shows sample spectra taken toward MM1a (upper spectrum) and MM1b (lower spectrum). The beam of the observed transition is shown in the bottom left side. All contours are 30%, 60%, and 90% of the peak emission.

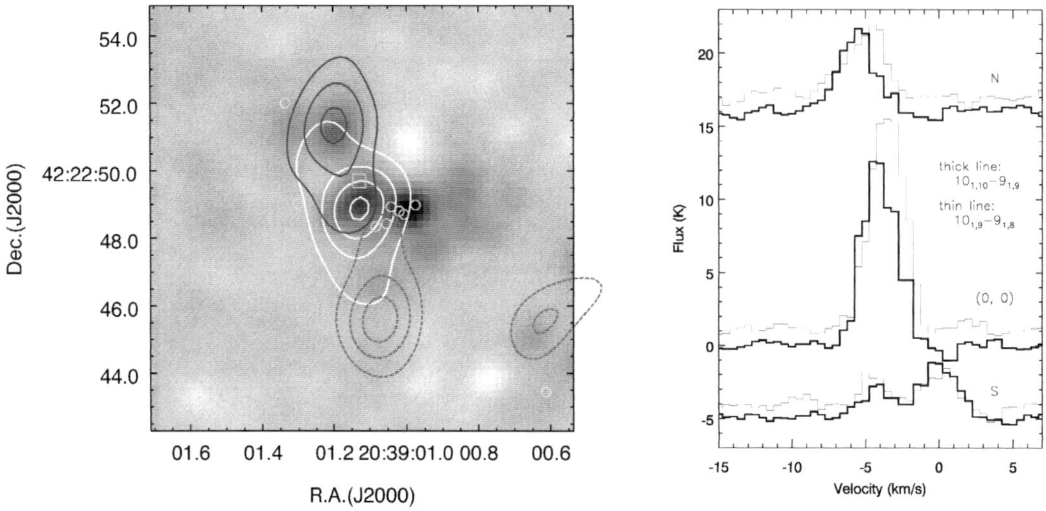

Figure 3. Integrated intensity map of the H_2CS $10_{1,10} - 9_{1,9}$ transition observed toward DR21(OH) [20] for the separate velocity ranges: (*blue*) $v_{lsr} = -10.0 \sim -5.55$ km s^{-1}, (*white*) $-5.5 \sim -2.0$ km s^{-1}, and (*red, dashed*) $-2.0 \sim 3.0$ km s^{-1}. Contour levels of the blueshifted and redshifted components increase by 1 Jy beam^{-1} km s^{-1} from 1.5 Jy beam^{-1} km s^{-1}. The white contours increase by 3 Jy beam^{-1} km s^{-1} from 3 Jy beam^{-1} km s^{-1}. The green box is the MM1 and circles are radio continuum sources [1, 18]. Sample spectra of the H_2CS $10_{1,10} - 9_{1,9}$ and $10_{1,9} - 9_{1,8}$ transitions taken at the emission peaks of the blue- and redshifted lobes and the center [20]. Flux of the spectra was transformed to the temperature unit.

time scale in the gas phase of $\sim 10^5$ years [6, 15], the observed H_2CS is highly likely to evaporate directly from the grain mantles [20].

Although there have been several suggestions on the sulfur on dust grains [35], the major form of solid sulfur is a long standing problem. H_2S has previously been suggested as a main reservoir of sulfur in grain mantles, but it was not found in the observations toward continuum sources on the grain mantles [11, 34]. OCS is the only sulfur-bearing species clearly detected on the grain surface, but it also has a very low fractional abundances $\sim 10^{-7}$ relative to hydrogen [25, 35]. H_2CS could be one of the sulfur-bearing solid species of the ice grain mantles, depending on the chemical environment of efficient hydrogenation.

2.2. G33.92+0.11

G33.92+0.11 is a massive star-forming region classified as a UC HII region at a distance of 7.1 kpc [8]. OB star-clusters are forming in this GMC and the total mass is about 10^5 $M_\odot$ in the central $\sim$5 pc [16]. Figure 4(*Left*) shows the overall density distribution of the cloud A of G33.92+0.11 [17, 22]. This region contains all representative features related to massive star formation, such as, hot cores, outflows, UC HII regions, etc., in addition to probable on-going accretion from the ambient gas [17]. In this region, two central clumps A1 and A2 are most prominent with masses of 100 – 300 $M_\odot$ [16, 36]. Near the A1 clump, there exists a bright UC HII region, where newly formed YSOs are embedded. Next generation stars are observed to be forming in nearby dense cores. The star formation activity seems to be enhanced by the interaction with accreting gas. The high angular resolution observations ($\sim 0.6'' - 0.8''$) made with ALMA reveal very complicated chemical gradients over the observed region.

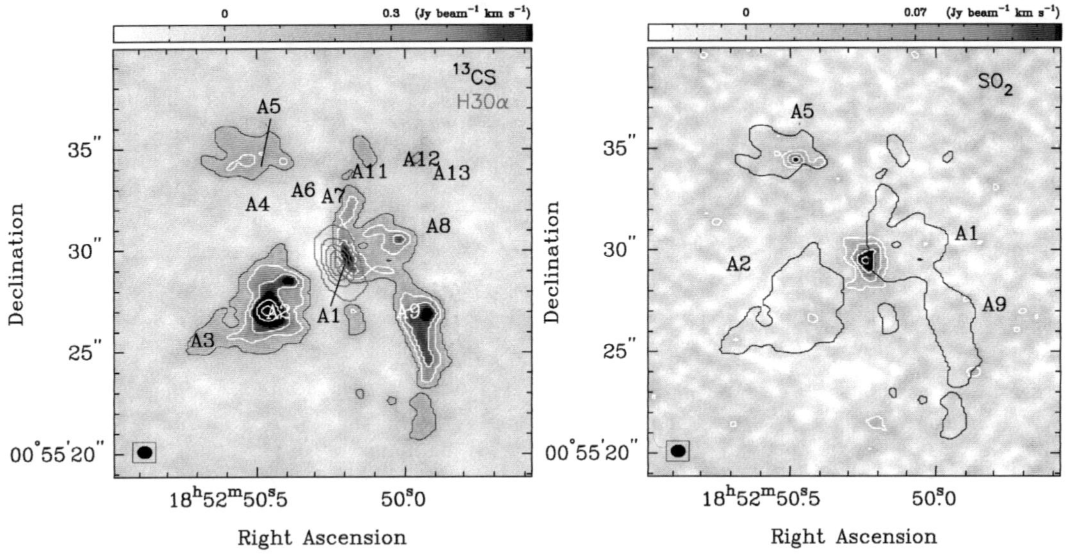

Figure 4. Integrated intensity map of the ^{13}CS $5-4$ transition (colored scale; *left*) observed toward G33.92+0.11 [22]. The outline contour in black is for the 0.15 Jy beam^{-1} km s^{-1} intensity. The magenta contour lines are for the H30α line with scale of 0.5, 1, 1.5, etc. Jy beam^{-1} km s^{-1}. The labels, A1 to A13, are the positions of the dust continuum peaks [17]. Integrated intensity map (colored image) of the SO_2 $14_{3,11}-14_{2,12}$ $1\nu_2$ transition (*right*). Its intensity scale is shown in the upper side of the panel. Contours in white is for the scale of $0.04\times[1, 2, 3, 4]$ Jy beam^{-1} km s^{-1}. Beam is shown in the bottom left corner. The contour in black is the outline of the ^{13}CS $5-4$ emission in the left panel.

Figure 4 shows the integrated intensity map of the SO_2 $14_{3,11}-14_{2,12}$ $1\nu_2$ transition. The energy of the upper level of this vibrational transition is about 250 K from the ground state. This high energy transition shows a very similar distribution with the H30α line of the UC HII region. This emission has also been detected toward the dust emission peak A5 with an intensity greater than 5σ of the observational uncertainty, where multiple SiO outflows have been detected [22]. The SO_2 emission is often observed together with UC HII regions in massive star-forming regions [19, 21]. This species seems to be formed in the high temperature gas phase from the atomic sulfur or sulfur-bearing species evaporated from the grain mantles.

Figure 5 shows the integrated intensity maps of the OCS and H_2S transitions observed toward G33.92+0.11. The OCS emission shows clumpy features around the dust continuum peaks of A1, A2, and A5, and also in the northern part of the A1 clump. These dust continuum peaks are the positions of hot cores associated with YSOs. The emission distribution of OCS is very similar to that of CH_3OH [22], which is an unambiguous first molecule evaporating in the dust mantles by heating from YSOs. As was reported previously, OCS, like CH_3OH, may have evaporated directly from the mantles.

On the other hand, H_2S shows a strong and extended emission in the A2 clump, but relatively weak at the A1 and A5 peak positions (Figure 5). Especially, the A5 peak is the position where new stars are forming and hot core features are prominent. In this position, the SiO emission also shows multiple outflows which must be driven by the embedded YSOs. H_2S shows a very similar emission distribution with ^{13}CS, but not with CH_3OH or OCS. This fact strongly suggests that H_2S is not a species evaporating directly from the mantles. The H_2S emission also shows an enhancement along the southern boundary of the A2 clump, which is probably resulted from

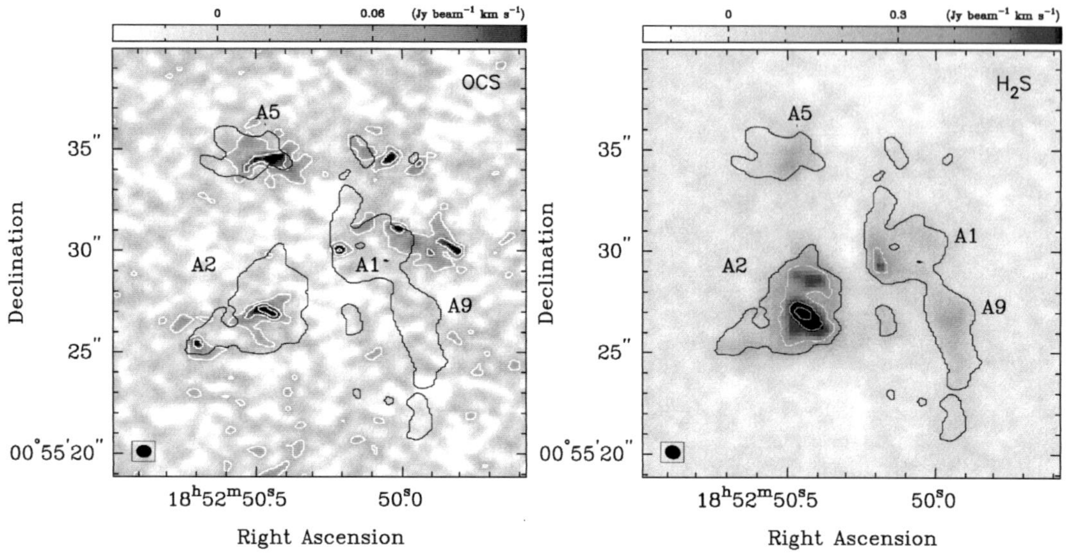

Figure 5. Integrated intensity maps of the OCS 19 − 18 (*left*) and H_2S $2_{2,0} - 2_{1,1}$ (*right*) transitions. The intensity scales are shown in the upper side of the panels. The white contour lines are for the scale of 0.05×[1, 2, 3] and 0.2×[2, 3, 4, ⋯] Jy beam^{-1} km s^{-1} for OCS and H_2S, respectively. Beams are shown in the bottom left corners. The boundary contour in black is the outline of the ^{13}CS 5 − 4 emission as included in Figure 4.

the interaction with the accreting gas to this source [17]. H_2S is thought to be formed in the warm and turbulent gas phase by using the evaporated sulfur or sulfur-bearing species from the grain mantles.

3. Summary on the chemical implications

Massive star-forming regions are observed to be extremely complicated in molecular emissions. There clearly exist small scale chemical differences within a cloud or a clump, which are probably resulted from the different evolutionary stages of massive star formation of the order of 10^4 years [21]. Recent high sensitive and high angular resolution observations are providing unprecedented amount of chemical data on the massive star-forming regions. If we can digest those huge amount of information properly, the complicated star formation process will certainly be better understood. Here we discuss on some observational results for the sulfur-bearing species observed toward two massive star-forming regions, DR21(OH) and G33.92+0.11 with SMA and ALMA, respectively.

Interstellar sulfur is largely depleted in the quiet and dense gas phase to the dust grains, and appears again in the gas phase in association with star formation activities. There exist some controversial issues on the sulfur species, such as the forms of solid state sulfur on the grain mantles or the evolution of the gas phase sulfur species before accretion. We confirm that H_2S is not a major solid sulfur species in the grain mantles. Instead, it seems to be forming in the dense turbulent gas phase from the evaporated sulfur-bearing species. On the other hand, OCS is believed to evaporate directly from the grain mantles, as evidenced by a good locational correlation with CH_3OH. OCS seems to be one of major forms of solid sulfur in the grain mantles, as suggested before. In addition, the bipolar-like outflow of the H_2CS emission observed in DR21(OH) suggests that H_2CS might be another solid sulfur-bearing species on the grain mantles. Depending on the chemical environment, the competition between hydrogenation

Journal of Physics: Conference Series **728** (2016) 052007 doi:10.1088/1742-6596/728/5/052007

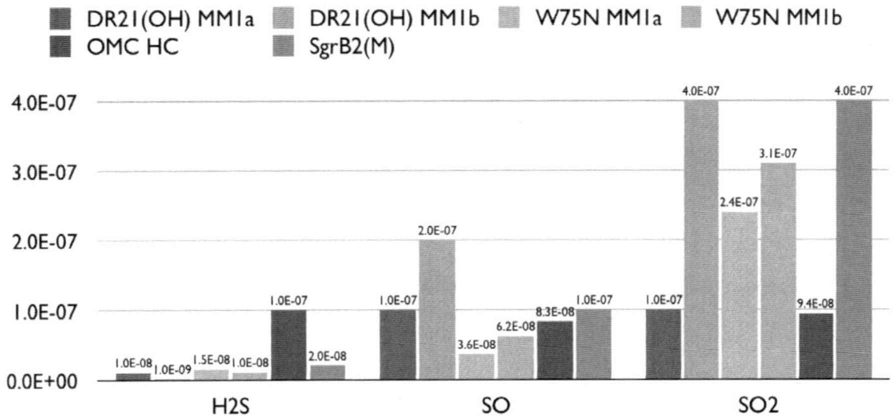

Figure 6. Fractional abundances of H_2S, SO, and SO_2 relative to the total H_2 abundances in selected massive star-forming regions [6, 19, 21, 31, 32].

and oxidization on the grain surface may result the formation of solid H_2CS or OCS on the grain surface [20, 25, 33].

The evaporated atomic sulfur and sulfur-bearing species from the dust grains will soon be transformed to other forms in the warm and dense gas phase. SO and SO_2 are the molecules frequently found in association with star formation activities, which show, in general, a good correlation with ionized gas, such as the UC HII regions. These species seem to be formed in the high temperature turbulent gas in a later stage of star formation after the hot core phase.

Figure 6 shows the abundance comparisons of some sulfur-bearing species in selected sources. Overall fractional abundances of H_2S relative to H_2 appear to be fairly consistent except toward OMC Hot Core, which is a nearby star-forming region consisted of various different emission sources. For SO and SO_2, MM1a is a little bit younger than MM1b in both sources DR21(OH) and W75N. MM1a contains hot cores and MM1b contains UC HII regions, which may result the relative abundances of SO and SO_2 enhanced in MM1b compared to MM1a. Giving consideration to the uncertainties involved in deriving the column densities, especially of total H_2 abundances, the fractional abundances of these sulfur-bearing species are observed to be relatively consistent in these sources. Although the high angular resolution observations show complicated locational variations even in a clump within a source, these sources may be experiencing very similar chemical and consequently physical evolutionary stages on the whole. From the careful investigation on the chemical evolution associated with the source structure, we may find a consistent and coherent evolutionary pattern of the massive star formation process in general.

References

[1] Araya E D, Kurtz S, Hofner P and Linz H 2009 *Astrophys. J.* **698** 1321
[2] Argon A L, Reid M J and Menten K M 2000 *Astrophys. J. Suppl. Ser.* **129** 159
[3] Bachiller R and Pérez Gutiérrez M 1997 *Astrophys. J.* **487** L93
[4] Batrla W and Menten K M 1988 *Astrophys. J.* **329** L117
[5] Caselli P, Hasegawa T I and Herbst E 1994 *Astrophys. J.* **421** 206
[6] Charnley S B 1997 *Astrophys. J.* **481** 396
[7] Codella C, Viti S, Williams D A and Bachiller R 2006 *Astrophys. J.* **644** L41
[8] Fish V L, Reid M J, Wilner D J and Churchwell E 2003 *Astrophys. J.* **587** 701
[9] Fish V L, Muehlbrad T C, Pratap P, Sjouwerman L O, Strelnitski V, Pihlström Y M and Bourke T L 2011 *Astrophys. J.* **729** 14

[10] Genzel R and Downes D 1977 *Astron. Astrophys. Suppl. Series* **30** 145
[11] Gibb E L *et al* 2000 *Astrophys. J.* **536** 347
[12] Hatchell J, Thompson M A, Millar T J and Macdonald G H 1998 *Astron. Astrophys.* **338** 713
[13] Kogan L and Slysh V 1998 *Astrophys. J.* **497** 800
[14] Kurtz S, Hofner P and Álvarez C V 2004 *Astrophys. J. Suppl. Series* **155** 149
[15] Langer W D, van Dishoeck E F, Bergin E A, Blake G A, Tielens A G G M, Velusamy T and Whittet D C B 2000 *Protostars and Planets IV* 29
[16] Liu H B, Jiménes-Serra I, Ho P T P, Chen H-R, Zhang Q and Li Z-Y 2012 *Astrophys. J.* **756** 10
[17] Liu H B, Galvań-Madrid R, Jimeńez-Serra I, Román-Zúñiga C, Zhang Q, Li Z and Chen H-R 2015 *Astrophys. J.* **804** 37
[18] Mangum J C, Wootten A and Mundy L G 1991 *Astrophys. J.* **378** 576
[19] Minh Y C, Su Y-N, Chen H-R, Liu S-Y, Yan C-H and Kim S-J 2010 *Astrophys. J.* **723** 1231
[20] Minh Y C, Liu S-Y, Chen H-R and Su Y-N 2011 *Astrophys. J.* **737** L25
[21] Minh Y C, Chen H-R, Su Y-N and Liu S-Y 2012 *J. Korean Astron. Soc.* **45** 157
[22] Minh Y C, Liu H B and Galvań-Madrid R 2016 *submitted*
[23] Motte F, Bontemps S, Schilke P, Schneider N, Menten K M and Broguière D 2007 *Astron. Astrophys.* **476** 1243
[24] Norris R P, Booth R S, Diamond P J and Porter N D 1982 *Mon. Not. R. Astron. Soc.* **201** 191
[25] Palumbo M E, Geballe T R and Tielens A G G M 1997 *Astrophys. J.* **479** 839
[26] Plambeck R L and Menten K M 1990 *Astrophys. J.* **364** 555
[27] Pineau des Forêts G, Roueff E, Schilke P and Flower D R 1993 *Mon. Not. R. Astron. Soc.* **262** 915
[28] Reipurth B and Yan C-H 2008 *Handbook of Star Forming Regions: Volume I, The Northern Sky* ed Reipurth B (San Francisco, CA: ASP) **4** 869
[29] Rygl K L J, Brunthaler A, Sanna A, Menten K M, Reid M J, van Langevelde H J, Honma M, Torstensson K J E and Fujisawa K 2012 *Astron. Astrophys.* **539** 79
[30] Schneider N, Bontemps S, Simon R, Jakob H, Motte F, Miller M, Kramer C and Stutzki J 2006 *Astron. Astrophys.* **458** 855
[31] Sutton E C, Jaminet P A, Danchi W C and Blake G A 1991 *Astrophys. J. Suppl. Series* **77** 255
[32] Sutton E C, Peng R, Danchi W C, Jaminet P A, Sandell G and Russell A P G 1995 *Astrophys. J. Suppl. Series* **97** 455
[33] Tielens A G G M and Hagen W 1982 *Astron. Astrophys.* **114** 245
[34] van der Tak F F S, Boonman A M S, Braakman R and van Dishoeck E F 2003 *Astron. Astrophys.* **412** 133
[35] Wakelam V, Caselli P, Ceccarelli C, Herbst E and Castets A 2004 *Astron. Astrophys.* **422** 159
[36] Watt S and Mundy L G 1999 *Astrophys. J. Suppl. Ser.* **125** 143
[37] Woody D P, Scott S L, Scoville N Z, Mundy L G, Sargent A I, Padin S, Tinney C G and Wilson C D 1989 *Astrophys. J.* **337** L41
[38] Zapata L A, Loinard L, Su Y-N, Rodríguez L F, Menten K M, Patel N and Galván-Madrid R 2012 *Astrophys. J.* **744** 86
[39] Zinnecker H and Yorke H W 2007 *Annu. Rev. Astron. Astrophys.* **45** 481

Journal of Physics: Conference Series **728** (2016) 062001 doi:10.1088/1742-6596/728/6/062001

Organic dust in galaxies

Takashi Onaka

Department of Astronomy, Graduate School of Science, The University of Tokyo, 7-3-1 Hongo, Bunkyo-ku, Tokyo 113-0033, Japan

E-mail: onaka@astron.s.u-tokyo.ac.jp

Abstract. Recent space infrared telescopes, *Infrared Space Observatory*, *Spitzer Space Telescope*, and *AKARI* have made significant progress in our understanding of organic dust in the Universe. In this review, we discuss recent observations with these space telescopes of the unidentified infrared emission (UIE) features in the near to mid-infrared, which come from very small organic dust, and the absorption features from 3 to 7 μm, which characterize large organic dust. They provide us with a new view of organic dust in galaxies. We also briefly discuss latest *AKARI* observations of H_2O and CO_2 ices in 2.5–5 μm in the Large Magellanic Cloud in comparison with observations in our Galaxy, which suggests the importance of dust surface chemistry in the formation of organic matters in the Universe.

1. Introduction

The major constituents of interstellar dust are thought to be silicates and carbonaceous grains [1]. However, their exact structures and compositions are still not completely understood. In particular, carbon atom can have various hybridized orbitals and constitute various types of organic dust. The advent of recent space infrared telescopes, such as *Infrared Space Observatory (ISO)*, *Spitzer Space Telescope*, and *AKARI* has made signifiant progress in our understanding of organic dust in the Universe. Here an overview of recent observations with space infrared space telescopes is presented with particular focuses on on emission features arising from small organic dust [2], absorption from large organic dust [3], and absorption features of ice species [4] in galaxies. Other carbon-bearing dust, such as C_{60} or related materials, are discussed by others in this conference and are not covered here.

2. Emission features

A family of prominent emission features are ubiquitously seen in the near- to mid-infrared in the general interstellar medium (ISM) and normal spiral galaxies [5, 6, 7, 8, 9, 10, 11]. The major features are observed at 3.3, 3.4–3.5, 6.2, 7.7, 8.6, and 11.3 μm associated with minor features. These features have been attributed to polycyclic aromatic hydrocarbons (PAHs) [12, 13, 14] or materials containing PAH components, such as QCC [15]. Appearance of the emission features is stable in the various environments, leading to an idea that there may be a small number of dominating PAH species, called grand PAH, which is yet to be identified, rather than a number of various PAH species [16]. Recently Kwok and Zhang [17, 18] proposed mixed aromatic-aliphatic organic nanoparticles (MAON) as a more likely candidate of the carriers because the features at 3.4–3.5 μm clearly suggest the presence of aliphatic CH bonds. The exact nature of the feature carriers is thus still under debate and it is called them the unidentified infrared emission (UIE) features in the following.

The UIE features are seen even in distant galaxies. The highest redshift at which the UIE feature is detected is $z = 4.055$ to date [19]. The intensity of the UIE feature shows a good correlation with the far-infrared intensity and thus the star-formation rate [19]. Since they are not seen in H II regions [20], the UIE features are suggested to be a good tracer for star-formation of B-type stars rather than massive star-formation [21]. The equivalent width of the UIE feature is also a good measure for the presence of AGNs [22]. Therefore, they are a useful probe to estimate the nature and star-forming activities of distant galaxies if we understand their properties correctly. The UIE features are absent in low-metallicity dwarf galaxies [11, 23, 24]. This can be interpreted in terms of efficient destruction and/or inefficient production of the feature carriers in these environments [25], delayed formation of carbonaceous dust [26], or carrier formation from fragmentation of carbonaceous dust [27]. It is not clear at present which process plays the most dominant role and the formation process of the carriers in galaxies is still not clearly understood (see § 2.2).

2.1. Band ratio diagnosis

The UIE features have several emission bands arising from different vibrational modes and thus their relative band ratios can be used to estimate the physical conditions of the band carriers [2, 28]. However, systematic variations in the band ratios have so far been seen only in a few cases. One of the clear cases is small ratios of the 7.7 μm to the 11.3 μm bands seen in elliptical galaxies (Fig. 1) [11, 29, 30, 31]. Similar small ratios of a lesser degree are also seen in the halo of disk galaxies [32, 33], which may be attributed to the dominance of neutral band carriers because the 7.7 μm band is thought to come predominantly from ionized carriers [28].

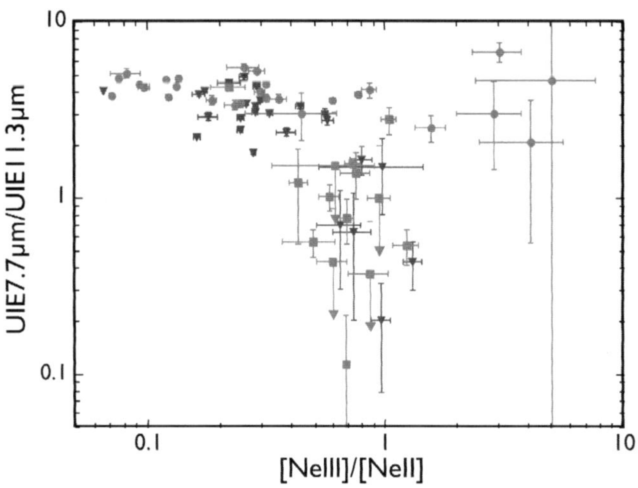

Figure 1. Intensity ratio of the UIE 7.7 to 11.3 μm bands against the line ratio of [Ne III]15.6 μm to [Ne II]12.8 μm [31]. The red circles indicate SINGS galaxies with H II -dominated nuclei and the blue triangles show those with low-luminosity AGNs [11]. Elliptical galaxies [29, 30] are indicated by the green squares. Note that galaxies with high [Ne III]/[Ne II] are low-metallicity dwarf galaxies, which show band ratios similar to normal star-forming galaxies.

The most clear systematic variation in the band ratios is found in AKARI/IRC observations of star-forming regions in the Large Magellanic Cloud (LMC) [34]. Figure 2 shows plots of several band ratios against the infrared color. The abscissa color indicates the strength of the star-forming activity and thus the intensity of the incident radiation field. The band ratios show correlations with the color for PDR targets (blue circles), which can be successfully interpreted in terms of an increase of the ionization fraction and the excitation of band carriers (see Fig. 3). Thus it provides the first clear evidence that the central exciting source becomes hotter with the increase of the intensity. On the other hand, H II region targets (red triangles), which show emission lines from ionized gas, do not follow the sequence. The variation of the band ratios

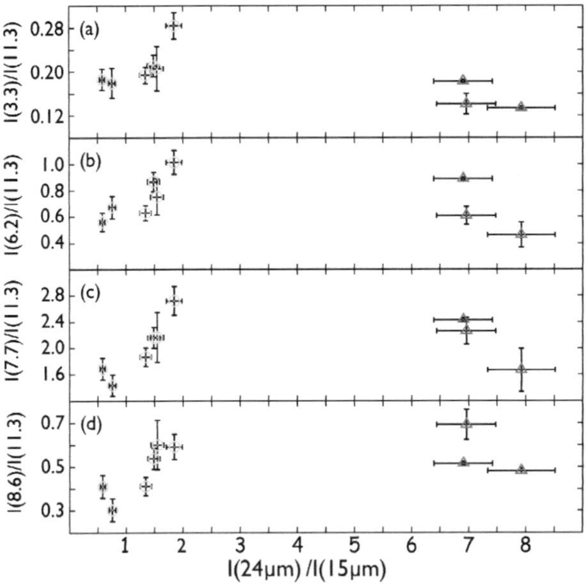

Figure 2. UIE band ratios versus the infrared color of 24 to 11 μm for the LMC star-forming regions [34]. (a) 3.3 μm band, (b) 6.2 μm band, (c) 7.7 μm band, and (d) 8.6 μm band to the 11.3 μm band intensities. The blue circles indicate the PDR targets, while the red triangles show the H II region targets. The infrared color indicates the strength of star-formation activity. The band ratios of the PDR targets show systematic variations with the infrared color, while those in H II region targets do not show a clear correlation.

can be seen more clearly in the two band-ratio diagram (Fig. 3), in which grids of simple model calculations are also overlaid [34]. As described above, the trend of the PDR targets can be accounted for by an increase of the effective temperature of the exciting source (T_*) and the ionization fraction (f_i). The different behavior of the H II region targets can be explained by a decrease of small-sized band carriers (an increase of the number of carbon atoms in the smallest carriers, $n_{\rm c,min}$), suggesting destruction of small carriers in ionized regions. This study strongly demonstrates the importance of the 3.3 μm band for the study of processing of the carriers and the UIE band diagnostics.

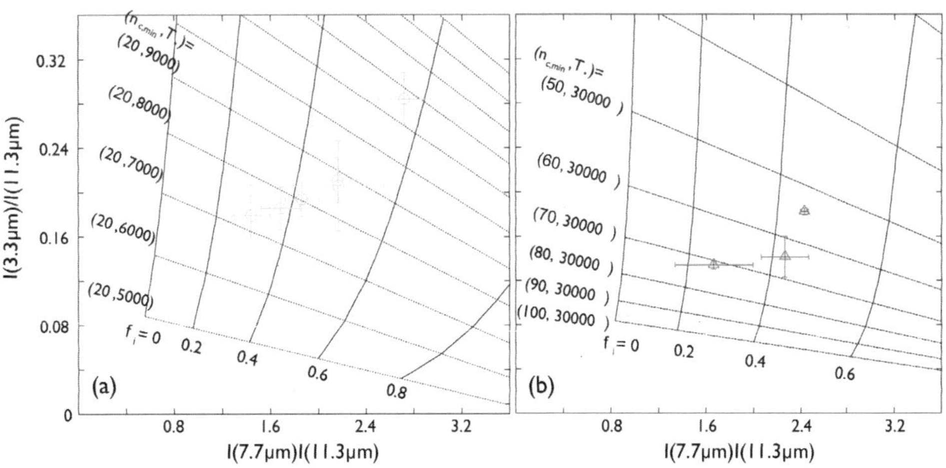

Figure 3. Two band-ratio diagram of the 3.3 to the 11.3 μm band ratio versus the 7.7 to the 11.3 μm band ratio for the LMC star-forming regions [34]. (a) PDR targets (blue circles) and (b) H II region targets (red triangles). In (a) a model grid of the ioniziation fraction f_i and the effective temperature T_* of the exciting sources is also plotted. The number of carbon atoms in the smallest band carriers $n_{\rm c,min}$ is fixed as 20. In (b), a model grid of f_i and $n_{\rm c,min}$ is plotted. The effective temperature T_* is fixed as 30000K in the grid of (b).

2.2. Formation of UIE band carriers in galaxies
Formation sites of the UIE band carriers in galaxies are not yet clearly identified. They may be formed either in stellar outflows [36, 37, 38] or molecular clouds [39, 40, 41]. Observations of *AKARI*/IRC also suggest that the band carriers may be formed by fragmentation of large carbonaceous grains in a supernova remnant [42] and a filament produced by a star-burst activity in a dwarf galaxy [43]. Figure 4 shows another interesting case. It shows the *AKARI*/IRC two-color artificial image of the interacting galaxy NGC 2782 [44]. This galaxy shows a very long tidal tail in the west with an extended structure in the east in the H I 21 cm (red contours), both of which are thought to have resulted from collision with a minor galaxy that has passed through the major galaxy [45]. Interestingly, a similar structure to H I in the east is seen in the 7 μm image of *AKARI* /IRC (Fig. 4b). Since the IRC 7 μm band efficiently probes the UIE 6.2 and 7.7 μm bands [43], the *AKARI* observations suggest that the UIE band carriers were formed in the collision event, possibly from fragmentation of larger carbonaceous grains. In fact, a simple SED analysis of this structure suggests a decrease of nanometer-sized dust [44]. These observations strongly suggest that fragmentation of large carbonaceous grains is one of the major processes to produce the UIE band carriers.

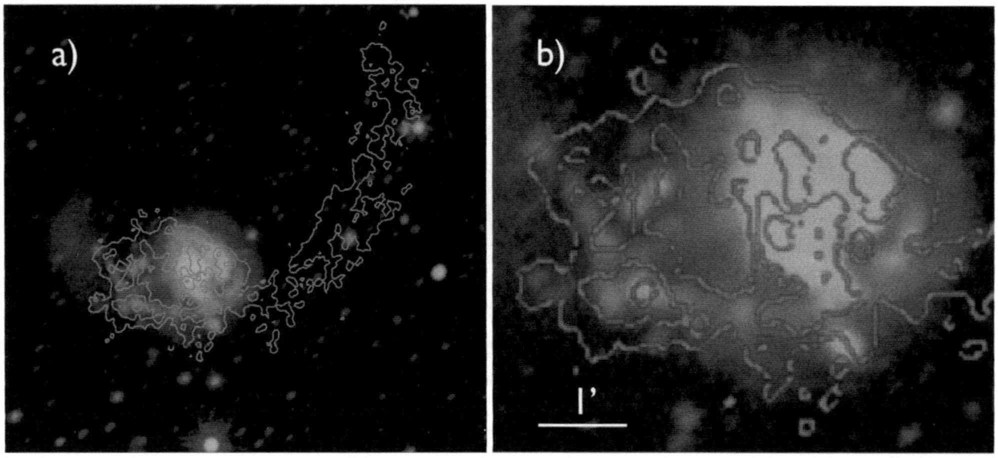

Figure 4. *AKARI*/IRC artificial two-color image of NGC 2782 [44]. The blue is *AKARI*/IRC 3.2 μm, the green is *AKARI*/IRC 7 μm, and the red contours indicate H I 21 cm intensity [45]. Similar extended structures are clearly seen both in H II and at 7 μm in the enlarged figure (b), suggesting that the UIE carriers may have been produced in the gas stripping process during the collision event.

3. Absorption features
CH stretching modes of carbonaceous dust have been observed at $\sim$ 3.4, 6.8, and 7.2 μm and CC vibrations at $\sim$ 6.2 μm in absorption in our Galaxy [3, 46]. The 3.4 μm features consist of several components, which correspond to sp^3 CH_2 and CH_3 as well as sp^2 CH vibrations. These features arise from large carbonaceous grains. Since they require a high column density to be observed, they are not easily seen in normal spiral galaxies, but are detected in dust torii around AGNs [47, 48]. Figure 5 shows the infrared spectrum of IRAS 08572+3915 ($z = 0.05835$) taken with *AKARI*/IRC with a fit with four Gaussian components, which is in good agreement with the results taken with the United Kingdom Infrared Telescope (UKIRT) [47].

These recent analyses of the absorption features both in our Galaxy and an AGN suggest that large carbonaceous grains are dominated by sp^2 carbon, but that most of hydrogen atoms

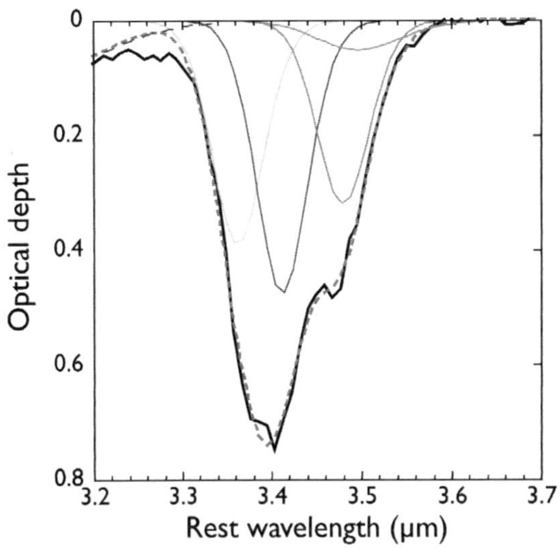

Figure 5. Optical depth of the 3 μm region absorption of IRAS 08572+3915 derived from near-infrared spectrum taken with *AKARI*/IRC plotted against the rest wavelength. The black solid line indicates the observed spectrum and the red dashed line shows the fitted spectrum. The green, blue, purple, and brown lines are 4 Gaussian components in the fit, corresponding to sp^3 CH_2 and CH_3 stretching vibrational modes. The dashed dark green line seen in the shorter wavelength side indicates absorption of H_2O ice. Due to the low resolution of the IRC spectrum, the decomposition of the longest two components has ambiguity, but it is basically in good agreement with the results of [47].

are associated with sp^3 C. The suggested structure of large carbonaceous grains can be shown in the ternary diagram of amorphous carbon (Fig. 6) [49]. Large carbonaceous grains both in our Galaxy and an AGN have similar structures within uncertainties. Note that the results of [47] have been revised based on the new absorption coefficient [3]. The results suggest that the backbone structure of large carbonaceous grains is hydrogen-poor and sp^2 carbon-dominant, while most hydrogen is attached in sp^3 carbon. Based on existing amorphous carbons, these results lead to a new view of organic dust in the ISM [3, 47]. Interstellar organic dust may consist of the hydrogenated amorphous carbon mantle on top of the pure aromatic carbon core. It produces PAHs or PAH-containing small dust by shattering [3]. This view is also consistent with the evolution of latest model of organic dust that suggests aromatization of organic dust in the ISM [50]. Latest observations with *AKARI* indicate a decrease of aliphatic to aromatic CH bonds with the radiation field intensity, supporting this new view of evolution of the cosmic organic dust [51].

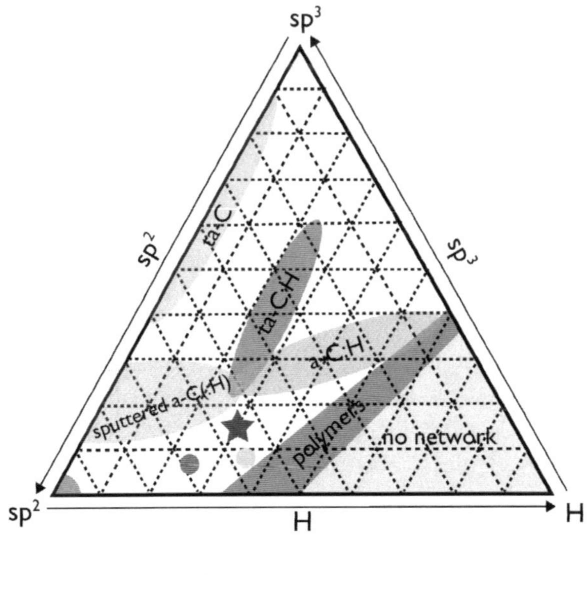

Figure 6. Ternary diagram of hydrocarbon materials, in which various forms of hydrocarbons are plotted for reference [49]. "ta-C" denotes tetrahedral amorphous carbon, while "ta-C:H" indicates hydrogenated tetrahedral amorphous carbon and "a-C:H" denotes hydrogenated amorphous carbon. The red and green circles indicate the hydrocarbons in our Galaxy with different absorption strengths [3]. The blue star symbol indicates the hydrocarbons in the AGN IRAS 08572+3915 [47] estimated with the revised absorption strengths of [3].

4. Ice absorption features

ISO, *Spitzer* and *AKARI* observations towards young stellar objects (YSOs) in our Galaxy show that H_2O ice is the most abundant ice species and that the column density of the second abundant CO_2 ice is very well correlated with that of H_2O ice [52, 53, 54]. The good correlation strongly suggests concurrent formation of both ice species on the grain surface [55, 56]. *AKARI*/IRC observations of M82 suggest a correlation between the ratio of the column densities of both ice species and the ultraviolet radiation intensity [57], while a similar correlation between the two ice species is not seen in a larger sample of galaxies observed with *AKARI* [58]. It may be due to various components on the line of sight that erase the intrinsic correlation.

AKARI observations indicate that CO_2 ice is clearly more abundant relative to H_2O ice towards massive YSOs in the LMC than those in our Galaxy [59, 60] (Fig. 7). It can be accounted for by the high dust temperature in the LMC [62], which enhances the mobility of CO and thus the formation of CO_2 ice on the dust surface [60]. Alternatively it can also be explained by the reduced H_2O ice abundance relative to CO_2 ice due to the strong ultraviolet radiation and the low dust-to-gas ratio at low-metallicity in the LMC [61]. Recent ground-based near-infrared observations towards these YSOs find a paucity of CH_3OH ice relative to H_2O ice compared to our Galaxy, which can also be interpreted consistently in terms of the high dust temperature in the LMC due to the rapid diffusion or evaporation of hydrogen atoms that reduces hydrogenation of CO to form CH_3OH ice [63]. CH_3OH ice is a key to the formation of complex organic molecules [64]. Hence, the paucity of CH_3OH ice predicts a difficulty of formation of complex organic matter in the LMC, suggesting an impact of dust surface chemistry on the formation of organic matter in the Universe.

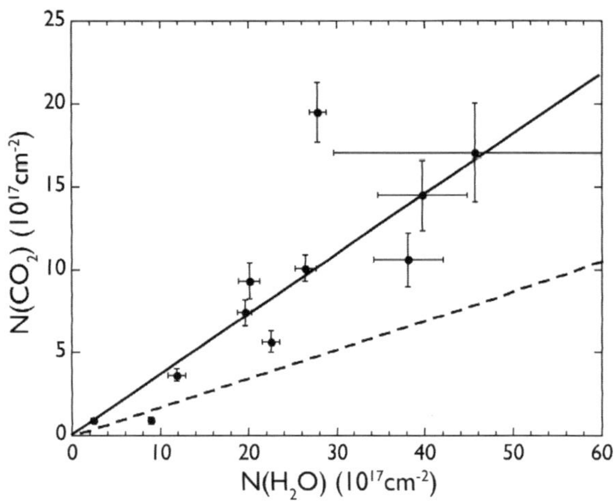

Figure 7. Correlation of column densities between CO_2 and H_2O ices towards massive YSOs in the LMC [59, 60]. The solid line indicates the correlation with the slope of 0.34 for the LMC YSOs, while the dashed line shows the correlation slope (0.17) for Galactic massive YSOs [52, 53]. The plot clearly shows the overabundance of CO_2 ice relative to H_2O ice in massive YSOs in the LMC compared to our Galaxy.

5. Summary

Recent observations of space infrared telescopes, *ISO*,, *Spitzer*, and *AKARI*, have made significant progress in our understanding of organic dust in the Universe. The UIE features give us information on small organic dust. Although the exact nature of the feature carriers remains unsettled, recent studies steadily unveil their properties. Observations of the LMC suggest possible processing of small organic dust in ionized regions and demonstrate the importance of the 3.3 μm band emission for the band diagnostics. Several observations of *AKARI*/IRC also suggest that fragmentation of large carbonaceous grains is one of the major processed for the formation of the UIE band carriers. Observations of absorption features in the near- to mid-infrared both for our Galaxy and AGNs suggest that large organic dust is sp^2 carbon-dominant and hydrogen-poor, and most hydrogen is attached to sp^3 carbon. These results lead us to a new view of organic dust and its evolution in the Universe. Observations of ice species in the low-metallicity LMC suggest a distinct difference from the ices in our Galaxy, indicating the importance of the environmental effect on the surface chemistry of dust grains, which affects the formation of complex organic molecules.

Acknowledgments

This work is based on observations with *AKARI*, a JAXA project with the participation of ESA. The author greatly thank all the members of the AKARI project and the members of the Interstellar and Nearby Galaxy team for their help and continuous encouragement. In particular he is grateful to Istuki Sakon, Ryou Ohsawa, Tamami Mori, Tomohiko Nakamura, Fumihiko Usui, Ronin Wu, Takashi Shimonishi, Yoko Okada, Daisuke Ishihara, and Hidehilro Kaneda for their contribution. This work is supported by a Grant-in-Aid for Science Research from the Japan Society for the Promotion of Science (no. 23244021).

References

[1] Draine B T 2003 *Ann. Rev. Astron. Astrophys.* **41** 241–289
[2] Onaka T 2013 *Proc. Int. Astron. Union* (S292) **8** 259–265
[3] Chiar J E, Tielens A G G M, Adamson A J and Ricca A 2013 *Astrophys. J.* **770** 78
[4] Gibb E L, Whittet D C B, Boogert A C A and Tielens A G G M 2004 *Astrophys. J. Suppl. Series* **151** 35–73
[5] Giard M, Serra G, Caux E, Pajot E and Lamarre J M 1988 *Astron. Astrophys.* **201** L1–L4
[6] Tanaka M, Matsumoto T, Murakami H, Kawada M, Noda M and Matsuura S 1996 *Publ. Astron. Soc. Japan* **48** L53–L57
[7] Onaka T, Yamamura I, Tanabé T, Roellig T L and Yuen L 1996 *Publ. Astron. Soc. Japan* **48** L59–L63
[8] Mattila K, Lemke D, Haikala L K, Laureijs R J, Leger A, Lehtinen K, Leiner C and Mezger P G 1996 *Astron. Astrophys.* **315** L353–L356
[9] Tsumura K, Matsumoto T, Matsuura S, Sakon I, Tanaka M and Wada T 2013 *Publ. Astron. Soc. Japan* **65** 120
[10] Helou G, Lu N Y, Werner M W, Malhotra S and Sibermann N 2000 *Astrophys. J. Lett.* **532** L21–L24
[11] Smith J D T, Draine B T, Dale D A, Moustakas J, Kennicutt Jr R C, Helou G, Armus L, Roussel H, Sheth K and Bendo G J 2007 *Astrophys. J.* **656** 770–791
[12] Leger A and Puget J L 1984 *Astron. Astrophys.* **137** L5–L8
[13] Allamandola L J, Tielens A G G M and Barker J R 1985 *Astrophys. J. Lett.* **290** L25–L28
[14] Tielens A G G M 2008 *Ann. Rev. Astron. Astrophys.* **46** 289–337
[15] Sakata A, Wada S, Tanabé T and Onaka T 1984 *Astrophys. J. Lett.* **287** L51–L54
[16] Berné O, Joblin C, Mulas G, Tielens A G G M and Goicoechea J R 2009 *SPICA joint European/Japanase Workshop* (Electronic online proceedings) ed A M Heras, B M Swinyard, K G Isaak and J R Goicoechea (Les Ulis and London: EDP Sciences) 03005
[17] Kwok S and Zhang Y 2011 *Nature* **479** 80–83
[18] Kwok S and Zhang Y 2013 *Astrophys. J.* **771** 5
[19] Riechers D *et al* 2014 *Astrophys. J.* **786** 31
[20] Onaka T, Mori T I and Okada Y 2015 *Planet. Sp. Sci.* **116** 84–91
[21] Peeters E, Spoon H W W and Tielens A G G M 2004 *Astrophys. J.* **613** 986–1003
[22] Armus L *et al* 2007 *Astrophys. J.* **656** 148–167
[23] Wu Y, Charmandaris V, Hao L, Brandl B R, Bernard-Salas J, Sppn H W W and Houck J R 2006 *Astrophys. J.* **639** 157–172
[24] Wu R, Hogg D and Moustakas J 2011 *Astrophys. J.* **730** 111
[25] O'Halloran B, Satyapal S and Dudik R P 2006 *Astrophys. J.* **641** 795–800
[26] Galliano F, Dwek E and Chanial P 2008 *Astrophys. J.* **672** 214–243
[27] Seok J Y, Hirashita H and Asano R S 2014 *Mon. Not. R. Astron. Soc.* **439** 2186–2196
[28] Draine B T and Li A 2001 *Astrophys. J.* **551** 807–827
[29] Kaneda H, Onaka T and Sakon I 2005 *Astrophys. J. Lett.* **632** L83–L86
[30] Kaneda H, Onaka T, Sakon I, Kitayama T, Okada Y and Suzuki T 2008 *Asrtrophys. J.* **684** 270–281
[31] Onaka T, Matsumoto H, Sakon I and Kaneda H 2008 *Proc. Int. Astron. Union* (S251) **4** 229–236
[32] Irwin J A and Madden S C 2006 *Astron. Astrophys.* **445** 123–141
[33] Galliano F, Madden S C, Tielens A G G M, Peeters E and Jones A P 2008 *Astrophys. J.* **679** 310–345

[34] Mori T I, Sakon I, Onaka T, Kaneda H, Umehata H and Ohsawa R 2012 *Astrophys. J.* **744** 68
[35] Onaka T *et al* 2007 *Pulb. Astron. Soc. Japan* **59** S401–S410
[36] Latter W B 1991 *Astrophys. J.* **377** 187–191
[37] Cherchneff I, Barker J R and Tielens A G G M 1992 *Astrophys. J.* **401** 269–287
[38] Paradis D *et al* 2009 *Astron. J.* **138** 196–209
[39] Herbst E 1991 *Astrophys. J.* **366** 133–140
[40] Greenberg J M, Gillette J S, Munñoz Caro G M, Mahajan T B, Zare R N, Li A, Schutte W A, de Groot M and Mendoza-Gómez C 2000 *Astrophys. J. Lett.* **531** L71–L73
[41] Sandstrom K M, Bolatto A D, Draine B T, Bot C and Stanimirović S 2010 *Astrophys. J.* **715** 701–723
[42] Seok J Y, Koo B-C and Onaka T 2012 *Astrophys. J.* **744** 160
[43] Onaka T, Matsumoto H, Sakon I and Kaneda H 2010 *Astron. Astrophys.* **514** A15
[44] Nakamura T, Onaka T, Kaneda H, Sakon I, Ohsawa R and Mori T 2013 *Proceedings of the Life Cycle of Dust in the Universe: Observations, Theory, and Laboratory Experiments PoS (LCDU2013)* ed A Andersen, M Baes, C Kemper and D Watson (Taipei: Proceedings of Science) 080
[45] Smith B J 1994 *Astron. J.* **107** 1695–1712
[46] Pendleton Y J, Sandford S A, Allamandola L J, Tielens A G G M and Sellgren K 1994 *Astrophys. J.* **437** 683–696
[47] Dartois E, Geballe T R, Pino T, Cao A-T, Jones A, Deboffle D, Guerrini V, Bréchignac Ph and D'Hendecourt L 2007 *Astron. Astrophys.* **463** 635–640
[48] Dartois E and Muñoz-Caro G M 2007 *Astron. Astrophys.* **476** 1235–1242
[49] Robertson J 2002 *Material Science Engineering R* **37** 129–281
[50] Jones A P, Fanciullo L, Köhler M, Verstraete K, Guillet V, Bocchio M and Ysard N 2013 *Astron. Astrophys.* **558** A62
[51] Mori T I, Onaka T, Sakon I, Ishiihara D, Shimonishi T, Ohsawa R and Bell A C 2014 *Astrophys. J.* **792** 80
[52] Nummelin A, Whittet, D C B, Gibb E L, Gerakines, P A and Chiar J E 2001 *Astrophys. J.* **558** 185–193
[53] Pontoppidan K M, Boogert A C A, Fraser H, van Dishoeck E F, Blake G A, Lahuis F, Öberg K, Evans II N J and Salyk C 2008 *Astrophys. J.* **678** 1005–1031
[54] Noble J A, Fraser H J, Aikawa Y, Pontoppidan K M and Sakon I 2013 *Astrophys. J.* **775** 85
[55] Oba Y, Watanabe N, Kouchi A, Hama T and Pirronello V 2010 *Astrophys. J. Lett.* **712** L174–L178
[56] Ioppolo S, van Boheemen Y, Cuppen H M, van Dishoeck E F and Linnartz H 2011 *Mon. Not. R. Astron. Soc.* **413** 2281–2287
[57] Yamagishi M, Kaneda H, Ishihara D, Oyabu S, Onaka T, Shimonishi T, Suzuki T and Minh Y C 2013 *Astrophys. J. Lctt.* **773** L37
[58] Yamagishi M, Kaneda H, Ishihara D, Oyabu S, Onaka T, Shimonishi T and Suzuki T 2015 *Astrophys. J.* **807** 29
[59] Shimonishi T, Onaka T, Kato D, Sakon I, Ita Y, Kawamura A and Kaneda H 2008 *Astrophys. J. Lett.* **686** L99–L102
[60] Shimonishi T, Onaka T, Kato D, Sakon I, Ita Y, Kawamura A and Kaneda H 2010 *Astron. Astrophys.* **514** A12
[61] Oliveira J M, van Loon J Th, Sloan G C, Indebetouw R, Kemper F, Tielens A G G M, Simon J D, Woods P and Meixner M 2011 *Mon. Not. R. Astron. Soc.* **411** L36–L40
[62] Sakon I *et al* 2006 *Astrophys. J.* **651** 174–189
[63] Shimonishi T, Dartois E, Onaka T and Boulanger F 2016 *Astron. Astrophys.* **585** A107
[64] Chang Q and Herbst E 2014 *Astrophys. J.* **787** 135

Laboratory study of carbonaceous dust and molecules of astrochemical interest

F Cataldo[1,2], D A Garcia-Hernandez[3,4], A Manchado[3,4,5] and S Kwok[6]

[1] Actinium Chemical Research srl, Via Casilina 1626A, 00133 Rome, Italy
[2] INAF-Osservatorio Astrofisico di Catania, Via S. Sofia 78, 95123 Catania, Italy
[3] Instituto de Astrofísica de Canarias, Via Lactea s/n, 38205 La Laguna, Spain
[4] Departamento de Astrofísica, Universidad de La Laguna (ULL), 38206 La Laguna, Spain
[5] Consejo Superior de Investigaciones Cientificas, Madrid, Spain
[6] Laboratory for Space Research, Faculty of Science, The University of Hong Kong, Hong Kong, China

E-mail: [1] `franco.cataldo@fastwebnet.it`

Abstract. In this paper are reviewed some research works dedicated to the study of carbonaceous dust and molecules of astrochemical interest. First of all it is discussed the carbon arc through which it is possible to produce carbon soot and fullerenes under helium but also many other different products just changing the arcing conditions. For example, when the carbon arc is struck in an hydrocarbon solvent it is possible to produce and trap polyynes in the solvent. Monocyanopolyynes and dicyanopolyynes can be produced as well by selecting the appropriate conditions. Amorphous carbon soot or partially graphitized carbon black can be produced with the carbon arc. Fullerenes were found in space thanks to the reference infrared spectra and the absorption cross sections which were determined in laboratory. Fullerenes are readily reactive with hydrogen yielding fulleranes the hydrogenated fullerenes. Furthermore fullerenes react with PAHs and with iron carbonyl yielding adducts. All these fullerene derivatives were synthesized and their reference spectra recorded in laboratory. It was proposed that petroleum fractions can be used as model substrates in the explanation of the carriers of the AIB (Aromatic Infrared Bands) observed in protoplanetary and planetary nebulae and the UIE (Unidentified Infrared Bands) found in the interstellar medium.

1. Introduction
In recent years we were fascinated and involved in the research on topics of astrochemical interest which developed on laboratory scale. In most cases the ambition was to reproduce as closer as possible the astrophysical conditions were carbonaceous dust and molecules are formed and detected for instance by radioastronomy or by infrared spectroscopy.

2. Carbon vapor as the key source of carbon-rich molecules and carbon dust
It is well known that huge amounts of carbon dust are released in the space by carbon rich stars [1-4]. The elemental carbon is accompanied by carbon rich molecules like polyynes and cyanopolyynes [1-4]. Please refer to the contributions of Millar and Chernicharo to this conference issue for further details and references. The key molecule responsible for the formation of carbon dust and other molecules in circumstellar environment is C_2. This molecule can be viewed as an acetylenic diradical •C≡C•and can be produced in laboratory for example with

a carbon arc [5] or by laser ablation of carbonaceous targets [6]. The simplest approach to produce C_2 in laboratory is through a carbon arc and C_2 can be recognized by its electronic absorption spectrum [7]. A carbon arc is ignited when two carbon electrode are put in contact and then slightly separated each other. The intensity of the arc is governed by the electric current flowing in the arc. In our experiments the minimal electric current used was about 10 A in the submerged electric arc [8], but it can reach 100 A in the production of fullerenes, when the arc is ignited in an helium atmosphere [9]. Higher arc regimes are needed in the production of exotic carbon nanostructures like carbon nanohorns [10]. The electronic absorption spectrum of the carbon arc reveals always the presence of the C_2 molecule and the mass spectrometry of the gas phase carbon arc reveals the tendency of the C_2 molecule to polymerize into longer carbon chains which are acetylenic in nature [11,12]. At this point is only matter of the environment surrounding the carbon arc as summarized in Fig.1. If the arc is struck in an helium atmosphere carbon dust will be produced almost exclusively. However, this carbon dust contains fullerenes and for this reason is also called "fullerene black" [9]. The amount of fullerenes (mainly C_{60} and C_{70} fullerenes with trace amounts of higher fullerenes) can reach up to 10% by weight of the "fullerene black" carbon dust and must be extracted with an appropriate fullerene solvent like for instance benzene or toluene to remove C_{60} and C_{70} fullerene from the carbon dust matrix [9]. Thus, what we are learning from this laboratory experiment? That C_{60} and C_{70} fullerenes cannot be easily produced in pure form in space but they are produced necessarily mixed and adsorbed in carbon dust. Of course, there is also the possibility of fullerene desorption from the carbon dust by sublimation and we know that temperatures as high as 700°C are necessary to cause the sublimation of C_{60} [13]. Helium is an essential gas in favoring the fullerene formation in high yields and this is attributed to its uniquely high thermal conductivity which favors the rapid temperature quench of the products [14]. In fact when the carbon arc is struck in argon or other inert gases having lower thermal conductivity than helium, once again the fullerene yield drops almost to zero and only the carbon dust is the reaction product [14].

Another important thing we learn from the study of the carbon arc is the inhibiting effect of hydrogen in fullerene formation. When the carbon arc is struck in a hydrogen atmosphere, the fullerene yield drops nearly to zero and instead the resulting products may be polycyclic aromatic hydrocarbons (PAHs), polyynes carbon dust and carbon nanotubes, depending on the conditions adopted [15]. Reactive gases in the arc are able to orient the reaction products toward new and unusual products. For example, as shown in Fig. 1, the presence of ammonia or acetonitrile leads to the formation of a mixture of polyynes and monocyanopolyynes [12,13]while when the carbon arc is struck in liquid nitrogen a mixture of dicyanopolyynes are obtained [16]. Another reactive gas is chlorine, its presence in the carbon arc favors the formation of chlorinated PAHs [9].

In summary, the general view is that C_2 polymerizes into longer polyynes chains. These chains are stabilized if end-capped by hydrogen atoms and nitrogen atoms yielding polyynes and cyanopolyynes. If instead there is an inert gas, C_2 may undergo a cyclization and crosslinking reaction producing fullerenes and carbon dust. In this process the eventual presence of hydrogen is crucial because it can orient the reaction products toward PAHs, polyynes, carbon nanotubes. The formation of carbon dust in an arc regime under He is the most common product. As already said, the dust consists -almost exclusively- of amorphous carbon (see Fig. 2 for a TEM image), in other words by carbon black with a certain amounts of fullerenes, the so-called "fullerene black", but in special carbon regimes also unique carbon nanostructures known as "carbon nanohorns" are produced (see Fig. 1) [10]. On the other hand, when acetylene or methane are cracked in a carbon arc, then a large amount of carbon soot is produced together with polyynes and PAHs [17]. In this case the carbon soot rather than amorphous it appears partially graphitized at the transmission electron microscope as shown in Fig. 2 (right).

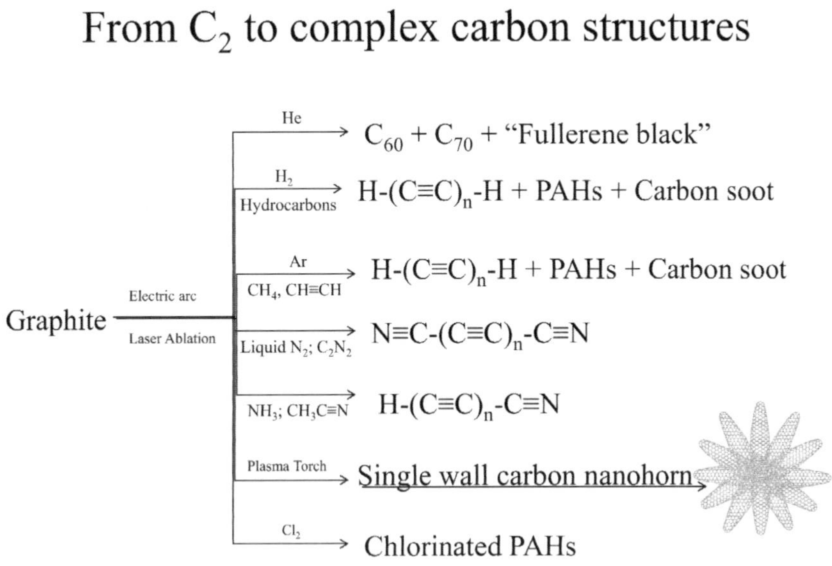

Figure 1. Different products obtainable from the carbon arc or by laser ablation of graphite under different conditions.

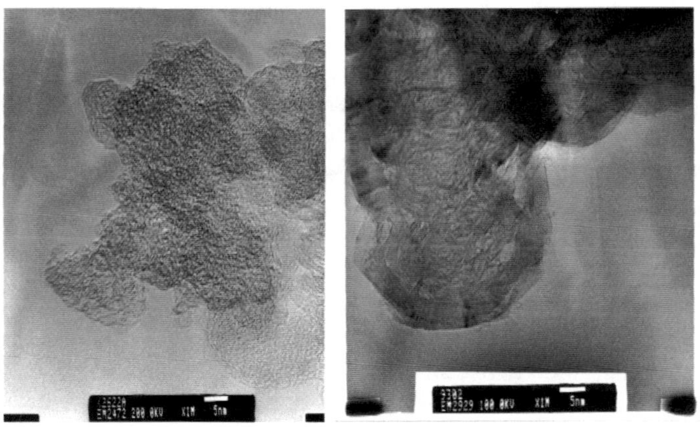

Figure 2. Transmission Electron Microscopy (TEM) images of amorphous carbon black (left) and partially graphitized carbon black from methane-acetylene arc (right).

3. Synthesis of polyynes, cyanopolyynes and other products through the submerged electric arc

Polyynes are acetylenic carbon chains derived for example from the polymerization of the C_2 unit in a carbon arc condition or in circumstellar medium conditions [5,11,12]. The hydrogen terminated carbon chains are the canonical polyynes but are known also nitrogen-terminated acetylenic chains known as monocyanopolyynes and dicyanopolyynes:

Journal of Physics: Conference Series **728** (2016) 062002 doi:10.1088/1742-6596/728/6/062002

$H-(C\equiv C)_n-H$	polyynes
$H-(C\equiv C)_n-C\equiv N$	monocyanopolyynes
$N\equiv C-(C\equiv C)_n-C\equiv N$	dicyanopolyynes

These molecules are highly unstable in the Earth and can be studied only in dilute solutions but are stable and can be detected by radioastronomy in the circumstellar medium of carbon rich stars and other astrophysical objects like hot molecular cores and in dense and dark molecular clouds like TMC-1, where very low gas density and/or low temperatures ensure their permanence [2-4,18,19]. Indeed, one of the largest organic molecule known in space is the monocyanopolyyne $H\text{-}(C\equiv C)_{10}\text{-}C\equiv N$. The synthesis of such molecules in laboratory represents a big challenge and was achieved only in 1970 by a step synthesis [5]. A much simpler, rapid approach in the polyynes synthesis is offered by the submerged carbon arc. A carbon arc between two graphite electrodes is struck at 10-20 Amperes inside an hydrocarbon solvent like n-hexane or n-heptane with external cooling (Fig. 3) [5]. All the polyynes family from $H-(C\equiv C)_2-H$ to $H-(C\equiv C)_{18}-H$ is produced in one shot and quenched in the solvent where it remains dissolved. Mixture of polyynes up to 10^{-3} M can be produced in hydrocarbon solvent and preserved for further studies [8]. The versatility of the submerged carbon arc synthesis can be pointed out by the fact that also monocyanopolyynes and dicyanopolyynes can be produced as well. The monocyanopolyynes can be produced by running the arc submerged in acetonitrile ($CH_3-C\equiv N$) solvent or in presence of ammonia, while the dicyanopolyynes are produced by arcing graphite electrodes submerged in liquid nitrogen and collecting the products in a solvent trap [11,12].

In each cases polyynes, monocyanopolyynes and dicyanopolyynes are produced all together as mixture of products. The HPLC (High Pressure Liquid Chromatography) analysis on a normal C8 column in reversed phase was an excellent tool in the separation of the polyynes and the collection of their electronic absorption spectra by a DAD (Diode Array Detector) coupled with the HPLC apparatus [5]. In this way we have available the individual electronic absorption spectra of each polyyne, monocyanopolyyne and dicyanopolyyne produced with the submerged arc approach [5].

From the electronic absorption spectra it is possible to determine the relative abundance of the polyynes produced in laboratory with the electric arc and expressed in terms of the molar ratio $C_8H_2/C_{10}H_2$, $C_{10}H_2/C_{12}H_2$ and so on. It was shown that the relative abundance of the polyynes generated with the arc is practically the same as that determined in the circumstellar environment by Chernicharo et al. [18] suggesting a similar mechanism of formation of these molecular species both in the electric carbon arc and in the circumstellar medium of carbon rich stars [11,12].

Thermodynamic calculations show that the Gibbs free energy of formation from the elements of polyynes becomes negative at the threshold temperatures above 3500 K as shown in Fig. 4. Such threshold temperature is even higher for monocyanopolyynes and dicyanopolyynes which is found respectively at 4100 K and at 4400 K. Thus, thermodynamics tells us that the reactions like $nC+H_2 \rightarrow H-(C\equiv C)_n-H$ or the analogous for cyanopolyynes: $nC+1/2H_2 + 1/2N_2 \rightarrow H-(C\equiv C)_n-C\equiv N$ and $nC+N_2 \rightarrow H-(C\equiv C)n-C\equiv N$ become possible only at very high temperatures which in laboratory can be reached quite easily with the carbon arc. Once the polyynes are formed in the high temperature of the arc, they are ejected in the solvent and quenched to much lower temperature and then preserved in the solvent .

The formation of polyynes in the submerged electric arc is always accompanied by the formation of a mixture of polycyclic aromatic hydrocarbons (PAHs) and by the formation of carbon soot [20]. In general the latter component is just a by-product while the PAHs formation is quite appreciable. There are available chemical methods to separate the polyynes from the PAHs in solution. One method consists in the precipitation of polyynes as copper or silver acetylides and, after filtration, their re-solubilization in a fresh solvent by the action of a mineral acid [5].

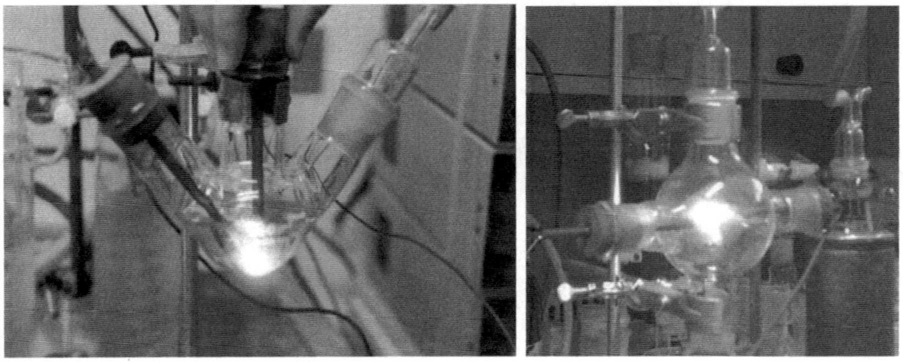

Figure 3. (Left) Submerged electric carbon arc: the two graphite electrodes are arced submerged in an hydrocarbon solvent. (Right) Gas phase carbon arc: a stream of reaction gas is blown through the carbon arc and the resulting products collected in an external cold trap.

Another interesting aspect about the polyynes chemical properties which may have astrochemical implications, regards the fact that they are reactive with atomic hydrogen. In laboratory the hydrogenation of polyynes is accessible under mild conditions and does not lead to fully hydrogenated products but to partially hydrogenated polyynes which are better described as ene-ynes, where the polyynes after the partial hydrogenation can be described as a sequence of conjugated acetylenic and ethylenic units displaying very rich and complex electronic absorption spectra [21].

4. Carbon arc in gas phase

Gas phase carbon arc is another interesting tool to simulate certain astrophysical conditions. Cracking of methane in a carbon arc (for example with a diluting inert gas like argon) is an excellent way to produce acetylene, polyynes and an important amount of carbon dust which we call acetylene black and whose microstructure is shown in Fig. 2 (left) [17]. Thermodynamics calculations show that above 1500K methane is unstable toward its reaction into acetylene and hydrogen: $2CH_4 \rightarrow HC{\equiv}CH + 3H_2$.

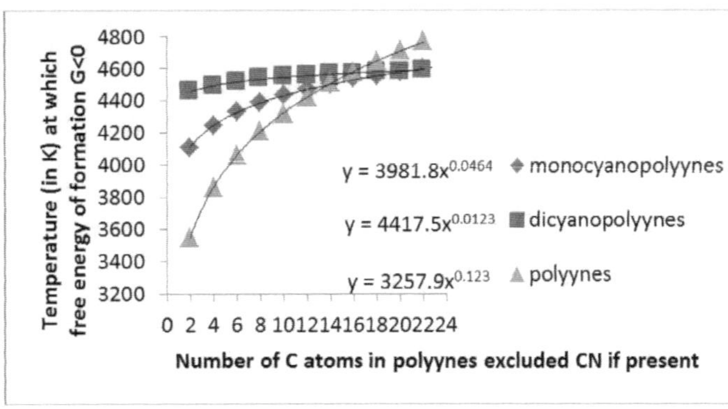

Figure 4. Thermodynamic calculations showing the temperature above which $\Delta G<0$ and the synthesis of polyynes and cyanopolyynes becomes allowed

Together with acetylene we have detected a mixture of polyynes as secondary products, while the acetylene black is formed for example by the thermal decomposition of acetylene into the elements which is thermodynamically allowed at temperature below 3500 K. Simple PAHs were found as well as secondary products and adsorbed on the carbon black, after solvent extraction [22]. The best way to produce in high yield carbon soot in a carbon arc is by arcing halomethanes like CCl_4, $CHCl_3$ and CH_2Cl_2 [23].

5. Fullerenes, fulleranes and other derivatives

C_{60} fullerene was expected to be present in space since the time of its discovery, but only in 2010 it was discovered through its infrared spectral signature in a young planetary nebula [24]. Later, C_{60} was found in other astrophysical environments including the interstellar medium [25,26]. Recently, C_{60} as radical cation was recognized as contributor to the DIBs in the gas phase [27]. The stability of C_{60} toward high energy radiation was demonstrated suggesting an high permanence of this molecule in the harsh space conditions [28]. In laboratory the infrared spectra of C_{60} and C_{70} fullerenes were recorded in a wide range of temperatures [29]. The position of certain infrared bands of fullerenes are sensitive to temperature and undergo a shift as function of temperature [29]. These properties are useful for searching and recognizing fullerenes in space as well as the determination of the absorption cross section of each absorption band which is a useful property for the determination of column density in space. Other spectral properties of fullerenes were determined in a strong oxidizing medium like oleum. In such an environment C_{60} and C_{70} fullerenes are losing one electron and oxidized to the radical cations [30,31]. The optical spectra of these species were recorded but unfortunately, because of the strong interaction between solute and solvent the electronic transitions were found blue shifted and broadened with respect to the radical cation band position in the gas phase [30,31].

C_{60} and C_{70} fullerenes are readily reactive with atomic hydrogen yielding fulleranes the hydrogenated fullerenes [32]. The fullerenes are so avid of hydrogen that the hydrogenation in the laboratory conditions leads directly to $C_{60}H_{18}$ and $C_{60}H_{36}$ [33]. Thus, it is completely reasonable to think that hydrogenated fullerenes may be present in space. Indeed a tentative detection of hydrogenated fullerenes was made by Kwok et al in the PPNe IRAS 01005+7910 [34]. Hydrogenated fullerenes are releasing molecular hydrogen when irradiated with UV light playing a role in the conversion of atomic hydrogen to molecular hydrogen in space [35]. Furthermore, $C_{60}H_{36}$ presents a unique absorption band at 217 nm resembling the UV "bump" absorption observed in the interstellar light extinction curve [35]. When irradiated with UV light $C_{60}H_{36}$ loses hydrogen and the maximum of absorption is gradually shifted towards longer wavelengths [35]. Thermal treatment of fulleranes causes the release of hydrogen and the restoration of the pristine C_{60} although the yield is not quantitative and also carbon soot as by-product is obtained [36].

Fullerenes were found in various space environments, sometime mixed with other complex molecules like PAHs or metals. Since it is known that fullerenes form adducts with PAHs through the Diels-Alder reaction, the synthesis of all the series of C_{60} and acenes (anthracene, tetracene and pentacene) was performed and the infrared spectra of the adducts were recorded together with the electronic absorption spectra (see Fig. 5)[37-40]. The reference spectra are now available for searching such adducts in space although the fullerene-acene adduct show just a few new infrared bands different than the two reactants which are of diagnostic utility for the identification.

In another work [41], C_{60} and C_{70} fullerenes were irradiated with visible light in presence of iron pentacarbonyl ($Fe(CO)_5$) yielding the respective iron tetracarbonyl adducts (see Fig. 5). Iron is one of the most abundant metal in the universe and CO is ubiquitous. Fe and CO can react in mild conditions to produce $Fe(CO)_5$ which is a highly volatile iron compound. When irradiated with C_{60} or C_{70} the iron carbonyl compound forms adducts of the type shown in

Fig. 5. The fullerene-iron carbonyl adduct present a moderate thermal stability but a very characteristic infrared spectra.

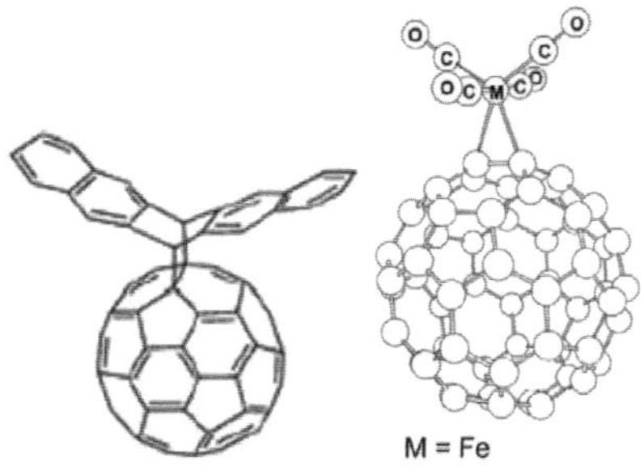

Figure 5. Fullerene C_{60}-pentacene adduct (left); Fullerene C_{60}-iron carbonyl adduct (right).

6. Petroleum fractions as model of the structures of mixed aliphatic-aromatic compounds observed in PPNe

The protoplanetary nebulae show a series of distinctive infrared emission features which are certainly due to aromatic components [2,4]. These emission features are known as AIB (Aromatic Infrared Bands) and are not seen in AGB stars [2]. This means that the carriers of the AIB are synthesized in the circumstellar envelope during the post-AGB phase. The AIB of protoplanetary nebulae suggest a mixed aliphatic aromatic structure of the carriers. Instead, in planetary nebulae the AIB suggest essentially aromatic carriers[2]. This implies an increased UV processing of the carriers of the AIB on the passage from protoplanetary to planetary nebulae with elimination of hydrogen and aliphatic chains [2]. Some of the AIB bands of PPNe are also found in the interstellar medium for example in the Orion Nebula or in various reflection nebulae [2]. The AIB bands suggest a mixture of different carriers and the individual molecules of these carrier should be complex. For these reasons it was proposed as model of the AIB the infrared spectra of certain heavy petroleum fractions [42-44]. These heavy petroleum fractions are the bottom residue of the petroleum distillation and are made by a mixture of relatively large and complex molecules which can be described as mixed aliphatic-aromatic-naphthenic, containing also some heteroatom as shown in Fig. 6 [45,46].

The petroleum fractions proposed as model spectra of protoplanetary nebulae can be distinguished into two classes: the lighter types which are shown in the left of Fig. 6, which are soluble in n-pentane. The "core" of these molecule is made by an aromatic and cycloaliphatic (naphthenic) ring with some side alkyl chains. To this type belong the petroleum fractions called DAE and RAE [45,46].

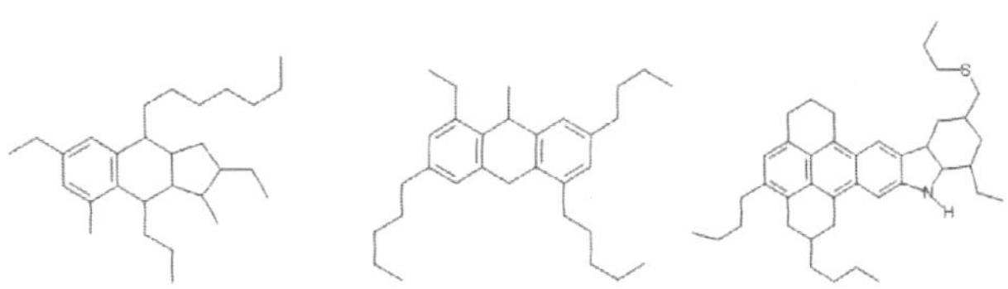

Figure 6. Examples of "average molecules" in heavy petroleum fractions and asphaltenes.

The other class is the heavier type which is insoluble in n-pentane and which belongs to the asphaltenes, represented by the average molecule in the right of Fig. 6 characterized by a "core" of the molecule made by two or three condensed aromatic rings [45,46]. To this type belongs the petroleum fraction called BQ-1. In Fig. 7 (left) it is shown that the infrared emission of the PPNe IRAS 22272+5435 is matched by the petroleum fraction DAE and MF2 and partially matched also by anthracite coal sample [42-44]. In Fig. 7 (right) it is shown that the infrared emission of the young PNe IRAS 212824+5050 is partially matched the BQ-1 petroleum fraction [46]. The model of petroleum fractions meets quite in full the model of MAONS (Mixed Aliphatic-Aromatic Organic Nanoparticles) proposed by S. Kwok and colleagues [4,47].

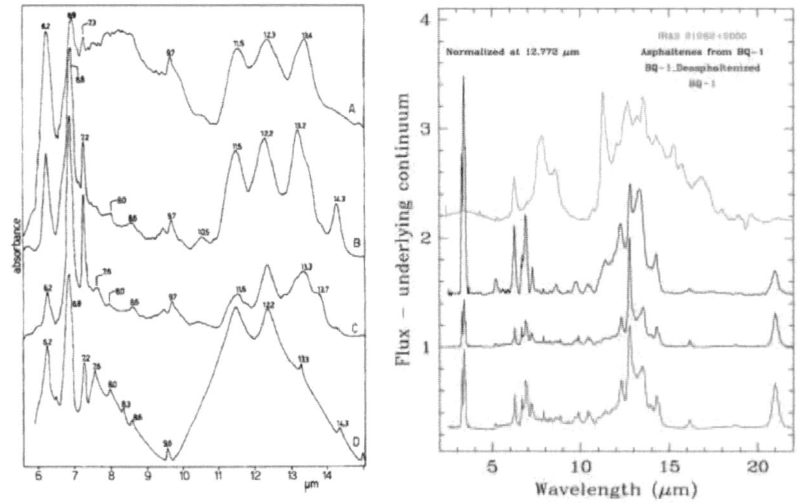

Figure 7. (Left) Infrared emission of the PPNe IRAS 22272+5435 (bottom) compared with the infrared spectra of petroleum fractions DAE and MF2 and by coal. (Right) Infrared emission of the PPNe IRAS 22272+5435 (top) compared with the spectra of petroleum fractions BQ-1 type.

References

[1] Hare J P and Kroto H W 1992 *Acc. Chem. Res.* **25** 106
[2] Kwok S 2011 *Organic Matter in the Universe.* (Weinheim V WileyVCH)
[3] Walsh C, Millar T J, Nomura H, Herbst E, Weaver S W, Aikawa Y, Laas J C and Vasyunin A I 2014 *Astron. Astrophys.* **563** A33
[4] Kwok S 2015 *Origins Life Evol. Biosph.* **45** 113.
[5] Cataldo F 2005 *Polyynes: Synthesis, Properties, and Applications* (Boca Raton -CRC Press)
[6] Tsuji M, Kuboyama S, Matsuzaki T and Tsuji T 2003 *Carbon* **41** 2141.
[7] Todorović-Marković B, Marković Z, Nikolić Z, Ristić Z and Nenadović T 2004 *Fullerenes Nanotubes Carbon Nanostructure* **12** 647
[8] Cataldo F 2004 *Carbon* **42** 129
[9] Hirsch A 1994 *The Chemistry of the Fullerenes* (Stuttgart: G. Thieme Verlag)
[10] Pagura C, Barison S, Mortalo C, Comisso N and Schiavon M 2012 *Nanosci. Nanotechnol. Lett.* **4** 160
[11] Cataldo F 2006 *Int. J. Astrobiology* **5** 37.
[12] Cataldo F 2006 *Origins Life Evol. Biosph.* **36** 467
[13] Cataldo F 2002 *Fullerenes, Nanotubes Carbon Nanostruct.* **10** 293
[14] Krätschmer W 1991 *Zeitschrift für Physik D Atoms, Molecules and Clusters* **19** 405
[15] Wang X K, Lin X W, Mesleh M, Jarrold M, Dravid V P, Ketterson J B and Chang R P H 1995 *J. Mater. Res.* **10** 1977
[16] Cataldo F 2004 *Polyhedron* **23** 1889
[17] Cataldo F 2007 *Fullerenes Nanotubes Carbon Nanostruct.* **15** 291

[18] Cernicharo J 2004 *Astrophys. J.* **608** L41
[19] Cherchneff I, Glassgold A E and Mamon G A 1993 *Astrophys. J.* **410** L188
[20] Cataldo F 2005 *Fullerenes Nanotubes Carbon Nanostruct.* **13** 21
[21] Cataldo F 2004 *Fullerenes Nanotubes Carbon Nanostruct.* **12** 765
[22] Cataldo F and Pontier-Johnson M A 2002 *Fullerenes, Nanotubes Carbon Nanostruct.* **10** 1
[23] Cataldo F 2005 *Fullerenes, Nanotubes, Carbon Nonstruct.* **13** 239
[24] Cami J, Bernard-Salas J, Peeters E and Malek S E 2010 *Science* **329** 1180
[25] Garciá-Hernández, D A, Villaver E, Garciá-Lario P, Acosta-Pulido J A, Manchado A, Stanghellini L, Shaw R A and Cataldo F 2012 *Astrophys. J.* **760** 107
[26] Sellgren K, Werner M W, Ingalls J G, Smith J D T, Carleton T M and Joblin C 2010 *Astrophys. J. Lett.* **722** L54
[27] Campbell E K, Holz M, Gerlich D and Maier J P 2015 *Nature* **523** 322
[28] Cataldo F, Strazzulla G and Iglesias-Groth S 2009 *Mon. Not. R. Astron. Soc.* **394** 615
[29] Iglesias-Groth S, Cataldo F and Manchado A 2011 *Mon. Not. R. Astron. Soc.* **413** 213
[30] Cataldo F, Iglesias-Groth S and Hafez Y 2013 *Eur. Chem. Bull.* **2** 1013 `http://www.eurchembull.com/index.php/ECB/article/view/1072`
[31] Cataldo F and Iglesias-Groth S 2014 *Eur. Chem. Bull.* **3** 920 `http://www.eurchembull.com/index.php/ECB/article/view/1497`
[32] Cataldo F and Iglesias-Groth S 2010 *Fulleranes: The Hydrogenated Fullerenes* (Heidelberg: Springer)
[33] Cataldo F 2003 *Fullerenes Nanotubes Carbon Nanostruct.* **11** 295
[34] Zhang Y and Kwok S 2013 *Earth, Planets and Space* **65** 1069
[35] Cataldo F and Iglesias-Groth S 2009 *Mon. Not. R. Astron. Soc.* **400** 291
[36] Luzan S M, Cataldo F, Tsybin Y O and Talyzin A V 2009 *J. Phys. Chem.* C **113** 13133
[37] Garciá-Hernández D A, Cataldo F and Manchado A 2013 *Mon. Not. R. Astron. Soc.* **434** 415
[38] Cataldo F, Garciá-Hernández D A and Manchado A 2014 *Eur. Chem. Bull.* **3** 740 `http://www.eurchembull.com/index.php/ECB/article/view/1456`
[39] Cataldo, F, Garciá-Hernández D A and Manchado A 2015 *Fullerenes Nanotubes Carbon Nanostruct.* **23** 818
[40] Cataldo F, Garciá-Hernández D A and Manchado A 2015 *Fullerenes Nanotubes Carbon Nanostruct.* **23** 760
[41] Garciá-Hernández D A, Cataldo F and Manchado A 2016 *Fullerenes Nanotubes Carbon Nanostruct.* **24** 225
[42] Cataldo F, Keheyan Y and Heymann D 2002 *Int. J. Astrobiology* **1** 79
[43] Cataldo F and Keheyan Y 2003 *Int. J. Astrobiology* **2** 41
[44] Cataldo F, Keheyan Y and Heymann D 2004 *Origins Life Evol. Biosph.* **34** 13
[45] Cataldo F, Angelini G, Garciá-Hernández D A and Manchado A 2013 *Spectrochim. Acta Part A: Molec. Biomolec. Spectrosc.* **111** 68
[46] Cataldo F, Garciá-Hernández D A and Manchado A 2013 *Mon. Not. R. Astron. Soc.* **429** 3025
[47] Kwok S and Zhang Y 2013 *Astrophys. J.* **771** 5

Franco Cataldo.

Theoretical infrared spectra of MAON molecules

SeyedAbdolreza Sadjadi, Sun Kwok and Yong Zhang

Laboratory for Space Research, Faculty of Science, The University of Hong Kong, Hong Kong, China

E-mail: ssadjadi@hku.hk, sunkwok@hku.hk, zhangy96@hku.hk

Abstract. Techniques of computational quantum chemistry (CQC) are used to explore the vibrational modes of complex organic molecules with mixed aromatic and aliphatic structures.

1. Introduction

Although the family of unidentified infrared emission (UIE) bands at 3.3 μm, 6.2 μm, 7.7 μm, 8.6 μm, 11.3 μm and 12.7 μm is commonly attributed to polycyclic aromatic hydrocarbon (PAH) molecules [1], there are other proposals suggesting the UIE bands originate from more complex organic compounds such as hydrogenated PAH [2, 3], hydrogenated amorphous carbon [4], soot and carbon nanoparticles [5], QQC [6], coal [7], oil fragments [8, 9], or mixed aromatic aliphatic organic nanoparticles (MAON) [10, 11].

The infrared spectra of complex amorphous hydrocarbons are poorly known theoretically as the vibrationally modes of large molecules have never been computed by chemists. The introduction of the Guassian functions by Boys [12] to overcome the difficulty of evaluation of integrals and the new formulations by Roothaan [13] opened the path to solve the Schrödinger equation for molecular systems. The reformulation of the mathematical framework into computer languages marked the first ab initio molecular quantum mechanics computer program in 1968 [14]. These pioneering works have led to a new discipline of computational quantum chemistry (CQC). With the availability of such modern powerful tools, it is now possible to explore the vibrational properties of large organic compounds. In this paper, we show briefly how CQC could help in the interpretation of astronomical IR spectra.

2. Infrared spectra of MAON

MAON's have mixed sp^2/sp^3 hybridizations and consist of isolated islands of benzene rings connected via the saturated hydrocarbon chains. An example of a MAON molecule $C_{155}H_{240}$ is shown in Figure 1. The molecular structure is assigned by applying the model(s) of chemical bond to the molecular geometry (defined by the list of Cartesian coordination of nuclei) [17]. The outcome of CQC calculations is the molecular geometry.

We applied the B3LYP/PC1 theoretical level implemented in PQS package [18] to calculate the local minimum geometry of $C_{155}H_{240}$, as well as the wavelength of $3N-6$ numbers of fundamental IR normal modes and their corresponding intensities. The blackbody thermal excitation model is then applied to these calculated data to simulate the IR emission spectra [19]. The calculated spectrum of this molecule is shown in Figure 2.

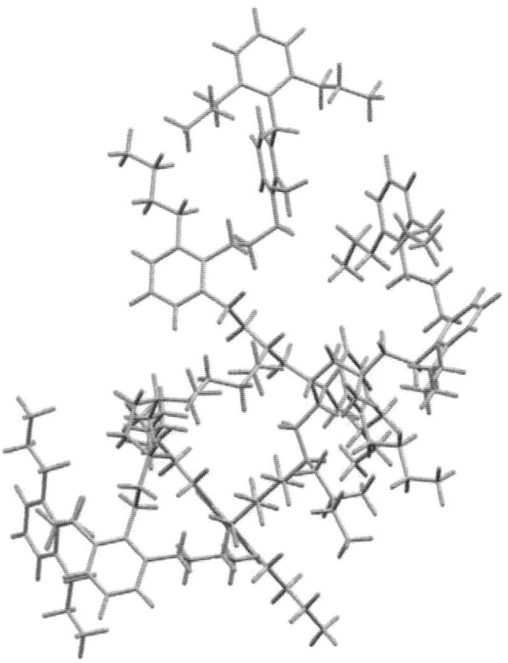

Figure 1. Structure of an example MAON molecule $C_{155}H_{240}$. The C atoms are shown in grey and H atoms in blue.

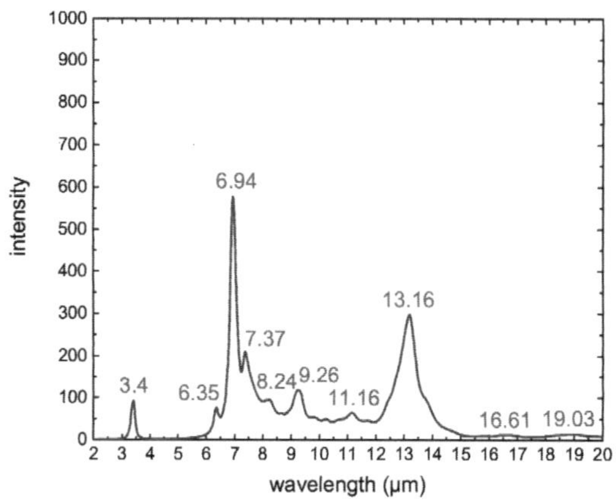

Figure 2. Theoretical spectra of the MAON molecule $C_{155}H_{240}$. Some of the peaks cannot be assigned to a single vibrational mode but are the result of coupled stretching and bending modes.

3. Vibrational analysis of MAON molecules

Unlike pre-quantum mechanical vibration models, there is no pre-defined modes of vibrations in CQC calculations. The molecular vibrational energy levels and intensities are calculated from the pure or mixed second derivatives of molecular energy. Since molecular energy is a hypersurface with $3N$ dimension (N is the number of atoms), the mathematical procedure is performed via special approximation called normal mode vibrational analysis. Besides the energy levels and intensities of vibrations, displacement vectors for all atoms in each normal mode are produced. When these vectors are animated alongside the molecular structure in graphical interface programs such as CHEMCRAFT [20] and GABEDIT [21], the three dimensional picture of molecular vibration can be observed. We can see which atoms, bonds or functional groups are participating in vibrations and how they vibrate.

In addition to visual examination of the movement of the displacement vectors, we have also developed a tool VIBANALYSIS [19] to quantify the contribution of each vibrational mode to a specific band. By such analysis, we can group the participant atoms into aliphatic and aromatic and calculate their contributions quantitatively.

4. The IR signatures of MAONs

The result of an analysis the vibrational bands of our sample MAON molecule is shown in Table 1. We can see that some vibrational modes are simple, e.g., the 3.4 μm band is due to sp^3 C–H stretching modes. However, some vibrational bands are coupled modes, e.g., the 6.35 μm band has major contributions from aromatic C–C stretching and C–H in-plane-bending (86%) coupled with 14% of aliphatic methylene twisting modes. The strongest band at 6.94 μm (Fig.2) has 85% contributions from sp^3 H–C–H scissoring mode. The second strongest band at 13.16

Table 1. Identification of the Vibartional Motions of Emission Bands of $C_{155}H_{240}$.

Bands (μm)	Vibrations	Ali/Ar[a]
3.4	methylene asymmetric C–H stretching	100/0
6.35	benzene C–C stretching + benzene C–H INP bending + methylene twisting	14/86
6.94	methyl and methylene scissoring+ benzene C–H INP bending	85/25
7.37	methyl umbrella motion+ methylene wagging	100/0.0
8.24	methyl deformation+ methylene wagging and twisting + benzene C–H INP bending+ benzene ring INP deformation	60/40
9.26	aliphatic chain vibration (methylene wagging + methyl deformation)+ benzene C-H INP bending	96/4
11.16	benzene C–H OOP + methylene wagging+ methyl deformation	85/15
13.16	benzene C–H OOP + methylene rocking	25/75
16.61	benzene ring OOP deformation+ aliphatic chains deformation	76/24
19.03	benzene ring OOP deformation + aliphatic chains deformation	68/32

[a] Aliphatic/Aromatic characters in precentage.

μm has 75% contribution from aromatic C–H out-of-plane bending mode. Deformation motions occur at longer wavelengths, e.g., in the 8.24, 9.26, 11.16, 16.61 and 19.03 μm bands. As shown in Figure 2, the intensities of such bands are significantly lower than other bands and they are broad and thus it may play role in the formation of plateaus in the astronomical emission spectra.

We have generated 40 MAON molecules with different molecular formulas, geometries and structures and calculated their vibrational spectra. Figure 3 shows that these molecules have similar IR signatures up to 10 μm and but different IR signatures beyond this region. This suggests the spectra with $\lambda < 10$ μm) are mostly determined by the chemical bonds within

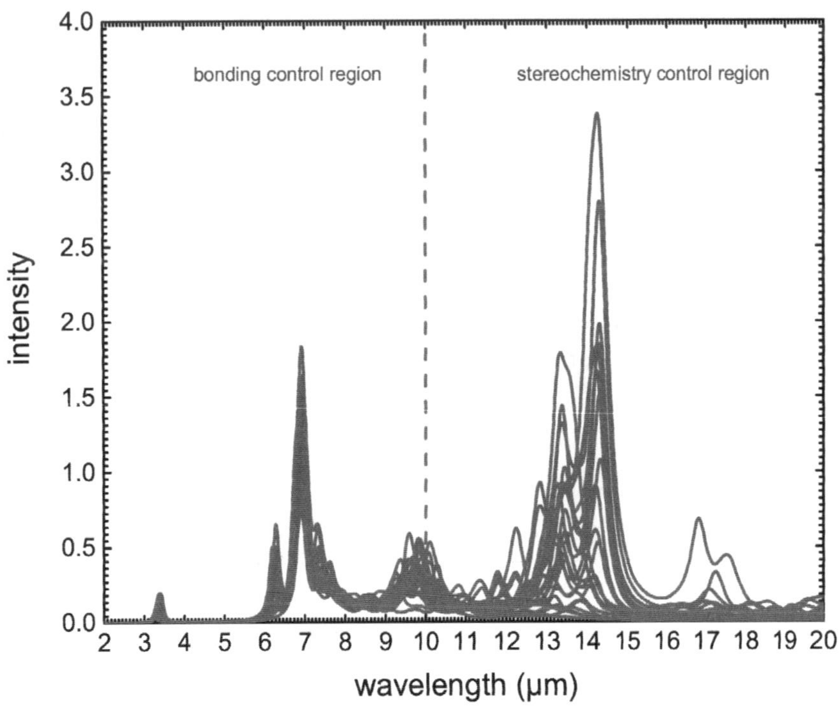

Figure 3. The IR spectra of 40 different types of MAON molecules.

the structure and the bands at $\lambda > 10$ μm are affected by molecular size and 3D structure (stereochemistry). The convergence of vibrational bands into certain specific wavelength bands gives us hope that the UIE bands can be naturally explained by the MAON structures.

5. Conclusion

Simple PAH molecules have very different individual spectral behavior in the infrared and the fitting of observed UIE spectra requires artificial combinations of PAH mixtures with many free parameters [22]. Preliminary results of infrared spectra of MAON molecules calculated with CQC presented here suggest that the spectra of MAON show consistent qualitative behavior. Further work on a larger group of MAON molecules is needed to confirm this.

Acknowledgments

The Laboratory for Space Research was established by a special grant from the University Development Fund of the University of Hong Kong. This work is also in part supported by a grant from the HKRGC (HKU 7027/11P).

6. References

[1] Allamandola L J, Tielens A G G M and Barker J R 1985 *Astrophys. J. Lett.* **290** L25
[2] Sandford S A, Bernstein M P and Materese C K 2013 *Astrophys. J. Suppl. Series* **205** 8
[3] Steglich M, Jäger C, Huisken F, Friedrich M, Plass W, Räder H J, Müllen K and Th H 2013 *Astrophys. J. Suppl. Series* **208** 26
[4] Jones A P, Duley W W and Williams D A 1990 *Quarterly Journal of the Royal Astronomical Society* **31** 567-582
[5] Hu A and Duley W W 2008 *Astrophys. J.* **677** L153-L156
[6] Sakata A, Wada S, Onaka T and Tokunaga A T 1987 *Astrophys. J.* **320** L63-L67
[7] Papoular R, Conrad J, Giuliano M, Kister J and Mille G 1989 *Astron. Astrophys.* **217** 204-208
[8] Cataldo F, Keheyan Y and Heymann D 2002 *Int. J. Astrobiology* **1** 79–86
[9] Cataldo F and Keheyan Y 2003 *Int. J. Astrobiology* **2** 41–50
[10] Kwok S and Zhang Y 2011 *Nature* **479** 80–83
[11] Kwok S and Zhang Y 2013 *Astrophys. J.* **771** 5
[12] Boys S F 1950 *Proceedings of the Royal Society of London A: Mathematical, Physical and Engineering Sciences* **200** 542–554
[13] Roothaan C C J 1951 *Rev. Modern Phys.* **23** 69–89
[14] Pople J A 1999 *Angewandte Chemie International Edition* **38** 1894–1902
[15] Schaefer H F 2015 *Nature* **517** 22–22
[16] Hehre W J, Radom L, Schleyer P V R and Pople J 1986 *Ab initio molecular orbital theory* (New York :: Wiley) `http://as.wiley.com/WileyCDA/WileyTitle/productCd-0471812412.html`
[17] Popelier P L A 2000 *Atoms in molecules : an introduction* (Harlow :: Prentice Hall)
[18] PQS 2013 Pqs version 4.0, parallel quantum solutions, 2013 green acres road, fayetteville, arkansas 72703 url: `http://www.pqs-chem.com` email:`sales@pqs-chem.com`
[19] Sadjadi S, Zhang Y and Kwok S 2015 *Astrophys. J.* **801** 34
[20] Zhurko G 2012 *http://www.chemcraftprog.com*
[21] Allouche A R 2011 *J. Computational Chem.* **32** 174–182
[22] Zhang Y and Kwok S 2015 *Astrophys. J.* **798** 37

Diamonds in space: a brief history and recent laboratory studies

Huan-Cheng Chang[1]

[1] Institute of Atomic and Molecular Sciences, Academia Sinica, Taipei 106, Taiwan

E-mail: [1] hchang@gate.sinica.edu.tw

Abstract. Stardust grains exist in outer space. However, due to their low abundance, structural heterogeneities, and lack of distinct spectroscopic features, unambiguous identification of their chemical composition has been a challenge. Diamond is a notable exception because it is composed of carbon and has several unique physicochemical properties that make the identification possible. Here, we provide a brief review on how diamonds were discovered in space based on the remarkable spectral matching between laboratory spectra and astronomical observations of the infrared emission at 3.43 and 3.53 μm, followed by a discussion of fluorescent nanodiamond as a possible carrier for extended red emission at 600 – 800 nm. Recent evidence to support the latter suggestion is provided.

1. Introduction

Carbon is the fourth most abundant element in the Galaxy. The presence of this element in the form of diamond in space was first proposed by Saslaw and Gaustad [1] in 1969 to account for the observed interstellar extinction curves in the far ultraviolet region of galactic radiation. Little attention was paid to this hypothesis until 1987 when Lewis et al. [2] reported the discovery of small diamond grains (with a mean size of ~3 nm and a concentration up to 400 ppm) in primitive meteorites from outside of the solar system. The search for diamond grains in the interstellar medium was actively pursued in the 1990s [3-6]. Allamandola et al. [3, 4], for example, made one of the first attempts to identify diamond dusts in dense molecular clouds based upon a broad infrared absorption band centered at 2880 cm^{-1} (3.47 μm), while Hill et al. [5] performed mid-infrared laboratory studies of meteoritic nanodiamonds. However, no satisfactory matches of the spectra were achieved.

An alternative approach of detecting nanodiamonds in space is to examine infrared emission bands from galactic nebulae and circumstellar medium. Among them, the emission from the HD 97048 star deserves special attention. It exhibits two prominent features at 3.43 and 3.53 μm, which are distinctly different than those from other nebulae. In a study of the nature of the emission from HD 97048 and Elias 1, Schutte et al. [7] noticed several anomalies in the observations. First, these two emission bands can be detected only under very special conditions (e.g. close to a hot stellar source). Second, the spatial variation of their intensities is fairly independent of other emission bands in the same frequency region. Third, these two bands always appear together, suggesting that they correlate spatially. An assignment of CVC overtones and the combination bands of highly excited large polycyclic aromatic hydrocarbon (PAH) molecules was proposed for these emission bands [7]. The assignment, however, did not present a satisfactory interpretation for the origin of these unidentified infrared emission (UIR) bands.

Extended red emission (ERE) is another type of mysterious emission observed in the sky. It has been observed in a wide variety of systems including reflection nebulae, planetary nebulae, HII regions, and the diffuse interstellar medium of our galaxy [8]. ERE was first discovered in 1980 by Schmidt et al. [9] in the spectra of the Red Rectangle nebula. ERE has a wavelength range of 500 – 900 nm and tends to exist in areas with abundant amounts of ultraviolet light, which suggests that it is a product of high-energy photons striking solid materials. No association of ERE with diamond had been made before 2000 except that of Duley [10] who proposed that the sharp emission lines observed at ~580 nm in the Red Rectangle spectra originate from the electronic transitions of structural defects similar to those found in terrestrial diamonds.

2. Unidentified infrared emission

Working independently, our group initiated a study on infrared spectroscopy and vibrational relaxation dynamics of CH_x ($x = 1V3$) on diamond nanocrystal and single-crystal surfaces in 1994 [11]. The single-crystal surfaces included H-terminated C(100), C(110), and C(111) (figure 1a). The aim of this project was to elucidate the formation mechanism of diamond by chemical vapor deposition (CVD). We hydrogenated diamond thin films made of 100 nm particles of irregular size and shape (figure 1b) by microwave plasma treatment and then obtained their absorption spectra at 3 – 5 μm with a Fourier transform infrared spectrometer. We observed surprisingly sharp absorption features at 2835 cm^{-1} and 2920 cm^{-1} (figure 1c), which were latter confirmed by single crystal experiments to arise from CH stretches on C(111) and C(100) facets [12]. The observation of such distinct and prominent absorption features for the hydrogenated nanodiamonds was attributed to the unique H-etching anisotropy of the diamond substrate [12]. Further studies of the thermal effect on the CH stretching vibrations showed a band redshift of ~5 cm^{-1} as the sample temperature was raised from 300 K to 1000 K (figure 1d) [13].

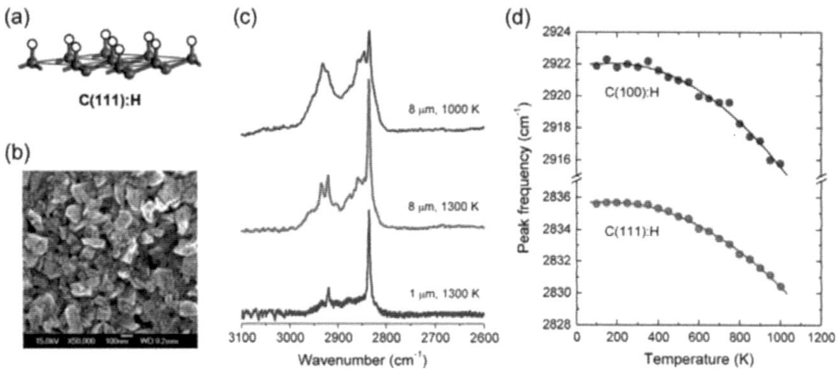

Figure 1. (a) Structure of a hydrogenated diamond C(111) surface. Grey spheres denote carbon atoms and white circles denote hydrogen atoms. (b) Scanning electron microscopy (SEM) image of diamond nanocrystallites used in the laboratory studies. (c) Laboratory absorption spectra of hydrogenated nanodiamond films prepared in different thickness (1 and 8 μm) and treated at different microwave plasma temperatures (1000 and 1300 K). The spectra were acquired at 300 K. (d) Temperature-dependent shifts of the absorption bands of the CH stretching vibrations on C(100) and C(111) facets of hydrogenated nanodiamonds.

In 1997, Geballe [14] provided a high-quality infrared emission spectrum of Elias 1 in the Taurus dark cloud from a ground-based observation. In this emission spectrum, the feature at

2830 cm^{-1} (3.53 μm) is much better resolved than before, with a full width at half maximum of only 11 cm^{-1}. Two years later, Guillois et al. [6] pointed out an astonishing match of the sharp UIR bands at 3.43 and 3.53 μm of Elias 1 (and also HD 97048) with the absorption bands of CH stretches on hydrogenated nanodiamond surfaces studied in our laboratories (figure 2). They identified these two peculiar emission bands as the CH stretching modes of H-terminated crystalline facets of diamonds, according to our assignments [11−13]. From the spectral redshift of 5 cm^{-1}, a temperature of 1000 K for these circumstellar nanodiamonds was estimated. Careful energy consideration was also given by this research team to understand the nature of the infrared emission process [6, 15].

One may ask: How can the circumstellar nanodiamonds display such distinct vibrational features? This is ascribable to these diamond grains being situated in close proximity to the star, from which they are exposed to intense UV photon fluxes and thus high temperatures can be reached. Such a high temperature and high H-flux condition is reminiscent of that of microwave plasma used in the diamond CVD synthesis. Conceivably, the temperatures reached in these two cosmic environments are sufficiently high to allow hydrogenation and H-etching of the nanodiamond surfaces to occur. The suggestions are indeed supported by the observations of H recombination lines in both Elias 1 and HD 97048 [16].

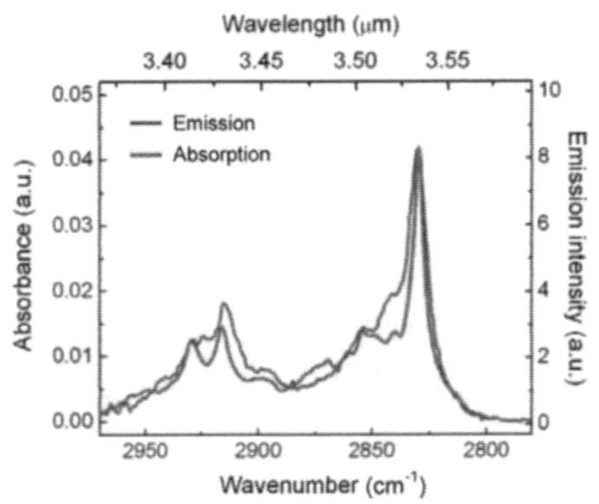

Figure 2. Comparison between the infrared emission spectrum of Elias 1 and the laboratory absorption spectrum of hydrogenated nanodiamonds measured at 300 K. Note that the absorption spectrum was redshifted by 5 cm^{-1} in the comparison.

Figure 2 represents one of the best matches ever obtained between laboratory spectra of solid-state candidates and infrared interstellar features. Taking advantage of the brightness of the emission, Habart et al. [17] and Goto et al. [18] were able to acquire spatially resolved spectra of the 3 gm bands at HD 97048 and Elias 1, respectively. It should be emphasized that CH stretches are the only distinct spectral features that can be clearly identified with nanodiamonds in the infrared. The corresponding CH bending vibrations could not be found in the 6 − 7 μm range [19] because of the weakness of the absorption and, possibly, severe band broadening resulted from vibrational coupling with diamond surface phonons resonant in the same frequency region. The absence of this spectral feature, in contrast to the prominence of the CH bending of PAH [20], represents another unique characteristic of H-terminated nanodiamonds.

The remarkable matches in the peak positions, bandwidths, and profiles of the two emission features at 3.43 and 3.53 μm in figure 2 verify the existence of diamonds in space. Another useful piece of information provided by the observations is the size of the UIR carriers. The information was deduced based on the finding that the 3.53 μm band coincides with the stretching resonance

Journal of Physics: Conference Series **728** (2016) 062004 doi:10.1088/1742-6596/728/6/062004

of CH monolayer on C(111) single-crystal surface [12]. Additionally, they have similar widths. Therefore, these nanodiamonds cannot be too small in size. Analysis of the infrared absorption spectra of meteoritic (~3 nm) [21] and synthetic (5 – 700 nm) [22] diamonds showed that the sharp 3.53 μm band can emerge only for particles with a mean size greater than 25 nm.

3. Extended red emission

In 2005, our group launched a new project to study the photoluminescence properties of nanodiamonds. The work was motivated by the findings that diamond can emit bright red photoluminescence from nitrogen-vacancy (NV) defects, which are potentially useful as photostable fluorophores for bioimaging applications [23]. The NV center in diamond is a crystallographic defect consisting of a nitrogen atom and a nearest-neighbour lattice vacancy (figure 3a). It can exist in two different forms: NV^0 (neutral) and NV^- (negatively charged). They both can be synthesized by bombarding synthetic diamond nanocrystallites (containing ~100 ppm N) with a high-energy (e.g. 3 MeV) proton beam from a particle accelerator, followed by annealing at elevated temperatures (e.g. 1000 K). Photoluminescence occurs as a response to excitation by green light (figure 3b). These two colour centers are characterized by the sharp zero-phone lines (ZPLs) at 575 nm and 638 nm, accompanied by broad phonon sidebands peaking at 610 nm and 680 nm, for NV^0 and NV^- respectively (figure 3c) [24]. Particles subjected to the irradiation and annealing treatment are called fluorescent nanodiamonds (FNDs). The typical concentration of NV centers in FND is 1 – 10 ppm with a photoluminescence quantum yield of 10 – 90% [25].

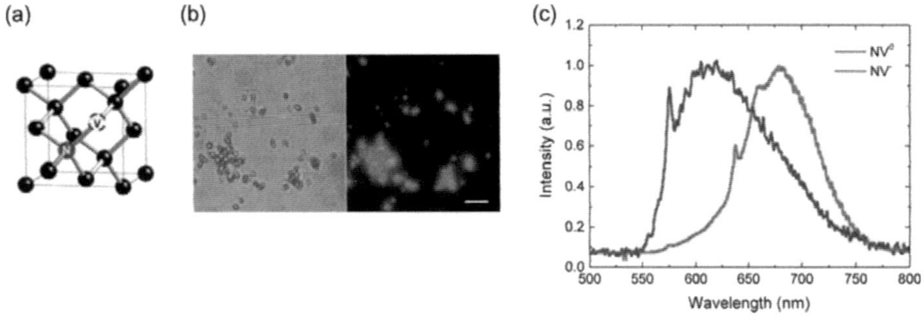

Figure 3. (a) Structure of a nitrogen-vacancy (NV) center in diamond. Black spheres denote carbon atoms in a unit cell. (b) Bright-field (left) and epifluorescence (right) images of FNDs and their aggregates on a glass slide. Scale bar: 10 gm. (c) Photoluminescence spectra of single NV^0 and NVV centers in nanodiamonds.

The laboratory studies of the photoluminescence properties of FNDs brought our attention to the ERE of reflection and planetary nebulae. Without much effort, we found a striking resemblance between the astronomical spectrum of the planetary nebula NGC 7027 [26] and the laboratory spectra of FNDs deposited on a glass slide (figure 4a) [27]. The close similarity between these two spectra led us to posit that nanodiamonds are the source of this red emission. The hypothesis had immediate backing: NGC 7027 is rich in carbon [28]. It was also supported by the fact that nitrogen atoms are abundant in the universe and they can be easily incorporated into the diamond lattice during synthesis. As vacancies in diamonds are created in laboratory by high-energy proton irradiation under vacuum, which can similarly occur in space, the FNDs are expected to form in the cosmic environments where ERE is observed.

Next, we compared the photoluminescence spectra of FNDs with the ERE bands of the Red Rectangle nebula, an object emitting ERE with exceptional strength. The high quality of its spectra allows for a stringent test of the theory that FND is a possible carrier of ERE. A notable feature of the Red Rectangle is that the spectral profile of the ERE varies substantially with distance from the central star HD 44179. The peak wavelengths of the spectra shift from 635 nm at 10" south to 660 nm at 6" south within the same object (figure 4b) [29, 30]. To account for the difference in the spectral profile, we consider that there exist two types of NV centers in FND: NV^0 and NVV. Spectra can thus be synthesized with data given in figure 3c for these two centers at different ratios. With the fluorescence intensity ratios of NV^0:NVV = 1:0 and 3:2, we can indeed properly reproduce the ERE bands at different offset distances (figure 4b).

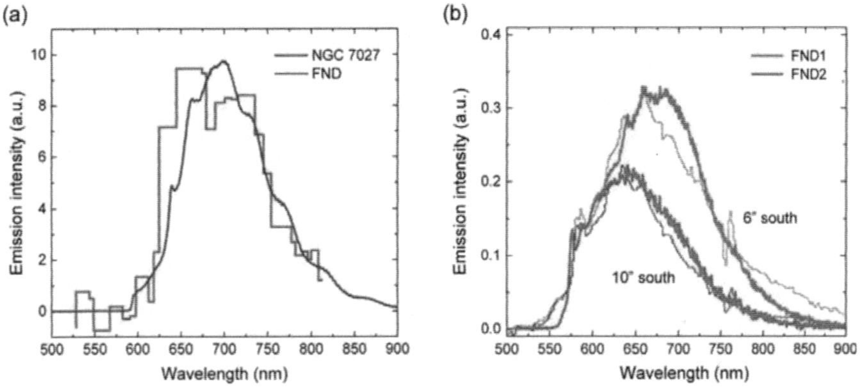

Figure 4. (a) Comparison between the observed ERE spectrum of NGC 7027 and the laboratory fluorescence spectrum of FND excited with 510 – 560 nm light. (b) Comparison between the observed ERE spectra of the Red Rectangle at 10 south and 6 south of HD 44179 and the laboratory fluorescence spectra of FNDs with different contents of NV^0 and NVV. The fluorescence intensity ratios of NV^0:NVV in samples FND1 and FND2 are roughly 3:2 and 1:0, respectively.

4. Conclusion

We have developed two spectroscopic approaches (infrared absorption and visible photoluminescence) to characterize nanodiamonds in laboratories and to assist their identification in space. The first approach is a proven success. However, only 2 star systems (Elias 1 and HD 97048) have been found so far to exhibit clear fingerprints of hydrogenated nanodiamonds in the infrared region. ERE is an appealing alternative to the 3.43- and 3.53-μm emission bands because it has been observed in a wide variety of dusty environments. The Red Rectangle is an ideal testing ground in this context and a proper explanation for the variation in the ERE spectra at different offset distances from the star HD 44179 is still lacking. To understand better this object as well as other ERE features in the galaxies, further studies of the thermal effect on the photoluminescence spectra of NV centers in the diamond grains are needed.

Acknowledgments

This work was supported by Academia Sinica, Taiwan. The author thanks Tom Geballe and Sun Kwok for insightful and helpful comments on the manuscript.

References

[1] Saslaw W C and Gaustad J E 1969 *Nature* **221** 160
[2] Lewis R S, Ming Y, Wacker J F, Anders E and Steele E 1987 *Nature* **326** 160

[3] Allamandola L J, Sandford S A, Tielens A G G M and Herbst T M 1992 *Astrophys. J.* **399** 134
[4] Allamandola L J, Sandford S A, Tielens A G G M and Herbst T M 1993 *Science* **260** 64
[5] Hill H G M, Jones A P and d'Hendecourt L B 1998 *Astron. Astrophys.* **336** L41
[6] Guillois O, Ledoux G and Reynaud C 1999 *Astrophys. J.* **521** L133
[7] Schutte W A, Tielens A G G M, Allamandola L J, Cohen M and Wooden D H 1990 *Astrophys. J.* **360** 577
[8] Smith T L and Witt A N 2002 *Astrophys. J.* **565** 304
[9] Schmidt G D, Cohen M and Margon B 1980 *Astrophys. J.* **239** L133
[10] Duley W W 1988 *Astrophys. Space Sci.* **150** 387
[11] Chang H-C, Lin J-C, Wu J-Y and Chen K-H 1995 *J. Phys. Chem.* **99** 11081
[12] Cheng C-L, Chang H-C, Lin J-C, Song K-J and Wang J-K 1997 *Phys. Rev. Lett.* **78** 3713
[13] Lin J-C, Chen K-H, Chang H-C, Tsai C-S, Lin C-E and Wang J-K 1996 *J. Chem. Phys.* **105** 3975
[14] Geballe T R 1997 *From Stardust to Planetesimals* vol 122 ed Y Pendleton and A G G M Tielens (San Franciso: ASP, Conference series) p 119
[15] Guillois O, Ledoux G, Nenner I, Papoular R and Reynaud C 1998 *Faraday Discuss.* **109** 335
[16] Van Kerckhoven C, Tielens A G G M and Waelkens C 1999 *The Universe as Seen by ISO* vol 1 ed E Cox and M F Kessler (Noordwijk: ESA SP-427) p 421
[17] Habart E, Testi L, Natta A and Carbillet M 2004 *Astrophys. J.* **614** L129
[18] Goto M, Henning Th, Kouchi A, Takami H, Hayano Y, Usuda T, Takato N, Terada H, Oya S, Jäger C and Andersen A C 2009 *Astrophys. J.* **693** 610
[19] Ando T, Ishii M, Kamo M and Sato Y 1993 *J. Chem. Soc. Faraday Trans.* **89** 1783
[20] Van Kerckhoven C, Tielens A G G M and Waelkens C 2002 *Astron. Astrophys.* **384** 568
[21] Jones A P, d'Hendecourt L B, Sheu S-Y, Chang H-C, Cheng C-L and Hill H G M 2004 *Astron. Astrophys.* **416** 235
[22] Sheu S-Y, Lee I-P, Lee Y T and H-C Chang 2002 *Astrophys. J.* **581** L55
[23] Yu S-J, Kang M-W, Chang H-C, Chen K-M and Yu Y-C 2005 *J. Am. Chem. Soc.* **127** 17604
[24] Rondin L, Dantelle G, Slablab A, Treussart F, Bergonzo P, Perruchas S, Gacoin T, Chang H-C, Jacques V and Roch J-F 2010 *Phys. Rev. B* **82** 115449
[25] Mohtashami A and Koenderink A F 2013 *New J. Phys.* **15** 043017
[26] Furton D G and Witt A N 1990 *Astrophys. J.* **364** L45
[27] Chang H-C, Chen K and Kwok S 2006 *Astrophys. J.* **639** L63
[28] Kwok S and Zhang Y 2011 *Nature* **479** 80
[29] Witt A N and Boroson T D 1990 *Astrophys. J.* **355** 182
[30] Ledoux G, Guillois O, Huisken F, Kohn B, Porterat D and Reynaud C 2001 *Astron. Astrophys.* **377** 707
[31] Chen X-D, Dong C-H, Sun F-W, Zou C-L, Cui J-M and Guo G-C 2011 *Appl. Phys. Lett.* **99** 161903
[32] Plakhotnik T and Gruber D 2010 *Phys. Chem. Chem. Phys.* **12** 9751
[33] Davies G 1979 *J. Phys. C: Solid State Physics* **12** 2551

Huan Cheng Chang giving his talk.

Left: Masatoshi Ohishi Right: Masaaki Otsuka

The diffuse interstellar bands - a brief review

T R Geballe

Gemini Observatory, 670 N. A'ohoku Place, Hilo, HI 96720 USA

E-mail: tgeballe@gemini.edu

Abstract.
The diffuse interstellar bands, or DIBs, are a large set of absorption features, mostly at optical and near infrared wavelengths, that are found in the spectra of reddened stars and other objects. They arise in interstellar gas and are observed toward numerous objects in our galaxy as well as in other galaxies. Although long thought to be associated with carbon-bearing molecules, none of them had been conclusively identified until last year, when several near-infrared DIBs were matched to the laboratory spectrum of singly ionized buckminsterfullerene (C_{60}^+). This development appears to have begun to solve what is perhaps the greatest unsolved mystery in astronomical spectroscopy. Also recently, new DIBs have been discovered at infrared wavelengths and are the longest wavelength DIBs ever found. I present the general characteristics of the DIBs and their history, emphasizing recent developments.

1. Introduction

The diffuse interstellar bands, which are commonly known as DIBs, are a class of absorption features that are observed in the spectra of objects in our galaxy and other galaxies. They arise in the interstellar medium, but are not due to atoms or simple molecules. When the first bands, at 5780 Å and 5797 Å, were discovered it was suspected, but not definitively shown that the absorptions were physically associated with interstellar medium; that became clear later. Originally the absorptions were only seen in optical spectra, but subsequently DIBs have been found at both longer and shorter wavelengths. They are most easily seen in the spectra of hot stars, because spectra of such stars contain fewer interfering photospheric lines than spectra of cooler stars.

Recent reviews of DIBs [1,2] are available and are considerably more comprehensive than the present paper. Readers are referred to them and references therein for more detailed information and discussion.

2. Description and Early history

The discovery of the first diffuse interstellar bands was made by a graduate student at Lick Observatory in California, Mary Lea Heger, almost a century ago during her search for "stationary" sodium D lines toward orbiting pairs of hot stars [3], whose intrinsic spectra shifted due to the changing orbital velocities. An excellent recounting of the discovery by McCall and Griffin is available [4]. The bands that Heger observed, as well as several additional ones discovered later by Merrill and his collaborators were shown by the latter investigators to arise in interstellar gas [5], not due to their stationary behavior, but because their strengths increased with increasing reddening of the stellar spectra. Although as discussed below the bands are

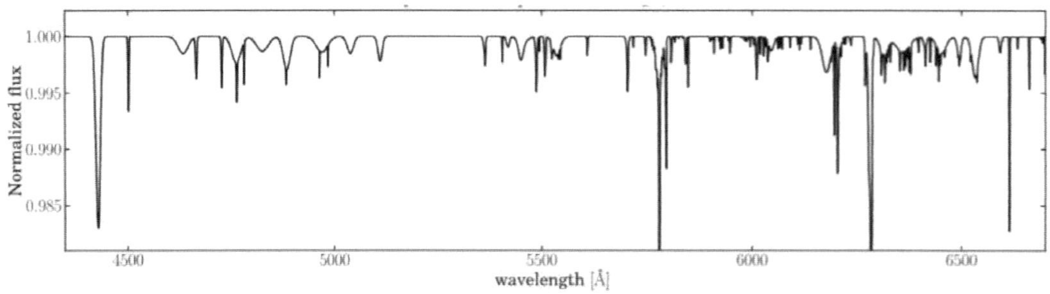

Figure 1. Synthetic DIB spectrum of Lan et al. [9] based on work by Jenniskens and Désert [10], at a resolution of 2 Å and for a reddening $E(B-V)$ of 0.1 mag. Note the small range of the vertical scale.

found predominantly in the diffuse (low density) interstellar medium, in the name DIB the term "diffuse" applies to their appearance on photographic plates where the first absorptions discovered are broader (more "diffuse") than interstellar absorption lines of atomic and molecular species.

In 1975 Herbig [6] listed 39 DIBs between 4400 Å and 6850 Å. With the advent of modern instrumentation and advances in spectroscopy, particularly in the near-infrared, the number has skyrocketed to the present value of ~500, through the work of many but in particular Hobbs and his collaborators [7,8]. However, as of the beginning of 2015, not a single DIB had been unequivocally identified with a specific carrier. Many have referred to the unknown identities of the DIBs as the greatest unsolved mystery in astronomical spectroscopy.

Figure 1 from Lan et al. [9] is a synthetic pure DIB spectrum for a reddening $E(B-V)$ of one-tenth mag, based on observations and analysis by Jenniskens and Désert [10] who averaged a large number of stellar spectra after removing the emission and/or absorption lines intrinsic to the stars themselves. For this reddening the strongest DIBs have optical depths less than 0.02. Most of the optical DIBs detected in the last decade are considerably weaker than those shown in Fig. 1 and are narrow (sub-Angstrom); detection and characterization of them requires high spectral resolution and high sensitivity [7,8].

3. Interstellar environment

Convincing evidence that most DIBs are formed predominantly in the diffuse interstellar medium comes from the strong correlations of their strengths vs $E(B-V)$ for small values of that parameter. Figure 2 from Lan et al. [9] shows this behavior for a number of DIBs, using as background sources objects both inside and outside of the Galaxy. Low values of $E(B-V)$ generally imply lines of sight passing through diffuse clouds rather than than dense clouds. As shown in these figures the correlation with $E(B-V)$ disappears at higher values of reddening, which correspond to sightlines that are more likely to include regions of denser and/or UV-shielded gas. Thus the flattening at these higher values suggests that most DIBs do not exist in the denser interstellar medium. Indeed it has also been found that on average DIB strengths are either uncorrelated with or anti-correlated with $N(H_2)$ [9].

However, there is convincing evidence that some DIBs are formed in denser or more shielded environments. The best known examples of this are the so-called C_2 DIBs, whose strengths roughly scale with $N(C_2)/E(B-V)$ [11]. Their carriers presumably are more susceptible to destruction by UV radiation than other DIBs carriers, and thus they exist mainly in regions where the molecular fraction is high, in either diffuse cloud cores or in dense clouds.

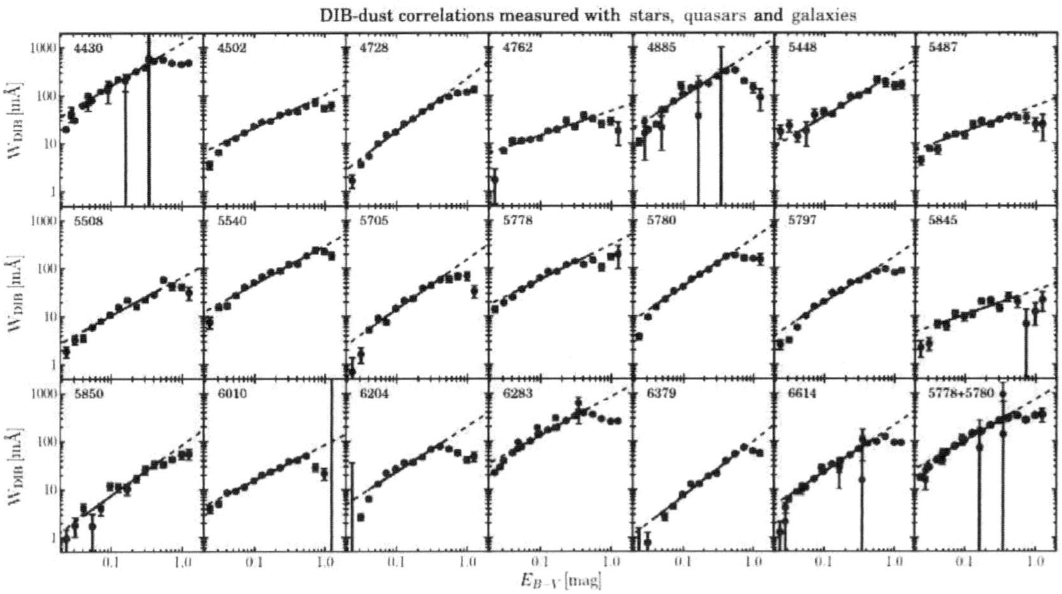

Figure 2. Dependence of DIBs equivalent widths on reddening for sightlines to Galactic stars, quasars, and galaxies (color-coded in original figure), from Lan et al. [9]. Wavelengths of DIBs are given in Angstroms.

4. DIB - DIB correlations

The large numbers of DIBs coupled with the absence of convincing identifications for them has led many to search for correlations between the strengths of individual DIBs as a way to constrain the numbers of carriers and possibly identify bands produced by the same carrier. Quoting Adámkovics, Blake and McCall [12], "Any group of features arising from a particular carrier, or set of chemically related carriers, must maintain the same relative intensities in all lines of sight." Although there are noteworthy counterexamples of this even among the stronger DIBs (a glaring instance is the aforementioned 5780 and 5797 Å bands [13], in most cases band strengths show a rough correlation from direction to direction, suggesting that most DIBs carriers tend to form under similar physical conditions . A very tight correlation between the strengths of two DIBs might suggest that each arises from the same carrier, but such correlations are very rarely the case. McCall [14] reported that in a study of 58 DIBs observed on 40 sightlines only 1.5% of the pairs had strength correlation coefficients of 0.95 or higher. This suggests that the DIBs must be produced by many species.

5. Free molecules or solids?

The strengths of many DIBs scale roughly with the visual interstellar extinction for small values of the latter. Because extinction is produced by dust particles, this behavior suggested to early investigators that DIBs carriers are physically associated with the dust particles. However, there are a number of arguments strongly suggesting otherwise and indicating that the carriers are free molecules rather than molecules bound to grains.

- Solid state absorption features tend to be much broader than the DIBs [15].
- The central wavelengths of the DIBs do not vary from sightline to sightline and the absorption profiles have been found to vary only in a very few cases [15]. In solids the

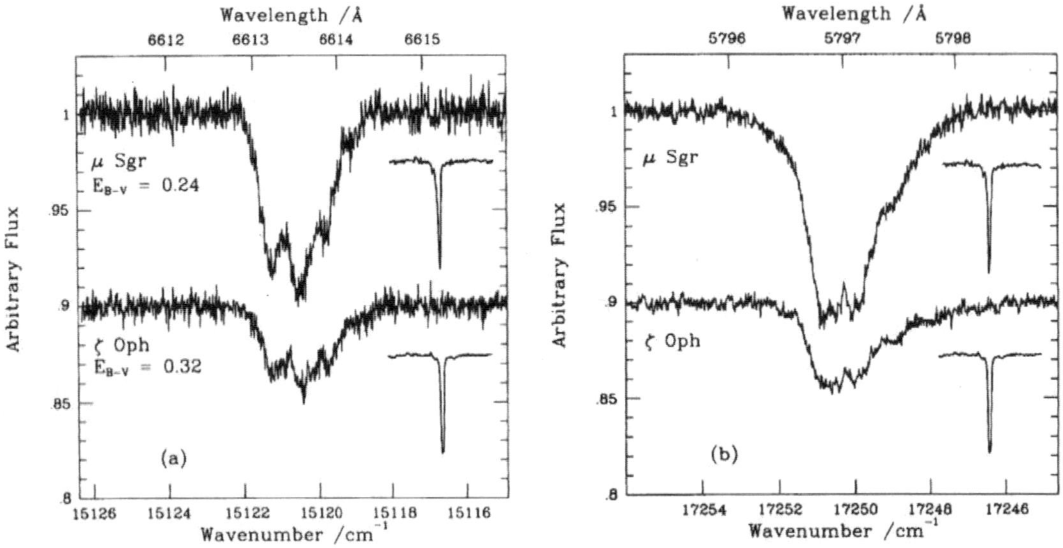

Figure 3. High resolution spectra of two DIBs toward two stars, showing resolved structure [16]. The interstellar lines of K I are shown to the right of each DIB spectrum

central wavelengths and shapes of absorption bands are influenced by the interaction of the absorbing molecules with their neighboring molecules, and significant differences are observed on different sightlines.

- At very high spectral resolution fine structure is observed in some DIBs [16,17] (see Fig. 3), which is suggestive of rotational structure and thus the absorbers being free molecules [17,18,19].
- Polarization studies of a few DIBs in a highly reddened star show that while the continuum is polarized, as one would expect due to magnetically aligned dust grains, there is no excess polarization at the wavelengths of the absorptions, which would occur if the absorbing material is on the aligned grains of standard sizes (~0.1 μm), regardless of whether the grains are silicates or are carbonaceous [20].

Thus it may be safely concluded that most if not all DIBs arise in free molecules. That the DIBS occur at optical and short infrared wavelengths implies that the transitions are vibronic in nature.

6. Identification

Identifications for DIBs have been proposed since the time that they were recognized as having an interstellar origin. The proposals can generally be separated into two groups: individual carriers, and classes of carriers. Proposed individual identifications typically have linked single or small numbers of DIBs with single species and in the first instance have usually appealed to wavelength matches. None of them have survived scrutiny. In most cases the wavelength matches were found to be imprecise, absorption features by the candidate species at other wavelengths were not seen in space, or astrophysical or astrochemical arguments precluded the candidates from being sufficiently abundant to account for the strengths of the observed features.

There have only been a few suggestions of classes of carriers. These have been prompted by arguments that the DIBs carriers are likely to be large, carbon-bearing molecules [21,22].

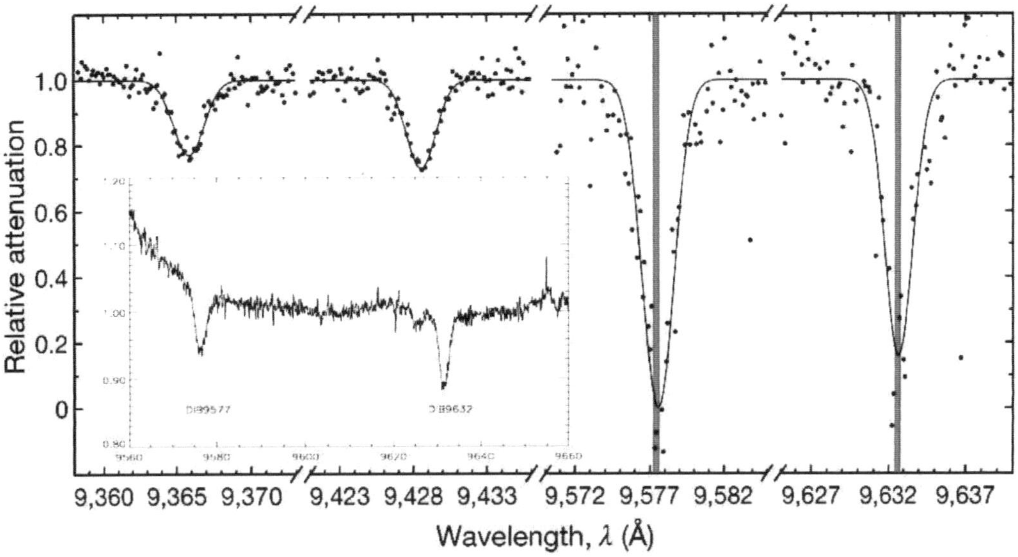

Figure 4. . Laboratory spectrum of C_{60}^{+} embedded in a He matrix (dots with continuous line fit) [32] and astronomical spectra of diffuse interstellar bands at 9577 and 9632 Å observed toward the star HD 183143 [28]. The vertical lines are at the observed central wavelengths of the two DIBs. The two weaker absorptions in the laboratory spectrum at shorter wavelengths have also been detected in space [33].

The most recent of these proposed candidates are polycyclic aromatic hydrocarbons (PAHs) [23,24,25], fullerenes [26,27] and fulleranes [28]. The arguments in their favor have not been based on wavelength correspondences. Instead the proposers have noted the stability of these classes of molecules against destruction, their compatibility as carbon-based compounds with abundance constraints, and either their known presence or likely presence, based on astrophysical and astrochemical considerations, in the diffuse interstellar medium. This approach has been necessary, as not only have accurate wavelengths of optical and near-IR transitions of these species as free molecules not been known (and indeed are extremely difficult to measure in the laboratory in the gas phase or in conditions approximating the gas phase – e.g., in a matrix of low mass inert atoms), but also because there are so many DIBs and vast numbers of PAHs, fullerenes, and fulleranes, each with its own unique spectrum.

Nevertheless, definitive proof of an association between a DIB and a carrier does require an accurate wavelength match. The lack of specific match of any DIB to a member of any of the above classes began to change in the early 1990s when laboratory spectroscopy of the fullerene C_{60}^{+} embedded in a Ne matrix revealed transitions at 9590 Å and 9640 Å [29]. It was expected that in the gas phase these lines would occur at slightly shorter wavelengths, but the precise wavelength shift was uncertain. Shortly thereafter two DIBs were found at wavelengths consistent with the expected shift and proposed to be due to C_{60}^{+} [30,31]. Finally, last year a laboratory spectrum of C_{60}^{+} trapped in dense, gaseous, and cold helium demonstrated that the central wavelengths of the two bands accurately match those observed in space, as shown in Fig. 4 [32]. Further confirmation has come with the detection in space of two additional DIBs, whose wavelengths match two weaker absorptions in the laboratory spectrum in Fig. 4 [33]. Based on these successes it seems plausible that fullerenes and their analogues produce many of the DIBs. Verifying that would likely require a large number laboratory studies.

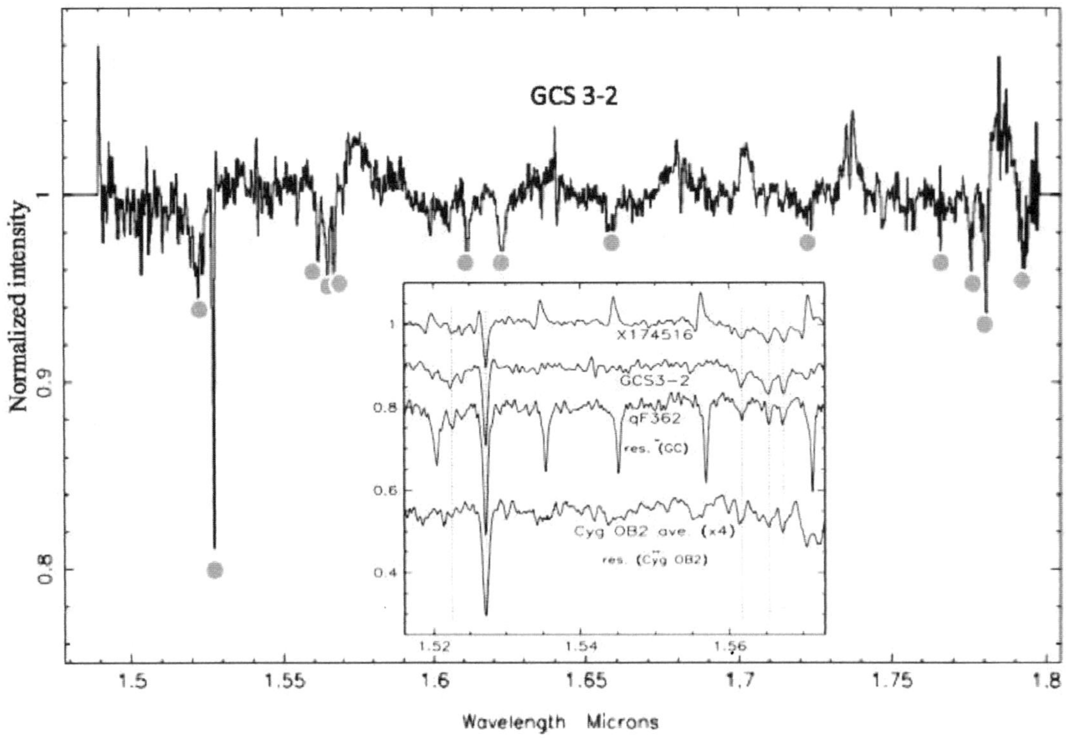

Figure 5. Thirteen diffuse interstellar bands, denoted by large dots, in the 1.5–1.8 μm interval toward the Galactic center source GCS3-2 and other stars [35]. The inset shows in more detail some of these DIBs toward three Galactic center sources and Cyg OB2.

7. New infrared DIBs

The vast majority of DIBs are at optical and near-infrared wavelengths. The four absorptions of C_{60}^{+} in Fig. 4 are not the longest wavelength DIBs. Prior to their discovery, in 1990, two DIBs, at 1.18 and 1.31 μm, had been found in spectra toward optically bright stars on known DIBs sightlines [34]. For two decades they were the longest wavelength DIBs known.

Recently a much longer wavelength set of DIBs were discovered at 1.5–1.8 μm toward stars in the Galactic center [35], for which spectroscopy at wavelengths below $\sim$1.0 μm is virtually impossible due to the extremely high extinction. These are shown in Fig. 5. Some of them, as well as several additional DIBs, mostly in the 1.0–1.3 μm region have also been found toward optically bright stars on known DIBs sightlines with much lower extinctions [35,36,37]. The infrared DIBs can be employed to investigate the diffuse interstellar medium in directions that are too highly obscured for optical spectroscopy. Recently the strongest of these DIBs, at 1.53 μm, has been used to map the Galaxy's interstellar medium [38].

8. Conclusion

Considerable progress has been made recently in understanding DIBs behavioral patterns, isolating DIBs families, and especially in convincingly identifying a few of the DIBS. It is possible that fullerenes and their analogues, and perhaps eventually PAHs, will be shown to be the keys to understanding a significant fraction of the DIBs. The recent success in linking four DIBs to C_{60}^{+} demonstrates that, while proposed identifications based on blind suggestions and wavelength coincidences alone are unlikely to be a successful way forward, educated guesses followed by laboratory spectroscopy can succeed. Clearly big challenges remain to be overcome in order to more fully solve this great mystery in astronomical spectroscopy.

9. Acknowledgements
This work was supported by the Gemini Observatory, which is operated by the Association of Universities for Research in Astronomy, Inc., on behalf of the international Gemini partnership of Argentina, Brazil, Canada, Chile, and the United States of America. The author thanks T. P. Snow for helpful comments on the original draft of this paper.

References
[1] Herbig G H 1995 *Ann. Rev. Astron. Astrophys.* **33** 19
[2] Snow T P 2013 *Proc. Int. Astron. Union* (S297) **9** 3
[3] Heger M L 1922 *Lick Obs. Bull.* **10** 146
[4] McCall B J and Griffin R E 2013 *Proc. Roy. Soc. A* **469** 2151
[5] Merrill P W and Wilson O C 1938 *Astrophys. J.* **87** 9 (see also references therein)
[6] Herbig G H 1975 *Astrophys. J.* **196** 129
[7] Hobbs L M *et al* 2008 *Astrophys. J.* **280** 1256
[8] Hobbs L M *et al* 2009 *Astrophys. J.* **705** 32
[9] Lan T-W, Ménard B and Zhu G 2015 *Mon. Not. R. Astron. Soc.* **452** 3629
[10] Jenniskens P and Désert F-X 1994 *Astron. Astrophys. Suppl. Series* **106** 39
[11] Thorburn J A, Hobbs L M, McCall B J, Oka T, Welty D E, Friedman S D, Snow T P, Sonnentrucker P and York D G 2003 *Astrophys. J.* **584** 339
[12] Adámkovics M, Blake, G A and McCall B J 2005 *Astrophys. J.* **625** 857
[13] Krelowski J and Walker G A H 1987 *Astrophys. J.* **312** 860
[14] McCall B J 2008 http://meetings.aps.org/link/BAPS.2008.DAMOP.Q2.3
[15] Smith W H, Snow T P and York D G *Astrophys. J.* **218** 124
[16] Sarre P J, Miles J R, Kerr T H, Hibbins R E, Fossey S J and Somerville W B 1995 *Mon. Not. R. Astron. Soc.* **277** L41
[17] Ehrenfreund P and Foing B H 1996 *Astron. Astrophys.* **307** L25
[18] Kerr T H, Hibbins, R E, Miles J R, Fossey S J, Somerville W B and Sarre P J 1996 *Mon. Not. R. Astron. Soc.* **283** 1105
[19] Bernstein L S, Clark F O, Lynch D K and Galazutdinov G A 2015 *Astrophys. J.* **801** 6
[20] Adamson A J and Whittet D C B 1995 *Astrophys. J.* **448** L49
[21] Cox N L J 2011 *EAS Publ. Series* **46** 349
[22] Snow T P and McCall B J 2006 *Ann. Rev. Astron. Astrophys.* **44** 367
[23] van der Zwet G P and Allamandola L J 1985 *Astron. Astrophys.* **146** 76
[24] Léger A and D'Hendecourt, L 1985 *Astron. Astrophys.* **146** 81
[25] Crawford M K, Tielens A G G M and Allamandola L J 1985 *Astrophys. J.* **293** L45
[26] Kroto H W 1988 *Science* **242** 1139
[27] Léger A, D'Hendecourt L, Verstraete L and Schmidt W 1988 *Astron. Astrophys.* **203** 145
[28] Webster A 1993 *Mon. Not. R. Astron. Soc.* **263** 385
[29] Fulara J, Jakobi M and Maier J 1993 *Chem. Phys. Lett* **211** 227
[30] Foing B H and Ehrenfreund P 1994 *Nature* **369** 296
[31] Foing B H and Ehrenfreund P 1997 *Astron. Astrophys.* **319** L59
[32] Campbell E K, Holz M, Gerlich D and Maier J P 2015 *Nature* **523** 322
[33] Walker G A H, Bohlender D A, Maier J P and Campbell E K 2015 *Astrophys. J.* **812** L8
[34] Joblin C, Maillard J P, D'Hendecourt, L and Léger A 1990 *Nature* **346** 729
[35] Geballe T R, Najarro F, Figer D F, Schlegelmilch, B W and de la Fuente D 2011 *Nature* **479** 200
[36] Cox N L J, Cami J, Kaper L, Ehrenfreund P, Foing B H, Ochsendorf B B, van Hooff S H M and Salama F 2014 *Astron. Astrophys.* **569** A117
[37] Hamano S *et al* 2015 *Astrophys. J.* **800** 137
[38] Zasowski G *et al* 2015 *Astrophys. J.* **798** 35

Tom Geballe listening to a question from the audience after his talk.

Carbon and silicate dust formation in V1280 Sco

I Sakon[1], S Sako[2], T Oanaka[1], T Nozawa[3], Y Kimura[4], T Fujiyoshi[5], T Shimonishi[6], F Usui[1], H Takahashi[2], R Ohsawa[2], A Arai[7], M Uemura[8], T Nagayama[9,10], B-C Koo[11] and T Kozasa[12]

[1] Department of Astronomy, Graduate Schools of Science, University of Tokyo, 7-3-1 Hongo, Bunkyo-ku, Tokyo 113-0033, Japan
[2] Institute of Astronomy, University of Tokyo, 2-21-1 Ohsawa, Mitaka, Tokyo 181-0015, Japan
[3] National Astronomical Observatory of Japan, Mitaka, Tokyo 181-8588, Japan
[4] Institute of Low Temperature Science, Hokkaido University, Sapporo 060-0819, Japan
[5] Subaru Telescope, National Astronomical Observatory of Japan, 650 North Aohoku Place, Hilo, HI 96720, USA
[6] Frontier Research Institute for Interdisciplinary Sciences, Tohoku University, Aramaki aza Aoba 6-3, Aoba-ku, Sendai 980-8578, Japan
[7] Koyama Astronomical Observatory, Kyoto Sangyo University, Motoyama, Kamigamo, Kita-ku, Kyoto, 603-8555, Japan
[8] Hiroshima Astrophysical Science Center, Hiroshima University, Kagamiyama 1-3-1, Higashi-Hiroshima 739-8526, Japan
[9] Department of Physics and Astronomy, Graduate School of Science and Engineering, Kagoshima University, 1-21-35 Korimoto, Kagoshima 890-0065, Japan
[10] Department of Astrophysics, Nagoya University, Furo-cho, Chikusa-ku, Nagoya, Aichi 464-8602, Japan
[11] Department of Physics and Astronomy, Seoul National University , 1 Gwanak-ro, Gwanak-gu, Seoul 151-742, Korea
[12] Department of Cosmosciences, Graduate School of Science, Hokkaido University, Sapporo 060-0810, Japan

E-mail: [1] isakon@astron.s.u-tokyo.ac.jp

Abstract. This study investigates the temporal evolution of the infrared emission from the dusty nova V1280 Sco over 2000 days from the outburst. We have revealed that the infrared spectral energy distributions at 1272, 1616 and 1947 days are explained by the emissions produced by amorphous carbon dust of mass $(6.6–8.7)\times10^{-8}$ $M_\odot$ with a representative grain size of 0.01 μm and astronomical silicate dust of mass $(3.4–4.3)\times10^{-7}$ $M_\odot$ with a representative grain size of 0.3–0.5 μm. Both of carbon and silicate dust travel farther away from the white dwarf without an apparent mass evolution throughout those later epochs.

1. Introduction

Classical novae are valuable phenomena which provide us unique opportunities to investigate the process of dust formation in the stellar ejecta. The composition of dust formed in the stellar ejecta (e.g., carbonaceous or silicate dust) depends on the C/O ratio in the envelope [1]. However, recent infrared observations of some dusty novae have revealed the presense of both carbonaceous and silicate dust: e.g., Nova V842 Cen [2]; Nova QV Vul [3], Nova V705 Cas [4]. Infrared monitoring observations of nearby dusty novae over several years from their outbursts are crucial in identifying the origin of such dual chemistry.

V1280Sco is a dusty nova discovered on 2007 February 4.86 by Y. Nakamura and Y. Sakurai [5]. It is located at the distance of 1.1 kpc [6]. Dust formation has been reported on 23 days after the discovery [7, 8]. Long-term photometric monitoring observations of V1280Sco in the optical and the near-infrared have shown that the light curve evolution of V1280Sco is extremely slow and, therefore, the white dwarf mass of V1280Sco is expected to be 0.6 $M_\odot$ or smaller [9, 6]. During the initial 150 days from the discovery, the VLTI/AMBER and MIDI interferometric observations of V1280Sco were carried out by Chesneau et al. [10] and linearly expanding dust shell with 0.35 mas day^{-1} was recognized.

In this proceedings paper, we report the results of spectral energy distribution (SED) analyses based on the late-time multi-epoch infrared observations of V1280 Sco (see [11] for details).

2. Dust Properties Formed in V1280 Sco

Mid-infrared imaging and spectroscopic observations of V1280 Sco were carried out on 2007 July 7 (Day 150) with Subaru/COMICS and on 2010 August 1 (Day 1272), 2011 July 10 (Day 1616), and 2012 June 6 (Day 1947) with Gemini-South/TReCS. The infrared spectral energy distributions (SEDs) at Days 150, 1272, 1616 and 1947 are shonw in Figure 1.

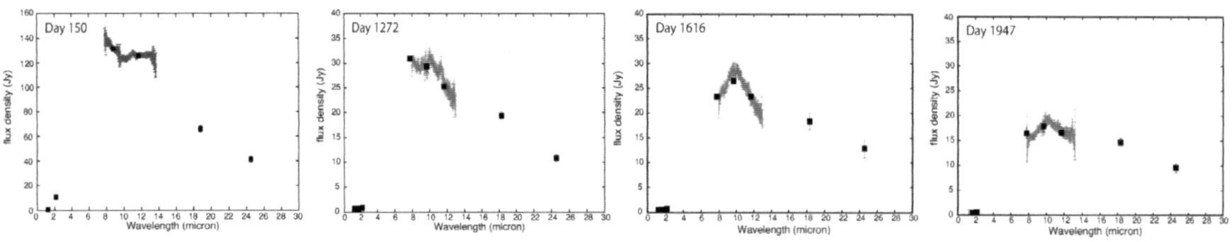

Figure 1. The infrared spectral energy distributions of V1280 Sco collected on Days 150, 1272, 1616 and 1947.

A notable point in the near- to mid-infrared SEDs of V1280 Sco at Days 1272, 1616 and 1947 is a clear appearance of the 9.7 μm and 18 μm amorphous silicate features in emission, which were not seen in the SED obtained at Day 150. The mid-infrared image collected at Day 1947 with Gemini-S/TReCS is consistent with the geometry of dusty nebulae ejected in the bipolar directions (see Figure 2), we assume a simple geometry in which both of the amorphous carbon and astronomical silicate components are distributed in bipolar cones with a common opening angle ϕ and is viewed edge on (see also [12]). We assume the density distributions of amorphous carbon (e.g., [13, 14]) [$n_{a.car.}(r)$] and astronomical silicate (e.g., [15, 14]) [$n_{a.sil.}(r)$] to be

$$n_{a.car.}(r) = \begin{cases} 0 & (r < r^{in}_{a.car.}) \\ n^{ini}_{a.car.} \times \left(\dfrac{r}{r^{in}_{a.car.}}\right)^2 & (r^{in}_{a.car.} < r < r^{out}_{a.car.}) \\ 0 & (r^{out}_{a.car.} < r) \end{cases} \tag{1}$$

and

$$n_{a.sil.}(r) = \begin{cases} 0 & (r < r^{in}_{a.sil.}) \\ n^{ini}_{a.sil.} \times \left(\dfrac{r}{r^{in}_{a.sil.}}\right)^2 & (r^{in}_{a.sil.} < r < r^{out}_{a.sil.}), \\ 0 & (r^{out}_{a.sil.} < r) \end{cases} \tag{2}$$

respectively, as a function of distance r from the white dwarf, where $r^{in}_{a.car.}$ and $r^{in}_{a.sil.}$ are the inner wall radii, $r^{out}_{a.car.}$ and $r^{out}_{a.sil.}$ are the outer wall radii, and $n^{ini}_{a.car.}$ and $n^{ini}_{a.sil.}$ are the initial number densities at the position of the inner wall for amorphous carbon and astronomical silicate, respectively. We assume the stellar parameters for the white dwarf at Days 1272, 1616, and 1947 as $T_{\rm WD} = 1.0 \times 10^5$ K and $L_{\rm WD} = 1.8 \times 10^4 L_\odot$ [16].

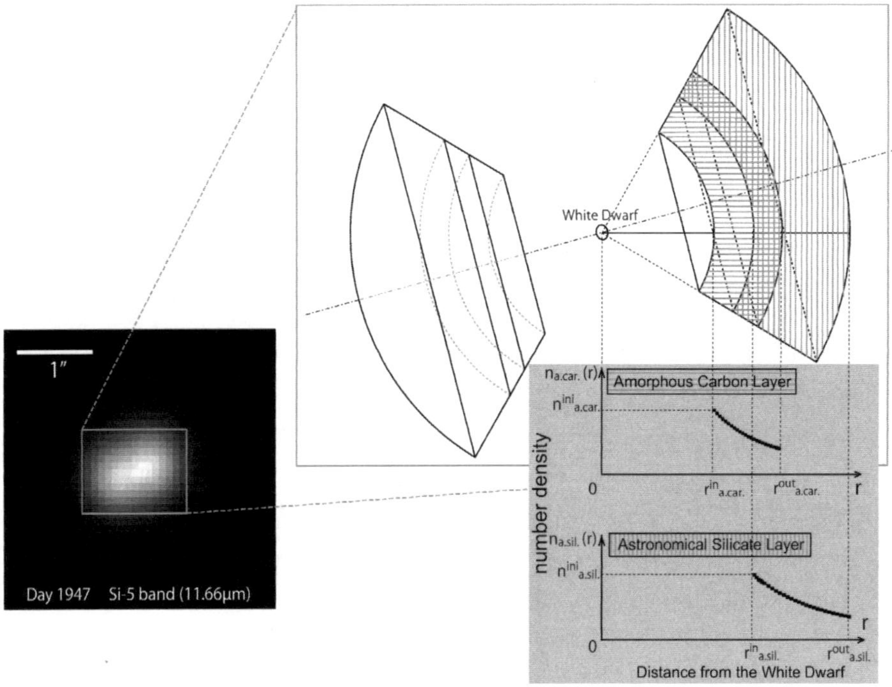

Figure 2. The 11.66 μm image of V1280 Sco at Day 1947 and the schematic view of dust geometry assumed in V1280 Sco at Days 1272, 1616, and 1947.

Based on above assumptions, an infrared spectrum produced by amorphous carbon and astronomical silicate in bipolar lobes of V1280 Sco is calculated for a geometric configuration characterized by 7 free parameters; $n^{ini}_{a.car.}$, $n^{ini}_{a.sil.}$, $r^{in}_{a.car.}$, $r^{in}_{a.sil.}$, $r^{out}_{a.car.}$, $r^{out}_{a.sil.}$, and ϕ. The optical depth at 9.7μm, $\tau_{9.7} = 0.24 \pm 0.01$, due to interstellar silicate towards V1280 Sco is fixed. Least-squares fitting is carried out to obtain a set of 7 best-fit parameters at Days 1272, 1616 and 1947 for each case of different grain sizes ($a_{a.car.}$,$a_{a.sil.}$)=(0.1, 0.1), (0.01, 0.01), (0.01, 0.1), (0.01, 0.2), (0.01, 0.3) and (0.01, 0.5) in μm. Particularly, inner wall radii $r^{in}_{a.car.}$ and $r^{in}_{a.sil.}$ are quite sensitive to the grain size. The values of $r^{in}_{a.car.}$ and $r^{in}_{a.sil.}$ obtained as a result of the fitting are shown in Figure 3. Because $r^{in}_{a.sil.}$ must be close to the semi-major axis of the deconvolved 7.73μm image of V1280 Sco at Day 1272 (i.e. 0".27) and $r^{in}_{a.sil.}$ must be smaller than the semi-major axis of the deconvolved 24.56μm image of V1280 Sco at Day 1947 (i.e. 0".45), the typical grain size of amorphous carbon is estimated as $a_{a.car.} = 0.01$ μm and that of astronomical silicate as 0.3 μm< $a_{a.sil.}$. We also find that $r^{in}_{a.sil.}$ is always larger than $r^{in}_{a.car.}$ when $a_{a.car.} = 0.01$ μm and $a_{a.sil.} < 0.5$ μm. As a conclusion, we obtained the typical size of amorphous carbon dust formed around V1280 Sco as $a_{a.car.}$ =0.01 μm and its mass as $M_{a.car.} \sim 8\times10^{-8}$ M$_\odot$, and the corresponding values for astronomical silicate as 0.3μm $< a_{a.sil} <$ 0.5μm and $M_{a.sil.} \sim 4\times10^{-7}$ M$_\odot$.

The dust formation scenario around V1280 Sco suggested from our analyses is that the amorphous carbon dust is formed in the nova ejecta followed by the formation of silicate dust in the expanding nova ejecta or as a result of the interaction between the nova wind and the circumstellar medium. Cool clumpy clouds produced via interactions between the pre-existing

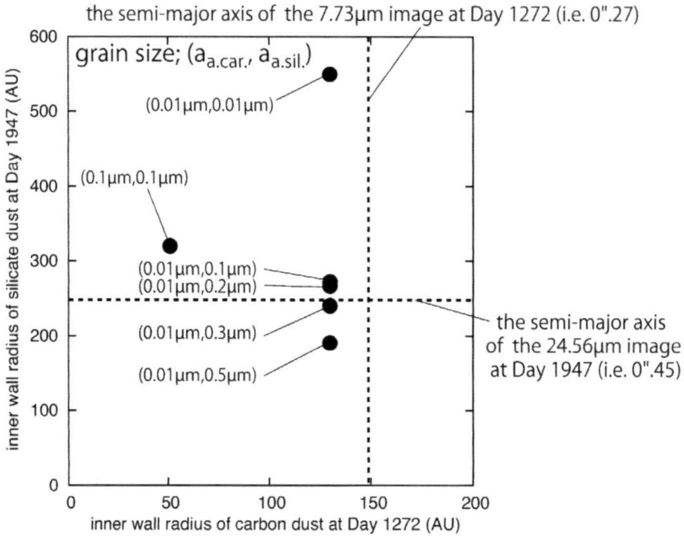

Figure 3. The plots of inner wall radii $r^{in}_{a.car.}$ at Day 1272 and $r^{in}_{a.sil.}$ at Day 1947 obtained as a result of the fitting for each case of different grain sizes $(a_{a.car.}, a_{a.sil.})$=(0.1, 0.1), (0.01, 0.01), (0.01, 0.1), (0.01, 0.2), (0.01, 0.3) and (0.01, 0.5) in μm.

circumstellar gas and high velocity nova ejecta may have provided a favorable condition for the silicate dust to attain a large grain size (i.e., $0.3\mu\text{m} < a_{a.sil} < 0.5\mu\text{m}$).

References

[1] Waters L B F M 2004 *ASP Conf. Series* **309** 229
[2] Smith C H, Aitken D K and Roche P F 1994 *Mon. Not. R. Astron. Soc.* **267** 225
[3] Gehrz R D, Jones T J, Woodward C E, Greenhouse M A, Wagner R M, Harrison T E, Hayward T L and Benson J 1992 *Astrophys. J.* **400** 671
[4] Evans A, Tyne V H, Smith O, Geballe T R, Rawlings J M C and Eyres S P S 2005 *Mon. Not. R. Astron. Soc.* **360** 1483
[5] Yamaoka H, Nakamura Y, Nakano S, Sakurai Y and Kadota K 2007 *IAU Circ.* **8803** 1
[6] Naito H *et al* 2012 *Astron. Astrophys.* **543** 86
[7] Das R K, Ashok N M and Banerjee D P K 2007 *CBET* **866** 1
[8] Rudy R J *et al* 2007 *Bul. Am. Astron. Soc.* **39** 817
[9] Hounsell R *et al* 2010 *Astrophys. J.* **724** 480
[10] Chesneau O *et al* 2008 *Astron. Astrophys.* **487** 223
[11] Sakon I *et al* 2016 *Astrophys. J.* **817** 145
[12] Chesneau O *et al* 2012 *Astron. Astrophys.* **545** A63
[13] Zubko V G 1996 *Mon. Not. R. Astron. Soc.* **283** 577
[14] Dwek E and Smith R K 1996 *Astrophys. J.* **459** 686
[15] Draine B T 1985 *Astrophys. J. Suppl. Series* **57** 587
[16] Kato M and Hachisu I 1994 *Astrophys. J.* **437** 802

Journal of Physics: Conference Series **728** (2016) 062007 doi:10.1088/1742-6596/728/6/062007

Comparison between 30 micron sources in different galaxies

Marcin Gładkowski[1], **Ryszard Szczerba**[1], **G. C. Sloan**[2], **Eric Lagadec**[3] **and Kevin Volk**[4]

[1] Nicolaus Copernicus Astronomical Center, ul. Rabiańska 8, 87 - 100 Toruń, Poland
[2] Cornell Center for Astrophysics and Planetary Science, Cornell University, Ithaca, NY 14853-6801, USA
[3] Laboratoire Lagrange, UMR 7293, Univ. Nice Sophia-Antipolis, CNRS, Observatoire de la Côte d'Azur, F-06300, Nice, France
[4] Space Telescope Science Institute, 3700 San Martin Drive, Baltimore, Maryland, U.S.A.

E-mail: [1] `seyfert@ncac.torun.pl`

Abstract. We present an analysis and comparison of the 30 μm dust feature in the spectra of carbon-rich objects located in the Milky Way, Magellanic Clouds and the Sagittarius dwarf spheroidal galaxy. These spectra were collected by Spitzer Space Telescope. All of these galaxies are characterized by the different metallicities. We expect that some physical and chemical processes related to the formation of the feature are a function of the metallicity. Our study should allow us to better understand the mass loss process and thus late stages of stellar evolution of carbon-rich stars in these four galaxies. Our analysis uses the "Manchester method" as a basis of estimating the temperature of dust for the carbon-rich stars and the planetary nebulae in our sample. In the case of post-AGB objects we changed the spectral ranges used for the temperature estimation, because of the presence of the 21 μm feature. We used a blackbody function with a single temperature deduced from the Manchester method to approximate the continuum under the 30 μm feature.

We have produced on-line catalogues of photometric and Spitzer IRS spectra for all objects that show the 30 μm feature. These resources are available on-line for use by the community.

1. Introduction

A broad emission dust feature peaking around 30 μm is seen in the spectra of some carbon-rich Asymptotic Giant Branch (AGB) stars, post-AGBs and planetary nebulae (PNe). This feature was observed for the first time by [1] who noticed it in the CW Leo and two other objects. Since the discovery this dust feature has been detected in many carbon-rich objects. In the carbon-rich post-AGB objects this feature is observed along with the 21 μm feature (the carrier of which is still unidentified), and is often extremely strong. The feature is generally weaker in PNe spectra.

[2] proposed solid magnesium sulfide (MgS) as the possible carrier of the 30 μm feature. Despite the MgS identification being widely accepted by the community, its identification remains unconfirmed. Other authors have proposed different materials as possible carriers of the 30 μm feature, such as hydrogenated amorphous carbon [3] or graphite [9]. It has been suggested that the feature arises from two-component grains rather than a simple homogeneous dust

particle. For instance, [8] suggested that MgS condenses as a coating on the top of amorphous carbon and SiC grains.

[5] have analysed a sample of 63 Galactic 30 μm feature spectra collected by Infrared Space Observatory. This is the biggest sample which has been analysed in a uniform way so far. The subsequent Spitzer Space Telescope mission (hereafter Spitzer), thanks to its sensitivity, was able to detect sources with 30 μm feature in nearby galaxies such as the Magellanic Clouds and the Sagittarius dwarf spheroidal galaxy (hereafter Sgr dSph).

2. The Toruń catalogues of 30 μm objects

The 30 μm feature spectra were identified by visual examination of all Spitzer *Infrared Spectrograph* (IRS) spectra for objects in the Magellanic Clouds, as well as spectra from programs observing carbon-rich objects in the Galaxy or Sgr dSph. We have a total 180 objects whose IRS spectra show the 30 μm feature, divided into three catalogues: 15 objects from the Small Magellanic Cloud, 100 objects from the Large Magellanic Cloud, and 65 objects divided between the Galaxy (60) and the Sgr dSph (5). The Sgr dSph sources are distinguished by special comments in the third catalogue. The catalogues are available on-line at:

- `http://www.ncac.torun.pl/postagb_30smc`,
- `http://www.ncac.torun.pl/postagb_30lmc`,
- `http://www.ncac.torun.pl/postagb_30galactic`.

The catalogues contain photometric data from Spitzer, WISE, 2MASS, 2MASS 6X, MCPS, AKARI and IRAS. The Galactic catalogue contains data from the GSC and USNO as well. Each catalogue also contains the Spitzer spectra.

3. Spectral analysis

3.1. The spectral sample

Our sample consists of archival data obtained by the Spitzer IRS instrument. We use optimally extracted low and high resolution spectra ([6]) from various observational programs. The spectral coverage is from 5 to 38 μm in the low resolution spectra. Discontinuities were removed between orders by multiplying each spectral segment by a scale factor to align them in the regions of overlap. For the most of objects the corrections are made up to the brightest segment, on the grounds that this segment is the one best centered in the slit.

3.2. Continuum

Our analysis is based on the "Manchester Method", which was introduced by [11, 10]. It uses two colour indices, [6.4]−[9.3] and [16.5]−[21.5]. The [6.4]−[9.3] colour is calculated by separately summing the total spectral emission from 6.25 to 6.55 μm and from 9.1 to 9.5 μm. [11] showed that the [6.4]−[9.3] colour provides a good estimate of the dust optical depth, whereas the [16.5]−[21.5] colour serves as an indicator of the dust temperature We summed over the regions 16–17 μm and 21–22 μm) to simulate the second colour value. The [6.4]−[9.3] colour shows a linear correlation with the measured mass-loss rates [4]. We used this method in the original form for the carbon-rich AGB objects and PNe.

In the case of post-AGB stars we defined three additional colour indices to estimate the dust temperature T_d, depending the strength of the 21 μm feature. Generally, if the 21 μm is not visible or is very weak we used the [18.4]−[22.45] colour (summing the emission from 18.1–18.7 and 22.3–22.6 μm) to estimate the dust temperature. When the 21 μm feature is relatively strong we took the [18.4]−[22.75] colour (summing the emission from 18.1–18.7 and 22.5–23 μm). For objects with a very strong 21 μm feature we used the [17.95]−[23.2] colour (summing the emission from 17.8–18.1 and 23–23.4 μm).

Journal of Physics: Conference Series **728** (2016) 062007 doi:10.1088/1742-6596/728/6/062007

To model the underlying continuum we use a blackbody emission with a single temperature derived from the colours discussed above. Then we fit the blackbody continuum level at specific wavelengths, which depend on the kind of object. The typical normalization range for the carbon-rich AGB stars is 16–22 μm, for carbon-rich post-AGB stars without the 21 μm feature we use 19.5–22.5 μm, and for the PNe we use 20–23 μm.

After subtraction of the continuum we measured the strength of the 30 μm feature, which is defined as the ratio between integrated flux (from 24 to 36 μm) of the feature and the integrated flux of the continuum (L/C).

4. Circumstellar properties

Fig. 1 shows the strength of the feature (L/C) as a function of the dust temperature. This feature tends to appear in the envelopes of objects with cool dust. In the case of AGB stars, the strength of the feature increases as the dust temperature decreases until about 400 K. After that, probably self-absorption reduces the strength of the 30 μm feature. The coolest objects with this feature are post-AGBs and PNe. However, it seems that there is no correlation between L/C and T_d. Fig. 2 shows the strength of the feature (L/C) as a function of [6.4]−[9.3] colour. The Galactic objects show a range of values of [6.4]−[9.3] colour. In the LMC the 30 μm emission become visible for the redder values of [6.3]−[9.4] colour ($\sim$ 0.5 mag). The SMC objects become visible around [6.4]-[9.3] $\sim$ 0.75 mag. This trend was also described by [10, 7]. This behaviour is consistent with a metallicity effect, because in the lower metallicity environments there is less Mg and S to form MgS, thus a higher dust production rate is needed to create the feature.

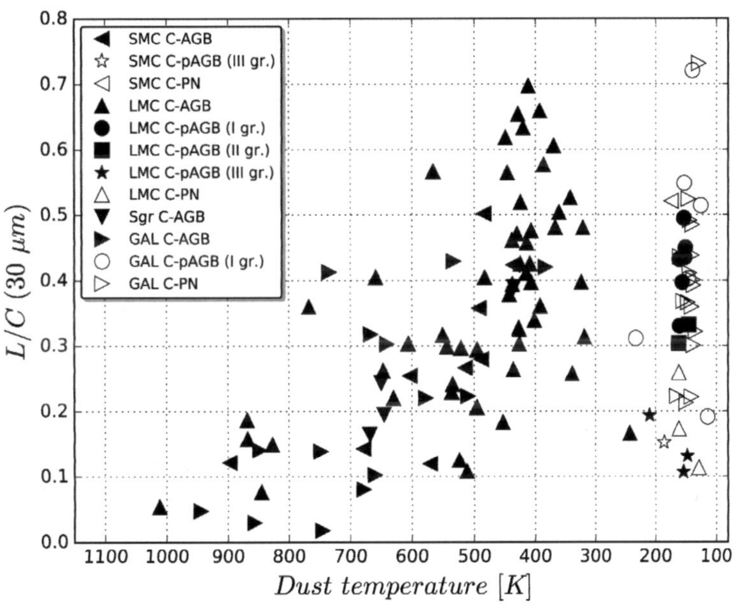

Figure 1. The strength of the 30 μm feature as a function of the dust temperature.

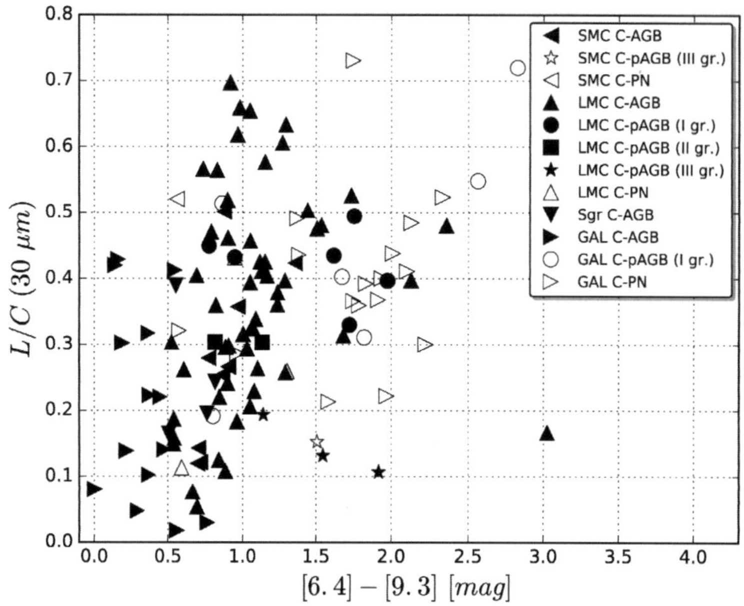

Figure 2. The strength of the feature as a function of [6.4]−[9.3] colour.

5. Summary/Conclusions

We present a large sample of 180 Spitzer IRS spectra of carbon-rich objects that exhibit the 30 μm feature, and a simple approach to determined the continuum under the 30 μm feature. The strength of the 30 μm feature shows a correlation with the T_d for AGB stars, and dependence on the metallicity of host galaxy. The full analysis of the 30 μm feature in our and nearby galaxies will appear elsewhere (Gładkowski et al. 2016, in preparation).

Acknowledgments

This work was financially supported by the MNiSW of Poland through grant No. 2014/15/N/ST9/04629. We have made extensive use of the SIMBAD and Vizier databases operated at the Centre de Données Astronomiques de Strasbourg. We also collected low and high resolution spectra from The Cornell Atlas of Spitzer/IRS Sources (CASSIS), which is a product of the Infrared Science Center at Cornell University, supported by NASA and JPL. Special thanks to Gregory Sloan for providing optimally extracted spectra for the most of objects from our sample.

References

[1] Forrest W J, Houck J R and McCarthy J F 1981 *Astrophys. J.* **248** 195
[2] Goebel J H and Moseley S H 1985 *Astrophys. J.* **290** L35
[3] Grishko V I, Tereszchuk K, Duley W W and Bernath P 2001 *Astrophys. J.* **558** L129
[4] Groenewegen M A T *et al* 2007 *Mon. Not. R. Astron. Soc.* **376** 313
[5] Hony S, Waters L B F M and Tielens A G G M 2002 *Astron. Astrophys.* **390** 533-53
[6] Lebouteiller V, Bernard-Salas J, Sloan G C and Barry D J 2010 *Publ. Astron. Soc. Pac.* **122** 188
[7] Leisenring J M, Kemper F and Sloan G C 2008 *Astrophys. J.* **681** 1557-73
[8] Lombaert R, de Vries B L, de Koter A, Decin L, Min M, Smolders K, Mutschke H and Waters L B F M 2012 *Astron. Astrophys.* **544** L18
[9] Otsuka M, Kemper F, Cami J, Peeters E and Bernard-Salas J 2014 *Mon. Not. R. Astron. Soc.* **437** 2577-93
[10] Sloan G C, Kraemer K E, Matsuura M, Wood P R, Price S D and Egan M P 2006 *Astrophys. J.* **645** 1118
[11] Zijlstra A A *et al* 2006 *Mon. Not. R. Astron. Soc.* **370** 1961

Journal of Physics: Conference Series **728** (2016) 072001 doi:10.1088/1742-6596/728/7/072001

Modelling the millimetre/submillimeter emission of the gas envelope of evolved axially symmetric stars

Pham Ngoc Diep*, Do Thi Hoai, Pham Tuyet Nhung, Nguyen Thi Phuong, Nguyen Thi Thao, Pham Tuan-Anh and Pierre Darriulat

Department of Astrophysics, VNSC/VAST, 18 Hoang Quoc Viet, Ha Noi, Viet Nam

E-mail: *pndiep@vnsc.org.vn

Abstract. We illustrate, with the examples of a few evolved stars, the results of considerations which we have developed with the aim of easing the modelling of their gas envelopes.

1. Introduction

The unprecedented quality of the observations available from ALMA calls for analysis methods making the best of them. We have addressed this issue in the context of the study of the millimetre/submillimeter emission of the gaseous circumstellar envelopes of evolved stars in [1], of which we give here a preliminary illustration The space reconstruction of their morphology and kinematics requires special attention to the evaluation of the orientation of the star axis in space and the differentiation between expansion along the star axis and rotation about it. We illustrate the arguments developed in [1] using concrete examples that contribute new results to earlier analyses of the CO emission of the CSE of three AGB stars. We use coordinates on the plane of the sky, y pointing east, z pointing north, $R=\sqrt{y^2+z^2}$, $r=\sqrt{x^2+R^2}$ and $\tan\varphi=z/y$. The x axis points along the line of sight away from the observer, with V_x the Doppler velocity. We call $f(y,z,V_x)$ the measured flux density with $F(y,z)=\int f(y,z,V_x)dV_x=\int \rho(x,y,z)dx$ where ρ is the effective density. The star axis makes an angle θ with the line of sight and its projection on the sky plane an angle ψ with north.

2. RS Cnc observed in CO(1-0) & CO(2-1) using IRAM telescopes [2, 3]

RS Cnc, a typical AGB, S star, having already experienced several thermal pulses with dredge up episodes, has a mass loss rate of $\sim$1.2 10^{-7} $M_\odot\,yr^{-1}$. The star axis has θ=38° and ψ=9°. It has a spherical CSE (Figure 1a,b) but displays a clear Doppler velocity asymmetry (Figure 1c,d) well described by a radial wind increasing from $\sim$2 km s^{-1} at the equator to $\sim$8 km s^{-1} at the poles. The P-V diagram (Figure 1e) shows clear expansion with no evidence for significant rotation. The angle θ is measured with a precision of 10° (Figure 1f). A small contribution of rotation about the star axis cannot be excluded and weak evidence is found in favour of stationary winds.

3. EP Aqr observed in CO(1-0) & CO(2-1) using IRAM telescopes [4, 5]

EP Aqr, a M star, joined the AGB later than RS Cnc and has a mass loss rate of 1.8 10^{-7} $M_\odot\,yr^{-1}$. It displays nearly spherical, but very lumpy, effective density and wind

Journal of Physics: Conference Series **728** (2016) 072001 doi:10.1088/1742-6596/728/7/072001

velocity maps. It was modeled earlier using a spherical r-dependent wind [4] and using a bipolar outflow with radial expansion velocity increasing from 2 km s^{-1} at equator to 10 km s^{-1} at the poles [5]. In order to obtain a unified picture, we first estimate the effective density averaged over stellar latitude in each φ-bin by solving the integral equation $F(y,z) = \int \rho(r,\alpha,\omega)dx$ with $\rho^*(r) = \rho_0^* r^{-n^*}$ (Figure 2a,b). We then use the relation $f(R,\varphi,V_x)dV_x = \rho(r,\alpha,\omega)dx$ to evaluate the expansion velocity V_{rad} in the r vs $\sin\alpha$ plane (Figure 2c,d), where α is the stellar latitude. The projections on $\sin\alpha$ (Figure 2e) and on r (Figure 2f) give evidence in favour of the second model.

4. π^1 Gru, an S-type star observed using the ALMA compact array in CO(3-2) [6]

π^1 Gru, an S star with a mass loss rate of 2.7 10^{-6} M$_\odot$ yr^{-1} and asymmetric Doppler velocities (Figure 3a), had been observed earlier in CO(2-1) using the SMA [7] and modelled as an expanding flared disk (Figure 3b). We analyse ALMA compact array data to show that the effective density and Doppler velocity vary smoothly from stellar equator to stellar poles as seen from the P-V diagram (Figure 3c) and the sine wave distributions of RF (Figure 3d) and $< V_x >$ (Figure 3e). A new model constructed along these lines gives a significantly improved description of the data and gives evidence for an oblate effective density distribution and a bipolar outflow (Figure 3f). We estimate the ratio between effective densities at stellar poles and equator not to exceed 10%.

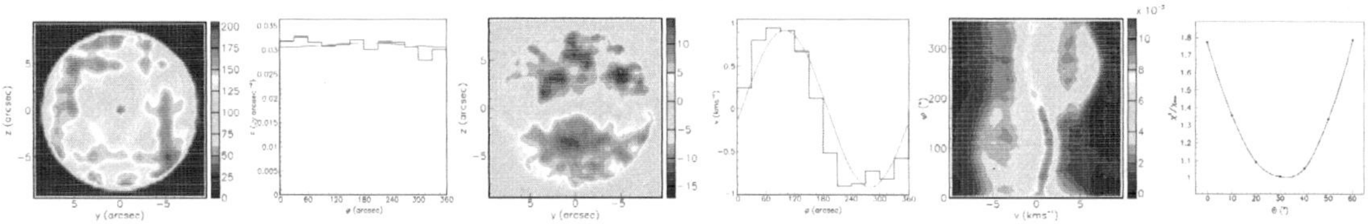

Figure 1. a) Map of RF; b) RF vs φ; c) Map of $< V_x >$; d) $< V_x >$ vs φ; e) P-V diagram; f) χ^2/χ^2_{min} vs θ.

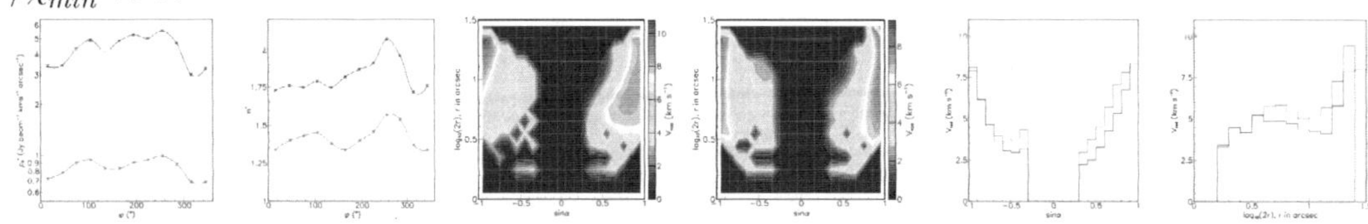

Figure 2. a) ρ_0^* vs φ; b) n^* vs φ; c,d) Maps of V_{rad}; e) V_{rad} vs $\sin\alpha$; f) V_{rad} vs r

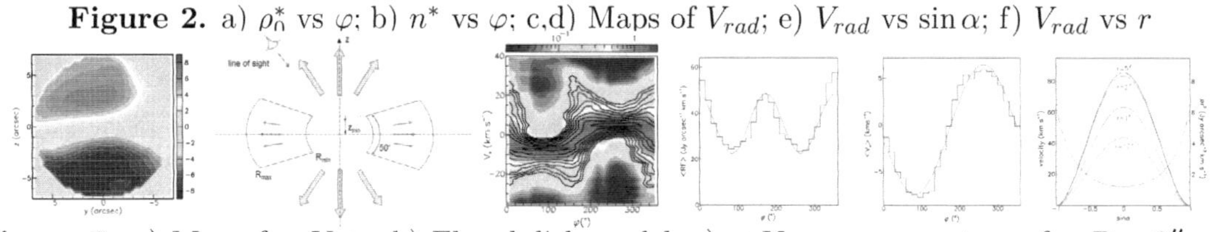

Figure 3. a) Map of $< V_x >$; b) Flared disk model; c) $< V_x >$ vs φ, contours for $R < 3''$ and colour for $R > 3''$); d) RF vs φ; e) $< V_x >$ vs φ; f) ρ and V_{rad} vs $sin\alpha$.

References

[1] Diep P N *et al* 2016 in preparation

[2] Hoai D T, Matthews L D, Winters J M, Nhung P T, Gerard E, Libert Y and Le Bertre T 2014 *Astron. Astrophys.* **565** A54

[3] Nhung P T, Thi Hoai D, Winters J M, Darriulat P, Gerard E and Le Bertre T 2015 *Research Astron. Astrophys.* **15** 713

[4] Winters J M, Le Bertre T, Pety J and Neri R 2007 *Astron. Astrophys.* **475** 559

[5] Nhung P T, Hoai D T, Winters J M, Le Bertre T, Diep P N, Phuong N T, Thao N T, Tuan-Anh P and Darriulat P 2015 *Astron. Astrophys.* **583** A64

[6] Nhung P T, Hoai D T, Diep P N, Phuong N T, Thao N T, Tuan-Anh P and Darriulat P 2015 arXiv:1601.04834

[7] Chiu P *et al* 2006 *Astrophys. J.* **645** 605

Journal of Physics: Conference Series **728** (2016) 072002 doi:10.1088/1742-6596/728/7/072002

Correlation of circumstellar SiO maser spot distribution with the stellar light curve

M. Oyadomari[1], H. Imai[1], T. Nagayama[2], T.Oyama[2], N. Matsumoto[2], J. Nakashima[3], S.-H. Cho[4]

[1] Kagoshima University, [2] National Astronomical Observatory of Japan, [3]Ural Federal University, [4]Korean Astronomy and Space Science Institute

E-mail: [1] `k7219031@kadai.jp`

Abstract. We have investigated the distributions of silicon monoxide (SiO) $v = 2$ and $v = 3$ $J = 1 \rightarrow 0$ masers around long-period variables (LPVs) in VLBI observations using the VLBI Exploration of Radio Astrometry (VERA) combined with the Nobeyama 45 m telescope. We find some examples of correlation of a maser spot distribution with the stellar light curve, which may provide a clue to elucidating the pumping mechanism of circumstellar SiO masers.

1. Introduction

There are three pumping models of circumstellar SiO masers: (1) stellar radiation pumping; (2) collisional pumping; (3) line overlapping [1, 2, 3]. The $v = 3$ $J = 1 \rightarrow 0$ SiO maser is expected as a unique probe to investigate the pumping mechanism of SiO molecules. According to the third model, the mid-infrared lines of H_2O molecules will pump SiO molecules from the $J = 0$ rotation level at lower vibration(v) levels to the $J = 1$ level at higher v-level, resulting in spatial correlation between the $v = 3$ and $v = 1/v = 2$ SiO masers. Multi-epoch VLBI mapping of these masers throughout different stellar light curve phase may enable us to discriminate a predominant pumping mechanism. In fact, it is suggested that the relative distributions of the $v = 2$ and $v = 3$ masers with respect to the central star may change with correlation with light curve [4].

2. Observations

We carried out VLBI observations of SiO $v = 2$ and $v = 3$ $J = 1 \rightarrow 0$ masers towards 12 LPVs (WX Psc, AP Lyn, U Ori, VY CMa, R Leo, RS Vir, W Hya, U Her, RU Her, V1111 Oph, V4120 Sgr, and T Cep) using four VERA 20 m telescopes and the 45 m telescope of Nobeyama Radio Observatory (NRO) on March 24–25 and May 21–22 in 2012.

3. Results and Discussion

The $v = 3$ maser emission was detected toward 5 stars. We succeeded in superimposition of $v = 2$ and $v = 3$ maser maps toward WX Psc, R Leo, W Hya (Fig.1), and T Cep (Fig.2). In T Cep, $v = 3$ masers well correlate with $v = 2$ masers within a few mas and 1 km s^{-1}, respectively. This result suggests that the H_2O–SiO line overlapping is predominant in T Cep. For T Cep and W Hya, we investigated correlation between the distributions of the two maser lines with the stellar light curve. With the visible light curves provided by the American Association of Variable Star Observers (AAVSO), we find that good spatial correlation between the SiO $v = 2$ and $v = 3$ masers is found only at $\phi \sim 0.2$, when the near infrared radiation may be at a maximum [5, 6]. The sequence of the line overlapping mechanism is as follows: (1) the $\sim 8\mu$m emission increases at $\phi \sim 0.2$; (2) SiO molecules are excited from $v = 1$ $J = 0$ to $v = 2$ $J = 1$ and from $v = 2$ $J = 0$ to $v = 3$ $J = 1$ by $11_{6,6}$ $\nu 2{=}1 \rightarrow 12_{7,5}$ $\nu 2{=}0$ and $5_{0,5}$ $\nu 2{=}2 \rightarrow 6_{3,4}$ $\nu 2{=}1$, respectively; (3) the $v = 3$ masers increase around $v = 2$ masers. We speculate that the line overlapping mechanism may be predominant only at $\phi \sim 0.2$ while the collisional pumping may be predominant in other phases.

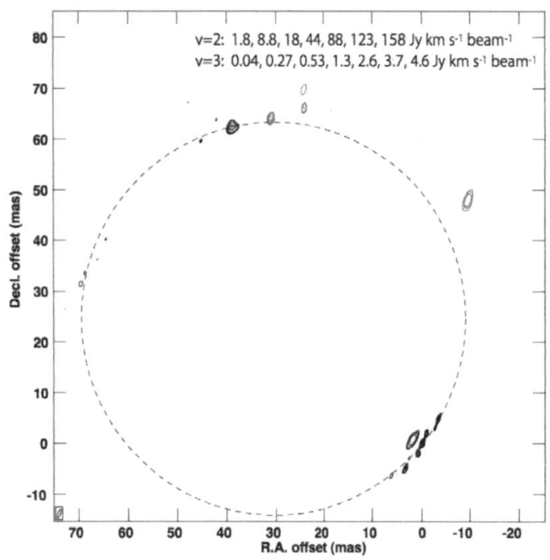

Figure 1. Composite map of $v = 2$ (black) and $v = 3$ (gray) maser lines toward W Hya. A dashed circle is drawn so as to fit to the $v = 2$ maser distribution. The $v = 3$ masers do not have any spatial correlation with $v = 2$. Interestingly, some of the $v = 3$ masers are located further away from the central star than the $v = 2$ masers.

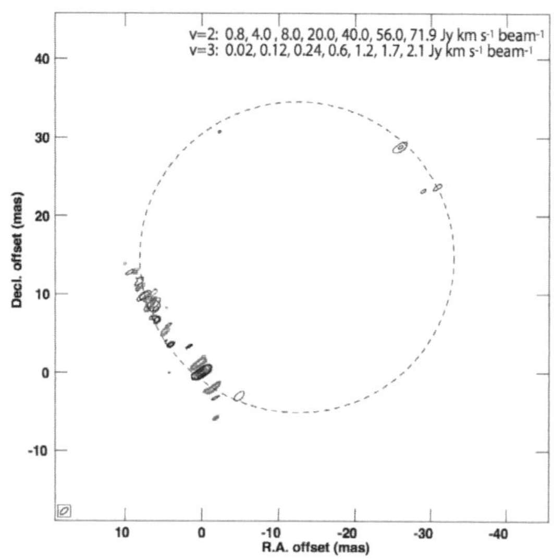

Figure 2. Same as Fig.1 but toward T Cep. These masers exhibit good correlation in position.

References

[1] Lockett P and Elitzur M 1992 *Astrophys. J.* **399** 704
[2] Bujarrabal V 1994 *Astron. Astrophys.* **285** 953
[3] Desmurs J-F, Bujarrabal V, Lindqvist M, Alcolea J, Soria-Ruiz R and Bergman P 2014 *Astron. Astrophys.* **565** A127
[4] Imai H, Nakashima J, Deguchi S, Yamauchi A, Nakagawa A and Nagayama T 2010 *Publ. Astron. Soc. Japan* **62** 431
[5] Smith B J, Price S D and Moffett A J 2006 *Astron. J.* **131** 612
[6] Zhao-Geisler R, Quirrenbach A, Köhler R, Lopez B and Leinert C 2011 *Astron. Astrophys.* **530** A120

Journal of Physics: Conference Series **728** (2016) 072003 doi:10.1088/1742-6596/728/7/072003

Rb and Zr abundances in massive Galactic AGB stars revisited

V. Pérez-Mesa[1,2], O. Zamora[1,2], D.A. García-Hernández[1,2], B. Plez[3], A. Manchado[1,2,4], A.I. Karakas[5] and M. Lugaro[6,7]

[1] Instituto de Astrofísica de Canarias, 38200 La Laguna, Tenerife, Spain

[2] Departamento de Astrofísica, Universidad de La Laguna, 38206 La Laguna, Tenerife, Spain

[3] Laboratoire Univers et Particules de Montpellier, Université Montpellier2, CNRS, 34095 Montpellier, France

[4] Consejo Superior de Investigaciones Científicas, 28006 Madrid, Spain

[5] Research School of Astronomy & Astrophysics, Australian National University, Canberra, ACT 2611, Australia

[6] Konkoly Observatory, Research Centre for Astronomy and Earth Sciences, Hungarian Academy of Sciences, 1121 Budapest, Hungary

[7] Monash Centre for Astrophysics, Monash University, VIC3800, Australia

E-mail: [1] `vperezme@iac.es`

Abstract. We report new abundances of Rb and Zr in a sample of massive Galactic asymptotic giant branch (AGB) stars that were previously studied with hydrostatic models by using more realistic dynamical model atmospheres. We use a modified version of the spectral synthesis code Turbospectrum, and consider the presence of a circumstellar envelope and a radial wind in the modelling of these Galactic AGB stars. The Rb and Zr are determined from the 7800 Å Rb I resonant line and the 6474 Å ZrO bandhead, respectively, and they are compared with the AGB nucleosynthesis theoretical predictions. The derived Rb abundances are much lower (~1-2 dex) with the new dynamical models, while the Zr abundances, however, are closer to the hydrostatic values. The new model atmospheres can help to resolve the problem of the mismatch between the observations and the nucleosynthesis theoretical predictions of massive AGB stars.

1. Observational data and models

We have used high-resolution optical echelle spectra (R ~ 50,000) in a sample of 21 massive Galactic AGB stars (see [1, 2]). The S/N ratios achieved in the reduced spectra strongly vary from the blue to the red orders (typically ~ 10–20 at 6000 Å and >100 at 8000 Å). By using new dynamical model atmospheres recently developed by us [3], we obtain reliable Rb (and Zr) photospheric abundances for 12 stars in our sample. The rest of stars displayed more than one circumstellar contribution, which cannot be reproduced by our present modified version of the Turbospectrum spectral synthesis code.

2. Results

The new Rb abundances derived from our dynamical models display an average decrease of 1 dex with respect to the static case. Fig. 1 shows that our dynamical atmosphere models reproduce the observed 7800 Å Rb I line much better than the classical hydrostatic models. However, the Zr abundances derived from dynamical models are similar to those obtained with the hydrostatic models because the 6474 Å ZrO bandhead is formed deeper in the atmosphere and is less affected than the Rb I line [3].

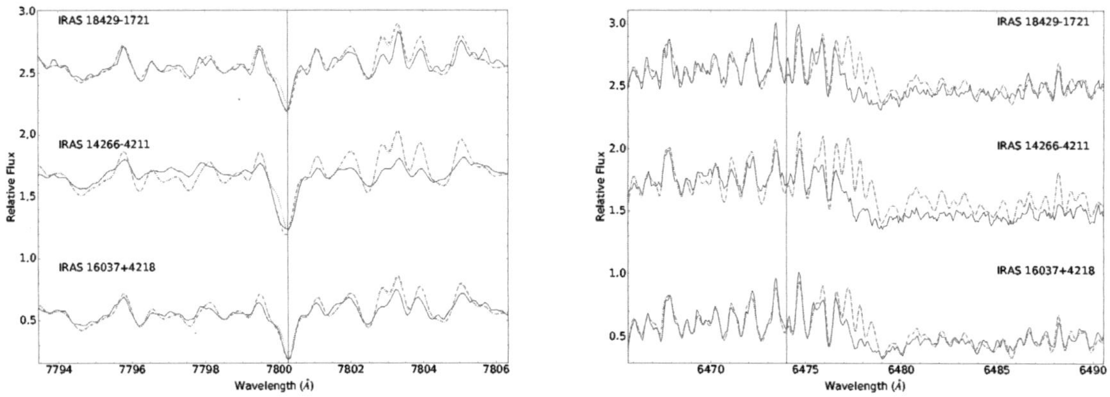

Figure 1. The Rb I 7800 Å (left panel) and ZrO 6474 Å (right panel) spectral regions in three massive Galactic AGB stars. The dynamical model best fits (dashed lines) to the observations (solid lines) in three sample stars are shown. For comparison, the hydrostatic models are also displayed (dotted lines). Note that the dynamical and hydrostatic models are identical in the right panel.

A comparison with standard nucleosynthesis models for massive AGB stars shows that the predicted Rb abundances range from [Rb/Fe] $\sim$ 0.00 up to 1.04 dex for solar metallicity stars as those studied here; the maximum Rb is found for a 6.5 $M_{\odot}$ star [4]. On the other hand, Karakas et al. [5] delayed the beginning of the superwind phase in massive AGB models and this increases the Rb production up to [Rb/Fe]=1.34 dex (for the 6 $M_{\odot}$ case).

The Rb abundances and [Rb/Zr] ratios previously derived with hydrostatic models [1, 2] represented a challenge for the theoretical models because both quantities are much larger than theoretically predicted values. The Rb abundances were correlated with the OH expansion velocity in the static case, while the Rb-V_{exp}(OH) distribution is now more flatter in the dynamical case. In short, the Rb abundances and [Rb/Zr] ratios derived here could resolve the previous mismatch between the observations of massive Galactic AGB stars and the theoretical predictions.

Acknowledgments
V.P.M., O.Z., D.A.G.H. and A.M. acknowledge support provided by the Spanish Ministry of Economy and Competitiveness (MINECO) under grant AYA-2014-58082-P. M.L. is a Momentum (Lendület-2014 Programme) project leader of the Hungarian Academy of Sciences.

3. References

[1] García-Hernández D A, García-Lario P, Plez B, D'Antona F, Manchado A and Trigo-Rodríguez J M 2006 *Science* **314** 1751
[2] García-Hernández D A, García-Lario P, Plez B, Manchado A, D'Antona F, Lub J and Halbing H 2007 *Astron. Astrophys.* **462** 711
[3] Zamora O, García-Hernández D A, Plez B and Manchado A 2014 *Astron. Astrophys.* **564** L4
[4] van Raai M A, Lugaro M, Karakas A I, García-Hernández D A and Yong D 2012 *Astron. Astrophys.* **540** 44
[5] Karakas A I, García-Hernández D A and Lugaro M 2012 *Astrophys. J.* **751** 8

Journal of Physics: Conference Series **728** (2016) 072004 doi:10.1088/1742-6596/728/7/072004

S-process nucleosynthesis in AGB stars with the full spectrum of turbulence scheme for convection.

A Yagüe[1,2,3], D A García-Hernández[2,3], P Ventura[1] and M Lugaro[4,5]

[1] Istituto Nazionale di Astrofisica - OAR, Monte Porzio Catone, Italy
[2] Instituto de Astrofísica de Canarias, La Laguna, Tenerife, Spain
[3] Departamento de Astrofísica, Universidad de La Laguna (ULL), La Laguna, Tenerife, Spain
[4] Konkoly Observatory, Hungarian Academy of Sciences, Budapest, Hungary
[5] Monash Centre for Astrophysics, Monash University, Australia

Abstract. The chemical evolution of asymptotic giant branch (AGB) stars models depends greatly on the input physics (e.g. convective model, mass loss recipe). Variations of hot bottom burning (HBB) strength, or third dredge-up (TDU) efficiency are among the main consequences of adopting different input physics in the AGB models. The ATON evolutionary code stands apart from others in that it uses the Full Spectrum of Turbulence convective model. Here we present the first results of a newly developed *s*-process nucleosynthesis module for ATON AGB models. Our results are compared also with observations and theoretical predictions of present AGB nucleosynthesis models using different input physics.

1. Motivation

Among the different asymptotic giant branch (AGB) stars evolutionary codes used in the literature, ATON[1] stands appart in that it uses the Full Spectrum of Turbulence convective model[2]. However, ATON lacked a way of simulating the nucleosynthesis due to slow neutron captures (the *s*-process). To study the dependence of the *s*-process nucleosynthesis on different input physics, we built an ATON module including 322 nuclear species up to Pb and Bi (Yagüe et al., in preparation). The nuclear network is the same as that of Lugaro et al.[3][4].

A critical part for the design is to settle in an overshooting (OV) strength parameter. This parameter represents the distance of exponential decay of velocities with which the mixing is performed. Different OV strengths would result in the *s*-process He-intershell overabundances being dredged up more efficiently to the surface.

2. Nucleosynthesis preliminary results

In order to set the OV strenght parameter, five simulations were run with an OV parameter in the range [0.04-0.12] for a 4 $M_{\odot}$ stellar model at solar metallicity. These simulations were later compared to observations and predictions from other nucleosynthesis codes.

As shown in Figure 1, the final surface abundances depend greatly on the OV parameter, up to an average difference of 0.5 dex among the three models with higher OV parameters.

3. Discussion and future work

Most of the theoretical predictions by other codes and observations lie between two (0.08 and 0.10) of our OV parameters, hinting at an intermediate value. In particular, the ATON

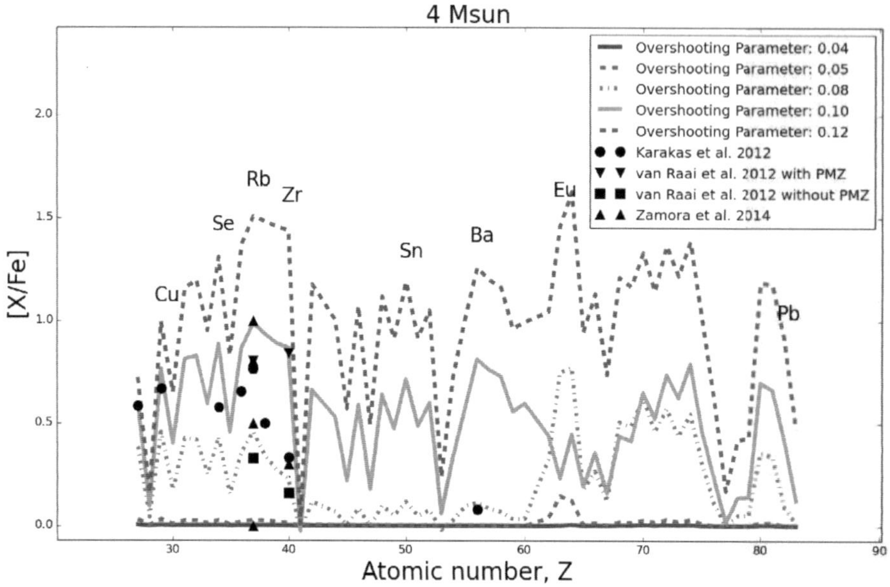

Figure 1. Final surface abundances ([X/Fe]) of elements heavier than Fe as a function of atomic number, Z, for a 4 $M_\odot$ solar metallicity model. Included are the predictions from Karakas et al.[5], for a 5 $M_\odot$ (equivalent to a 4 $M_\odot$ star in ATON) model with a delayed superwind, and from van Raai et al.[6] for a 5 $M_\odot$ model after a synthetic extension with and without a partial mixing zone (PMZ). The Rb and Zr abundances in massive Galactic AGB stars derived by Zamora et al.[7] are also shown. We are currently working on the anomalous Eu overabundance seen in these results.

simulations with OV parameter 0.10 predict higher Rb abundances than other calculations but in nice agreement with the maximum Rb content observed in massive Galactic AGB stars.

Also interesting are future numerical simulations for lower mass stars. In the 4 $M_\odot$ models presented here the *s*-process is driven both by the ^{22}Ne neutron source, which is strongly activated in the more massive AGB stars[8], and the ^{13}C neutron source. For a lower mass star, however, the ^{13}C neutron source is the main provider of neutrons. This source by itself can be even more sensitive to the OV parameter because the ^{13}C-pocket size is affected by it.

Acknowledgments

AYL and DAGH acknowledge MINECO support from grants RYC-2013-14182 and AYA-2014-58082-P. ML is a Momentum ("Ledület-2014" Programme) project leader of the Hungarian Academy of Sciences.

References

[1] Ventura P, D'Antona F and Mazzitelli I 2008 *Astrophys. Space Sci.* **316** 93
[2] Mazzitelli I, D'Antona F and Ventura P 1999 *Astron. Astrophys.* **348** 846
[3] Lugaro M, Karakas A I, Stancliffe R J and Rijs C 2012 *Astrophys. J.* **747** 2
[4] Lugaro M, Tagliente G, Karakas A I, Milazzo P M, Käppeler F, Davis A M and Savina M R 2014*Astrophys. J.* **780** 95
[5] Karakas A I 2010 *Mon. Not. R. Astron. Soc.* **403** 1413
[6] van Raai M A, Lugaro M, Karakas A I, García-Hernández D A and Yong D 2012 *Astron. Astrophys.* **540** A44
[7] Zamora O, García-Hernández D A, Plez B and Manchado A 2014 *Astron. Astrophys.* **564** L4
[8] García-Hernández D A, García-Lario P, Plez B, D'Antona F, Manchado A and Trigo-Rodríguez J M 2006 *Science* **314** 1751

Identification of oxygen-rich late/post-asymptotic giant branch stars and water fountains via maser and infrared criteria

Bosco H K Yung[1,2], Jun-ichi Nakashima[1,3], Christian Henkel[4,5] and Chih-Hao Hsia[6,7]

[1] Department of Physics, The University of Hong Kong, Pokfulam, Hong Kong
[2] Nicolaus Copernicus Astronomical Center, Rabiańka 8, 87-100 Toruń, Poland
[3] Department of Astronomy and Geodesy, Institute of Natural Science, Ural Federal University, Lenin Ave. 51, 620000, Ekaterinburg, Russia
[4] Max-Planck-Institut für Radioastronomie, Auf dem Hügel 69, D-53121 Bonn, Germany
[5] Astronomy Department, King Abdulaziz University, P.O. Box 80203, Jeddah 21589, Saudi Arabia
[6] Laboratory for Space Research, Faculty of Science, The University of Hong Kong, Pokfulam, Hong Kong
[7] Space Science Insitute, Macau University of Science and Technology, Avenida Wai Long, Taipa, Macau

E-mail: [2] byung@ncac.torun.pl

Abstract. The transitional phase between the asymptotic giant branch (AGB) and post-AGB phases holds the key to our understanding of the late-stage metamorphosis of intermediate-mass stars. In particular, high velocity jets forming during this phase are suggested to contribute significantly to the shaping of planetary nebulae. For oxygen-rich stars, the rare "water fountains (WFs)" have been regarded as representative objects in this phase, and it is important to identify more of them for further studies. Here we briefly report the results of our latest OH and H_2O maser surveys in which a new WF candidate (IRAS 19356+0754) was found. We also performed radiative transfer modelling on the spectral energy distributions (SEDs) of all known WFs. It is concluded that WFs might in fact not be the transitional objects, as opposed to previous belief. WFs could be AGB or post-AGB stars with no obvious similarities amongst their SEDs. Further efforts are still needed to improve the identification criteria.

1. Introduction

High velocity jets associated with asymptotic giant branch (AGB) and post-AGB stars could play an important role in the shaping of planetary nebulae. This process is often thought to mainly occur during the transitional phase between the AGB and post-AGB phases. In the case of oxygen-rich envelopes, the "water fountains (WFs)" — objects associated with bipolar jets traced by H_2O maser emission — have been suggested as possible representative objects for this short phase. Such objects are very rare, up to date there are only 16 confirmed members. In order to have a better understanding of the WFs, as well as their role during the late stages of stellar evolution, more samples are required. Hence, we explore better methods for identifying this particular type of AGB/post-AGB stars.

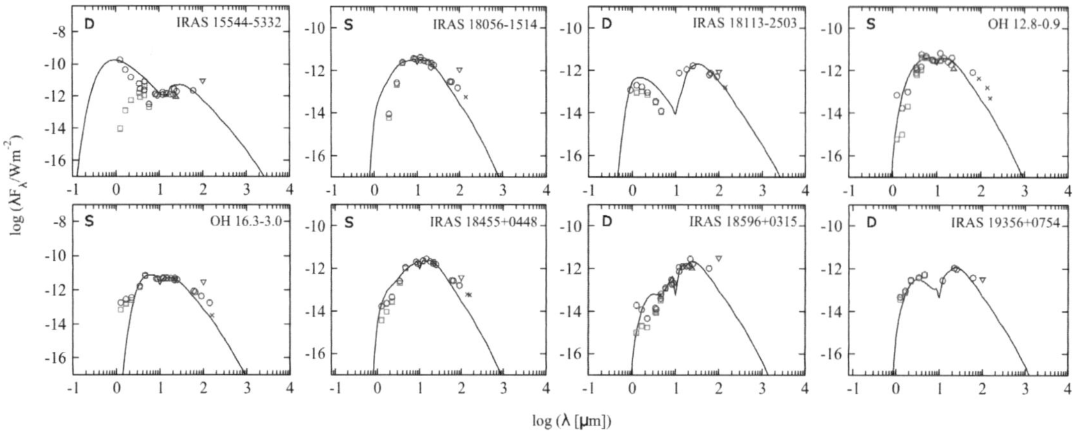

Figure 1. SEDs of WFs with the best fit DUSTY model curves. Only those with a good fit are shown here. "S" or "D" represent single-peaked or double-peaked profiles, respectively. Calibrated (○) and original (□, for $\lambda < 8$ μm) photometric data points are shown. Unreliable data points (×), upper (▽) and lower (△) intensity limits are also indicated, when applicable.

2. Maser survey and spectral energy distribution modelling

Our approach is to perform high sensitivity maser surveys, and to look for characteristics from the infrared spectral energy distributions (SEDs). We performed an OH and H_2O maser survey with the Effelsberg 100 m radio telescope in late 2012. The 1612, 1665, 1667 MHz OH maser lines and the 22 GHz H_2O maser line were observed with root-mean-square noise levels down to about 1–10 mJy. A total of 108 objects were covered (but not all of the objects were observed in all maser lines). A new WF candidate, IRAS 19356+0754, was identified. This object demonstrates large spectral velocity coverage in both the 1612 MHz OH ($\sim$67 km s^{-1}) and H_2O ($\sim$119 km s^{-1}) maser spectra, which is a clear sign of a high velocity outflow (see [1] for details).

We also constructed the SEDs of all known WFs from archival infrared photometric data. The SEDs were then modelled by the one-dimensional radiative transfer code DUSTY [2]. We found that WFs exhibit various types of SED profiles, and not all of them were well fit by the DUSTY models. Good fits could be obtained for eight WFs. Their SEDs and the corresponding best fit model curves are shown in Figure 1. Amongst these cases, four have a single-peaked SED profile resembling those from AGB stars, while the other four have a double-peaked profile and they are more likely post-AGB stars.

To conclude, WFs are difficult to be identified even with such a high sensitivity maser survey. We suggest that WFs are not necessarily the transitional objects between AGB and post-AGB phases, because no obvious common characteristics are found from their SEDs. However, if WFs could spread across various evolutionary phases, then it remains unclear why they are so rare.

Acknowledgments

This work is supported by a grant awarded to JN from the Research Grants Council of Hong Kong (project code: HKU 703308P; HKU 704209P; HKU 704710P).

References

[1] Yung B H K, Nakashima J and Henkel C 2014 *Astrophys. J.* **794** 81
[2] Ivezić Ž, Nenkova M and Elitzur M 1999 *User Manual for DUSTY, University of Kentucky Internal Report*

KaVA ESTEMA project

Miyako Oyadomari[1], Hiroshi Imai[1], Se-Hyung Cho[2], Yoshiharu Asaki[3], Yoon-Kyong Choi[2], Jaeheon Kim[2], Youngjoo Yun[2], Naoko Matsumoto[4], Cheul-Hong Min[4], Tomoaki Oyama[4], Sung-Chul Yoon[5], Dong-Hwan Yoon[2], Dong-Jin Kim[2], Richard Dodson[6], Maria Rioja[6], Ross Burns[1], Gabor Orosz[1], Akiharu Nakagawa[1], James Chibueze O.[7], Jun-ichi Nakashima[8] and Andrey Sobolev[8]

[1] Kagoshima University, 1-21-35 Korimoto, Kagoshima 890-0065, Japan
[2] Korean VLBI Network, Korea Astronomy and Space Science Institute, 776, Daedeokdae-ro, Yuseong-gu, Daejeon 305-348, Republic of Korea
[3] Chile Observatory, National Astronomical Observatory of Japan, Calle Joaquin Montero 3000, Oficina 702, Vitacura, Santiago, Chile
[4] Mizusawa VLBI Observatory, National Astronomical Observatory of Japan, 2-21-1 Osawa, Mitaka, Tokyo 181-8588, Japan
[5] Seoul National University, 1 Gwanak-ro, Gwanak-gu, Seoul 151-742, Korea
[6] The University of Western Australia, 35 Stirling Hwy, Crawley, WA 6009, Australia
[7] The University of Nigeria, Carver Building, 1 University Road, Nsukka, Nigeria
[8] Ural Federal University, Lenin Avenue 51, Ekaterinburg 620000, Russia

E-mail: [1] `hiroimai@sci.kagoshima-u.ac.jp`

Abstract. The ESTEMA (Expanded Study on Stellar Masers) project is one of three Large Programs of the KaVA (the combined array of the Korean VLBI Network and Japanese VLBI Exploration of Radio Astrometry), and conducted in 2015–2016. It aims to publish a database of the largest sample of VLBI images of circumstellar water (H_2O) and silicon-monoxide (SiO) maser sources towards circumstellar envelopes (CSEs) of 80 evolved stars in late AGB to early post-AGB phase. Here we present the specifications of the ESTEMA observations and the planned scientific goals in order to share the basic information of the ESTEMA with astronomical community and encourage future collaborations with the ESTEMA and future follow-up observations for the targeted stars.

Circumstellar maser sources are important tools of diagnostics of the dynamical and physical conditions of CSEs. These masers are spatially resolved into compact clumps of maser emission in VLBI, whose three-dimensional relative motions are also traceable though monitoring VLBI observations. Different species of maser lines are located at different distances from the central stars and, due to the maser actions, it is difficult to directly derive the physical parameters of the maser regions in CSEs without the knowledge of the maser properties. Therefore, simultaneous observations of these masers associated with common stars are crucial. Moreover, CSEs are affected by stellar pulsation and inhomogeneous mass loss; such performance of dynamic CSEs will be elucidated by monitoring VLBI observations.

Using the unique specifications of the Korean VLBI Network (KVN) and Japanese VLBI Exploration of Radio Astrometry (VERA), which enable, respectively, simultaneous observations of the masers at four frequency bands (around 22, 43, 86 and 129 GHz) and high-precision astrometry using the dual-beam receiving system, as well as the form of KaVA for high

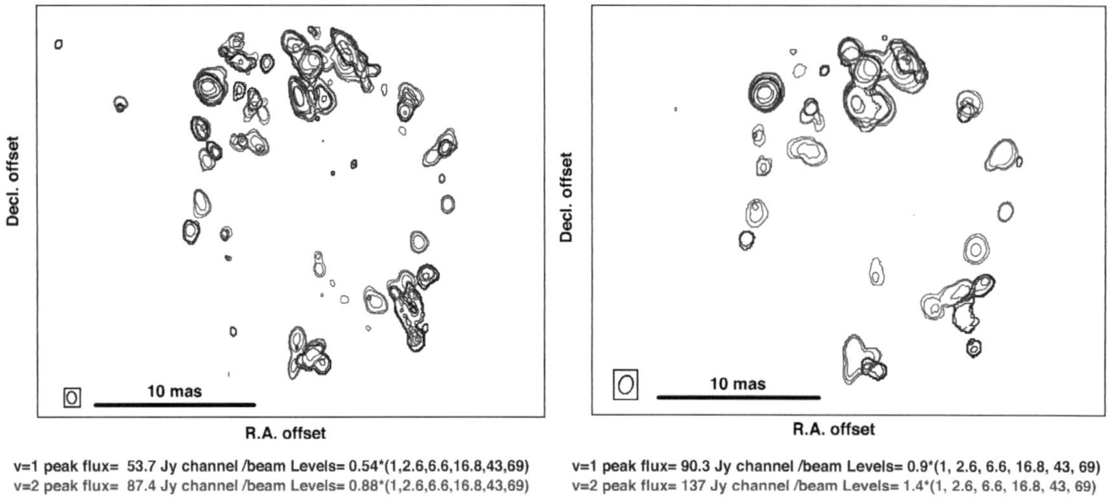

Figure 1. KaVA ESTEMA imaging demonstration for the SiO masers around BX Cam. The maser source was scanned in 17 times for 8 minutes each. The $v = 1$ and $v = 2$ $J = 1 \rightarrow$ masers brightness distributions are displayed in dark and light grey contours, respectively. *Left:* Map obtained by using all available visibility data of the KaVA observation. *Right:* Same as the left panel, but using the data expected to be obtained in an ESTEMA observation mode.

quality imaging [1], we have planned KaVA Large Projects to conduct more systematic legacy observations projects for circumstellar masers. We finally aims to monitor ~20 stars in VLBI intensively (every 2–12 weeks depending on stellar pulsation period) throughout a few pulsation cycles (2–10 years). The first stage of the Large Projects named ESTEMA, whose name comes from Spanish "estema" (English "stemma", meaning "a family line"), will finally select such 20 stars from the snapshot imaging surveys for 80 stars (adding a few sources observed before the ESTEMA in KaVA commissioning [2]), whose large fraction of maser lines will be mapped through the whole stellar pulsation cycle. Using the whole ESTEMA maps, which are expected to be yielded as shown in Figure 1, it enables statistical analyses of the masers observed among different types of stars at a variety of pulsation phases, from microscopic (individual maser spots) to macroscopic (CSE) views. In the ESTEMA, for each star the relative locations and distributions of different maser lines are directly compared. Thus it enables us to analyze the dependence of maser pumping mechanisms on stellar type and pulsation phase, plus the evolution of asymmetric stellar mass loss found through biased H_2O maser spatio-kinematic structures in the CSE with respect to the SiO maser location, pinpointing the central star.

Because the ESTEMA is a risk-shared project, there are some problems in array operations and data processing in the observations. Nevertheless, a large fraction of the data will be yielded as planned, and the ESTEMA outputs will become a legacy reference for planning any long-term intensive monitoring campaigns of evolved stars exhibiting high mass loss rates, with current large and forthcoming radio astronomy facilities including KaVA and ALMA.

Acknowledgments
The ESTEMA project is conducted with support by the KASI Commissioning Program.

References
[1] Matsumoto N et al 2014 *Astrophys. J.* **789** L1
[2] Yun Y J et al 2016 *Astrophys. J.* **822** 3

Near IR observations of η Car: Reaching its critical rotation?

Kazunori Ishibashi[1], **Andrea Mehner**[2] **and Takahiro Nagayama**[3]

[1] Graduate School of Science, Nagoya University, Japan
[2] ESO European Organisation for Astronomical Research in the Southern Hemisphere, Chile
[3] Department of Physics and Astronomy, Graduate School of Science and Engineering, Kagoshima University, Japan

E-mail: [1] bish@nagoya-u.jp

Abstract. We report a preliminary result in the monitoring of η Carinae in JHKs bands through the "spectroscopic" event of 2014.5 at InfraRed Survey Facility (IRSF) located in South African Astronomical Observatory (SAAO). The latest photometric data, combined with the data taken with the MK-II photometer [1], show a sign of the same cyclic variation in the J-H vs. H-K diagram reported by Mehner et al.[2]. The change can still be attributed to an apparent increase in black-body temperature, potentially reaching 3,000 to 6,000K as of March 2015.

NIR Monitoring of η Carinae

Near-Infrared monitoring of η Carinae continued throughout the spectroscopic event of 2014.5 (for those who are not familiar with the star and its 5.5yr cyclic and other secular changes, please see [2] and references therein). As the last event occurred in August 2014, some observations were proven difficult to conduct (η Carinae being too close to the Sun for many observatories), though we managed to obtain a few crucial data points during the phase of expected rapid changes in JHK colors.

Figure 1 shows the H-K vs. J-H color-color diagram of η Carinae, corrected for its long secular change in reddening. With those few data points in hand (marked in light green), the same rapid, cyclic change is confirmed to recur at the last event, as well as the increasing trend in apparent black-body temperature (of η Car's stellar atmosphere). While depending on an assumption on intrinsic reddening correction, it seems that the apparent plasma temperature has reached 6,000K (or as low as 3,000K), a jump in temperature at 5,000K (or 2,700K) in 2009.0, respectively.

Is η Carinae Spinning Up?

A cause of the cyclic changes is not clearly understood, though naturally we suspect it to be driven by interaction between η Carinae and its massive companion. As for the apparent jump in temperature, a tidal interaction between the two stars may play a significant role. As the companion passes its periastron, it swings by the primary at the speed of $\sim$300 km s^{-1}. It may lead to a spin-up of the rotation speed of the primary; in turn it leads to an appreciable expansion in its dense atmosphere, which will eventually allows ionizing radiation from its core to permeate much easily.

Here, two concerns are that how long this trend may continue and what would happen at its end point? Nothing is certain, but the history suggests that an eruption may be a way to carry away any excessive angular momentum once it reaches a limit in critical rotational stability.

Easy to speculate; harder to prove.

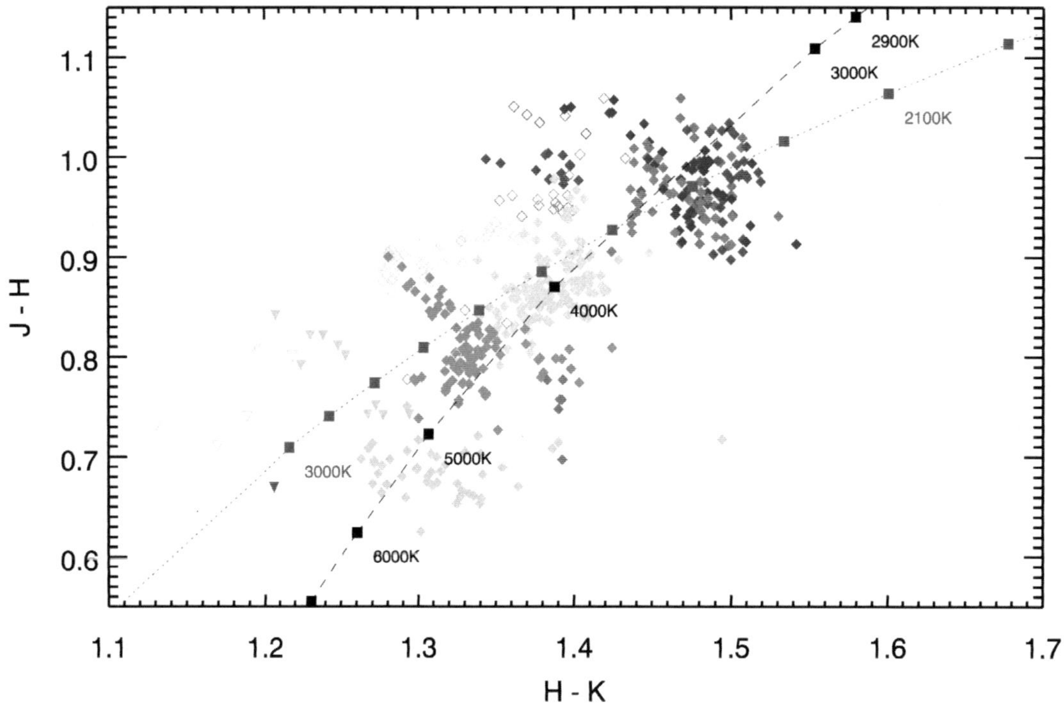

Figure 1. Historical Color-Color diagram of η Carinae in Near IR: data points are color-coded by 5.5yr phases. Note that the lone data point taken after the 2014.5 event shown here in gray (marked with reversed filled triangle) near H-K$\sim$1.2 and J-H$\sim$0.68. See [2] for more details.

Acknowledgments

We acknowledge Drs. Sugitani, Murata and many other observers to support this monitoring program. Ishibashi also acknowledges that this work was supported by JSPS KAKENHI Grant Number 26400227.

References

[1] Whitelock P A, Feast M W, Marang F and Breedt E 2004 *Mon. Not. R. Astron. Soc.* **352** 447–456

[2] Mehner A, Ishibashi K, Whitelock P, Nagayama T, Feast M, van Wyk F and de Wit W J 2014 *Astron. Astrophys.* **564** A14

Journal of Physics: Conference Series **728** (2016) 072008 doi:10.1088/1742-6596/728/7/072008

Discovery of new planetary nebulae in the Small Magellanic Cloud

D. Drašković[1], Q. A. Parker[2,4], W. A. Reid[1,3] and M. Stupar[3]

[1] Department of Physics and Astronomy, Macquarie University, NSW, Australia
[2] Department of Physics, The University of Hong Kong, Pokfulam Road, Hong Kong
[3] School of Computing, Engineering and Mathematics, Western Sydney University, Australia
[4] Australian Astronomical Observatory (AAO), North Ryde, NSW, Australia

E-mail: [1] danica.draskovic@mq.edu.au

Abstract. We present six new planetary nebulae (PNe) discovered in the Small Magellanic Cloud (SMC) from deep UK Schmidt telescope (UKST) narrow band Hα and broad-band short-red "SR" continuum images and confirmed spectroscopically. These 6 preliminary discoveries provide a 6% increase to the previously known SMC PN population of ~100. Once spectroscopic follow-up of all our newly identified candidates is complete, we expect to increase the total number of known SMC PNe by up to 50%. This will permit a significant improvement to determination of the SMC PN luminosity function (PNLF) and enable further insights into the chemical evolution and kinematics of the SMC PN population.

1. Introduction

Planetary nebulae (PNe) play a crucial role in understanding mass loss for low and intermediate mass stars, they influence chemical evolution of galaxies through interstellar enrichment and are a major Galactic dust factory. Their strong emission lines are laboratories for plasma physics and provide accurate radial velocities for Galactic kinematics while the PNLF in external galaxies is a powerful cosmological distance indicator.

The SMC is a gas rich, late type dwarf irregular galaxy with an average gas/dust ratio a factor of ~ 30 below the Galactic value [1]. The SMC's accurately determined distance of 61 ± 1 kpc [2] allows derivation of true PNe physical parameters from optical spectra. With a small angular extent of $5°20' \times 3°05'$, it can be easily studied in its entirety. SMC reddening and extinction is relatively low and uniform, enabling absolute nebula luminosity estimates.

2. Discovery technique

We have carefully examined UKST Hα and SR SMC image data for ~120 deg^2. We adopted the same powerful and proven colour merging technique successfully applied to the LMC by [3, 4, 5]. Using the KARMA visualisation software package, we surveyed these fields and created merged false-colour images, combining Hα (coloured in blue) and SR (coloured in red). All images were systematically visual scanned, looking for a faint, compact or barely resolved objects. By carefully choosing the image combination parameters we could balance the intensity of Hα and SR red matching images, allowing only specific features of one or other pass-band to be seen. In our scheme emission objects such as PNe and H II regions appear with a uniform strong blue

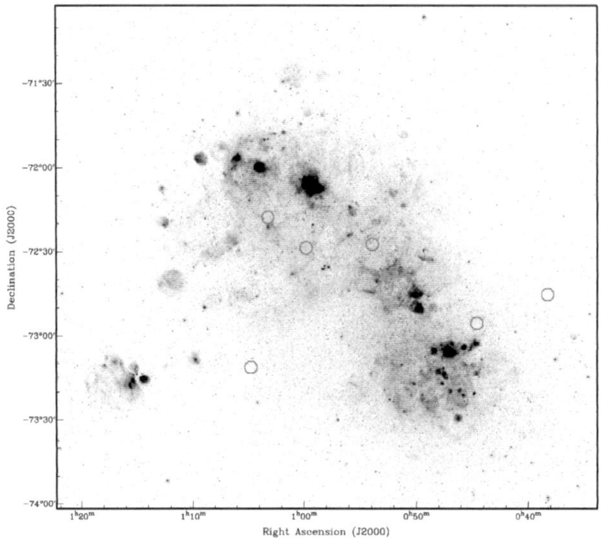

Figure 1. Positions of six new PNe discovered in the SMC.

colour, whereas normal continuum stars are a uniform pink-purple colour. Emission-line stars usually have a strong continuum component and so are easily detected by the narrow extent of their blue halos around a strong pink core [3, 4].

3. Conclusions

Confirmatory spectra of six preliminary PNe candidates are quite homogenous. Each was confirmed as an emission-line source with no obvious stellar continuum. This eliminates confusion with emission-line stars for compact sources. Each candidate also gave high [N II] to Hα ratios ($\sim$ 2 to 7) [6] not observed in H II regions, which are the most likely mimics [7]. The presence of only narrow diagnostic lines also removes the confusion with supernova remnants. These preliminary results demonstrate the strength and promise of our techniques for providing high-quality PNe candidates.

We expect to increase SMC PNe by 20 to 50% after spectroscopic follow-up of our candidates and refinement of identification of all known SMC emission-line sources. This will include multi-wavelength analysis and spectroscopic confirmation (or rejection) of objects previously listed as SMC PNe. Our survey will permit more extreme ends of the PNLF to be explored, especially at the faint end, and enable us to study underrepresented evolutionary stages of SMC PNe. This sample will have a major impact on improving rates of elemental enhancement in lower mass stars within a low-metallicity environment. Other advances will come from the improved SMC kinematical data from over 300 emission-line objects uncovered.

References

[1] Stanimirovic S, Staveley-Smith L, van der Hulst J M, Bontekoe T R, Kester D J M and Jones P A 2000 *Mon. Not. R. Astron. Soc.* **315** 791–807
[2] Hilditch R W, Howarth I D and Harries T J 2005 *Mon. Not. R. Astron. Soc.* **357** 304–324
[3] Reid W A and Parker Q A 2006 *Mon. Not. R. Astron. Soc.* **365** 401–413
[4] Reid W A and Parker Q A 2006 *Mon. Not. R. Astron. Soc.* **373** 521–550
[5] Reid W A and Parker Q A 2013 *Mon. Not. R. Astron. Soc.* **436** 604–624
[6] Drašković D, Parker Q A, Reid W A and Stupar M 2015 *Mon. Not. R. Astron. Soc.* **452** 1402–1411
[7] Frew D J and Parker Q A 2010 *Publ. Astron. Soc. Australia* **27** 129–148

Coaxial rings and H_2 knots in Hubble 12

Chih-Hao Hsia[1,2], Sun Kwok[1], Wayne Chau[1,2] and Yong Zhang[1,2]

[1] Laboratory for Space Research, Faculty of Science, The University of Hong Kong, Hong Kong, China

[2] Department of Physics, The University of Hong Kong, Hong Kong, China

E-mail: [1,2] xiazh@hku.hk

Abstract. Hubble 12 (Hb 12) is a young planetary nebula (PN) exhibiting nested shells. We present new near-infrared narrow-band imaging observations of Hb 12 using the Canada-France-Hawaii Telescope (CFHT). A number of co-axial rings aligned with the bipolar lobes and two pairs of separate H_2 knots with different orientations are detected.

1. Introduction

Morphological transformation from spherical geometry to bipolar is believed to occur shortly after the asymptotic giant branch (AGB). Recent near-infrared observations in the H_2 2.122 μm line [1] have shown that molecular gas is widely observable in PNs, providing us important information about the mass-loss history during AGB and shaping mechanism of PNs. A few hourglass PNs (e.g. MyCn 18, M 2-9, and Hen 2-104) are discovered to have high-speed collimated outflows [2], which could be produced through binary interactions. In this paper, we study a young PN Hubble 12 (Hb 12) with the goal of determining its molecular gas structures and properties of the central source.

2. Observations and Results

Deep near-infrared images of Hb 12 were taken using the Wide-field InfraRed Camera (WIRCam) on the *Canada-France-Hawaii Telescope* (CFHT). The nebula was imaged with three narrow band filters: H_2 (λ_c = 2.122 μm), Brγ (λ_c = 2.166 μm), and K continuum (Kc; λ_c = 2.218 μm). The H_2 *CFHT* image of Hb 12 is shown in Figure 1. Three pairs of bipolar lobes and a number of co-axial rings aligned with the outermost lobe can be seen in the image. The nested structures may have a common formation mechanism, most likely involving mass-loss processes influenced by a binary core [2]. These rings are thought to be the manifestation of a time-variable, collimated fast wind of bipolar lobes interacting with surrounding AGB medium. Moreover, two pairs of separate H_2 knots with different position angles (N1-S1: 175°±2° and N2-S1: 171°±2°) are found in the wide-field H_2 *CFHT* image (Figure 2). The pair of knots N1-S1 has a projected length of ~ 133 arcsec and N2-S2 has a projected size of ~ 353 arcsec, respectively. This could be related to the binary nature of the nucleus of Hb 12 [3], and the formation of bipolar, rotating, episodic jet. A bipolar precessing jet model for these knots is needed to better understand their nature.

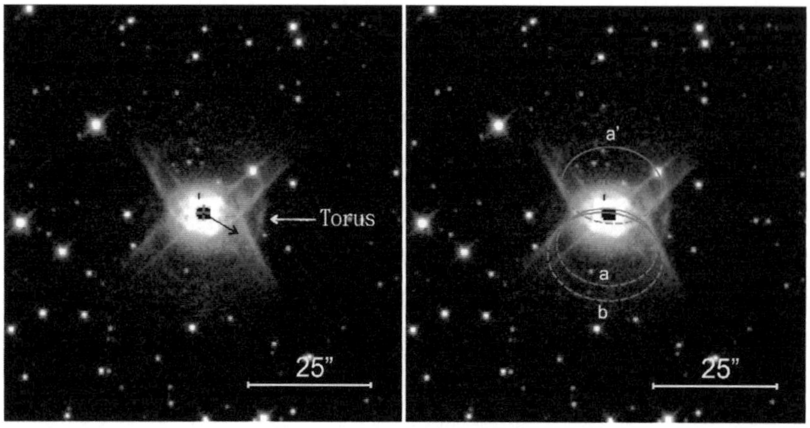

Figure 1. CFHT WIRCam H_2 image of Hb 12. (Left) Black arrow at the central part represent the direction of proper motion. (Right) This panel is the same with the left panel but with schematic outlines of three coaxial rings with inclination angle of 45° in outer lobes of Hb 12.

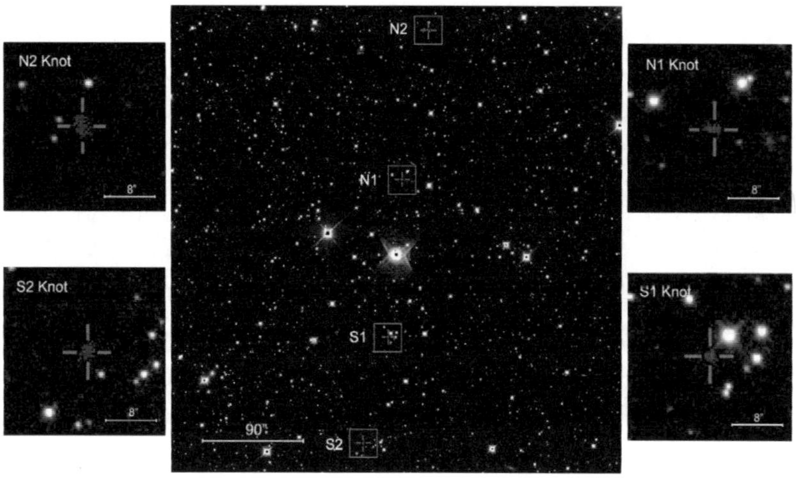

Figure 2. A wide-field CFHT H_2 image of Hb 12, whose central part is the same as Figure 1. The upper right, upper left, lower right, and lower left insets indicate the positions of N1, N2, S1, and S2 knots, respectively. The knots N1 and S1 might be the H_2 counterparts of the knots seen at [NII] [4].

Acknowledgments
Financial support for this work was provided by the Research Grants Council of the Hong Kong Special Administrative Region, Project no. HKU 7031/10P and HKU 7062/13P.

References

[1] Guerrero M A, Villiaver E, Manchado A, Garcia-Lario P and Prada F 2000 *Astrophys. J. Suppl. Series* **127** 125
[2] Corradi R L M, Livio M, Balick B, Munari U and Schwarz H E 2001 *Astrophys. J.* **553** 211
[3] Hsia C H, Ip W H and Li J Z 2006 *Astron. J.* **131** 3040
[4] Vaytet N M H, Rushton A P, Lloyd M, López J A, Meaburn J, O'Brien T J, Mitchell D L and Pollacco D 2009 *Mon. Not. R. Astron. Soc.* **398** 385

Near-infrared views of dust in planetary nebulae: *AKARI*/IRC NIR spectral atlas of Galactic PNe

Ryou Ohsawa[1], Takashi Onaka[2], Itsuki Sakon[2], Tamami Mori[2], Mikako Matsuura[3], Issei Yamamura[4], and Hidehiro Kaneda[5]

[1]Institute of Astronomy, Graduate School of Science, The University of Tokyo, 2-21-1 Osawa, Mitaka, Tokyo 181-0015, Japan
[2]Department of Astronomy, Graduate School of Science, The University of Tokyo, 7-3-1 Hongo, Bunkyo-ku, Tokyo 113-0033, Japan
[3]School of Physics and Astronomy, Cardiff University, Queen's Buildings, 5 The Parade, Roath, Cardiff CF24 3AA, United Kingdom
[4]Japan Aerospace Exploration Agency, Institute of Space and Astronautical Science, 3-1-1 Yoshinodai, Chuo-ku, Sagamihara, Kanagawa 252-5210, Japan
[5]Division of Particle and Astrophysical Science, Graduate School of Science, Nagoya University, Furo-cho, Chikusa-ku, Nagoya, Aichi 464-8602, Japan

E-mail: [1] ohsawa@ioa.s.u-tokyo.ac.jp

Abstract. Asymptotic Giant Branch (AGB) stars and post-AGB stars are thought to be major dust sources in galaxies. Investigation of dust in planetary nebulae (PNe), their successors, reveals properties of dust grains injected into the interstellar medium. The near-infrared (2.5–5.0 μm) emission of PNe consists of hot dust continuum, emission lines of ionized gases, and emission bands of carbonaceous dust grains. This wave range, however, has not been well investigated due to the lack of instruments with a high sensitivity. We conducted spectroscopic observations of PNe with the *AKARI*, obtaining 2.5–5.0 μm spectra of 72 Galactic PNe with a sensitivity of a few mJy. The spectra and the intensities of emission features are compiled as a catalog. The characteristics of the catalog are described.

1. Introduction

Cosmic dust is provided by Asymptotic Giant Branch (AGB) stars and post-AGB stars. Investigation of circumstellar matters in planetary nebulae (PNe), their successors, provides an effective tool to understand properties of dust grains injected into the interstellar medium. Near-infrared (2.5–5.0 μm) spectra are rich in emission features, which are useful to diagnose the physical conditions and radiation field in the nebula.

We conducted spectroscopic observations of Galactic PNe with the Infrared Camera (IRC) aboard the *AKARI*. We obtained 72 spectra of PNe with a sensitivity of a few mJy. The emission feature intensities were measured, which are compiled into a catalog (hereafter the PNSPC catalog) as well as the well-calibrated spectra [1]. We introduce details of the observations and characteristics of the PNSPC catalog.

2. Observations

Targets were selected from the Strasbourg-ESO Catalogue of Galactic Planetary Nebulae [2] by their magnitudes and sizes. Observations were carried out in the Post-Helium phase. Spectra

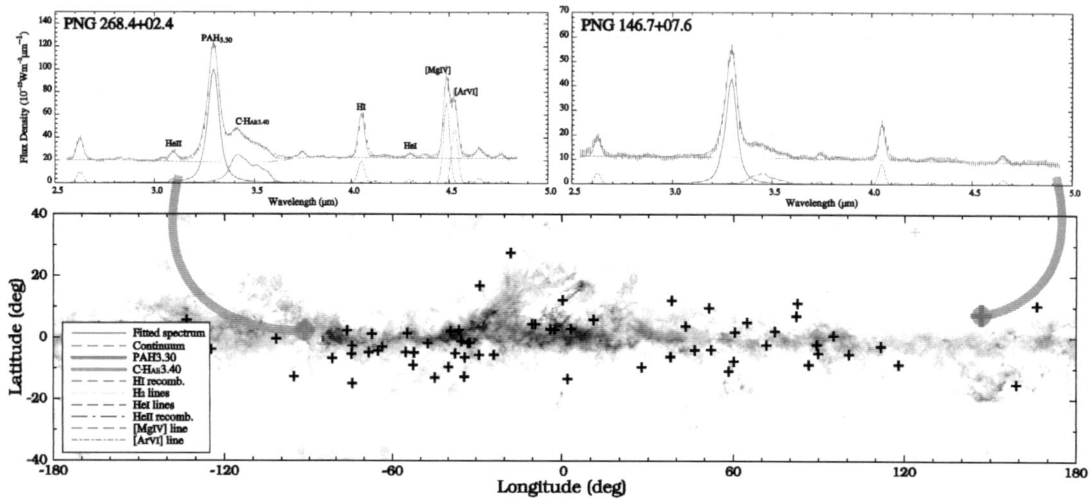

Figure 1. Typical near-infrared spectra in the PNSPC catalog. Top two panels show the spectra of PN G268.4+02.4 (left) and G146.7+07.6 (right). The bottom panel shows the locations of the PNe in the catalog overlaid on the Galactic line-of-sight extinction map [4].

were obtained in the slit-less spectroscopy mode with the grism. The achieved spectral resolution ($\lambda/\Delta\lambda$) is typically $\sim$100, which is degraded down to $\sim$40 for the most extended object. The distribution of the targets is illustrated in the bottom panel of Figure 1.

3. Results and Summary

The top two panels of Figure 1 show typical near-infrared spectra in the PNSPC catalog. The achieved sensitivity is about $0.7\times10^{-15}\,\mathrm{W\,m^{-2}\,\mu m^{-1}}$ at $3.0\,\mu$m, corresponding to $\sim$2 mJy. The absolute flux density is consistent with the WISE All-Sky Source Catalog [3] with 20% accuracy. The PNSPC catalog also tabulates the intensities of emission features (Hydrogen and Helium recombination lines, MgVI and ArIV fine-structure lines, the $3.3\,\mu$m PAH feature, and the 3.4–$3.5\,\mu$m aliphatic feature complex, *see* Figure 1). Emission features brighter than $2\times10^{-16}\,\mathrm{W\,m^{-2}}$ were detected with a signal-to-noise ratio > 3. The PNSPC catalog provides unique information for the investigation of 2.5–$5.0\,\mu$m emission of Galactic PNe.

Acknowledgments

This research is based on observations with *AKARI*, a JAXA project with the participation of ESA. This work is supported in part by Grant-in-Aid for JSPS fellows (25-8492).

References

[1] Ohsawa R, Onaka T, Sakon I, Matsuura M and Kaneda H 2016 *Astron. J.* **151** 93
[2] Acker A, Marcout J, Ochsenbein F, Stenholm B and Tylenda R 1992 *Strasbourg-ESO Catalogue of Galactic Planetary Nebulae, Parts I, II*
[3] Cutri R M, Wright E L, Conrow T, Bauer J, Benford D, Brandenburg H and Dailey J 2012 *VizieR Online Data Catalog*
[4] Dobashi K, Uehara H, Kandori R, Sakurai T, Kaiden M, Umemoto T and Sato F 2005 *Publ. Astron. Soc. Japan* **57** 1

A multiwavelength study of the Stingray Nebula; properties of the nebula, central star, and dust

Masaaki Otsuka[1], **Mudumba Parthasarathy**[2], **Akito Tajitsu**[3], **and Swetlana Hubrig**[4]

[1]ASIAA, 11F of Astronomy-Mathematics Building, AS/NTU. No.1, Sec. 4, Roosevelt Rd, Taipei 10617, Taiwan, R.O.C.
[2]Indian Institute of Astrophysics, II Block Koramangala, Bangalore 560034, Karnataka, India
[3]Subaru Telescope, 650 N Aohoku Place, Hilo, HI 96720, U.S.A.
[4]Leibniz-Institut fuer Astrophysik Potsdam (AIP) An der Sternwarte 12, 14482 Potsdam, Germany

E-mail: otsuka@asiaa.sinica.edu.tw

Abstract. We performed a detail chemical abundance analysis and photo-ionization modeling of the Stingray Nebula (Hen3-1357, Parthasarathy et al. 1993[1]) to more characterize this PN. We calculated nine elemental abundances using collisionally excited lines (CELs) and recombination lines (RLs). The RL C/O ratio indicates that this PN is O-rich, which is supported by the detection of the broad amorphous silicate features at 9 and 18 μm By photo-ionization modeling, we investigated properties of the central star and derived the gas and dust masses. The nebular elemental abundances, the core-mass of the central star, and the gas mass are in agreement with the AGB model for the initially 1.5 $M_\odot$ stars with the $Z = 0.008$.

1. Nebular elemental abundances

We performed a chemical abundance analysis using the MPG ESO 2.2-m/FEROS 0.36-0.9 μm high-dispersion spectrum taken on 2006 April and the *Spitzer*/IRS spectrum taken on 2005 March. The result is summarized in Table 1. The RL C/O ratio (0.20) indicates that this PN is O-rich. The nebular He/C/N/O/Ne abundances are similar to the predictions of the AGB star model by [2] for the initially 1.5 $M_\odot$ stars with the metallicity $Z = 0.008$.

2. Physical properties of the central star and the dusty nebula

Using the photo-ionization code CLOUDY [3], we investigated properties of the central star and the dusty nebula by fitting the near-UV FEROS to the Far-IR *AKARI*/FIS data.

Effective temperature ($T_{\rm eff}$) and surface gravity ($\log g$) When we adopted the model atmosphere for yr 2006 with $T_{\rm eff} = 55\,000$ K and $\log g = 6.0$ cm s^{-2} by [4] and the distance of 1.6 kpc [4], CLOUDY overestimated the fluxes of higher excitation lines such as [Ne III] and [O III]. $T_{\rm eff}$ could be cooler than 55 000 K; we estimated $T_{\rm eff}$ to be 50 500 K using the nebular [O III]/Hβ line ratio. We utilized TLUSTY O-star atmosphere [5] and searched for $T_{\rm eff}$ and $\log g$ to match the observations. We set $T_{\rm eff} = 45\,800$ K and $\log g = 4.55$ cm s^{-2}.

Distance We calculated the post-AGB age of $\sim$379 $D_{\rm kpc}$ yrs using the expansion velocity (21 km s^{-1}, from the Hβ line) and the nebula's outer radius (1.7$''$) measured from the

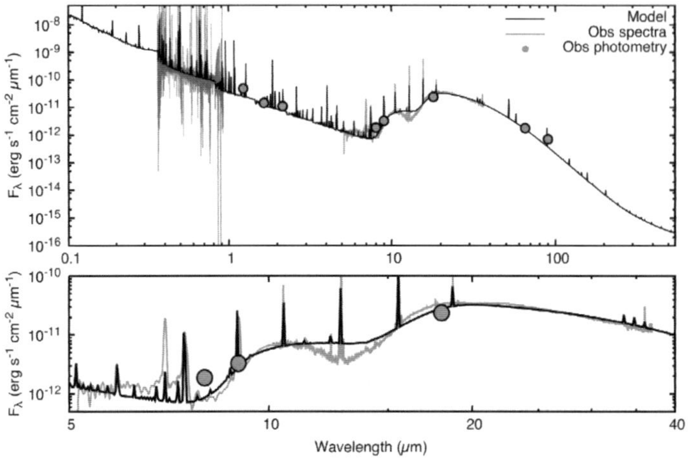

Figure 1. (*upper panel*) Comparison between the CLOUDY model and observational data of Hen3-1357. (*lower panel*) Closed-up plots for mid-IR wavelength. The mid-IR *Spitzer*/IRS spectrum shows the amorphous silicate broad features at 9 and 18 μm. We derived the dust mass of $2.2\times10^{-4}\,M_\odot$ and the temperature 50-176 K (grain radius a = 0.01-0.25 μm and $a^{-3.5}$ size distribution).

Table 1. Elemental abundances ($\log_{10}\epsilon(H) = 12$). The fourth and eighth columns are the predictions of the AGB star model by [2] for the initially $1.5\,M_\odot$ stars with the $Z = 0.008$.

X	ϵ(X)	[X/H]	Model	X	ϵ(X)	[X/H]	Model
He	11.04	0.11	10.98	Ne	8.09	0.22	8.18
C(RL)	8.09	-0.30	8.06	S	6.71	-0.48	7.16
N(CEL)	7.78	-0.05	7.70	Cl	5.12	-0.38	
N(RL)	7.81	-0.02	7.70	Ar	6.16	-0.39	
O(CEL)	8.65	-0.04	8.36	Fe	5.06	-2.41	
O(RL)	8.79	0.10	8.36				

HST/WFPC2 F487N image taken in 1998. Supposing that the central star was initially $1.5\,M_\odot$ and the current $T_{\rm eff}$ is $\sim$45 000-55 000 K, the post-AGB age is $\sim$1000-3000 yrs from the predictions of [6]. Thus, we estimated the distance of 2.7-6.7 kpc. Here we set 5 kpc.

Result We compare the observed SED plots and the SED predicted by the model in Figure 1. The calculated gas mass ($0.07\,M_\odot$) and the core-mass ($0.62\,M_\odot$) are consistent with [2], who predict that the initially $1.5\,M_\odot$ stars with $Z = 0.008$ will eject $\sim0.09\,M_\odot$ during the last thermal pulse and end as $\sim0.63\,M_\odot$ stars.

References

[1] Parthasarathy M, Garcia-Lario P, Pottasch S R, Manchado A, Clavel J, de Martino D, van de Steene G C M and Sahu K C 1993 *Astron. Astrophys.* **267** L19–L22

[2] Karakas A and Lattanzio J C 2007 *Publ. Astron. Soc. Australia* **24** 103–117

[3] Ferland G J, Porter R L, van Hoof P A M, Williams R J R, Abel N P, Lykins M L, Shaw G, Henney W J and Stancil P C 2013 *Revista Mexicana de Astronomia y Astrofisica* **49** 137–163

[4] Reindl N, Rauch T, Parthasarathy M, Werner K, Kruk J W, Hamann W R, Sander A and Todt H 2014 *Astron. Astrophys.* **565** A40

[5] Lanz T and Hubeny I 2003 *Astrophys. J. Suppl. Series* **146** 417–441

[6] Vassiliadis E and Wood P R 1994 *Astrophys. J. Suppl. Series* **92** 125–144

Journal of Physics: Conference Series **728** (2016) 072012 doi:10.1088/1742-6596/728/7/072012

New DSH planetary nebulae and candidates from optical and infrared surveys

Matthias Kronberger[1], Quentin A Parker[2], George H Jacoby[1,3], Agnes Acker[4], Filipe Alves[1], Ivan Bojicic[2], Paul Eigenthaler[5], David J Frew[2], Dianne Harmer[6], Dana Patchick[1], Warren Reid[7,8] and Johannes Schedler[9]

[1] Deep Sky Hunters collaboration
[2] Department of Physics, The University of Hong Kong, Hong Kong
[3] Giant Magellan Telescope/Carnegie Observatories, Pasadena, CA 91101, USA
[4] Observatoire astronomique de Strasbourg, 11 rue de l'Université, F-67000 Strasbourg, France
[5] Instituto de Astrofísica, Pontificia Universidad Católica de Chile, Santiago, Chile
[6] National Optical Astronomy Observatory, 950 N. Cherry Avenue, Tucson, AZ 85719, USA
[7] Department of Physics and Astronomy, Macquarie University, Sydney, NSW 2109, Australia
[8] University of Western Sydney, Locked Bag 1797, Penrith South DC, NSW 1797, Australia
[9] CHART32 Team

E-mail: [1] `matthias.kronberger@gmx.at`

Abstract. To date, the planetary nebula (PN) survey of the Deep Sky Hunters collaboration has led to the detection of more than 250 previously unknown candidate planetary nebulae (PNe). About 60% of them were found during the past two years and are expected to be true, likely or possible PNe because careful vetting has already thrown out more doubtful objects. The majority of the new PN candidates are located within the boundaries of the SHS and IPHAS Hα surveys and were discovered by combining MIR data from the WideField Infrared Survey Explorer (WISE) with optical data from the IPHAS, SHS and DSS surveys, and UV data from the Galaxy Evolution Explorer (GALEX).

Since the discovery of M 27 by the French astronomer Charles Messier about 250 years ago, ≈ 3500 more Galactic PNe have been discovered [1]. However, this number is still short of the total number of Galactic PNe estimated from local volume counts ($N_{tot=11,000}$ [2]). To a great extent, this difference can be attributed to the circumstance that, until recently, PN surveys were conducted mainly at optical wavelengths and over a relatively limited latitude range of the Galactic plane. Consequently, a considerable gain in the number of Galactic PNe can be expected from surveys in the MIR range, allowing for the discovery of highly obscured dust-rich and high excitation PNe that are invisible at optical wavelengths.

The Deep Sky Hunters [3] (DSH) have been scanning the Milky Way for new clusters and nebulae for more than decade now. Until fairly recently, our survey for PN candidates was restricted to medium Galactic latitudes ($5° \leq |b| \leq 20°$), and based on broadband Digitized Sky Survey (DSS) images [2,4,5] as well as on [S II]/Hα/[O III] narrowband images of the Milky Way taken with modestly-sized amateur telescopes equipped with modern CCD imagers [6]. Since then, the focus has shifted to fields closer to the Galactic Equator, and from optical to MIR and UV wavelengths, thanks to the availability of high resolution imagery from the WISE and GALEX missions. This has allowed us to detect 162 true, likely and possible PNe during the past

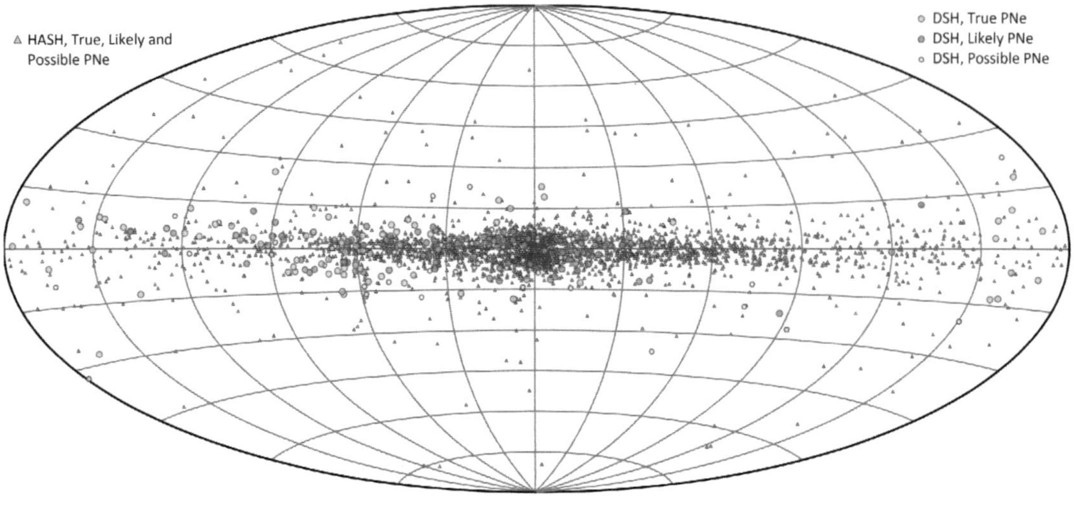

Figure 1. Aitoff-Hammer projection of the Galactic Plane, showing the distribution of all True, Likely and Possible PNe of the DSH sample.

two years, increasing the census of DSH PNe and candidates by more than 150%. It is worth noting that many of the new PN candidates, although originally discovered in the MIR, are clearly visible on SHS [7] and IPHAS [8] survey images but were overlooked thus far, most likely because of their faintness, because of insufficient contrast between Hα and their comparatory broadband imagery, or because of their proximity to bright sources in the field. Object sizes of the new PN candidates found at optical wavelengths range from essentially unresolved to several arcmin. All good quality DSH candidates and all those confirmed spectroscopically will be ingested into the new HASH database [9].

Figure 1 shows the distribution of all True, Likely and Possible PNe of the DSH sample in an Aitoff-Hammer projection of the Galactic Plane. Objects listed as True, Likely and Possible PNe in the HASH database are shown for comparison. Gaps in the DSH candidate distribution correspond to unsurveyed Milky Way regions. We plan to obtain confirmatory spectroscopy of the Possible and Likely PNe during the next years, and Hα + [O III] images of another ≈ 130 objects that we consider to be PN candidates based alone on their MIR characteristics.

References

[1] Parker Q A, Bojicic I, Frew D J, Acker A, Ochsenbein F and MASPN Database Team 2015 *American Astronomy Society Meeting* **225** 108.06
[2] Jacoby G H *et al* 2010 *Publ. Astron. Soc. Australia* **27** 156
[3] Kronberger M *et al* 2006 *Astron. Astrophys.* **447** 921
[4] Kronberger M *et al* 2012 *Proc. Int. Astron. Union* (S283) **7** 414
[5] Kronberger M *et al* 2014 *Asymmetrical Planetary Nebulae VI* 48
[6] Acker A, Boffin H M J, Outters N, Miszalski B, Sabin L, Le Dû P and Alves F 2012 *Revista Mexicana de Astronomia y Astrofisica* **38** 223
[7] Parker Q A *et al* 2005 *Mon. Not. R. Astron. Soc* **362** 689
[8] Drew J E *et al* 2005 *Mon. Not. R. Astron. Soc* **362** 753
[9] Parker Q A *et al these proceedings*

Improving the distances of post-AGB objects in the Milky Way

Shane B Vickers[1], David J Frew[2], Matt S Owers[1,3], Quentin A Parker[2,3], and Ivan S Bojičič[2]

[1]Department of Physics and Astronomy, Macquarie University, NSW 2109, Australia
[2]Department of Physics, The University of Hong Kong, Pokfulam Road, Hong Kong, China
[3]Australian Astronomical Observatory, PO Box 915, North Ryde, NSW 1670, Australia

E-mail: [1] shane.vickers@students.mq.edu.au

Abstract.
Post-AGB (PAGB) stars are short-lived, low-intermediate mass objects transitioning from the asymptotic giant branch (AGB) to the white dwarf (WD) phase. These objects are characterised by a constant, core-mass dependent luminosity and a large infrared excess from the dusty envelope ejected at the top of the AGB. PAGB stars provide insights into the evolution of their direct descendants, planetary nebulae (PNe). Calculation of physical characteristics of PAGB are dependent on accurately determined distances scarcely available in the literature. Using the Torun catalogue for PAGB objects, supplemented with archival data, we have determined distances to the known population of Galactic PAGB stars. This is by modelling their spectral energy distributions (SED) with black bodies and numerically integrating over the entire wavelength range to determine the total integrated object flux. For most PAGB stars we assumed their luminosities are based on their positional characteristics and stellar evolution models. RV Tauri stars however are known to follow a period-luminosity relation (PLR) reminiscent of type-2 Cepheids. For these variable PAGB stars we determined their luminosities via the PLR and hence their distances. This allows us to overcome the biggest obstacle to characterising these poorly understood objects that play a vital part in Galactic chemical enrichment.

1. Introduction
Pre-planetary nebulae (PPNe) are a very brief phase in the late-stage evolution of mid-mass stars ($\sim 1 - 8\,\mathrm{M}_\odot$) between the asymptotic giant branch (AGB) and the PN phase. The tenuous envelope ejection in the final superwind of AGB evolution reaches rates of $10^{-4}\ \mathrm{M}_\odot\ \mathrm{yr}^{-1}$, and increases the effective temperature of the central star. The rate of increase is a strong function of core mass [1,2]. RV Tauri stars are F to K type variable stars with alternating deep and shallow minima and are believed to be luminous population II Cepheids that follow a period-luminosity relationship similar to type-2 Cepheids [3]. Understanding these objects depends on accurate distances that are not yet available for the more poorly-quantified objects. We aim to rectify this by determining accurate distances to the known Galactic PAGB star population [6].

2. Sample
We use the Toruń catalogue of PAGB and related objects [4,5] and flux data for known Galactic objects that includes 209 likely PPN, 112 RV Tauri, 72 R CrB/eHe/LTP and 87 possible PAGB

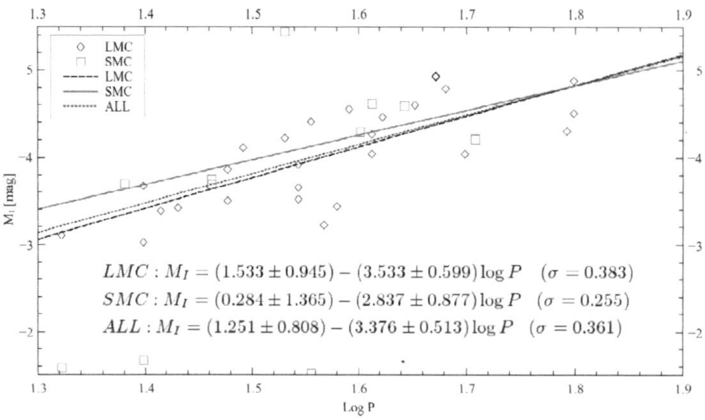

Figure 1. I-Band PLR for LMC and SMC RV Tauri variables taken from [7,8].

objects. We determine RV Tauri star distances via a newly defined period luminosity relation. These data are supplemented with publicly available data from other surveys from the optical to the sub-mm to build SEDs for all objects.

3. Method

We use the SED fitting method described in [6], with the luminosity calculation for RV Tauri stars described as follows. Using the data from [4,5] as well as publicly available photometry, we build an SED that is fit with a number of black body functions to calculate the total bolometric flux. This is used with the luminosity to determine the distance. Light curves and photometry from [7,8] are used to define a new period-luminosity relation for RV Tauri stars in both the LMC and the SMC to determine luminosities for their Galactic counterparts. Using these data we have defined a new period luminosity relations for RV Tauri stars in multiple wavebands the I-band PLR is shown in Figure 1.

4. Summary and future work

With a new catalogue of distances to the different PAGB objects we can work on determining the progenitors, amongst other physical parameters, for the different classes. With distances we can compare the z-heights for the different object classes, shining light on their progenitors. This will help uncover the shaping mechanisms responsible for the observed complex morphologies. This is the first ever distance catalogue for these objects.

References

[1] Schönberner D 1983 *Astrophys. J.* **272** 708
[2] Vassiliadis E and Wood P R 1994 *Astrophys. J. Suppl. Series* **92** 125
[3] Alcock et al 1998 *Astron. J.* **115** 1921
[4] Szczerba R, Siódmiak N, Stasińska G and Borkowski J 2007 *Astron. Astrophys.* **469** 799
[5] Szczerba R, Siódmiak N, Stasińska G, Borkowski J, García-Lario P, Suárez O, Hajduk M and García-Hernández D A 2012 *Proc. Int. Astron. Union* (S283) **7** 506
[6] Vickers S B, Frew D J, Parker Q A and Bojičič I S 2015 *Mon. Not. R. Astron. Soc.* **447** 1673
[7] Matsunaga N et al 2006 *Mon. Not. R. Astron. Soc.* **370** 1979
[8] Ripepi V et al 2015 *Mon. Not. R. Astron. Soc.* **446** 3034
[9] Vickers S B, Frew D J, Owers M S, Parker Q A and Bojičič I S 2016 *Mon. Not. R. Astron. Soc.* in preparation

Raman-scattered O VI λ1032 and He II λ1025 and Bipolar Outflow in the Symbiotic Star V455 Sco

Jeong-Eun Heo[1], Rodolfo Angeloni[2], Francesco Di Mille[3], Tali Palma[4], Seok-Jun Chang[1], Chae-Lin Hong[1] and Hee-Won Lee[1]

[1] Department of Physics and Astronomy, Sejong University, Korea
[2] AURA-GEMINI Observatory, Chile
[3] Las Campanas Observatory, Chile
[4] Pontificia Universidad Católica de Chile, Chile

E-mail: jeung6145@gmail.com

Abstract. Raman-scattering by atomic hydrogen is a unique spectroscopic process that may probe the mass transfer and mass loss phenomena in symbiotic stars(SSs). In the optical high-resolution spectra of the S-type SS V455 Sco, we note the presence of two Raman-scattered features, one at around 6825 Å with a triple-peak profile formed from Raman-scattering of O VI λ1032 and the other Raman-scattered He II λ1025 at around 6545 Å. Adopting an accretion flow model with additional contribution from a collimated bipolar outflow, we propose that the blue and central peaks are contributed from the accretion flow and the bipolar flow is responsible for the remaining red peak. With the absence of [N II] λ6548, the Raman-scattered He II λ1025 at around 6545 Å is immersed in the broad Hα wings that appear to be formed by Raman-scattering of far-UV continuum near Lyman series.

1. Raman O VI and Bipolar Outflow

A number of symbiotic stars(SSs) exhibit broad emission features at around 6825 Å and 7082 Å which are formed through Raman-scattering of O VI $\lambda\lambda$1032, 1038 by atomic hydrogen.[1] Raman O VI features in SSs are known to exhibit complicated profiles including double-peak profiles and triple-peak profiles and strong polarization.[2] They are unique and useful tools to probe the wind accretion process as some fraction of material from the giant is captured by the white dwarf. [3-5]

In Fig. 1 we present the optical high-resolution spectrum of the S-type SS V455 Scorpii taken with the Echelle spectrograph on the Magellan telescope in March 2015, in which we find that the Raman-scattered O VI λ1032 feature exhibits a triple-peak profile. We transform the Raman O VI feature to the rest frame determined by the optical emission line He I λ7065 to find that the line center of the Raman O VI at 6825 Å feature falls on the dip dividing the blue peak and the central peak. In the Doppler factor(ΔV) space (upper x-axis of Fig.1), which is measured by an imaginary observer that is at rest with respect to the giant, the first dip is found to correspond to 0 km s^{-1} and the two main peaks are ~ -30 km s^{-1} and $+40$ km s^{-1}.

Based on the accretion flow model [3], the accretion flow is divided into the 'Blue Emission Region(BER)' and the 'Red Emission Region(RER)', where the BER approaches and the RER recedes from the imaginary observer located at the giant. In this picture, we propose that the blue peak and central peak are contributed from the accretion flow around the white dwarf.

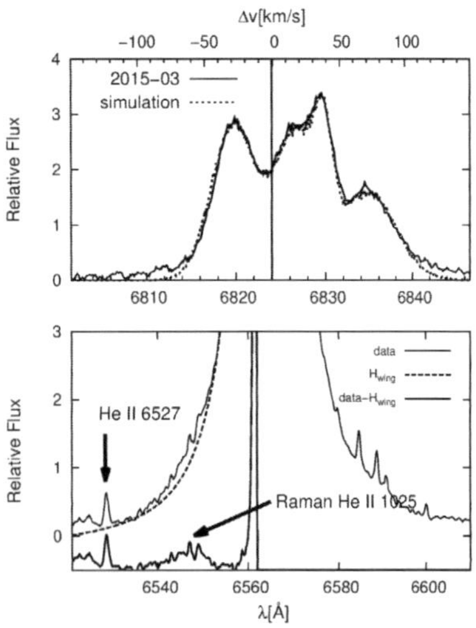

Figure 1. MIKE spectrum around the Raman-scattered O VI λ1032 at 6825 Å (upper panel) and He II λ 1025 at 6545 Å (lower panel)

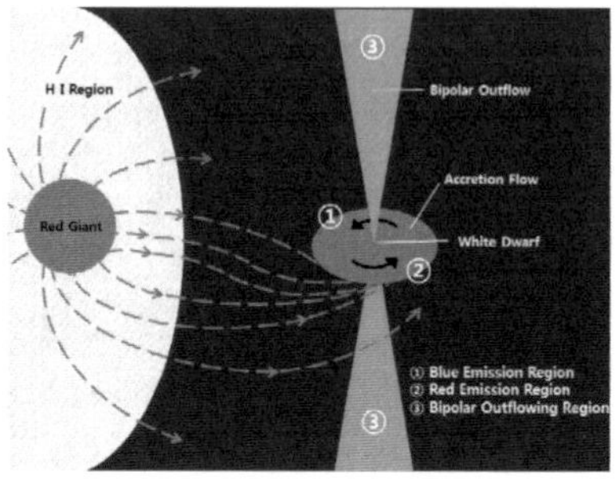

Figure 2. A schematic illustration of the accretion and bipolar outflow in V455 Sco

Based on the separation of two peaks, we suggest that the O VI emission region lies within 1AU from the white dwarf. We also propose that the remaining red peak is formed from O VI coming from the bipolar outflowing region which moves away with a velocity $\sim +80$ km s^{-1}.(e.g., [6])

Assuming the O VI emission region including the accretion flow and bipolar outflow as shown in Fig. 2, we show the profile of Raman O VI 6825 feature produced by our Monte-Carlo code as shown by dotted line in Fig. 1.

2. Raman He II λ1025

Raman He II features are useful as a spectroscopic diagnostic to estimate the mass loss rate of the scattering region. However, all the Raman He II λ1025 at 6545 Å features discovered so far suffer from heavy blending with the forbidden line [N II] λ6548 rendering it very difficult to obtain the line profile with sufficient quality.

He II λ4686 and He II λ6527 emission lines in V455 Sco spectrum provide strong hints pointing out the existence of the Raman He II features. We find the Raman-scattered He II 6545 feature by subtracting Hα wings which are excellently fitted by a $\Delta\lambda^{-2}$ profile as shown in the lower panel of Fig. 1. It is quite noticeable that the Raman He II feature is free of blending with [N II] λ6548 which is absent in the spectrum.

Acknowledgments

This research was supported by the Basic Science Research Program of the National Research Foundation (NRF-2014R1A1A2054887) and the Korea Astronomy and Space Science Institute under the R&D program (Project No. 2015-1-320-18).

References

[1] Schmid H M 1989 *Astron. Astrophys.* **211** L31
[2] Harries T J and Howarth I D 1997 *Astron. Astrophys. Suppl. Series* **121** 15
[3] Lee H-W and Park M-G 1999 *Astrophys. J.* **515** L89
[4] Kang E-H and Lee H-W 2008 *J. Korean Astron. Soc.* **41** 49
[5] Heo J-E and Lee H-W 2015 *J. Korean Astron. Soc.* **48** 105
[6] Angeloni R, Di Mille F, Bland-Hawthorn J and Osip D J 2011 *Astrophys. J. Lett.* **743** L8

Journal of Physics: Conference Series **728** (2016) 072015 doi:10.1088/1742-6596/728/7/072015

Subsynchronous sdBV binaries

Andrzej Baran[1]

[1] Uniwersytet Pedagogiczny w Krakowie, ul. Podchorążych 2, 30-552 Kraków, Poland

E-mail: [1] sfbaran@cyf-kr.edu.pl

Abstract. We found subsynchronicity of sdB stars in binaries. An assumption of synchronous rotation in such binary systems may lead to incorrect conclusions.

1. Introduction

Subdwarf B stars (hereafter: sdB) are descendants of roughly one solar masses main sequence stars. After the hydrogen is burnt out in their cores, they move towards the red giant branch to ignite helium. While burning helium, sdB stars settle down on the hot-end of the horizontal branch since, unlike most red giant stars, sdB stars lose their hydrogen envelopes before the helium flash. A population study including several different envelope ejection mechanisms was presented by [1]. The mechanisms that are producing sdB in binaries include short-period binaries, e.g. sdB+white dwarf or sdB+main sequence star, which are the products of common envelope ejection, long-period binaries, sdB+F/G/K, which are products of stable Roche lobe overflow, as well as single sdB stars, which are mergers of two objects. Some of the sdB star show stellar oscillations (sdBV).

2. Binaries

Analyzing Kepler data of sdB binaries, we found two interesting features, Doppler boosting and subsynchronicity. Doppler boosting is a relativistic effect in which a brightness of an object is magnified when heading towards an observer and dimmed when moving away. The brightness change is tiny, typically below 1ppt, and we have never detected this effect in data of sdB in binaries collected from the ground. Doppler boosting permits an estimate of the radial velocity without resorting to spectroscopic data. The second feature, subsynchronicity, was accidentally detected and never considered to be present in sdB binaries, particularly those, in tied orbits.

3. Synchronization

Subsynchronicity was detected in all Kepler sdBV with rotational periods derived. We can divide this sample in two groups; three sdBs have rotational period of days and orbital period of hours (sdB+dM) while the other four have rotational period of months and orbital period of days (sdB+WD). Based on [2, 3] showed that for stars with radiative envelopes the synchronization time scales with the orbital period to the power 17/3. Indeed, when comparing the seven subsynchronous Kepler sdBV binaries to those of comparable binaries which are considered to rotate synchronously, the expected synchronization time scales are several hundred times longer. Therefore the seven binaries provide no good cases to constrain the formalism in [3] for sdB stars, since these systems are not close enough to be fully synchronized. Their life on

Journal of Physics: Conference Series **728** (2016) 072015 doi:10.1088/1742-6596/728/7/072015

Table 1. The orbital and rotational periods [in days] of selected sdBV binaries.

KIC	Orbital	Rotational	companion
2991403	0.44	10.5	dM
11179657	0.39	7.2	dM
2438324	0.40	9.6	dM
7664467	1.56	35	WD
10553698	3.4	43	WD
11558725	10	45	WD
7668647	14.2	47	WD

the horizontal branch will be too short to allow for synchronization. However, the fact that they are so subsynchronous is remarkable, and must mean that many of the sdB binaries are 'born' subsynchronous on the EHB, and that the RGB progenitor has effectively lost a lot of rotational angular momentum while being converted into an sdB star. The fact that the Kepler sdBV+dM systems, with rotations derived seismically, rotate about a factor of 4 faster than the Kepler sdBV+WD systems and than the apparently single Kepler sdBVs, all with seismic rotation measurements, may point out that the synchronization mechanism is, as expected, more effective in the former group as they have shorter orbital periods. Alternatively, the rotational discrepancy may point at the possibility that sdBV+dM systems are born on the horizontal branch with higher sdB rotation rates than newborn sdB+WD systems or newborn single sdBs have.

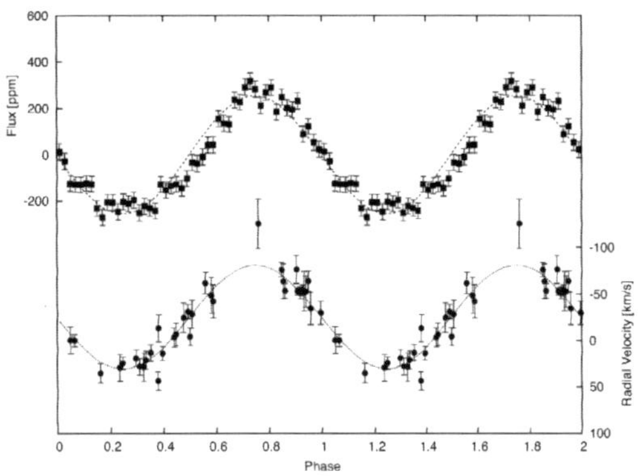

Figure 1. Doppler boosting in KIC7664467.

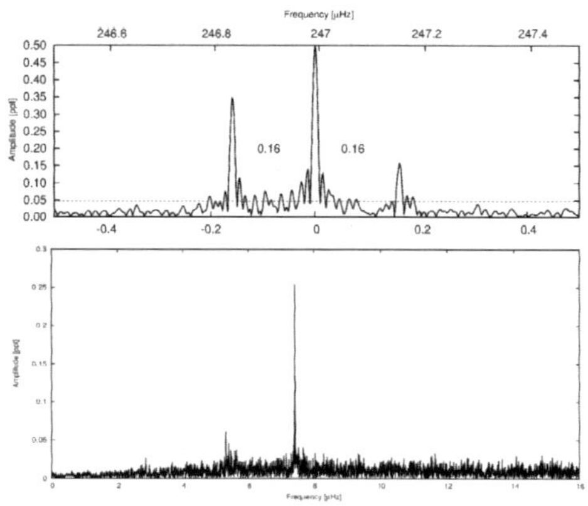

Figure 2. Rotation and orbital periods of KIC7664467.

Acknowledgments

This work was supported by Polish NCN grant No. UMO-2011/03/D/ST9/01914.

References

[1] Han Z, Podsiadlowski Ph, Maxted P F L, Marsh T R and Ivanova N 2002 *Mon. Not. R. Astron. Soc.* **336** 449
[2] Claret A and Cunha N 1997 *Astron. Astrophys.* **318** 187
[3] Zahn J-P 1975 *Astron. Astrophys.* **42** 329

Journal of Physics: Conference Series **728** (2016) 072016 doi:10.1088/1742-6596/728/7/072016

Formation of broad Balmer wings in symbiotic stars

Seok-Jun Chang, Jeong-Eun Heo, Chae-Lin Hong and Hee-Won Lee

Department of Physics and Astronomy, Sejong University, Korea

E-mail: csj607@gmail.com

Abstract. Symbiotic stars are binary systems composed of a hot white dwarf and a mass losing giant. In addition to many prominent emission lines symbiotic stars exhibit Raman scattered O VI features at 6825 and 7088 Å. Another notable feature present in the spectra of many symbiotics is the broad wings around Balmer lines. Astrophysical mechanisms that can produce broad wings include Thomson scattering by free electrons and Raman scattering of Lyβ and higher series by neutral hydrogen. In this poster presentation we produce broad wings around Hα and Hβ adopting a Monte Carlo techinique in order to make a quantitative comparison of these two mechanisms. Thomson wings are characterized by the exponential cut-off given by the termal width whereas the Raman wings are dependent on the column density and continuum shape in the far UV region. A brief discussion is provided.

1. Broad Balmer Wings in Symbiotic Stars

Consisting of a hot white dwarf and a mass losing giant, symbiotic stars are important objects to study the mass transfer through gravitational capture of the slow stellar wind from the giant component. Notable features in the spectra of symbiotic stars include broad wings around Balmer emssion lines and Raman scattered O VI features at 6825 Å and 7082 Å([1]). The astrophysical mechanisms responsible for the formation of broad wings include Thomson scattering by free electrons in the H II region around the white dwarf([3] and [2]) and Raman scattering of far UV radiation around Lyman series by atomic hydrogen in the vicinity of the giant([1] and [4]). In this work we adopt a Monte Carlo technique to produce broad Balmer wings via Thomson scattering and Raman scattering in order to make a quantitative comparison.

2. Monte Carlo Results

In Fig. 1, we show our Monte Carlo results. Raman scattering wings are shown in the upper panel with a solid line and the dotted line is the best fitting Lorentzian function. The H I column density is chosen to be $N_{HI} = 10^{20}$ cm^{-2}. The lower panel shows Thomson scattering wings (soild line) with a Gaussian fitting function (dotted line).

Raman scattering wings delineate approximately the total scatttering cross section multiplied by the branching ratio([5]). In turn, the cross section is approximated by a Lorentzian function, whereas the branching ratio is a smooth function of the wavelength. Therefore, Raman wings are fitted well by a Lorentzian function as illustrated in Figure 1. Two factors, the H I column density and covering factor, control the width and flux of the wings in ref [5].

Thomson scattering wings are characterized by the column density and the temperature of free electrons, where the former affects the total Thomson wing flux and the latter determines the wing width. Consequently, Thomson scattering wings are well fitted by a Gaussian function as shown in the lower panel of Figure 1. A typical Thomson scattering wing width is ~ 500 km/s for the temperature of $\sim 10^4$ K. Under the condition of the Thomson scattering optical depth less than unity, the wing flux is nearly proportional to the optical depth in ref [2].

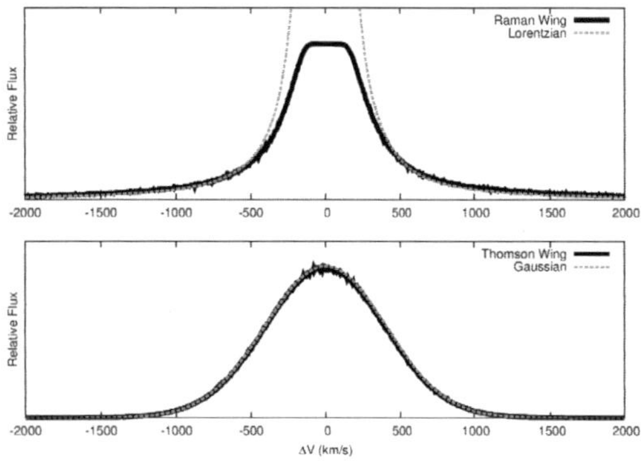

Figure 1. Raman scattering wings (upper panel) and Thomson scattering wings (lower panel) obtained from our Monte Carlo simulations. The solid lines show the Monte Carlo results. The dotted line in the upper panel represents the best fitting Lorentzian function and the one in the lower panel is the best fitting Gaussian function.

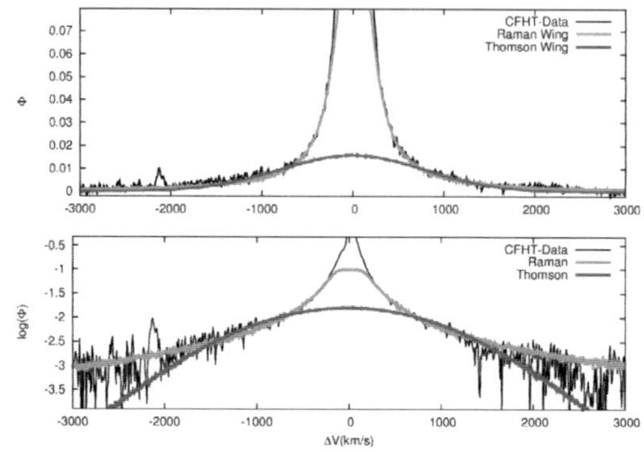

Figure 2. Overplot of the Thomson and Raman wings against the observed data of Z And obtained with the Canada-France-Hawaii Telescope in 2014. The lower panel is the same plot in the logarithmic scale. Apparently the Raman scattering wings provide a better fit to the observed data than the Thomson wings.

3. Discussion

In order to make a clear and quantitative comparison, we plot the two Monte Carlo wings with the observational data of the symbiotic star Z And obtained with the Canada-France-Hawaii Telescope in 2014. The lower panel is the same plot shown in a logarithmic scale, in which we easily verify the exponential cut-off shown in the Thomson scattering wings. In the lower panel, we find that the Raman scattering wings provide a better fit to the observed data than the Thomson wings. Due to significant difference in the cross section around Lyβ and Lyγ for Raman scattering, it is expected that Hβ wings are relatively much weaker than Hα wings if they are formed through Raman scattering. Our future work will include the intercomparisons of Hα wings and Hβ wings simultaneously.

This research was supported by the Basic Science Research Program of the National Research Foundation (NRF-2014R1A1A2054887) and the Korea Astronomy and Space Science Institute under the R&D program (Project No. 2015-1-320-18).

References

[1] Schimid H M 1989 *Astron. Astrophys.* **211** L31-L34
[2] Kim H J, Lee H-W and Kang S 2007 *Mon. Not. R. Astron. Soc.* **374** 187-195
[3] Sekeras M and Skopal A *Mon. Not. R. Astron. Soc.* **427** 979-987
[4] Heo J-E and Lee H-W 2015 *J. Korea Astron. Soc.* **48** 105-112
[5] Chang S-J, Heo J-E, Di Mile F, Angeloni R, Palma T and Lee H-W 2015 *Astrophys. J.* **814** 98

Journal of Physics: Conference Series **728** (2016) 072017 doi:10.1088/1742-6596/728/7/072017

Hot subdwarf stars and their connection to thermonuclear supernovae

S. Geier[1,2], T. Kupfer[3], E. Ziegerer[2], U. Heber[2], P. Németh[2], A. Irrgang[2] and MUCHFUSS team

[1] Department of Physics, University of Warwick, Coventry CV4 7AL, UK
[2] Dr. Karl Remeis-Observatory & ECAP, Astronomical Institute, Friedrich-Alexander University Erlangen-Nuremberg, Sternwartstr. 7, D 96049 Bamberg, Germany
[3]Division of Physics, Mathematics, and Astronomy, California Institute of Technology, Pasadena, CA 91125, USA

E-mail: [1]s.geier@warwick.ac.uk

Abstract. Hot subdwarf stars (sdO/Bs) are evolved core helium-burning stars with very thin hydrogen envelopes, which can be formed by common envelope ejection. Close sdB binaries with massive white dwarf (WD) companions are potential progenitors of thermonuclear supernovae type Ia (SN Ia). We discovered such a progenitor candidate as well as a candidate for a surviving companion star, which escapes from the Galaxy. More candidates for both types of objects have been found by crossmatching known sdB stars with proper motion and light curve catalogues. The Gaia mission will provide accurate astrometry and light curves of all the stars in our hot subdwarf sample and will allow us to compile a much larger all-sky catalogue of those stars. In this way we expect to find hundreds of progenitor binaries and ejected companions.

1. Introduction

Hot subdwarf stars (sdO/Bs) are evolved core helium-burning stars with very thin hydrogen envelopes, which can be formed by common envelope ejection. Close sdB binaries with massive C/O-WD companions are candidates for supernova type Ia (SN Ia) progenitors, because mass-transfer can lead to the thermonuclear explosion of the WD. The project Massive Unseen Companions to Hot Faint Underluminous Stars from SDSS (MUCHFUSS) aims at finding the sdB binaries with the most massive compact companions like massive white dwarfs, neutron stars or black holes.

We selected and classified about ~ 1400 hot subdwarf stars from the Sloan Digital Sky Survey (SDSS DR7). Stars with high velocity variations have been reobserved and analysed. In total 177 radial velocity variable subdwarfs have been discovered and 1914 individual radial velocities measured. We constrain the fraction of close massive companions of H-rich hot subdwarfs to be smaller than $\sim 1.3\%$ [1]. Orbital parameters as well as minimum companion masses have been derived from the radial velocity curves of 30 sdB binaries [2].

2. Discovery of an SN Ia progenitor and an ejected companion

We detected high RV-variability of the bright sdB CD$-30°$11223. Photometric follow-up revealed both shallow transits and eclipses, allowing us to determine its component masses and fundamental parameters. The binary system, which is composed of a C/O-WD ($\sim 0.76\,M_{\odot}$)

and an sdB ($\sim 0.51\,M_\odot$) has a very short orbital period of ~ 0.049 d. In the future mass will be transfered from the helium star to the white dwarf. After a critical amount of helium is deposited on the surface of the white dwarf, the helium is ignited. Modelling this process shows that the detonation in the accreted helium layer should be sufficiently strong to trigger the explosion of the core. Thermonuclear supernovae have been proposed to originate from this so-called double-detonation of a WD [3, 4]. The surviving companion star will then be ejected with its orbital velocity. The properties of such a remnant match the hypervelocity star US 708, a helium-rich sdO star moving with $\sim 1200\,\mathrm{km\,s^{-1}}$, exceeding the escape velocity of our Galaxy by far and making it the fastest unbound star known in our Galaxy [5].

3. Finding more progenitor and ejected companion candidates

Since the properties of the ejected companions, especially the ejection velocity, allow us to constrain the properties of the binary progenitors such as the orbital period and the companion mass right at the moment of the explosion, we will gain an unprecedented insight into the formation of SN Ia and learn about other acceleration mechanisms for hypervelocity stars, if more such objects can be found and studied. The distribution of orbital periods and WD companion masses of progenitor binaries will help us to constrain SN Ia progenitor models. While binaries with periods longer than about 2 hr will merge as double degenerates, closer binaries might be progenitors for the helium double-detonation channel.

To search for ejected companions we compiled a catalogue of all known sdO/B stars from the literature and our own database (~ 4500 stars) and crossmatch it with proper motion catalogues. Candidates with high velocities are followed-up with spectroscopy (Keck/ESI, VLT/XSHOOTER, SOAR/Goodman, CAHA/TWIN, WHT/ISIS) to measure spectroscopic distances and derive kinematics. Several good candidates for unbound hypervelocity sdO/Bs have already been found.

We found that hot subdwarf binaries with massive WDs in close orbits are quite rare. To find more of those objects, we crossmatch the hot subdwarf catalogue with light curve catalogues (e.g. CRTS, PTF, SWASP, GALEX gPhoton, Kepler K2) and search for the characteristic sinusoidal variations caused by the ellipsoidal deformation of the sdB. Several candidates have been found and will be followed-up spectroscopically and photometrically.

Eventually, the Gaia mission will provide accurate astrometry and light curves of all the stars in our hot subdwarf sample and will allow us to compile a much larger all-sky catalogue of those stars. In this way we expect to find hundreds of progenitor binaries and ejected companions.

References

[1] Geier S *et al* 2015 *Astron. Astrophys.* **577** A26
[2] Kupfer T, Geier S, Heber U, Østensen R H, Barlow B N, Maxted P F L, Heuser C, Schaffenroth V and Gänsicke B T 2015 *Astron. Astrophys.* **576** A44
[3] Fink M, Röpke F K, Hillebrandt W, Seitenzahl I R, Sim S A and Kromer M 2010 *Astron. Astrophys.* **514** A53
[4] Geier S *et al* 2013 *Astron. Astrophys.* **554** A54
[5] Geier S *et al* 2015 *Science* **347** 1126–1128

Journal of Physics: Conference Series **728** (2016) 072018 doi:10.1088/1742-6596/728/7/072018

Spiral-shells and nascent bipolar outflow in CIT 6: hints for an eccentric-orbit binary?

Hyosun Kim[1,2,4], Sheng-Yuan Liu[1], Naomi Hirano[1], Ronny Zhao-Geisler[3,1], Alfonso Trejo[1], Hsi-Wei Yen[1], Ronald E. Taam[1], Francisca Kemper[1], Jongsoo Kim[2], Do-Young Byun[2] and Tie Liu[2]

[1] Academia Sinica Institute of Astronomy and Astrophysics, Taiwan
[2] Korea Astronomy and Space Science Institute, Korea
[3] National Taiwan Normal University, Department of Earth Sciences, Taiwan
[4] EACOA fellow

E-mail: [1] hkim@asiaa.sinica.edu.tw

Abstract. We present the essential results pointed out in a recently published paper, Kim et al. 2015, Astrophys. J., 814, 61. The carbon star CIT 6 reveals evidences for a binary in a high-resolution CO line emission map of its circumstellar envelope taken with the Submillimeter Array. The morphology of the outflow described by the spiral-shell pattern, bipolar (or possibly multipolar) outflow, one-sided interarm gaps, and double spiral feature point to a plausible scenario that CIT 6 is a binary system in an eccentric orbit with the mass losing star evolving from the AGB.

Many planetary nebulae (PNe) and pre-PNe simultaneously present both outer ring-like patterns and inner asymmetric (bipolar or multipolar) structures. This coexistence is a result of shape transition from spherically-symmetric envelopes of asymptotic giant branch (AGB) stars to highly-asymmetric PNe. The bipolar outflows are hypothesized as the outcome of a *close* binary interaction, while the typical dynamical timescales of the observed ring patterns are consistent with the orbital periods of *wide* binaries. This puzzle may be resolved by binaries in highly eccentric orbits whose close periastron passages facilitate strong binary interactions and consequent accretion onto the companion stars.

In a high-resolution CO J=2–1 line observation of a carbon star CIT 6 with the Submillimeter Array, we have found various features expected in a highly eccentric binary system [1]. The circumstellar envelope of CIT 6 unveils its spiral-shell pattern (Figure 1) and a bipolar (or multipolar) structure (Figure 2) in the CO maps, in a similar manner to the pre-PNe observed in the optical images. The spiral-shell pattern was predicted earlier [2] based on the HC_3N J=4–3 segments [3], which are now well connected by the CO pattern. The tiny bipolar outflow was observed in an optical polarization image [4].

In addition, the CO maps further reveal new interesting aspects. The double spiral feature in the innermost region (Figure 3) is expected in a slow wind condition, where the flow is captured by the gravitational potential of the companion [5]. The existence of the double spiral in CIT 6 may imply that the companion of the carbon star lies in, or passes through, the wind acceleration zone of the AGB star. A typical size of the acceleration zone is several stellar radii and, therefore, the binary orbit is possibly highly eccentric. The one-sided lowering of interarm emission (Figure 4) can be also reproduced by an eccentric orbit binary model [6].

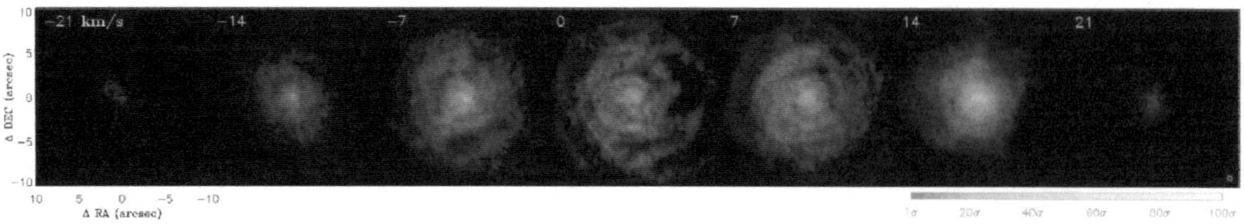

Figure 1. SMA CO J=2–1 channel maps of CIT 6 unveiling the circumstellar spiral-shell pattern. The central channel velocity with respect to the systemic velocity is marked at the top left corner of each panel. The color bar labels the line emission intensities of the CO map in units of the noise level. Its synthesized beam size is denoted at the bottom right corner.

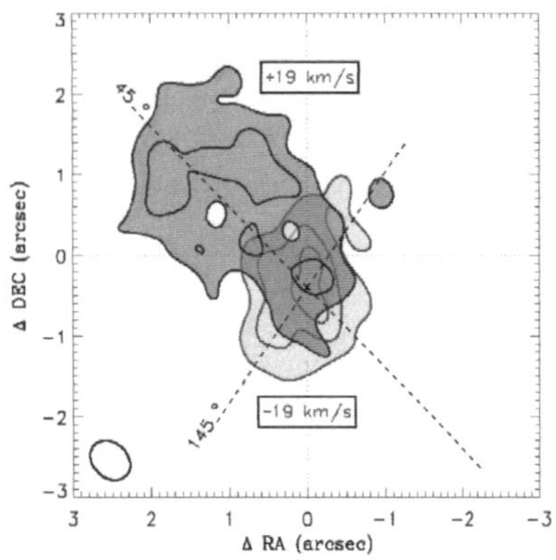

Figure 2. Innermost region of CO line-edge channels present the asymmetric (bipolar, or possibly multipolar) outflow of CIT 6. The outflow directions are marked by dashed lines. The beam size is given by an ellipse at the bottom left. Contour levels are 8σ, 12σ, and 16σ with the noise level σ.

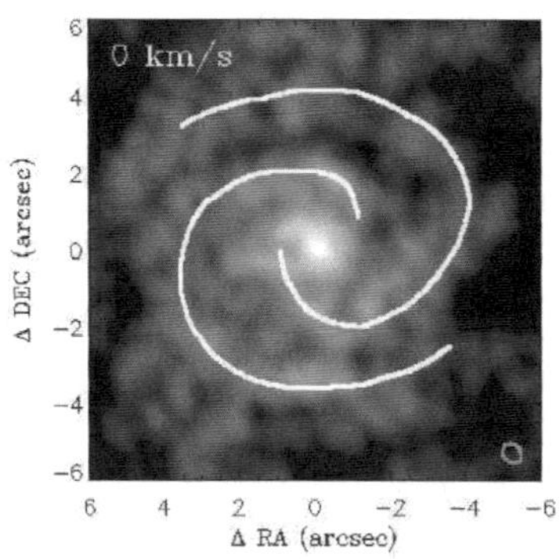

Figure 3. Inner region in the systemic velocity channel of CO emission line reveals the double-spiral feature of CIT 6. This is the same as the middle panel of Figure 1, but for about half the field and with the overlying hand-drawn lines.

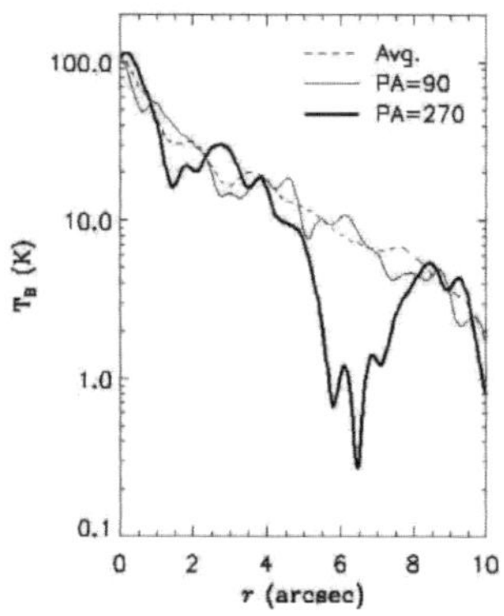

Figure 4. CO brightness temperature of CIT 6 as a function of radius to the east (PA=90°), to the west (PA=270°), and in azimuthal average. The absence of interarm emission to the west is significant and interpreted as a hint of eccentric binary.

Acknowledgments
We thank the organizing committee of the 11th Pacific Rim Conference on Stellar Astrophysics and the participants for valuable discussion during the conference. We would like to thank the SMA support staff, for their help in executing the observations. H.K. acknowledges support through the EACOA Fellowship.

References
[1] Kim H *et al* 2015 *Astrophys. J.* **814** 61
[2] Kim H, Hsieh I-T, Liu S-Y and Taam R E 2013 *Astrophys. J.* **776** 86
[3] Claussen M J, Sjouwerman L O, Rupen M P, Olofsson H, Schöier F L, Bergman P and Knapp G R 2011 *Astrophys. J. Lett.* **739** L5
[4] Schmidt G D, Hines D C and Swift S 2002 *Astrophys. J.* **576** 429
[5] Mohamed S and Podsiadlowski Ph 2012 *Baltic Astronomy* **21** 88
[6] He J H 2007 *Astron. Astrophys.* **467** 1081

The binary progenitors of core-collapse supernovae

Hanindyo Kuncarayakti[1,2]
[1] Millennium Institute of Astrophysics, Chile
[2] Departamento de Astronomía, Universidad de Chile, Casilla 36-D, Santiago, Chile
E-mail: hanin@das.uchile.cl

Abstract. Core-collapse supernovae are thought to be the death explosions of massive stars. Mass and metallicity have been regarded as the primary drivers in stellar evolution, including in the evolution of the progenitor stars of supernovae. However, the diversity of supernova explosions cannot seem to be explained by these two parameters only, furthermore there have been evidences that some supernovae are produced by binary progenitors. This supports the current paradigm that binarity plays an important role in the evolution of massive stars.

1. Introduction

Among core-collapse supernovae (SNe), type-Ib and Ic SNe are known to lack hydrogen in the spectra. While SNe Ib appear to still retain helium, SNe Ic does not. This leads to the notion that the progenitors of these SNe must have lost their outer envelope by some mechanism before the terminal explosion. The progenitors of these SNe are expected to bear similarities with Wolf-Rayet (WR) stars, which are devoid of hydrogen. These stars lost hydrogen and helium in their outer layers via strong metallicity-dependent winds. Single stellar evolution models predict that stars $\gtrsim 25\ M_{\odot}$ at solar metallicity will evolve into WR stars and eventually explode as type-Ib/c SNe [1]. However, observations suggest that massive single star evolution to WR stars may not be the only channel in the production of hydrogen-poor core-collapse SNe. Detections of SN progenitors in pre-explosion *Hubble Space Telescope (HST)* images have successfully identified the progenitor stars of some type-II SNe, nevertheless for the type-Ib/c SNe there has been no firm detection [2]. Furthermore, initial mass estimates from SN parent stellar population age indicate that there exist populations of SN Ib/c progenitors of both $\lesssim 25\ M_{\odot}$ and $\gtrsim 25\ M_{\odot}$ [3], which is also supported by the calculation of the stellar initial mass function (IMF) [4].

An alternative to single massive WR progenitor channel in SNe Ib/c is the massive binary interaction [5]. In this scenario, the progenitor star of SN Ib/c lost its outer envelope in a Roche lobe overflow mechanism during close binary interaction. As such, the the progenitor star may be less massive than $\sim 25 M_{\odot}$ as it does not need to have such high mass and metallicity in order to produce strong wind that propels the outer envelope.

2. Probing the binary system

SN 1993J was the first SN to have its binary companion discovered [6]. The presence of the binary component was predicted through stellar evolution modeling of the progenitor, and eventually it was discovered spectroscopically in late-time *HST* observations. Direct detection of the binary companion of SN 2011dh has been achieved also with late-time *HST* imaging after the SN has faded significantly [7]. The colors are consistent with an early B-type star. These two SNe are currently the strongest cases where a binary companion to the SN progenitor has been confirmed. There have been other examples of hints of the progenitor companion such as in SNe 2008ax [8] and 2013df [9]. Interestingly, all of these SNe with binary companions are of type IIb which initially showed hydrogen lines in the early spectra but gradually morph into type Ib.

The only type-Ib SN with possible archival progenitor detection thus far, iPTF13bvn [10], has been suggested to have resulted from a progenitor in a binary system from the light curve analysis [11] and nebular spectroscopy [12]. Future observations will put this deduction into test, as well as providing the necessary confirmation whether the progenitor candidate has actually disappeared. As shown in the case of iPTF13bvn [12], nebular spectroscopy provides an estimate to the core mass, hence the initial mass of the progenitor star. In the same work it is shown that the distribution of the [O I]/[Ca II] line ratio, which is indicative of the core mass, suggests that there are two distinct populations for hydrogen-poor SNe with different mass ranges as opposed to a more uniform distribution in SNe II. This may be interpreted as the two channels of SN progenitors: those coming from massive WR stars and those from relatively less massive close binaries. It is to be noted that the majority of massive stars undergo binary interaction at some point during their lifetime [13]. As a result, this implies that the majority of core-collapse SN progenitors should also experience the same evolution, and this may be a driving factor resulting in the loss of envelope in some of the progenitors and the diversity of the SN explosion.

Acknowledgments
Support for HK is provided by the Ministry of Economy, Development, and Tourism's Millennium Science Initiative through grant IC120009, awarded to The Millennium Institute of Astrophysics, MAS. HK acknowledges support by CONICYT/FONDECYT grant 3140563.

References

[1] Crowther P A 2007 *Annu. Rev. Astron. Astrophys.*, **45** 177–219
[2] Eldridge J J, Fraser M, Smartt S J, Maund J R and Crockett R M 2013 *Mon. Not. R. Astron. Soc.* **436** 774–795
[3] Kuncarayakti H, Doi M, Aldering G, Arimoto N, Maeda K, Morokuma T, Pereira R, Usuda T and Hashiba Y 2013 *Astron. J.* **146** 30
[4] Smith N, Li W, Filippenko A V and Chornock R 2011 *Mon. Not. R. Astron. Soc.* **412** 1522–1538
[5] Podsiadlowski P, Joss P C and Hsu J J L 1992 *Astrophys. J.* **391** 246–264
[6] Maund J R, Smartt S J, Kudritzki R P, Podsiadlowski P and Gilmore G F 2004 *Nature* **427** 129–131
[7] Folatelli G, Bersten M C, Benvenuto O G, Van Dyk S D, Kuncarayakti H, Maeda K, Nozawa T, Nomoto K, Hamuy M and Quimby R M 2014 *Astrophys. J. Lett.* **793** L22
[8] Folatelli G, Bersten M C, Kuncarayakti H, Benvenuto O G, Maeda K and Nomoto K 2015 *Astrophys. J.* **811** 147
[9] Maeda K *et al* 2015 *Astrophys. J.* **807** 35
[10] Cao Y *et al* 2013 *Astrophys. J. Lett.* **775** L7
[11] Bersten M C, Benvenuto O G, Folatelli G, Nomoto K, Kuncarayakti H, Srivastav S, Anupama G C, Quimby R and Sahu D K 2014 *Astron. J.* **148** 68
[12] Kuncarayakti H *et al* 2015 *Astron. Astrophys.* **579** A95
[13] Sana H, de Mink S E, de Koter A, Langer N, Evans C J, Gieles M, Gosset E, Izzard R G, Le Bouquin J B and Schneider F R N 2012 *Science* **337** 444

Journal of Physics: Conference Series **728** (2016) 072020 doi:10.1088/1742-6596/728/7/072020

Mass-transfer in close binary and their companions

Wenping Liao[1,2], Shengbang Qian[1,2,3], Liying Zhu[1,2,3] and Linjia Li[1,2]

[1] Yunnan Observatories, Chinese Academy of Sciences, P.O. Box 110, 650216 Kunming, P.R. China

[2] Key Laboratory for the Structure and Evolution of Celestial Objects, Chinese Academy of Sciences, P.O. Box 110, 650216 Kunming, P. R. China

[3] University of the Chinese Academy of Sciences, Yuquan Road 19#, Shijingshan Block, 100049 Beijing, P. R. China

E-mail: [1] liaowp@ynao.ac.cn

Abstract. Secular and/or cyclical orbital period variations of close binaries can be derived by analyzing the (O-C) diagram. The secular variations are usually explained as mass transfer between components, while the most plausible explanation of the cyclic period changes is the light-travel time effect (LTTE) through the presence of a third body. Mass transfer and additional companions in close binary systems are important for understanding the formation and evolution of the systems. Here, UV light curves of several close binaries based on the Lunar-based Ultraviolet Telescope (LUT) observations are presented and analyzed with the Wilson-Devinney (W-D) method. Then, based on those light-curve solutions and new analysis of the orbital period variations, the multiplicity, geometrical structure and evolution state of targets are discussed.

1. Introduction

Three Algol-type binaries (AI Dra, TW Dra and V548 Cyg) were observed by using the LUT. The light variability of AI Dra was suspected by Schilt & Hill [1] and its Algol type eclipsing nature was confirmed by Geyer et al. [2]. Many observations and detailed photometric and spectroscopic analysis have been carried out since its discovery. By using the LUT, AI Dra was monitored in October 2014. The variability of TW Dra was discovered in 1910 [3]. It is an Algol-type system with a delta scuti-like oscillating primary component [4], also the A-component of the visual binary ADS 9706. Detailed spectroscopic and photometric investigations of TW Dra have been published. A historical summary of investigation was given in Tkachenko et al. [5] and Zejda et al. [6]. By using the LUT, near-UV light curve of TW Dra is reported. The light variability of V548 Cygni was discovered by Hoffmeister [7]. Until 1983, several studies have been published. After that, the binary has generally been neglected in photometric analysis.

LUT is one of the payloads on the Chang'e-III lunar lander, it works in the near ultraviolet band ($245 \sim 345\,\mathrm{nm}$) [8]. It is a 15-cm telescope with a Nasmyth focus that has a field view of $1.^{\circ}27 \times 1.^{\circ}27$ and can observe objects as low as 13.5 mag [9, 10]. It is very useful to monitor eclipsing binaries and variable stars without interruption because of the slow rotation of the Moon [8]. The data processing was done with a special data processing method developed by Meng et al. because LUT works in lunar daytime for sufficient power supply [10]. The calibration of those photometric data was given by Wang et al [11].

2. LUT light curves and photometric analysis

Complete LUT light curves of AI Dra, TW Dra and V548 Cyg are displayed in figures 1-3. As shown in these figures, the light curves are all EA-types. We analysed light curves by using W-D 2013 method. The solid lines in figures show the theoretical light curves, open circles are the observations obtained with LUT. Our photometric solutions show that these three binaries are all semi-detached binaries with the secondary components fills their critical Roche lobes, which indicates that a mass transfer from the secondary component to the primary one should happen.

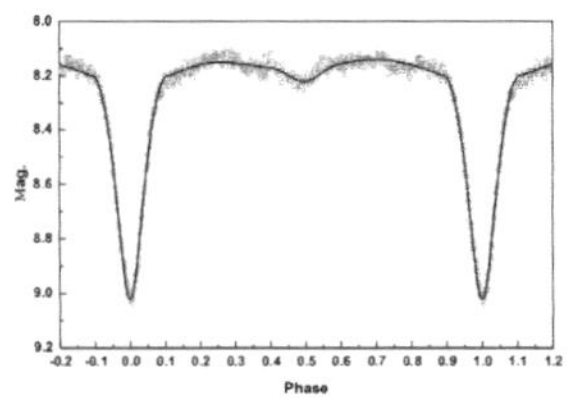

Figure 1. Light curve of AI Dra.

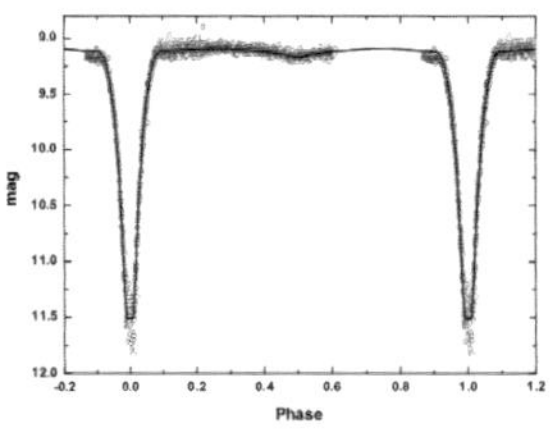

Figure 2. Light curve of TW Dra.

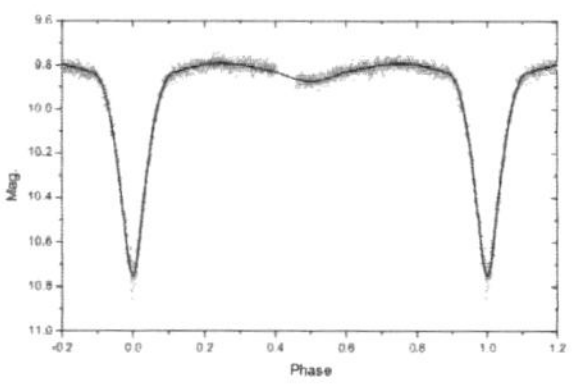

Figure 3. Light curve of V548 Cyg.

Based on all available data, the orbital period changes of three binaries were carried out. A secular increase + two cyclical variations and a secular increase + three cyclical variations are found in the O-C diagram of AI Dra and TW Dra, respectively. Three cyclic variations in the O-C diagram are obtained in V548 Cyg. The secular increase in the orbital period of the system can be interpreted as being the result of mass transfer from the secondary component to the primary one, which is consistent with the results of light-curve solutions. Cyclic variations can be interpreted by the LTTE via the presence of additional companions [12]. No obvious long-term changes in the orbital period of V548 Cyg were found indicating that the contributions of the mass transfer and the mass loss due to magnetic braking to the period variations are comparative.

Acknowledgments

This work is supported by the Chinese Natural Science Foundation (Nos. 11403095, 11133007, 11325315), Yunnan Natural Science Foundation (2014FB187), and the Strategic Priority Research Program "The Emergence of Cosmological Structures" of the Chinese Academy of Sciences, Grant No. XDB09010202.

References

[1] Schilt J and Hill S J 1938 *Contributions from the Rutherford Observatory of Columbia University New York* **31** 1
[2] Geyer E, Reim W, Remus G and Plattner D 1955 *Kleine Veroff. Remeis Sternw* 12
[3] Pickering E C 1910 *Harvard College Obs. Circ.* **159** 3
[4] Zhang X B, Luo C Q and Fu J N 2013 *Astrophys. J.* **777** 77
[5] Tkachenko A, Lehmann H and Mkrtichian D 2010 *Astron. J.* **139** 1327
[6] Zejda M, Mikulášek Z and Wolf M 2008 *Astron. Astrophys.* **489** 321
[7] Hoffmeister C 1949 *Erg. Hefte z.d. AN* **12** 1, 7
[8] Cao L *et al* 2011 *Science China Physics, Mechanics and Astronomy* **54** 558
[9] Wen W-B *et al* 2014 *Research Astron. Astrophys.* **14** 1674
[10] Meng X-M, Cao L, Qiu Y-L, Wu C, Wang J, Han X-H, Deng J-S, Xin L-P, Cai H-B and Wei J-Y 2015 *Astrophys. Space Sci.* **358** 47
[11] Wang J *et al* 2015 *Research Astron. Astrophys.* **15** 1068
[12] Liao W-P and Qian S-B 2010 *Mon. Not. R. Astron. Soc.* **405** 1930

Journal of Physics: Conference Series **728** (2016) 072021 doi:10.1088/1742-6596/728/7/072021

Cool and luminous transients from mass-losing binary stars

Ondřej Pejcha[1,2], Brian D. Metzger[3] and Kengo Tomida[4]

[1] Department of Astrophysical Sciences, Princeton University, 4 Ivy Lane, Princeton, NJ 08540, USA
[2] Hubble and Lyman Spitzer Jr. Fellow
[3] Columbia Astrophysics Laboratory, Columbia University, New York, NY 10027, USA
[4] Department of Earth and Space Science, Graduate School of Science, Osaka University, 1-1 Machikaneyama, Toyonaka, Osaka 560-0043, Japan

E-mail: [1] pejcha@astro.princeton.edu

Abstract. Motivated by the recently established link between luminous red novae (LRN) and catastrophic phases of binary star evolution, we perform smoothed particle hydrodynamic calculations of outflows from binary stars with realistic equation of state and opacities. We focus on the case of mass loss from the outer Lagrangian point (L_2), where the resulting spiral stream experiences tidal torques from the binary and becomes unbound. As the individual spiral arms merge and collide near the binary, the outflow thermalizes about 5% of its kinetic energy. For reasonable binary parameters, the outflow can produce luminosities up to $10^6\,L_\odot$ with effective temperatures between 500 and 6000 K, depending on the optical depth through the outflow. This is compatible with many examples of the LRN such as V838 Mon and V1309 Sco. The luminosity and the expansion velocity are correlated, as is roughly observed in the known LRN. The outflow readily forms dust, leading to great variations of the appearance of the transient as a function of the viewing angle. Our results are relevant for a more general class of equatorial outflows with asymptotic velocity and heating rate near the binary proportional to its orbital speed.

1. Introduction

The discovery of a contact binary with rapidly decreasing orbital period followed by an outburst in V1309 Sco [1] established a connection between catastrophic phases of binary star evolution and a group of transients best represented by objects such as V838 Mon, M31 RV, and M85 OT and characterized by their red color and luminosity in the gap separating classical novae and supernovae. However, these objects, often labeled as luminous red novae or intermediate luminosity optical transients, exhibit considerable diversity in their properties [2, 3].

2. Hydrodynamic simulations

To better understand the emission mechanism and the underlying physics of LRN, we explored the dynamics of outflows from mass-lossing binary stars using smoothed particle hydrodynamic simulations, which employ realistic low-temperature equation of state, radiative diffusion and cooling with realistic opacities including dust, and irradiation by the central binary [3]. We inject the particles at the outer Lagrangian point of the binary and leave the initially corotating gas evolve in the time-changing gravitational field of the central binary. For a set of binary

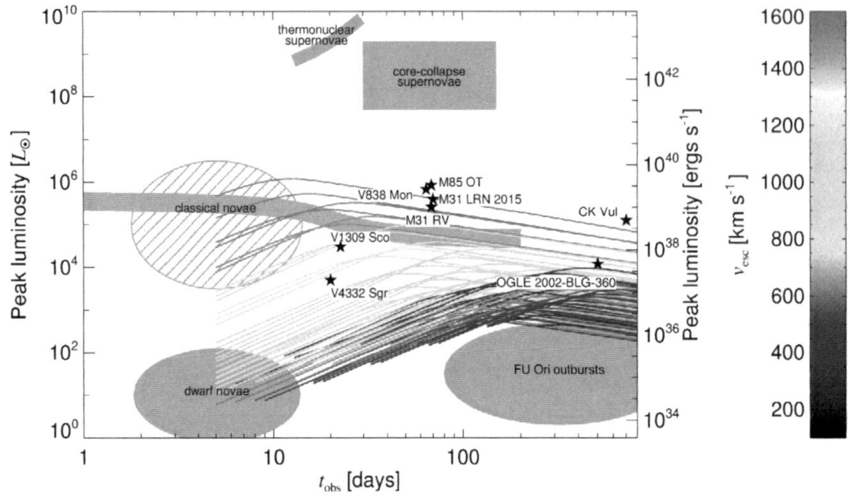

Figure 1. Peak luminosities as a function of transient duration [4]. Coloured lines mark our predictions for mass-losing binaries, where each line corresponds to a single set of binary parameters and a range of time-scales (left as a free parameter). The colour marks the escape velocity of the binary, as denoted in the colour bar on the right. Positions of red transients are shown with black stars. Approximate regions occupied by dwarf and classical novae, FU Ori outburst, and thermonuclear and core-collapse supernovae are shown in grey. More details are provided in [3].

parameters, the resulting equatorial and primarily radial unbound outflow exhibits internal shocks, which thermalize part of the kinetic energy. We track the luminosity and effective temperature of the resulting emission, and construct a semi-analytic model capturing the salient features of the simulations.

3. Results

In Figure 1 we show that our results can broadly explain the known transients associated with mass-losing binaries. Brighter transients come from binaries with higher orbital speed, and that fainter or longer transients are cooler, with effective temperatures $\lesssim 500\,$K possible. The group of long and extremely cold transients recently identified with Spitzer [5] might thus share common mechanism with LRN. We predict correlation between the observed expansion velocity and the transient luminosity. Due to copious dust formation in the equatorial outflow, the transient appearance might vary greatly depending on the orientation of the binary orbital plane.

Acknowledgments

Support for OP was provided by NASA through Hubble Fellowship grant HST-HF-51327.01-A awarded by the Space Telescope Science Institute, which is operated by the Association of Universities for Research in Astronomy, Inc., for NASA, under contract NAS 5-26555.

References

[1] Tylenda R *et al* 2011 *Astron. Astrophys.* **528** 114
[2] Pejcha O 2014, *Astrophys. J.* **788** 22
[3] Pejcha O, Metzger B D and Tomida K 2016 *Mon. Not. R. Astron. Soc.* **455** 4351
[4] Kasliwal M M 2011 PhD thesis, California Institute of Technology
[5] Kasliwal M M *et al* 2014 *Astron. Telegram*, 6644

Journal of Physics: Conference Series **728** (2016) 072022 doi:10.1088/1742-6596/728/7/072022

Study of neglected (near-)contact binaries using small telescopes

Miloslav Zejda[1], Jia Zhang[2], Shengbang Qian[2], Liying Zhu[2] and Zdeněk Mikulášek[1]

[1] Department of Theoretical Physics and Astrophysics, Masaryk University, Kotlarska 2, 61137 Brno, Czech Republic

[2] Yunnan Observatories, Chinese Academy of Sciences, Kunming, China

E-mail: [1] zejda@physics.muni.cz

Abstract. We present the first results of our new program aimed at relatively bright neglected (near-)contact binaries. The targets can be observed using small telescopes from several to several tens of centimeters in diameter. Light ephemeris determined by phenomenological modeling and physical parameters of the system obtained by Wilson-Devinney method are given for the first stars V373 Dra, OQ UMa from our program.

1. Introduction

Eclipsing binaries (EBs) serve very often as independent and accurate source of information about stellar masses and radii and thus they can be used for testing of stellar evolution models [1] and determination of their distances. At present, there is a wide collection of photometric surveys, thus it is expected that most of known eclipsing binaries have their models at least made by any (semi-)automatic tools based on WD code or EBOP and others. However, there are more than 1500 eclipsing variable stars in SIMBAD brighter than 12 mag (V) with number of references lower than 6. Such systems are easily observable by the small telescopes common also in amateur community. Research of these variables could be a suitable subject for pro-am cooperation.

Shaw [2] defined Near Contact Binaries (NCBs) as a subclass of close binaries with both components filling or nearly filling their critical Roche Lobes. These close binaries show β Lyrae type variations. According to the geometric definition of this subclass, NCBs actually comprise semidetached, marginal contact and marginally detached systems. Close configurations of NCBs bring strong tidal interaction and mass transfers become the main character of them. On the other hand, contact binaries may undergo oscillations around the state of marginal contact. NCBs thus comprise the bridge between detached and contact binaries. So using NCBs, we can study not only mass transfer in close binaries, but also investigate the formation of contact binaries. However, NCB targets are relatively rare, because this stage is a very short compared to the life time of close binaries [3, 4]. More samples are needed to reveal their physical characters.

2. Methods

Collecting available measurements from all surveys and our own observations we create aggregate of data. As a first step of data processing, a phenomenological model is created. This allow us to solve shifts in phases and magnitudes as well as to detrend the datasets and remove outliers. We use general model function in nonlinear LSM described in [5]. Using the mathematical model we were able to improve light ephemeris substantially. New light ephemeris were then used as input for determination of system parameters in WD or Phoebe respectively.

3. The first results

To show the first results of our program, we opted V373 Dra and OQ UMa, the given parameters were calculated using WD code (see table 1). Both systems include relatively cool red dwarves without any striking surface activity.

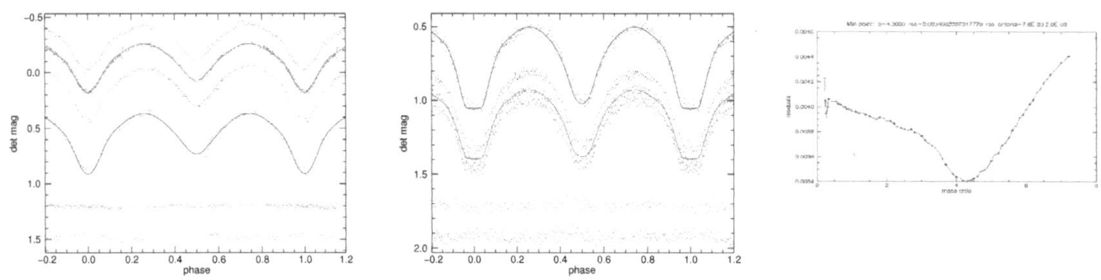

Figure 1. CCD observations of V373 Dra (left), OQ UMa (middle), search for mass ratio m_2/m_1 of OQ UMa.

Table 1. Parameters of both studied stars.

Star	V373 Dra		OQ UMa	
Component	primary	secondary	primary	secondary
Inclination	67.5(3)°		82.6(6)°	
Mass ratio	3.80(19)		4.30(5)	
Temperature	5000 K (fixed)	4565(20) K	4800 K (fixed)	4620(20) K
Radius: pole	0.261(3)	0.476(16)	0.2466(8)	0.479(4)
side	0.273(3)	0.516(23)	0.2571(9)	0.520(6)
back	0.314(3)	0.543(31)	0.2926(14)	0.544(7)

4. Summary

We started the data mining of (near-)contact binaries among neglected eclipsing binaries with the goal to increase the sample of known (N)CBs. Afterwards using statistically larger dataset we will be able to better understand and interpret the behaviour of close binaries in the limiting stage of their evolution when both components are close to the critical Roche lobes.

Acknowledgments

The study is supported by the grants of Ministry of Education of the Czech Republic LH14300, grant of Masaryk University MUNI/A/0930/2013, and by the Chinese Natural Science Foundation (Nos. 11133007, and 11325315). This research has made use of the SIMBAD database, operated at CDS, Strasbourg, France and VSX database, operated by AAVSO, USA.

References

[1] Torres G, Andersen J and Giménez A 2010 *Astron. Astrophys. Rev.* **18** 67
[2] Shaw J S 1990 *Proc. NATO Advanced Study Institute on Active Close Binaries* **319** 241
[3] Zhu L Y and Qian S B 2006 *Mon. Not. R. Astron. Soc.* **367** 423
[4] Zhu L Y, Qian S B, Mikulášek Z, Zejda M, Zvěřina P and Diethelm R 2009 *Astron. J.* **140** 215
[5] Mikulášek Z 2015 *Astron. Astrophys.* **584** A8

Substellar objects around the sdB eclipsing Binaries

Liying Zhu[1,2,3], Shengbang Qian[1,2,3], Wenping Liao[1,2], Ergang Zhao[1,2,3] and Linjia Li[1,2]

[1] Yunnan Observatories, Chinese Academy of Sciences (CAS), P.O. Box 110, 650011 Kunming, P. R. China

[2] Key laboratory of the structure and evolution of celestial objects, Chinese Academy of Sciences, P.O. Box 110, 650011 Kunming, P. R. China

[3]University of the Chinese Academy of Sciences, Yuquan Road 19#, Sijingshang Block, 100049 Beijing, P. R. China

E-mail: [1] `zhuly@ynao.ac.cn`

Abstract. The sdB-type eclipsing binary consists a very hot subdwarf B (sdB) type primary and a low mass secondary with short period. They are detached binaries and show very narrow eclipse profiles, which benefits the determination of the precise eclipse times. With the precise times of light minimum, we can detected small mass objects around them by analyzing the observed-calculated (O-C) curve based on the light time effect. For searching the substellar objects orbiting around the binaries, we have monitored sdB-type eclipsing binaries for decades. A group of brown dwarfs and planets have been detected since then. In the present paper, we focus on the target NSVS07826147, which may be another exoplanet host candidate among the group of the sdB-type eclipsing binaries.

1. Introduction

The hot sdB-type components in sdB eclipsing binaries are located on the extreme horizontal branch (EHB) in Hertzsprung-Russell diagram. They burn helium in their cores and have very thin hydrogen envelopes [1]. The sdB eclipsing binaries are believed to be formed from binary systems through common envelope(CE) ejection. Their progenitors should be solar-type binary with initial separations less than 5AU, and will evolve into cataclysmic binaries (CVs) [2] when the secondary components filling their critical Roche Lobe. Since 2009, more and more sdB eclipsing binaries have been found to be the host stars of substellar objects by our group and other investigators. Such as HS 0705+6700 [3][4],HW Vir [5], NSVS14256825 [6][7], NY Vir [8][9] and SDSSJ0820+0008 [10] etc. The discovery of the circumbinaries planets and brown dwarfs orbiting these evolved binaries not only increases our knowledge about the diversity of the exoplanets host stars, but also has very important implications for the formation of sdB stars and the fates of low-mass companion systems. Here we report some new results of one sdB eclipsing binary NSVS 07826147, which we have monitored for years.

2. The progress on NSVS07826147

NSVS 07826147 was classified as sdB eclipsing binary by Kelley & Shaw [11]. For et al. [12] gave the detailed absolute parameters using their photometric and spectroscopic observations. In 2009, We improved its period as 0.16177046 days based on all available times of light minimum at that time [13]. After that, we continue to monitor this target for years with some small telescopes in china, such as 1m and 60cm telescopes in YNO and 85cm telescope in Xinglong station. Some of the eclipse profiles obtained with these telescopes are displayed on the figures 1-2. With the high precise timings derived by our observations, we constructed the O-C curve of this system, see Figure 3. The O-C curve shows the trend of the periodic variation.

The presence of a tertiary body in the eclipsing binaries can cause the relative distance changes of the eclipsing pair as it orbits the barycenter of the triple system. Seen by a distant observer, the light-travel time of the eclipsing pair will change corresponding to the change of the relative distance. Then the change between the observed and calculated eclipsing timings will follow the strict period variation. By analyzing the O-C curve of NSVS 07826147, we detected a small amplitude cyclic variation. The preliminary result suggests that there is a Jupiter mass planet orbiting around NSVS 07826147 at the distant of 3 AU. The O-C curve and the corresponding fit (solid line) are shown in Figure 3. Thus, NSVS 07826147 may be another exoplanet host star among the group of the sdB-type eclipsing binaries.

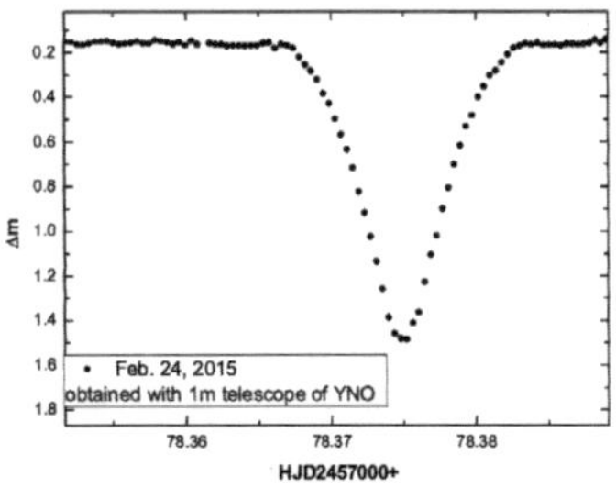

Figure 1. Obtained with 1m telescope.

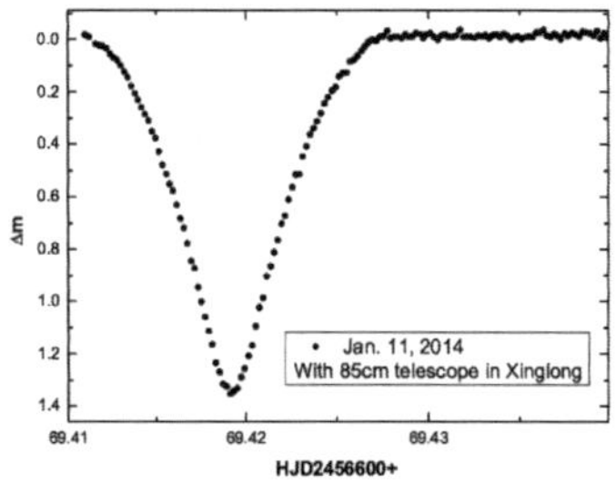

Figure 2. Obtained with 85cm telescope.

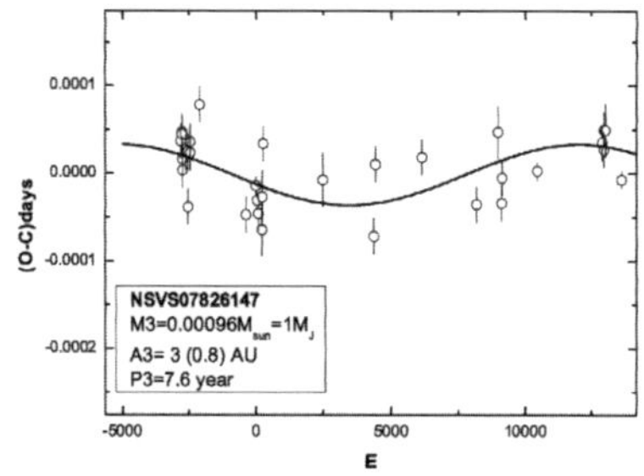

Figure 3. The O-C curve and the corresponding fit of NSVS 07826147.

Acknowledgments

This work is supported by the Chinese Natural Science Foundation (Nos.11133007, 11325315 and 11573063), Yunnan Natural Science Foundation (No.2013FB084), and the Strategic Priority Research Program "The Emergence of Cosmological Structures" of the Chinese Academy of Sciences (No.XDB09010202). New CCD photometric observations of NSVS 07826147 were obtained with the 1.0-m and 60-cm telescopes at Yunnan observatories (YNO) and 85cm telescope at xinglong station of NAO in China.

References

[1] Heber U 2009 *Annu. Rev. Astron. Astrophys.* **47** 211
[2] Zorotovic M and Schreiber M R 2013 *Astron. Astrophys.* **549** 95
[3] Qian S-B, Zhu L-Y, Zola S, Liao W-P, Liu L, Li L-J, Winiarski M, Kuligowska E and Dreiner J M 2009 *Astrophys. J* **695** L163
[4] Qian S-B *et al* 2013 *Mon. Not. R. Astron. Soc.* **436** 1408
[5] Lee J W, Kim S-L, Kim C-H, Koch R H, Lee C-U, Kim H-I and Park J-H 2009 *Astron. J.* **137** 3181
[6] Zhu L, Qian S, Liu L, Liao W P, He J J, Li L J, Zhao E G, Dai Z B, Zhang J and Li K 2011 *9th Pacific Rim Conference on Stellar Astrophysics* (ASP Conf. Series vol 451) p 155
[7] Almeida L A, Jablonski F and Rodrigues C V 2013 *Astrophys. J.* **766** 11
[8] Qian S-B, Zhu L-Y, Dai Z-B, Fernández-Lajús E, Xiang F-Y and He J-J 2012 *Astrophys. J.* **745** L23
[9] Lee J W, Hinse T C, Youn J H and Han W Y 2014 *Mon. Not. R. Astron. Soc.* **45** 2331
[10] Geier S *et al* 2011 *Astrophys. J* **731** L22
[11] Kelley N and Shaw J S S 2007 *J. Southeastern Association for Research in Astronomy* **1** 13
[12] For B-Q, Green E M and Fontaine G 2010 *Astrophys. J* **708** 253
[13] Zhu L and Qian S 2010 *Astrophys. Space Sci.* **329** 107

Journal of Physics: Conference Series **728** (2016) 072024 doi:10.1088/1742-6596/728/7/072024

Comparison of multi-band period-luminosity relations for classical Cepheids in the Magellanic Clouds

Chow-Choong Ngeow[1] **and Shashi M. Kanbur**[2]

[1] Graduate Institute of Astronomy, National Central University, Jhongli 32001, Taiwan
[2] Department of Physics, SUNY Oswego, Oswego, NY 13126, USA

E-mail: [1] cngeow@astro.ncu.edu.tw

Abstract. The period-luminosity (PL) relation for classical fundamental mode Cepheids (hereafter Cepheids) is an important astrophysical tool in distance scale applications. Because of this, we initiated a program to derive multi-band PL relations with Cepheids in the Large and Small Magellanic Cloud (hereafter LMC and SMC, respectively), as there are $\sim 10^3$ Cepheids found in these two nearby galaxies. When compared the slopes of the multi-band PL relations for Cepheids in the LMC and SMC, we found that these PL slopes agree with each others except in the V and J band. We also found an excellent agreement of the PL slopes in Wesenheit function, hence we calibrated the Period-Wesenheit (PW) relation by combining the data from both Clouds, together with an accurate LMC distance based on measurement from late-type eclipsing binaries. Our calibrated Wesenheit function is $M_W = -3.314\log(P) - 2.601$.

1. Data and Methods

We used almost the same data sources to derive the multi-band PL relations in LMC and SMC. Catalogs of Cepheids in the LMC and SMC were adopted from the third phase of Optical Gravitational Lensing Experiment (OGLE-III), which provide the position, pulsation period and the mean VI band photometry (see [1] & [2]). The OGLE-III catalogs were cross-matched to the 2MASS point source catalogs ([3]) to derive the mean JHK band photometry, as well as to the Spitzer IRAC band data available from SAGE ([4]) and SAGE-SMC ([5]) projects. Extinction corrections to these multi-band photometry were done using the Zaritsky's extinction maps ([6] & [7]). We also applied almost identical algorithms in our analysis when deriving the multi-band PL relations in both Magellanic Clouds. For examples, we applied the same procedures for random-phase corrections as presented in [8] to JHK-band data taken from the 2MASS catalog, and employed the same iterative 2.5σ-clipping regression when fitting the PL relations. We also included the extinction free Wesenheit function in the form of $W = I - 1.55(V - I)$.

2. Results: The Calibrated Wesenheit Function

Details of our work and results from this program can be found in [9] and [10] for the LMC and SMC PL relations, respectively. Since the data sources and algorithms used in our program were almost identical, we can compare the slopes of the multi-band PL relations for Cepheids in these two Clouds. These PL relations are shown in Figure 1 & 2 for the LMC and SMC Cepheids,

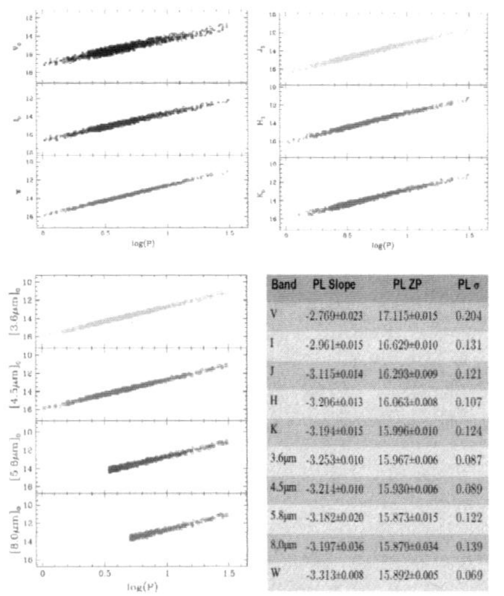

Band	PL Slope	PL ZP	PL σ
V	-2.769±0.023	17.115±0.015	0.204
I	-2.961±0.015	16.629±0.010	0.131
J	-3.115±0.014	16.293±0.009	0.121
H	-3.206±0.013	16.063±0.008	0.107
K	-3.194±0.015	15.996±0.010	0.124
3.6μm	-3.253±0.010	15.967±0.006	0.087
4.5μm	-3.214±0.010	15.930±0.006	0.089
5.8μm	-3.182±0.020	15.873±0.015	0.122
8.0μm	-3.197±0.036	15.879±0.034	0.139
W	-3.313±0.008	15.892±0.005	0.069

Figure 1. The multi-band PL relations for LMC Cepheids.

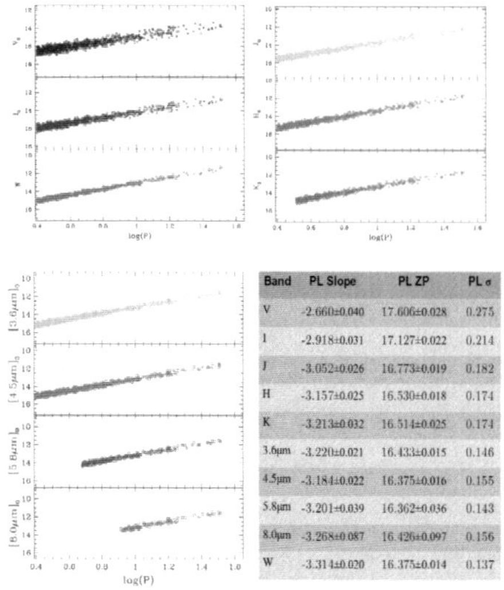

Band	PL Slope	PL ZP	PL σ
V	-2.660±0.040	17.606±0.028	0.275
I	-2.918±0.031	17.127±0.022	0.214
J	-3.052±0.026	16.773±0.019	0.182
H	-3.157±0.025	16.530±0.018	0.174
K	-3.213±0.032	16.514±0.025	0.174
3.6μm	-3.220±0.021	16.433±0.015	0.146
4.5μm	-3.184±0.022	16.375±0.016	0.155
5.8μm	-3.201±0.039	16.362±0.036	0.143
8.0μm	-3.268±0.087	16.426±0.097	0.156
W	-3.314±0.020	16.375±0.014	0.137

Figure 2. The multi-band PL relations for SMC Cepheids.

respectively. We found that the PL slopes in V and J band do not agree with each others by using the statistical t-test (see [10] for more details). On the other hands, slopes of the PW relation are almost identical (-3.313 ± 0.008 in LMC vs. -3.314 ± 0.020 in SMC). Therefore, the difference in PW intercepts directly gives the relative distance modulus (μ) of the two Clouds, which is found to be $\Delta\mu = 0.483 \pm 0.015$. Combining the LMC & SMC Wesenheit function using the derived $\Delta\mu$, together with absolute distance to LMC from [11] (with $\mu_{LMC} = 18.493$), then the absolute calibration of the Period-Wesenheit relation is found to be:

$$M_W = -3.314 \log(P) - 2.601, \quad \sigma = 0.099.$$

Acknowledgments
CCN thanks the support from grant MOST 104-2112-M-008-012-MY3.

References
[1] Soszynski I, Poleski R, Udalski A, Szymanski M K, Kubiak M, Pietrzynski G, Wyrzykowski L, Szewczyk O and Ulaczyk K 2008 *Acta Astronomica* **58** 163
[2] Soszynski I, Poleski R, Udalski A, Szymanski M K, Kubiak M, Pietrzynski G, Wyrzykowski L, Szewczyk O and Ulaczyk K 2010 *Acta Astronomica* **60** 17
[3] Cutri R M *et al* 2003 *VizieR On-line Data Catalog: II/246* **2246** 0
[4] Meixner M *et al* 2006 *Astron. J.* **132** 2268
[5] Gordon K D and SAGE-SMC Spitzer Legacy Team 2010 *Bulletin of the American Astronomical Society* **42** 489
[6] Zaritsky D, Harris J, Thompson I B, Grebel E K and Massey P 2002 *Astron. J.* **123** 855
[7] Zaritsky D, Harris J, Thompson I B and Grebel E K 2004 *Astron. J.* **128** 1606
[8] Soszyński I, Gieren W and Pietrzyński G 2005 *Publ. Astron. Soc. Pac.* **117** 823
[9] Ngeow C-C, Kanbur S M, Neilson H R, Nanthakumar A and Buonaccorsi J 2009 *Astrophys. J.* **693** 691
[10] Ngeow C-C, Kanbur S M, Bhardwaj A and Singh H P 2015 *Astrophys. J.* **808** 67
[11] Pietrzynski G *et al* 2013 *Nature* **495** 76

Journal of Physics: Conference Series **728** (2016) 072025 doi:10.1088/1742-6596/728/7/072025

Cometary grains in the HD 32297 debris disk

Y.-G Yang[1], and Aigen Li[2]

[1] Information College/School of Physics and Electronic Information, Huaibei Normal Univeristy, Huaibei 235000, Anhui Province, China

[2] Department of Physics and Astronomy, University of Missouri, MO 65211, USA

E-mail: [1]yygcn@163.com; [2]lia@missouri.edu

Abstract. HD 32297 is a young A-type star with a bright edge-on debris disk. The dust thermal emission spectral energy distribution and scattered starlight spectrum are simultaneously modeled in terms of porous cometary grains. Our modeling suggests that, similar to the solar system, the debris disk around HD 32297 may have an inner warm ring and an outer cold disk which are seen in other young debris disks as well.

1. Introduction

Dusty debris disks around main sequence (MS) stars [1], are freshly produced through the attrition of planetesimals. Direct imaging of circumstellar disks offers us insight into the origin, evolution and distribution of dust in planetary systems. These disks are typically dominated by cool dust, which emits at the far infrared (IR) wavelengths although some much warmer dust is often also seen (e.g., see [2]). Much less common are debris disks where most of the IR emission appears in the mid-IR and is emitted by warm dust particles with temperatures of ~ 200 K (e.g., see [3]).

HD 32297, an A-type MS star at a distance of about 112 pc, is surrounded by a debris disk. This disk was discovered through scattered starlight based on the 1.1 μm HST/NICMOS coronagraph [4]. It was subsequently resolved by HST/NICMOS at 1.6 μm and 2.05 μm [5], by Keck/NIRC2 and VLT/NACO at 2.16 μm [6, 7, 8], and by LBTI/LMIRcam at 3.8 μm [9]. This disk has also been detected through dust thermal emission from the mid-IR to millimeter by WISE, Spitzer/IRS, Spitzer/MIPS, IRAS, Gemini, IRAM, SMA, as well as Herschel (see [10] and references therein). Donaldson et al. (2013) [10] modeled the IR emission of the HD 32297 disk using the cometary dust model of Li & Greenberg (1998) [11], with the dust being a porous mixture of silicate, carbon, water ice and vacuum. However, Rodigas et al. (2014) [9] argued that this model fails to explain the disk's surface brightness and scattered light spectrum. They found that the best fit was given by pure water ice grains. In this work, we re-consider the cometary dust model with an aim of simultaneously reproducing the observed scattered starlight and the thermal emission spectral energy distribution (SED).

2. Method

The cometary grain model consists of a porous mixture of amorphous silicate, amorphous carbon, and water ice [11]. This model has been successfully applied to a number of dust disks (e.g., see [12, 13, 14, 15, 16, 17]).

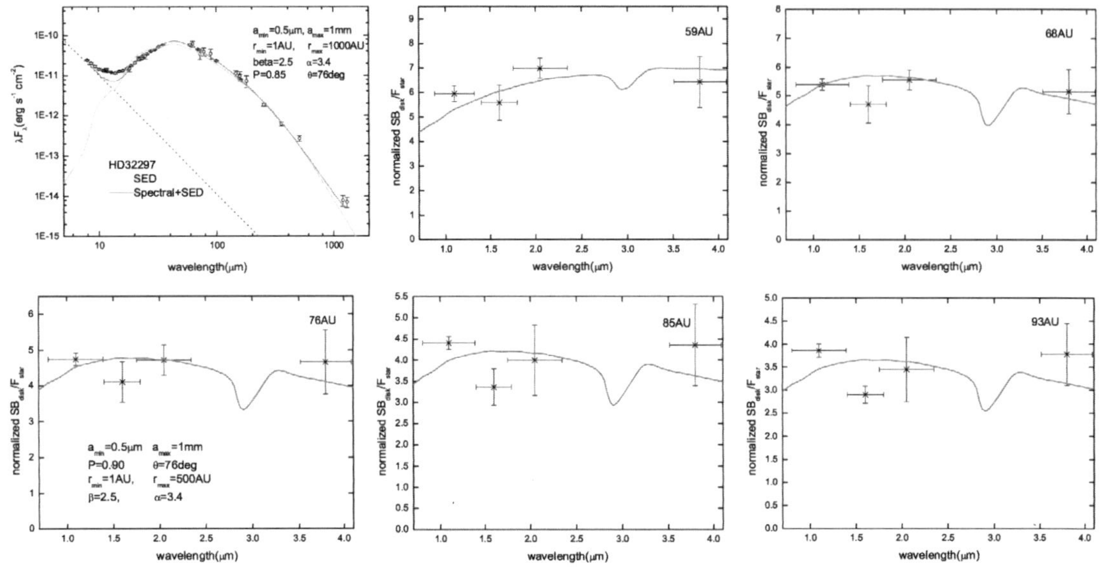

Figure 1. Comparison of the model predictions with the dust thermal emission SED (*upper left panel*) and the observed scattered light spectrum (*other panels*) of HD 32297.

3. Results

As shown in Figure 1, cometary grains closely reproduce both the dust thermal emission SED and the observed scattered-light spectrum. The best fit is given by grains with a porosity of $P = 0.85$, a power-law size distribution of $dn/da \propto a^{-3.4}$ for $0.5\,\mu\text{m} < a < 1\,\text{mm}$, and a spatial distribution of $d/dr \propto r^{-2.5}$ for $1 < r < 1000\,\text{AU}$. Figure 1 also shows that the model emission is a little bit deficient at $\sim 10\,\mu\text{m}$. This indicates the possible existence of an inner warm ring [18]. Also, the model fails to reproduce the scattered light dip at $\sim 1.5\,\mu\text{m}$, possibly due to a weak ice resonant feature.

Acknowledgments

This research is supported partly by NSFC grants 11473009 and U1131102. YYG thanks AL for inviting him to collaborate on protoplanetary and debris disks at the University of Missouri as a visiting scholar from September 1, 2014 to February 28, 2015.

References

[1] Aumann H H, Beichman C A, Gillett F C, de Jong T, Houck J R, Low F J, Neugebauer G, Walker R G and Wesselius P R 1984 *Astrophys. J.* **278** 23
[2] Morales F Y, Rieke G H, Werner M W, Bryden G, Stapelfeldt K R and Su K Y L 2011 *Astrophys. J.* **730** 29
[3] Melis et al. 2010???
[4] Schneider G, Silverstine M D, Murray D and Hines D C 2005 *Astrophys. J.* **629** L117
[5] Debes J H, Weinberger A J and Kuchner M M 2009 *Astrophys. J.* **702** 318
[6] Currie T *et al* 2012 *Astrophys. J.* **757** 28
[7] Boccaletti A, Augereau J-C, Lagrange A-M, Milli J, Baudoz P, Mawet D, Mouillet D, Lebreton J and Maire A-L 2012 *Astron. Astrophys.* **544** 85
[8] Esposito T M, Fitzgerald M P, Graham J R and Kalas P 2014 *Astrophys. J.* **780** 25
[9] Rodigas T J *et al* 2014 *Astrophys. J.* **783** 21
[10] Donaldson J K, Lebreton J, Roberge A, Augereau J-C and Krivov A V 2013 *Astrophys. J.* **772** 17
[11] Li A and Greenberg J M 1998 *Astron. Astrophys.* **331** 291
[12] Li A and Lunine J I 2003 *Astrophys. J.* **590** 368
[13] Li A and Lunine J I 2003 *Astrophys. J.* **594** 987
[14] Li A, Lunine J I and Bendo G J 2003 *Astrophys. J.* **598** 51
[15] Köhler M, Mann I and Li A 2008 *Astrophys. J.* **686** 95

[16] Seok J Y and Li A 2015 *Astrophys. J.* **809** 22
[17] Seok J Y and Li A 2016 *Astrophys. J.* **818** 2
[18] Yang Y G and Li A 2016 in preparation